Handbook of Animal Experimentation

Handbook of Animal Experimentation

Edited by Arthur Yates

SYRAWOOD
PUBLISHING HOUSE
New York

Published by Syrawood Publishing House,
750 Third Avenue, 9th Floor,
New York, NY 10017, USA
www.syrawoodpublishinghouse.com

Handbook of Animal Experimentation
Edited by Arthur Yates

International Standard Book Number: 978-1-68286-554-5 (Hardback)

Cataloging-in-Publication Data

Handbook of animal experimentation / edited by Arthur Yates.
p. cm.
Includes bibliographical references and index.
ISBN 978-1-68286-554-5
1. Animal experimentation. 2. Dairy farming. 3. Livestock--Research. I. Yates, Arthur.
SF241 .H36 2018
637--dc23

Printed in China.

TABLE OF CONTENTS

PREFACE

Animal experimentation is a field of research that performs tests on animals for scientific and commercial purposes. Research in this field usually analyzes the behavior of animals under controlled environments. Animal experimentation and research is extremely useful for industries such as cosmetics, livestock management, etc. This book attempts to understand the multiple branches that fall under this discipline and how such concepts have practical applications. It will prove immensely beneficial to students and researchers in the field.

This book is a comprehensive compilation of works of different researchers from varied parts of the world. It includes valuable experiences of the researchers with the sole objective of providing the readers (learners) with a proper knowledge of the concerned field. This book will be beneficial in evoking inspiration and enhancing the knowledge of the interested readers.

In the end, I would like to extend my heartiest thanks to the authors who worked with great determination on their chapters. I also appreciate the publisher's support in the course of the book. I would also like to deeply acknowledge my family who stood by me as a source of inspiration during the project.

Editor

Noisy Galvanic Vestibular Stimulation Promotes GABA Release in the Substantia Nigra and Improves Locomotion in Hemiparkinsonian Rats

Ghazaleh Samoudi[1], Hans Nissbrandt[1], Mayank B. Dutia[2], Filip Bergquist[1,3]*

1 Department of Pharmacology, Institute of Neuroscience and Physiology, The Sahlgrenska Academy, University of Gothenburg, Göteborg, Sweden, **2** Centre for Integrative Physiology, University of Edinburgh, Edinburgh, United Kingdom, **3** Department of Neurology, Sahlgrenska University Hospital, Göteborg, Sweden

Abstract

Background: The vestibular system is connected to spinal, cerebellar and cerebral motor control structures and can be selectively activated with external electrodes. The resulting sensation of disturbed balance can be avoided by using stochastic stimulation patterns. Adding noise to the nervous system sometimes improves function. Small clinical trials suggest that stochastic vestibular stimulation (SVS) may improve symptoms in Parkinson's disease. We have investigated this claim and possible mechanisms using the 6-hydroxydopamine (6-OHDA) hemilesion model of Parkinson's disease.

Methodology/Principal Findings: Animals were tested in the accelerating rod test and the Montoya staircase test of skilled forelimb use. In 6-OHDA hemilesioned animals, SVS improved rod performance by 56±11 s. At group level L-DOPA treatment had no effect, but positive responders improved time on rod by 60±19 s. Skilled forelimb use was not altered by SVS. To investigate how SVS may influence basal ganglia network activity, intracerebral microdialysis was employed in four regions of interest during and after SVS. In presence of the γ-amino buturic acid (GABA) transporter inhibitor NNC 711, SVS induced an increase in GABA to 150±15% of baseline in the substantia nigra (SN) of unlesioned animals, but had no effect in the pedunculopontine nucleus (PPN), the striatum or the ventromedial thalamus (VM). Dopamine release remained stable in all areas, as did GABA and amine concentrations in the SN of unstimulated controls. Following SVS, a sustained increase in GABA concentrations was observed in the ipsilesional, but not in the contralesional SN of 6-OHDA hemilesioned rats. In contrast, L-DOPA treatment produced a similar increase of GABA in the ipsi- and contra-lesional SN.

Conclusions/Significance: SVS improves rod performance in a rat model of Parkinson's disease, possibly by increasing nigral GABA release in a dopamine independent way. We propose that SVS could be useful for treating symptoms of Parkinson's disease.

Editor: Mathias Toft, Oslo University Hospital, Norway

Funding: This study was sponsored by the Swedish Research Council, grant no 2009-2618, Parkinsonfonden (Swedish Parkinson Foundation), Svenska läkaresällskapet (Swedish Medical Association), and Jeansons stiftelse and Göteborgs läkaresällskap (Gothenburg Medical Association). The funders had no role in study design, data collection and analysis, decision to publish, or preparation of the manuscript.

Competing Interests: The authors have declared that no competing interests exist.

* E-mail: filip.bergquist@gu.se

Introduction

Standard Parkinson's disease treatments with dopaminergic drugs, and sometimes deep brain stimulation, often fail to alleviate axial rigidity and gait problems. Axial symptoms are strongly related to the risk of falls and better treatment options could be of great benefit to a large group of patients with Parkinson's disease [1].

Recent findings suggest that a sufficient level of noise may be necessary for normal function of the central nervous system [2,3,4,5,6]. Under some circumstances, adding noise to a system will result in improved signal detection or linearity of responses [7,8,9,10]. This phenomenon can be predicted from mathematical models and is often referred to as stochastic resonance, or in a wider sense, noise benefit [11]. Several recent studies describe cross-modal noise benefit in cognitive tasks in healthy subjects [12,13,14,15,16], indicating that noise benefit is not restricted to signal detection but can be observed also when the outcome is a complex and abstract behavior. In a computational model of the degenerating brain, higher levels of external noise were needed for optimal function under conditions representing aging and loss of plasticity [4]. This could imply that the effects of neurodegeneration can be counteracted to some extent by introducing more noise in the central nervous system.

Noisy sensory signals can be transmitted to the central nervous system through any sensory pathway, but vestibular pathways have some characteristics that may be particularly useful for modulating motor deficits. Neurons in the vestibular brainstem nuclei influence axial motor functions, eye movements, autonomic cardiovascular reflexes, as well as spatial perception. Several cortical regions respond to vestibular stimulation [17] indicating widespread indirect effects. Galvanic stimulation of the vestibular system using DC current has been used for over a hundred years and is safe, but leads to disturbed posture and balance. However, by using stochastic current patterns, it is possible to activate the vestibular system without such adverse effects [18,19,20,21]. Such stochastic vestibular stimulation (SVS) can improve balance in healthy subjects [22,23]. In patients with neurodegenerative disease, like Parkinson's disease, there is some evidence that SVS improves motor functions [19,24], autonomic reflexes and cognitive executive control [19]. Static postural sway is also reduced by SVS in Parkinson disease patients [25]. It is not known if there is a general mechanism for noise benefit in higher functions. A model involving altered dopamine release has been proposed [26], but experimental evidence to support that is lacking [27].

We used the 6-OHDA hemilesioned rat model to determine if SVS can improve motor functions in dopamine-deficient rats. Furthermore, we investigated how SVS might influence basal ganglia networks under normal and dopamine deficient conditions by measuring amino acid and dopamine release in intact and 6-OHDA hemilesioned rats.

Four nuclei that relay or mediate basal ganglia signaling were investigated in intact rats: the SN, the striatum, the PPN and the VM. Dopamine release in the striatum and SN is known to facilitate movement initiation and to occur during motor activity [28,29,30]. The output from the basal ganglia is inhibitory and mainly mediated by GABA-ergic neurons projecting from the internal globus pallidus (entopeduncular nucleus in rodents) and the SN *pars reticulata* to thalamus and brainstem nuclei including the PPN. The PPN is of particular interest as it responds to a large number of sensory stimuli and is involved in gait initiation [31]. Furthermore cholinergic/glutamatergic PPN neurons project to the SN and influence motor functions [32].

In line with previous clinical studies, we report improved locomotion in hemilesioned rats during SVS, but we were unable to demonstrate a change in skilled forepaw use. In unlesioned rats, SVS increased extracellular GABA concentrations only in the SN. The other investigated neurotransmitters were not significantly altered in the SN, striatum, PPN and VM. In hemilesioned animals, L-DOPA injections induced similar GABA increases in the ipsi- and contralesional SN, whereas SVS induced differential effects in the two SN, suggesting that altered nigral GABA release has a role for the behavioral effects of SVS in 6-OHDA hemilesioned rats, possibly by decreasing the inhibitory activity of the ipsilesional SN *pars reticulata* neurons.

Results

Effects of SVS and L-DOPA on rotarod performance in 6-OHDA and sham hemilesioned rats

Unilateral 6-OHDA lesion resulted in reduced performance on the rotarod. The total time on the rod decreased by 48% from 472±80 s before the lesion to 266±86 s after (−206±45 s, paired t-test, $p<0.01$). In contrast, the sham procedure did not significantly reduce the total time on the rod (pre-sham performance: 544±135 s, post-sham performance: 430±112 s, paired t-test $p=0.1004$).

Because individual animals differ slightly in performance from day to day the effect of treatments were evaluated in comparison to a baseline performance obtained immediately before intervention. Treatment or sham treatment was administered in a balanced pseudo-randomized order so each animal received either sham treatment or active treatment first, and the remaining intervention in a separate trial. Baseline performances immediately before SVS and sham SVS did not differ (272±75 s and 279±77 s, respectively, paired t-test, $p=0.526$), and this was the case also with baseline performances immediately before L-DOPA and NaCl administration (265±62 s and 244±57 s, respectively, paired t-test, $p=0.511$).

6-OHDA hemilesioned animals increased the time spent on rod during SVS compared to baseline significantly more than during sham SVS ($\Delta t=56\pm11$ s vs. $\Delta t=24\pm11$ s, paired t-test, p = 0.011, Fig. 1). The improvement observed during SVS corresponded to an increase in performance of 24±6% of the immediately preceding baseline performance. As changes in neurotransmission were sustained (see below), we also analyzed if the order of treatment had an effect. Analysis of variance with treatment and test day order as independent factors revealed no effect of test day order, $F(1,8)=0.0003$, p = 0.987, failing to show a carry-over effect of SVS to the following test session. At group level, L-DOPA treatment did not improve the mean time spent on the rod relative to baseline and was in that respect not better than sham treatment with a saline injection (paired t-test, $p=0.818$, Fig. 1). At the individual level, however, animals could be categorized as positive responders to L-DOPA (n = 3) or negative responders (n = 3). The positive responders increased time on rod by 60±19 s compared to the preceding baseline and this was nearly significant (p = 0.067, paired t-test) compared to saline injection (−5±7 s). In contrast, in negative responders, time on rod changed by −63±15 s compared to the preceding baseline in response to L-DOPA and by 11±30 s in response to vehicle (not significant).

Sham-lesioned animals (n = 5) did not perform better during SVS than no SVS (paired t-test for change from preceding baseline, $p=0.8206$, n = 5. Fig. 1). Similar to L-DOPA treated hemilesioned animals the response to SVS was variable in this group. One animal improved by 192 s, and another deteriorated by 138 s during SVS. The other three retained stable performance.

Effects of SVS on skilled forelimb use

Forelimb use was evaluated with the Montoya staircase test. As expected there was a significant decrease in the ratio of pellets picked up with the contralesional forelimb following a 6-OHDA

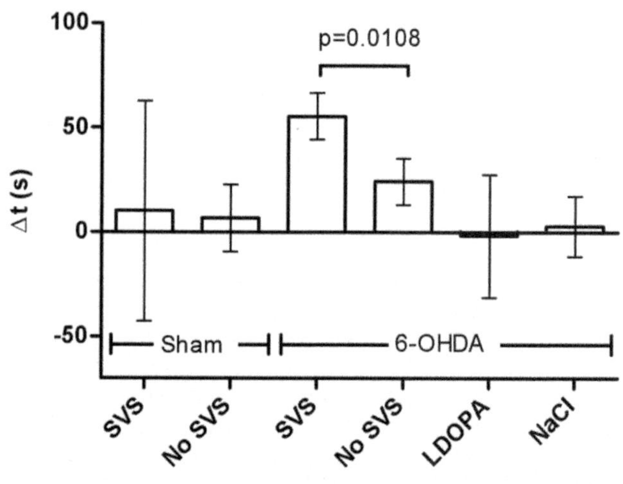

Figure 1. Rotarod performance after 6-OHDA or sham-hemilesions. Δt represents the change in time on rod (s, mean±SEM) compared to baseline performance on the day of the experiment. Stochastic vestibular stimulation (SVS) or no stimulation (No SVS) was administered in a counterbalanced order. In the following week a repeat experiment was performed with 6-OHDA hemilesioned animals and the effect of L-DOPA treatment or a vehicle injection (NaCl) was evaluated in counterbalanced order. P-value for paired t-test.

hemilesion (0.93±0.1 prelesion vs. 0.65±0.1 postlesion, paired t-test, $p = 0.035$). No statistical difference was found between the total number of pellets retrieved before and after the 6-OHDA lesion. Unlike 6-OHDA hemilesioned animals, the sham-lesioned animals retrieved a higher number of pellets after the lesion procedure than before, suggesting continued performance improvement, and they did not develop a side preference (Table 1). For technical reasons we were unable to administer SVS during the skilled forelimb task, and we did not observe any change in the total number of retrieved pellets or the ratio retrieved with the contralesional forelimb when animals were tested immediately after a 30 minutes SVS-session (Table 1).

Effects of SVS on dopamine and amino acids in un-lesioned animals

We used un-lesioned animals to broadly investigate if SVS induces any detectable changes in neurotransmitter release in four key brain regions in, or connected to, the basal ganglia (Fig. 2). Extracellular concentrations of dopamine, dopamine metabolites and amino acids were sampled from microdialysis probes in the SN, the striatum, the PPN and the VM before, during and after SVS. The GABA transporter inhibitor NNC 711 was retrodialysed to amplify synaptic release of GABA as we have previously demonstrated that rapid increases in GABA transmission can otherwise go undetected [33].

Following SVS in un-lesioned animals, dialysate GABA concentrations from the SN increased to 150±15% of baseline (at T = 150 min), while GABA concentrations in control animals that received no SVS were stable (Fig. 3A, Two way ANOVA, main effect of treatment $F(1,26) = 14.41$, $p = 0.0022$, time $F(2,26) = 3.83$, $p = 0.035$ and interaction treatment×time $F(2,26) = 3.53$, $p = 0.044$). The deadspace of tubings introduced a delay of approximately 20–30 minutes in the microdialysis measurements. This means that the observed increase in GABA started early during SVS-stimulation and persisted for at least 30 minutes after stimulation was terminated. DA concentrations remained stable in the SN and striatum of SVS-treated animals during SVS and the following 60 min (Fig. S1).

DA concentrations were below detection levels (0.05 nM) in the PPN and VM throughout the experiment. No significant changes in GABA concentrations were measured in the PPN, VM, or the striatum following SVS in un-lesioned animals (Fig. 3B). Concentrations of glutamate, aspartate, glycine, taurine, serine, alanine, DOPAC and HVA were not significantly affected by SVS in any of the investigated regions (not shown), but in the SN, there were large variations in glycine and glutamate concentrations after SVS (Fig. S2).

Table 1. Montoya staircase performance before and two weeks after a hemilesion procedure.

		Pre lesion	Post les	*p-value*	No SVS	SVS	*p-value*
6-OHDA	*Total*	51±2	45±3	*0.377*	45±3	46±3	*0.753*
(n = 5)	*Ratio*	0.93±0.1	0.65±0.1	*0.035*	0.58±0.1	0.54±0.1	*0.618*
Sham	*Total*	40±3	50±3	*0.005*	50±4	46±3	*0.497*
(n = 3)	*Ratio*	1.23±0.2	1.05±0.1	*0.239*	1.25±0.3	1.10±0.2	*0.520*

Total number of pellets retrieved and the contralesional/ipsilesional ratio is given (mean±SEM) Three weeks post lesion, the effect of stochastic vestibular stimulation (SVS) or no stimulation (No SVS) was evaluated on different days in a counterbalanced order. P-values for paired t-tests.

Effects of SVS and L-DOPA treatment on dopamine and GABA in 6-OHDA hemilesioned animals

Based on our findings in unlesioned animals, dopamine and amino acid concentrations were measured in the bilateral SN of some hemilesioned 6-OHDA animals, during and after SVS and L-DOPA treatments. These animals were not tested behaviorally, but the degree of dopamine lesion was assessed after the experiment. The effects of treatments on the nigral concentrations of DA, GABA and glutamate were evaluated and compared between the ipsilesional and contralesional SN.

In the 6-OHDA hemilesioned animals there was a trend towards lower baseline GABA concentrations in the ipsilesional SN at the beginning of sampling, but the difference was not significant ($ipsi_{n=6}$: 24±6 nM vs. $contra_{n=6}$: 50±20 nM, $p = 0.353$, paired t-test. In unlesioned animals, n = 15, absolute baseline concentrations were 50±17 nM).

Following SVS, the absolute GABA concentration changed differently in the two SN. In the ipsilesional SN GABA concentrations increased and in the contralesional SN they tended to fall. This differential effect is expressed as a significant interaction between time and SN side in a two way ANOVA using the absolute GABA concentrations at t = 90 to t = 150 minutes as the independent variable (Fig. 4A, $F(2,20) = 4.89$, $p = 0.019$).

In contrast, L-DOPA treatment on day two produced roughly parallel changes in absolute GABA concentrations in the two SN, despite apparent asymmetries in the dopamine concentrations in the two SN following the L-DOPA injection (Fig. 4B). The relative changes in GABA concentrations in the bilateral SN following L-DOPA treatment (Fig. 5) was of similar magnitude to that observed after SVS in intact animals (Figs. 3, 5) and in the ipsilesional SN, GABA concentrations (percent of baseline) were significantly increased compared to untreated control animals at T = 120 (Fig. 5). Due to tubing deadspace, that timepoint corresponds to approximately 30–40 minutes after L-DOPA administration.

Like in unlesioned animals SVS did not alter DA concentrations in the bilateral SN of 6- OHDA hemilesioned animals (Fig. 4C). After the L-DOPA administration the DA concentrations increased in both ipsi- and contralesional SN and peaked in the first succeeding sample (T = 90, Fig. 4D). This indicates that DA levels peaked within 10 minutes after the L-DOPA injection. The increase in DA following L-DOPA treatment was in absolute terms more than two times larger in the ipsilesional SN (ipsin = 4: +2.0±0.3 nM, $contra_{n=5}$: +0.9±0.2 nM, p = 0.028, unpaired t-test), but peak values were not significantly higher in the ipsilesional SN (ipsin = 4: 2.3±0.4 nM, $contra_{n=5}$: 1.5±0.2 nM, p = 0.119). For technical reasons however, only two measurements could be obtained from the ipsilesional SN at T = 90, so peak values from the ipsilesional SN may be underestimated.

Concentrations of the amino acid glutamate in dialysates from the two SN of 6-OHDA hemilesioned animals were also analyzed. Following SVS, as well as L-DOPA injection, no significant changes or differences between the two sides were observed (Fig. S3.).

Discussion

This is to our knowledge the first study to evaluate motor effects of stochastic vestibular stimulation (SVS) in rodents, and the first to explore the neurochemical effects of SVS *in vivo*. Stochastic resonance in central nervous system function has previously mainly been studied as a phenomenon, without trying to understand how activity patterns in the brain may change. This

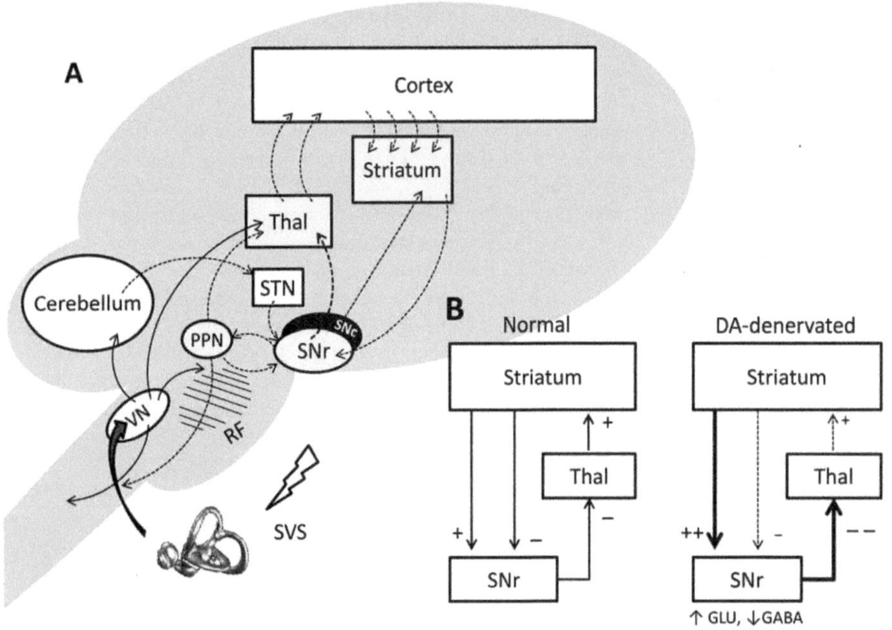

Figure 2. Vestibular pathways that may influence basal ganglia transmission. A) SVS is expected to activate multiple pathways from the vestibular nuclear complex (VN). Of particular interest may be pathways that connect the cerebellum and the basal ganglia over the subthalamic nucleus and thalamus [48,49]. We chose to sample the SN, the striatum, the PPN and the VM for dopamine and amino acid concentrations before, during and for 60 minutes after stochastic vestibular stimulation in unlesioned animals. Panel B indicates the activity of the direct and indirect loop projections to the SN in normal intact rats and after nigral dopamine cell degeneration. Loss of nigrostriatal dopamine neurons lead to hyperexcitation of SN *pars reticulata* neurons, which can be counteracted by increased GABA release following L-DOPA treatment. SVS also increases nigral GABA concentrations, but the pathways involved in this effect remains to be elucidated. RF: reticular formation, SNc: Substantia nigra *pars compacta*, SNr: Substantia nigra *pars reticulata*, STN: subthalamic nucleus, SVS: stochastic vestibular stimulation, Thal: thalamus.

study demonstrates that SVS improves rod performance in hemilesioned rats. SVS was associated with increased GABA concentrations in the SN, but not with altered dopamine transmission in normal animals. In hemilesioned rats SVS increased GABA in the ipsilesional SN by a magnitude similar to that observed after L-DOPA treatment. As described previously [34], L-DOPA treatment increased nigral DA release. In contrast, SVS was not associated with altered dopamine release in lesioned animals.

For technical reasons we used two different stimulation setups and stimulation currents were not identical (Fig. S4). The current used in behavioral experiments and microdialysis experiments in hemilesioned animals contained more frequencies over 30 Hz than the current used in microdialysis in unlesioned animals. Although vestibular neurons can support activity frequencies well over 100 Hz, the vestibular system is optimized for frequencies up to ~30 Hz [35]. The low frequency content of the signal can therefore be expected to have a higher impact on brainstem activity. The neurochemical effects of the two currents were however similar, and importantly both protocols stimulated vestibular afferents in a way that did not cause observable nystagmus or balance deficits.

Behavioral effects of SVS

Stochastic resonance (SR) is characterized by a biphasic response curve shaped as an inverted U [5]. The design of this study did not include more than one noise level per animal, so we cannot know if the observed improvement in rod performance involves SR. A SR phenomenon has been demonstrated previously for balance improvement with SVS [22], and adding noise with small amplitude, both electrical and mechanical, to proprioceptive pathways improves balance [36,37]. The observation that 6-OHDA hemilesioned, but not sham lesioned animals, improved rod performance can be interpreted as support for the notion that brains with impaired function display more noise benefit than the normally working brain [4,26,27]. The lack of improvement in sham lesioned animals is not likely to be only a ceiling effect, as the group response included both large improvement and impairment in response to SVS. Although SR can serve as a theoretical framework for understanding the effect of noisy sensory stimulation like SVS, it remains an unproved hypothesis that the effects of external noise on central nervous system functions involve SR. SVS could alternatively be viewed as a method to specifically activate the vestibular system without the adverse effects associated with stimulation with regular pattern.

SVS improved rod performance which is an integral measure of balance and locomotion. Improvements in time on an accelerating rod indicate not only improved endurance but also an increase in motor and balance functions, because the difficulty of the task increases gradually. The effect of SVS on rod performance was of similar magnitude as the effect of a single L-DOPA injection in positive responders, suggesting that the efficacy of SVS in terms of improved rod performance is not marginal. Furthermore, unlike L-DOPA treatment, SVS improved the rod performance in all lesioned animals.

The vestibular pathways are mainly influencing axial motor systems and a stronger effect of SVS on balance and locomotion than on appendicular motor control can therefore be expected. As there are reports of cross modal function improvements following noisy sensory stimulation [16,38], we hypothesized that skilled forelimb use would also improve in response to SVS. We were,

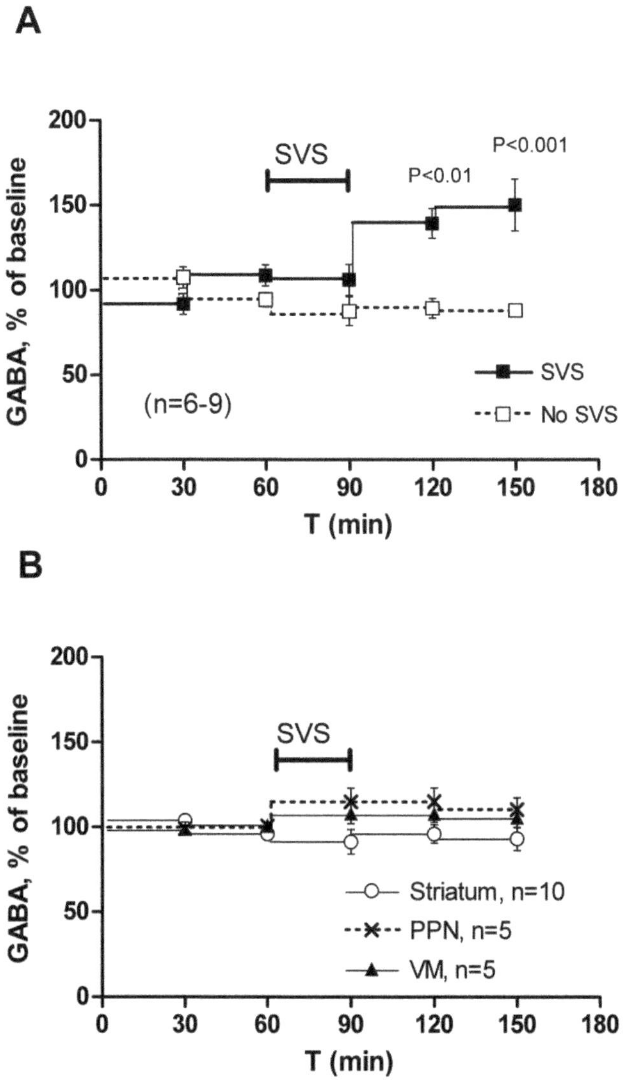

Figure 3. Microdialysate concentrations of GABA during stochastic vestibular stimulation (SVS) in intact animals and untreated intact controls. Percent of baseline, mean±SEM is shown. The GAT-inhibitor NNC 711 was included in the dialysates throughout the experiment. The SVS period is indicated by a horizontal bar. A. Nigral GABA concentration in SVS treated intact animals (n = 9) and untreated controls (n = 6). P-values from Bonferroni corrected post hoc tests following two-way ANOVA for timepoints T = 90–T = 150 minutes, indicating significant interaction between treatment and time $F(2,26) = 3.53$, $p = 0.044$. B. GABA concentrations in the PPN, VM and striatum of intact animals following SVS.

however, unable to demonstrate any effect of SVS in the Montoya stair case test. It therefore appears that noise benefit did not spread to forepaw motor function. Because SVS could not be administered in the Montoya box, an alternative explanation is that the effect of SVS wanes off rapidly. However, microdialysis measurements show that the increase in nigral GABA release is sustained for at least 30 minutes after SVS, making that explanation for the lack of observed effects on skilled forelimb use less likely. Future studies should also evaluate neurotransmission in the entopeduncular nucleus, the rodent equivalent of globus pallidus *pars interna*, an output structure which is analogue to SN *pars reticulata* and in more direct control of appendicular motor programs in humans [39].

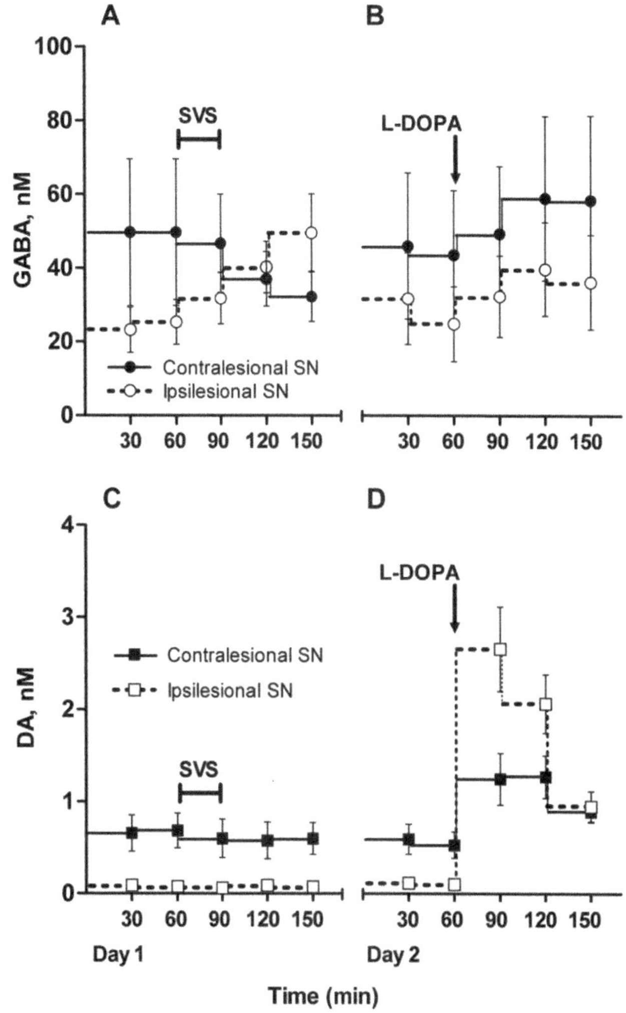

Figure 4. Absolute concentrations of GABA and dopamine in the ipsi- and contralesionalSN of hemilesioned 6-OHDA animals. Panel A–B shows the GABA concentrations and panel C–D the simultaneous dopamine (DA) concentrations following stochastic vestibular stimulation (SVS) and L-DOPA treatment (nM, mean±SEM). Panel A and C are measurements from day 1 and B and D from day 2. NNC 711 (30 µM) was present throughout the experiment and left in the microdialysis tube that was re-sealed over night. SVS treatment is indicated by a horizontal bar and the L-DOPA injection by an arrow.

A positive effect of SVS on axial muscle functions may have some clinical implications because axial rigidity and imbalance in Parkinson's disease are often difficult to control with L-DOPA and subthalamic DBS [40,41].

The effects of SVS on neurotransmission

The detection of rapid increases in GABA release with microdialysis sampling is improved by using a GABA re-uptake inhibitor (NNC 711) in the perfusion fluid [33]. One concern with this approach is that the re-uptake inhibitor may prolong and extend the increase in GABA concentrations and that this could lead to secondary changes in network activity. The increase in nigral GABA release following SVS is, however, not a long term effect of NNC 711 retrodialysis, because GABA concentrations remained stable in retrodialysed animals that did not receive SVS but were only retrodialysed with NNC 711 (Fig. 3A). Furthermore,

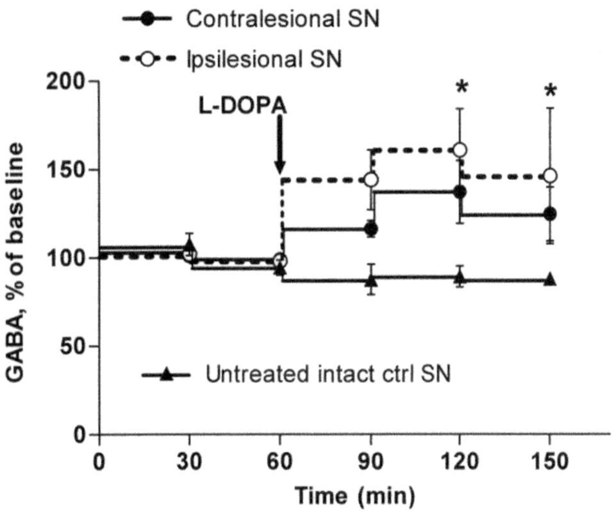

Figure 5. Relative changes in GABA concentrations in the ipsi- and contralesional substantia nigra of hemilesioned 6-OHDA animals following L-DOPA treatment. The microdialysate concentrations of GABA (percent of baseline, mean±SEM, n = 5–7) in the bilateral substantiae nigrae (SN) of 6-OHDA hemilesioned rats following an i.p. injection of L-DOPA are compared to GABA concentrations in the SN of untreated control animals (n = 6). Two-way repeated measure ANOVA for t = 90–t = 150 with L-DOPA treatment and time as main factors indicated a significant effect of treatment $F(2,30) = 4.68$, $p = 0.026$. * $p<0.05$ in post hoc Bonferroni corrected t-tests comparing ipsilesional SN in treated animals to control SN.

the increase occurred selectively in the SN and not in the other investigated nuclei following the vestibular stimulation.

The microdialysis measurements did not reveal any other significant changes in neurotransmitter concentrations following SVS than the increase in nigral GABA. Because microdialysis measurements of amino acids do not necessarily reflect synaptic release activity [42], at least not without using a reuptake inhibitor or stimulated release, this does not exclude altered activity and release of neurotransmitters other than dopamine and GABA in the investigated regions. In microdialysates from the SN, mean values of glutamate and glycine increased following SVS, but the change was not significant. It may be possible to determine if this represents an increase in afferent activity by measuring stimulated release or reuptake inhibitors. Immediate early gene expression may also be informative regarding by which mechanism nigral GABA release increases following SVS.

Nigrostriatal dopamine lesions lead to increased activity of the indirect loop and decreased activity of direct loop [43,44], and a decrease in GABA release in the SN can therefore be expected (Fig. 2). Increasing GABA release in the SN disinhibits the activation of motor programs and counteracts Parkinsonism. This is illustrated by the increase in nigral GABA concentrations that was observed following L-DOPA treatment, a treatment that is known to ameliorate Parkinsonism. There are obvious similarities between the nigral GABA increases observed after SVS in intact animals and the nigral increases in GABA after L-DOPA treatment of hemilesioned rats, not least regarding the timing of the GABA increase. Because SVS was not associated with altered dopamine release in striatum (or SN) it appears that the SVS-induced increase in nigral GABA release is not mediated by the action of dopamine on striatonigral GABA neurons. It is beyond the scope of this article to determine which pathways may mediate the observed nigral GABA increase following SVS, but as indicated in Fig. 2, vestibular neurons project to the cerebellum and the *formatio reticularis*, and possibly the PPN; all structures that may directly or indirectly alter neurotransmission in the SN. The subthalamic nucleus was not investigated in this study but may be of particular interest as DBS of the subthalamic nucleus has well documented antiparkinsonian effects, and increases nigral GABA and glutamate release [45].

Surprisingly, SVS did not induce similar changes in GABA release in the ipsi- and contralesional SN. Despite the bilateral nature of SVS the changes in GABA release in the two SN were in opposite direction, towards a more balanced GABA release between the two SN. Although we have no explanation for this phenomenon, it may have some relevance for the consistent improvement in rod performance that was observed after SVS, but not after L-DOPA treatment. Because L-DOPA increases nigral GABA release bilaterally, it may not reduce the imbalance in locomotor functions that is detrimental for rod performance. This could perhaps also explain why the initial laterality of Parkinsonism remains clearly observable throughout the course of the disease also in optimally treated Parkinson's disease patients.

Conclusion

SVS improves rod performance in 6-OHDA hemilesioned rats. We observed a prominent and sustained increase in GABA release in the SN following SVS, but no change in dopamine release in the SN or striatum. We propose that vestibular stimulation leads to a dopamine independent disinhibition of basal ganglia output that could promote movement initiation. We suggest that it should be evaluated as a treatment alternative for Parkinson's disease, in particular with prominent axial involvement.

Materials and Methods

Animals

Experimental design and procedures on unlesioned animals were carried out in compliance with the UK Animal (Scientific Procedures) Act 1986 (unlesioned animals) and approved by the UK Home Office under project license no 60/3334. Experimental design and procedures for experiments involving hemilesioned and sham lesioned animals were carried out in compliance with the European Communities Council Directive of November 24th, 1986 and approved by *Göteborgs djurförsöksetiska nämnd*, the local ethics committee in Gothenburg, Sweden, under project license no. 331/10. Experiments on unlesioned animals were performed using 18 male Lister-Hooded rats (120–150 g, age 5–6 weeks, Charles River Ltd, UK), 10 of these received microdialysis probes in striatum/SN and 8 received probes in PPN/VM. Experiments with 6-OHDA and sham-lesioned animals were performed using 32 female Sprague-Dawley rats (150–200 g, age 6–8 weeks, Charles River, Germany). For the microdialysis tests 12 of these were lesioned and 4 were sham lesioned. The other 16 animals were used for behavioral tests subsequent to a hemilesion in 11 of them and a sham-hemilesion in 5 of them. Animals were housed 4–6 per cage under standard controlled environmental conditions. After implantation of microdialysis guides they were single-caged for the remainder of the experiment.

Surgical procedures

All procedures were performed in deep surgical anesthesia (1.7–2.5% isoflurane). The surgical area was shaved and disinfected before lidocaine (1%) was infused subcutaneously for pre-emptive analgesia. Ketoprofen (5 mg/kg, s.c.) was administered as needed for post surgical analgesia.

6-hydroxydopamine hemilesion

The animal was placed in a stereotactic frame with bregma and lambda in the horizontal plane. After exposure of the skull, a hole was drilled over the medial forebrain bundle (4.2 mm posterior and +/− 1.2 mm lateral to bregma). A 150 µm diameter fused silica capillary was lowered 8 mm below the dura mater and 10 µg of 6-OHDA (Sigma Aldrich) dissolved in 0.9% NaCl, 0.3% ascorbate, 5 µg/µl, was injected over 2 minutes. The capillary was slowly removed after another minute and the brain surface was covered with periost membrane before closing the wound. Sham-treated animals received the saline ascorbate vehicle only.

Implantation of vestibular electrodes

Bilateral vestibular stimulation electrodes were attached by securing the 1 mm peeled end of a teflon coated stainless steel wire (0.2 mm diameter) over the horizontal canals of the two labyrinths by pushing short wire loops through the most ventral ends of the bilateral petrosal crests. The electrode wires were fixed to an acrylic cement foundation on the skull and externalized. Animals receiving microdialysis implants were implanted with electrodes in the same surgical session.

Microdialysis probe implantation

The unlesioned animals were stereotactically implanted with guide cannulae two days before microdialysis experiments. The skull was exposed and holes were drilled over two of the target nuclei. Targets were identified in the Paxinos and Watson rat brain atlas [46] and the coordinates relative to bregma were for the striatum: +1.0 mm anterior, 2.8 mm lateral, for the SN: 5.3 mm posterior, 2.1 mm contralateral, for the PPN: 8.0 posterior, 1.8 mm lateral, and for the VM: 2.6 posterior, 1.5 mm contralateral). Holes were also drilled in the parietal bones to fasten two stainless steel jewellers' screws. The dura was opened and the underlying pial membrane was opened by pulling it gently with fine forceps. Microdialysis guides (MAB 4.15.IC, Microbiotech, Sweden) were slowly implanted to a depth immediately above the target nucleus (6.0 mm from the brain surface for the PPN and VM, 3.2 mm and 6.2 mm for the striatum and the SN, respectively). The guide cannulae were secured to the skull with acrylic dental cement.

Some 6-OHDA hemilesioned animals were chronically implanted with in-house constructed microdialysis probes positioned bilaterally in the SN as described previously [47] and provided with vestibular electrodes as described above. Animals were allowed to recover for 24–48 h before the first microdialysis experiment.

Microdialysis

Two days after implanting the microdialysis guide cannulae, the unlesioned animals were briefly anaesthetized to insert bilateral microdialysis probes (Cuprophane membranes extending 1 mm, or with striatum guides, 3 mm, below the guide, MAB 4.15, Microbiotech, Sweden). The probes were continuously perfused with a Ringer solution containing 140 mM NaCl, 3.0 mM KCl, 1.2 mM CaCl2 and 1.0 mM MgCl2 and a GABA reuptake inhibitor, NNC 711 (30 µM, Tocris Cookson Ltd, UK). The animals were allowed to recover and were then kept in a standard microdialysis bowl, with the fluid lines from the microdialysis probes connected to the external equipment via a swiveled balance arm. The perfusion rate was kept at 4.0 µl/min for one hour and was then lowered to 1.5 µl/min. Baseline sampling commenced after 2 hours of perfusion. Microdialysis samples were collected every 30 minutes. After two baseline samples, stochastic vestibular stimulation commenced. Microdialysis perfusion continued throughout the 30 minute stimulation period, and for another 60 minutes, providing a total of 5 microdialysis samples from each probe. The samples were split and analyzed for amino acid content as described previously [33], and for amine content by two-dimensional HPLC with electrochemical detection as described by Lindgren et al. [34].

The 6-OHDA hemilesioned animals were subjected to a repeat microdialysis experiment on the following day. The experiment protocol was as described above, but instead of a 30 minutes stimulation period, they were injected with L-DOPA and benserazide (6 mg/kg and 12 mg/kg, respectively, i.p.) after the second sampling period.

After the last microdialysis experiment animals were euthanized and the brain was removed and fixed in paraformaldehyde. The brains were sectioned and probe locations were determined with reference to the Paxinos and Watson rat brain atlas [46]. Results from probes with less than 50% of the active membrane in the target nucleus were excluded. This reduced the final n of investigated brain areas to 5 for PPN and VM and 10 for striatum, 9 for SN respectively in intact animals. In 6-OHDA lesioned animals, a total of 6 probes were included in the lesioned SN, and 6 in the unlesioned. The positions of the active membranes of all included probes are indicated in Fig. S5.

Stimulation protocol

Current was delivered through the bilateral electrodes using two setups. Unlesioned animals were stimulated using two stimulus isolators (NeuroLog NL800, Digitimer Ltd. Hertfordshire, UK) connected to a biphasic pulse buffer unit (NeuroLog NL512) and a stimulus pattern fed from a voltage source pulse generator or a DAT recorder with a prerecorded stochastic stimulation (<30 Hz) pattern. For lesioned animals a bipolar analogue stimulus isolator (model 2200 A-M Systems, Sequim, Washington, U.S.A) was used, and the stimulus pattern was fed from a low pass (<30 Hz) filtered voltage source generator (BitScope 100 with BitGen coprocessor, Metachip Pty. Ltd., Australia), which was programmed to generate sinusoid or white noise patterned voltages.

Bipolar 1 Hz sinusoid stimulation pattern was used to determine an appropriate maximum current. The amplitude was slowly increased until a 1 Hz rocking of the animals head was clearly visible. The lowest amplitude A that reproducibly produced head

Table 2. Maximum stimulation amplitudes (mA, mean±SEM) in the different experiments.

Experiment	Animals	Mean±SEM (mA)
Rotarod	6-OHDA hemiles (n = 6)	0.30±0.02
	Sham hemiles (n = 5)	0.27±0.03
Montoya staircase	6-OHDA hemiles (n = 5)	0.23±0.05
	Sham hemiles (n = 3)	0.27±0.03
Microdialysis	SN/ST, unlesioned (n = 9)	0.48±0.02
	PPN/VM, unlesioned (n = 5)	0.41±0.03
	Bilat SN, 6-OHDA hemiles (n = 7)	0.47±0.08

Stimulation amplitudes were determined by subjecting the animals to a sinusoid current that was increased until head rocking was just visible. During stochastic stimulation the amplitude never exceeded ± this amplitude. A different current delivery system was used for the microdialysis experiments in unlesioned animals. 6-OHDA 6-hydroxy-dopamine, PPN pedunculopontine nucleus, ST striatum, SN substanta nigra, VM ventromedial thalamus.

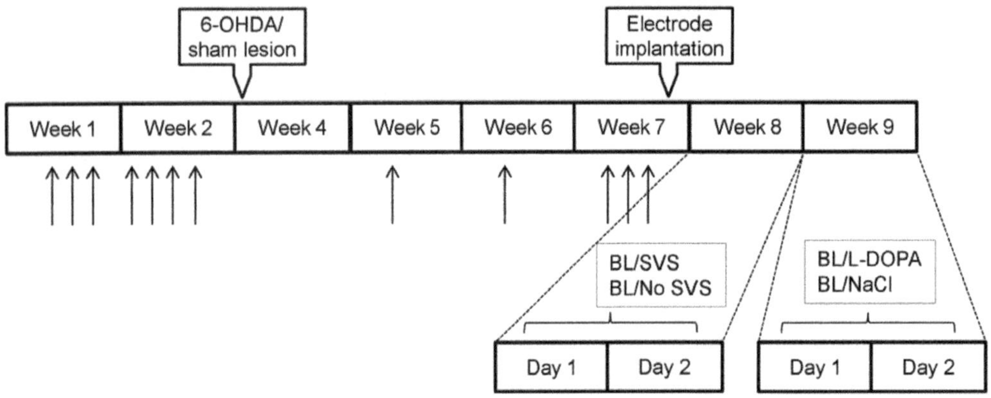

Figure 6. Schematic illustration of the experimental design for the Rotarod testing procedure. The black arrows represent training episodes for each animal. Rod performance during stochastic vestibular stimulation (SVS) or no SVS was compared to the baseline measurement (BL, 4×10 min before either condition) on the day of the experiment. The stimulated condition (30 min SVS in cage followed by 4×10 min testing during SVS) or non stimulated condition (30 min rest in cage followed by 4×10 min testing during no SVS) was carried out in a counterbalanced order after the BL measurements. L-DOPA testing was carried out in the beginning of the following week in counterbalanced order; BL testing followed by either an L-DOPA or NaCl injection, rest in cage for 30 min followed by testing for 4×10 min.

rocking was used as maximum amplitude during the stochastic vestibular stimulation period of the experiment.

The mean maximum stimulation amplitudes are given in Table 2, and were typically between 0.2 and 0.6 mA. The stochastic stimulation protocol consisted of a 30 minutes sequence containing bipolar current with random amplitude between (−A) and (+A) and random frequencies <30 Hz. Examples of the stimulation currents are given in Fig. S4.

Motor behavior – Rotarod

Animals were trained to run on a 6 cm (ø) accelerating rod. Training and a set of prelesion baseline tests took place before any surgical procedure, and animals were re-tested three weeks after sham/6-OHDA lesion procedures. Each baseline test and experimental test episode consisted of four 10 minute long accelerations (5–40 rpm). The mean time on rod during each test episode was determined as previously described [28]. Three to five days after implantation of bilateral vestibular electrodes the animals were baseline tested and after 30 minutes tested again with or without vestibular stimulation.

Stimulation was started 30 minutes before the first of four rod session, and continued throughout testing. This procedure was repeated for two days in a counterbalanced order so that each animal was tested one day with and one day without stimulation. The same procedure was carried out over an additional two day period, this time with either an L-DOPA and benserazide intraperitoneal injection (6 mg/kg and 12 mg/kg respectively) or a sham injection (NaCl 0,9%) given in a counterbalanced order (Fig. 6). The change in performance invoked by the SVS or no SVS condition was calculated with reference to the baseline performance on the same day. 6-OHDA-lesioned animals with less than 70% dopamine depletion (n = 5), as revealed by port-mortem analysis, were excluded from the Rotarod data analysis.

Motor behavior - Montoya staircase test

Two weeks before the hemilesion procedure, rats were food restrained to 90–95% of free feeding weight and trained to retrieve sugar pellets (45 mg each; BioServ, Frenchtown, NJ, U.S.A.) in a Montoya staircase apparatus (9×6×30 cm, Campden Instruments Ltd, Loughborough, UK). Training took place between 9am–1pm, over a period of 5 days and consisted of 2×15 min sessions in the box with 1 h 45 min rest between sessions. The box was covered with a dark cloth to maintain a dark and constant environment throughout training and testing. Animals which after initial training retrieved at least 9 pellets per side were included in the experiment. Seven days after implantation of vestibular electrodes the rats were food restrained and baseline tested. They were tested again over the following two days with and without vestibular stimulation in a counterbalanced order. Stimulation was applied for 30 minutes immediately before the animal was transferred to the test box. 6-OHDA lesioned animals with less than 70% dopamine depletion in the post mortem analysis were excluded from analysis.

Statistical analysis

When appropriate, microdialysate concentrations were transformed to percent of baseline to reduce the effects of recovery variations. The treatment effect on neurotransmitter concentrations was evaluated by repeated measure two-way ANOVA with treatment and time as independent factors or with one sample t-tests with 100% as the theoretical mean, as appropriate. Locomotion effects were evaluated with paired t-tests comparing the change in time (s) on rod observed during stimulation or no-stimulation conditions. On the Montoya staircase, the overall number of sugar pellets consumed, total number of pellets on each side as well as the ratio between number of pellets eaten from impaired (contralesional) side and non-impaired (ipsilesional) side after vestibular stimulation or after no stimulation, were evaluated by paired t-tests. Data are given as mean±SEM, and $p \leq 0.05$ was considered statistically significant.

Supporting Information

Figure S1 Dopamine concentrations in dialysates from the substantia nigra and striatum of normal animals. Figure shows dopamine (DA) concentrations (percent of baseline, mean±SEM) in dialysates from the SN and the contralateral striatum of 9 un-lesioned animals subjected to stochastic vestibular stimulation for 30 minutes (horizontal bar). Concentrations remained stable throughout the stimulation period and the following hour. For clarity, measurements from untreated control animals (n = 6) are omitted from the figure. NNC 711 (30 mM) was present in the perfusate throughout the experiment.

Figure S2 Amino acid concentrations in dialysates from the substantia nigra of normal animals. Panel A shows the relative concentrations (percent of baseline, mean±SEM) of glutamate and glycine from the SN of un-lesioned animals subjected to stochastic vestibular stimulation (SVS, horizontal bar, $n = 9$) or no SVS (ctrl, $n = 6$) for 30 minutes. Over this time there were no significant difference in relative concentrations (Two-way repeated measure ANOVA for $t = 90$–$t = 150$ with SVS treatment and time as main factors, Glutamate: $F_{treat}(1, 24) = 2.76$, $p = 0.12$, $F_{time}(2, 24) = 0.5$, $p = 0.61$, $F_{interact}(2,24) = 1.01$, $p = 0.38$, Glycine $F_{treat}(1, 26) = 1.36$, $p = 0.26$, $F_{time}(2, 26) = 1.63$, $p = 0.21$, $F_{interact}(2,26) = 1.34$, $p = 0.28$). The higher mean values and large variability in samples $t = 120$ and $t = 150$ coincided with increased exploratory behavior in the cage. Taurine and glutamine levels remained stable following SVS (Panel B, two-way repeated measure ANOVA for $t = 90$–$t = 150$ with SVS treatment and time as main factors, Taurine: $F_{treat}(1, 26) = 1.88$, $p = 0.19$, $F_{time}(2, 26) = 1.63$, $p = 0.22$, $_{interact}(2,26) = 0.44$, $p = 0.65$, Glutamine $F_{treat}(1, 26) = 1.71$, $p = 0.21$, $F_{time}(2, 26) = 1.81$, $p = 0.18$, $F_{interact}(2,26) = 0.30$, $p = 0.75$). NNC 711 (30 mM) was present in the perfusate throughout the experiment.

Figure S3 Glutamate concentrations in dialysates from the bilateral substantiae nigrae of 6-OHDA hemilesioned rats. Figure shows glutamate (GLU) concentrations (percent of baseline, mean±SEM) in dialysates from the ipsi- and contralesional SN before, during and after stochastic vestibular stimulation (SVS, horizontal bar, 30 minutes) in panel A and in response to L-DOPA treatment (6 mg/kg, i.p., arrow) in panel B. NNC 711 (30 mM) was present in the perfusate throughout the experiment.

Figure S4 Stochastic stimulation current patterns. Panel A shows the stimulation current used in normal animals and panel B the current used in hemilesioned rats. The time base is 200 ms per division, and the upmost part of each panel shows the frequency distribution.

Figure S5 Placement of microdialysis probes in intact and hemilesioned animals. Unlesioned animals are shown schematically in panel A (striatum and substantia nigra) and panel B (ventromedial thalamus and the pedunculopontine nucleus). Panel C shows the bilateral nigral locations in hemilesioned animals. Solid lines indicate the estimated location of the active dialysis membrane.

Author Contributions

Conceived and designed the experiments: FB HN MBD. Performed the experiments: GS FB. Analyzed the data: GS FB. Contributed reagents/materials/analysis tools: FB HN MBD. Wrote the paper: GS FB HN MBD.

References

1. Bloem BR, Hausdorff JM, Visser JE, Giladi N (2004) Falls and freezing of gait in Parkinson's disease: a review of two interconnected, episodic phenomena. Mov Disord 19: 871–884.
2. Aihara T, Kitajo K, Nozaki D, Yamamoto Y (2010) How does stochastic resonance work within the human brain? - Psychophysics of internal and external noise. Chemical Physics 375: 616–624.
3. Hospedales TM, van Rossum MC, Graham BP, Dutia MB (2008) Implications of noise and neural heterogeneity for vestibulo-ocular reflex fidelity. Neural Comput 20: 756–778.
4. Li SC, von Oertzen T, Lindenberger U (2006) A neurocomputational model of stochastic resonance and aging. Neurocomputing 69: 1553–1560.
5. McDonnell MD, Abbott D (2009) What is stochastic resonance? Definitions, misconceptions, debates, and its relevance to biology. PLoS Comput Biol 5: e1000348.
6. Yu XL, Lewis ER (1989) Studies with spike initiators: linearization by noise allows continuous signal modulation in neural networks. IEEE Trans Biomed Eng 36: 36–43.
7. Funke K, Kerscher NJ, Worgotter F (2007) Noise-improved signal detection in cat primary visual cortex via a well-balanced stochastic resonance-like procedure. Eur J Neurosci 26: 1322–1332.
8. Wells C, Ward LM, Chua R, Timothy Inglis J (2005) Touch noise increases vibrotactile sensitivity in old and young. Psychol Sci 16: 313–320.
9. Zeng FG, Fu QJ, Morse R (2000) Human hearing enhanced by noise. Brain Res 869: 251–255.
10. Collins JJ, Imhoff TT, Grigg P (1996) Noise-enhanced tactile sensation. Nature 383: 770–770.
11. Kosko B (2006) Noise. New York: Viking Adult. 272 p.
12. Smith PF, Geddes LH, Baek JH, Darlington CL, Zheng Y (2010) Modulation of memory by vestibular lesions and galvanic vestibular stimulation. Front Neurol 1: 141.
13. Usher M, Feingold M (2000) Stochastic resonance in the speed of memory retrieval. Biological Cybernetics 83: L11-16–L11-16.
14. Wilkinson D, Ko P, Kilduff P, McGlinchey R, Milberg W (2005) Improvement of a face perception deficit via subsensory galvanic vestibular stimulation. Journal of the International Neuropsychological Society: JINS 11: 925–929.
15. Wilkinson D, Nicholls S, Pattenden C, Kilduff P, Milberg W (2008) Galvanic vestibular stimulation speeds visual memory recall. Exp Brain Res 189: 243–248.
16. Wilkinson D, Zubko O, Degutis J, Milberg W, Potter J (2010) Improvement of a figure copying deficit during subsensory galvanic vestibular stimulation. J Neuropsychol 4: 107–118.
17. Lobel E, Kleine JF, Bihan DL, Leroy-Willig A, Berthoz A (1998) Functional MRI of galvanic vestibular stimulation. J Neurophysiol 80: 2699–2709.
18. Mian OS, Day BL (2009) Determining the direction of vestibular-evoked balance responses using stochastic vestibular stimulation. J Physiol 587: 2869–2873.
19. Yamamoto Y, Struzik ZR, Soma R, Ohashi K, Kwak S (2005) Noisy vestibular stimulation improves autonomic and motor responsiveness in central neurodegenerative disorders. Ann Neurol 58: 175–181.
20. Pavlik AE, Inglis JT, Lauk M, Oddsson L, Collins JJ (1999) The effects of stochastic galvanic vestibular stimulation on human postural sway. Exp Brain Res 124: 273–280.
21. Scinicariello AP, Inglis JT, Collins JJ (2002) The effects of stochastic monopolar galvanic vestibular stimulation on human postural sway. J Vestib Res 12: 77–85.
22. Mulavara AP, Fiedler MJ, Kofman IS, Wood SJ, Serrador JM, et al. (2011) Improving balance function using vestibular stochastic resonance: optimizing stimulus characteristics. Exp Brain Res 210: 303–312.
23. Scinicariello AP, Eaton K, Inglis JT, Collins JJ (2001) Enhancing human balance control with galvanic vestibular stimulation. Biol Cybern 84: 475–480.
24. Pan W, Soma R, Kwak S, Yamamoto Y (2008) Improvement of motor functions by noisy vestibular stimulation in central neurodegenerative disorders. J Neurol 255: 1657–1661.
25. Pal S, Rosengren SM, Colebatch JG (2009) Stochastic galvanic vestibular stimulation produces a small reduction in sway in Parkinson's disease. J Vestib Res 19: 137–142.
26. Sikström S, Söderlund G (2007) Stimulus-dependent dopamine release in attention-deficit/hyperactivity disorder. Psychol Rev 114: 1047–1075.
27. Pålsson E, Söderlund G, Klamer D, Bergquist F (2010) Noise benefit in prepulse inhibition of the acoustic startle reflex. Psychopharmacol 214: 675–685.
28. Andersson D, Nissbrandt H, Bergquist F (2006) Partial depletion of dopamine in substantia nigra impairs motor performance without altering striatal dopamine neurotransmission. Eur J Neurosci 24: 617–624.
29. Bergquist F, Shahabi HN, Nissbrandt H (2003) Somatodendritic dopamine release in rat substantia nigra influences motor performance on the accelerating rod. Brain Res 973: 81–91.
30. Lappin JM, Reeves SJ, Mehta MA, Egerton A, Coulson M, et al. (2009) Dopamine release in the human striatum: motor and cognitive tasks revisited. J Cereb Blood Flow Metab 29: 554–564.
31. Winn P (2008) Experimental studies of pedunculopontine functions: are they motor, sensory or integrative? Parkinsonism Rel Disord 14 Suppl 2: S194-198–S194-198.
32. Andersson DR, Bjornsson E, Bergquist F, Nissbrandt H (2010) Motor activity-induced dopamine release in the substantia nigra is regulated by muscarinic receptors. Exp Neurol 221: 251–259.
33. Bergquist F, Ludwig M, Dutia MB (2008) Role of the commissural inhibitory system in vestibular compensation in the rat. J Physiol 586: 4441–4452.
34. Lindgren HS, Andersson DR, Lagerkvist S, Nissbrandt H, Cenci MA (2010) L-DOPA-induced dopamine efflux in the striatum and the substantia nigra in a rat model of Parkinson's disease: temporal and quantitative relationship to the expression of dyskinesia. J Neurochem 112: 1465–1476.

35. Dakin CJ, Son GML, Inglis JT, Blouin J-S (2007) Frequency response of human vestibular reflexes characterized by stochastic stimuli. J Physiol 583: 1117–1127.
36. Gravelle DC, Laughton CA, Dhruv NT, Katdare KD, Niemi JB, et al. (2002) Noise-enhanced balance control in older adults. Neuroreport 13: 1853–1856.
37. Priplata AA, Patritti BL, Niemi JB, Hughes R, Gravelle DC, et al. (2006) Noise-enhanced balance control in patients with diabetes and patients with stroke. Ann Neurol 59: 4–12.
38. Manjarrez E, Mendez I, Martinez L, Flores A, Mirasso CR (2007) Effects of auditory noise on the psychophysical detection of visual signals: cross-modal stochastic resonance. Neurosci lett 415: 231–236.
39. Baker KB, Lee JY, Mavinkurve G, Russo GS, Walter B, et al. (2010) Somatotopic organization in the internal segment of the globus pallidus in Parkinson's disease. Exp Neurol 222: 219–225.
40. Ostergaard K, Sunde NA (2006) Evolution of Parkinson's disease during 4 years of bilateral deep brain stimulation of the subthalamic nucleus. Mov Disord 21: 624–631.
41. Umemura A, Oka Y, Okita K, Toyoda T, Matsukawa N, et al. (2010) Predictive factors affecting early deterioration of axial symptoms after subthalamic nucleus stimulation in Parkinson's disease. Parkinsonism Rel Disord 16: 582–584.
42. Timmerman W, Westerink B (1997) Brain microdialysis of GABA and glutamate: what does it signify? Synapse 27: 242–261.
43. DeLong MR, Wichmann T (2007) Circuits and Circuit Disorders of the Basal Ganglia. Arch Neurol 64: 20–24.
44. Kravitz AV, Freeze BS, Parker PR, Kay K, Thwin MT, et al. (2010) Regulation of parkinsonian motor behaviours by optogenetic control of basal ganglia circuitry. Nature 466: 622–626.
45. Windels F, Bruet N, Poupard A, Urbain N, Chouvet G, et al. (2000) Effects of high frequency stimulation of subthalamic nucleus on extracellular glutamate and GABA in substantia nigra and globus pallidus in the normal rat. Eur J Neurosci 12: 4141–4146.
46. Paxinos G, Watson C (1998) The Rat Brain in Stereotaxic Coordinates, Fourth Edition. Waltham, MA: Academic Press. 256 p.
47. Bergquist F, Jonason J, Pileblad E, Nissbrandt H (1998) Effects of local administration of L-, N-, and P/Q-type calcium channel blockers on spontaneous dopamine release in the striatum and the substantia nigra: a microdialysis study in rat. J Neurochem 70: 1532–1540.
48. Bostan AC, Dum RP, Strick PL (2010) The basal ganglia communicate with the cerebellum. Proc Natl Acad Sci U S A 107: 8452–8456.
49. Hoshi E, Tremblay L, Feger J, Carras PL, Strick PL (2005) The cerebellum communicates with the basal ganglia. Nat Neurosci 8: 1491–1493.

2

DPP4 Deficiency Preserved Cardiac Function in Abdominal Aortic Banding Rats

Hui-Chun Ku, Ming-Jai Su*

Institute of Pharmacology, College of Medicine, National Taiwan University, Taipei, Taiwan

Abstract

Dipeptidyl peptidase-4 (DPP4) enzyme inhibition has been reported to increase plasma glucagon-like peptide-1 (GLP-1) level for controlling postprandial glucose concentration. A prominent GLP-1 level in DPP4-deficient rats contributed to the resistance of endotoxemia and myocardial infarction. DPP4 deficiency also increased the capability against H_2O_2-induced stress in cardiomyocyte. However, long term effect of loss DPP4 activity on cardiac performance remained unclear. We used abdominal aortic banding (AAB) to induce pressure overload in wild-type and DPP4-deficient rats, and investigated the progression of heart failure. Cardiac histology and function were determined. Blood sample was collected for the plasma biochemical marker measurement. Heart weight to body weight ratio increased 1.2-fold after 6 weeks of AAB surgery. Cardiac function was compensated against pressure overload after 6 weeks of AAB surgery, but progressed to deterioration after 10 weeks of AAB surgery. AAB induced cardiac dysfunction was alleviated in DPP4-deficient rats. DPP4 activity increased significantly in wild-type rats after 10 weeks of AAB surgery, but remained unchanged in DPP4-deficient rats. In contrast, GLP-1 concentration was elevated by AAB after 6 weeks of surgery in DPP4-deficient rats, and remained high after 10 weeks of surgery. Ang II level markedly increased after 6 weeks of AAB surgery, but were less in DPP4-deficient rats. Massive collagen deposits in wild-type rat hearts appeared after 10 weeks of AAB surgery, which were alleviated in DPP4-deficient rats. Long term deficiency of DPP4 activity improved cardiac performance against pressure overload in rat, which may be attributed to a great quantity of GLP-1 accumulation during AAB.

Editor: Konrad Urbanek, Second University of Naples, Italy

Funding: This work was supported by National Science Council, Taiwan (NSC 101-2325-B-002 -058, NSC 102-2325-B-002 -056, and NSC 102-2325-B-002 -086). The funders had no role in study design, data collection and analysis, decision to publish, or preparation of the manuscript.

Competing Interests: The authors declare that they have no competing interests.

* E-mail: mingja@ntu.edu.tw

Introduction

Heart failure remains a leading cause of death in patients. Sustained pressure overload stimulates pathological cardiac hypertrophy, which may lead to heart failure [1]. Insulin resistance has been reported to occur with increased left ventricular mass and cardiac hypertrophy [2], and also lead to high incidence of heart failure [2,3]. Diabetes impairs insulin sensitivity which results in a shift of myocardial substrate usage towards even higher fatty acid utilization [4], which contributes to myocardial dysfunction [5]. Changes in myocardial structure, function and metabolism have been described in diabetes, which link to high risk factor of cardiovascular disease, so the medication that decreases cardiovascular risk factor in the diabetic subject is an important issue. However, the efficacy of current therapy to prevent cardiovascular disease in diabetes is still limited.

Incretin hormones have recently been developed as a new target for the maintenance of systemic glucose homeostasis. Glucagon-like peptide-1 (GLP-1), one of the incretin hormones, is secreted from the gastrointestinal tract after food intake and stimulates glucose-dependent insulin secretion by activating GLP-1 receptor [6]. When cleaved by dipeptidyl peptidase-4 (DPP4), GLP-1 is characterized by a complete lack of insulinotropic activity [7]. DPP4 is an enzyme that catalyzes the release of dipeptides from the N terminus which contain Pro or Ala in the third amino acid position [7]. GLP-1-based therapies have been approved recently as diabetes treatment, such as GLP-1 analogue and DPP4 inhibitor [8–10]. In addition, GLP-1 receptor is expressed in the cardiovascular system [11,12], and modulates cardiac function. GLP-1 signaling has been demonstrated to improve cardiac function in several animal model experiments, such as endotoxemia, myocardial infarction, and diabetic cardiomyopathy [11,13–18]. All these experiments evaluate that the cytoprotective effect of GLP-1 is mediated mainly by the activation of survival pathway, inhibition of apoptosis pathway, and improvement of glucose utilization to against oxidative and inflammatory stress in heart.

Accumulating studies have indicated that DPP4 inhibitors also play an important role on cardiovascular system. Lack of DPP4 activity acts to decrease myocardial infarct size, stabilize the cardiac electrophysiological state during myocardial ischemia, reduce ischemia/reperfusion injury, and prevent left ventricular remodeling after myocardial infarction [14,15,19,20]. In the cell study, DPP4 deficiency exerts protective effect against H_2O_2-induced stress in isolated cardiomyocytes [21]. Long term deficiency of DPP4 induces changes of cellular function, and increase the capability against oxidative stress via enhancing AKT signaling and glucose uptake in association with preserving catalase activity, via both GLP-1 receptor dependent and independent signaling [21]. Moreover, DPP4 inhibitors also protect the vasculature through their anti-inflammatory, anti-atherosclerotic effects, and promote vascular relaxation. [19].

Several clinical studies were also performed to investigate the safety of DPP4 inhibitors on diabetes with cardiovascular disease. DPP4 inhibitors improve several risk factors in patients with T2DM, and the positive effects on heart have also been described in patients with ischemic heart disease or congestive heart failure [10,22,23]. However, a controversy study was also found. Saxagliptin did not affect the rate of ischemic events, but increased the rate of hospitalization for heart failure [24]. The potential explanation for the differential studies is the exposure time of the drug may not long enough and the difference of glycated hemoglobin levels between patients [24].

Accordingly, long term safety of DPP4 inhibitor remains unclear. Abdominal aortic banding is one of the most widely used animal models in induction of heart failure [25]. As the animal grows, the aortic outflow is increasingly impeded, and hypertension is developed gradually. In this context, we aimed to investigate the progression of cardiac response after pressure overload surgery in both wild-type and DPP4-deficient rats.

Materials and Methods

1. Experimental Animals

The research was performed in accordance with the Guide for the Care and Use of Laboratory Animals published by the US National Institutes of Health (NIH publication no. 85–23, revised 1996), and was approved by the Institutional Animal Care and Use Committee of the National Taiwan University. Adult male Fischer 344 and DPP4-deficient mutant (DchcHsd-DPP IV) rats were purchased from National Laboratory Animal Center of Taiwan and Harlan laboratories, Inc, respectively. Fischer 344 is the control strain for the DPP4-deficient rats. Rats were maintained under a 12-h light/dark cycle at a controlled temperature (21±2°C) with free access to food and tap water.

2. Experimental Model of Pressure-overload in vivo

Pressure-overload was induced by abdominal aortic banding (AAB) as described previously [26]. Briefly, rats (8 week old) were anesthetized with pentobarbital (60 mg/kg, i.p.). After rats appeared calm, the abdominal aorta was isolated adjacent to renal arteries, and constricted by a 4–0 silk suture ligature tied firmly against a 22-gauge needle. The needle was removed immediately and performed a constriction of 0.7 mm in diameter. Sham-operated rat underwent a similar surgical procedure without aortic constriction. After operation, rats were placed on an electric blanket to maintain body temperature until awake, and monitor after 6 or 10 weeks of surgery.

3. Determination of Cardiac Function in vivo

Cardiac function was estimated in vivo by a pressure-volume catheter (1.9 F; Scisense Instruments, Ontario, Canada) as described previously [14,15,27]. After 6 or 10 weeks of surgery, rats were anesthetized, and the right carotid artery was cannulated with a catheter, which connected to a control box with signal conditioning and amplification circuitry (112B-C009, Scisense, Canada). Then the catheter was advanced into the left ventricle for the measurement of pressure and volume. PV loops were recorded in basal condition, and several parameters were analyzed. PV loops were then measured under the condition of preload change, which induced by inferior vena cava constriction as described previously [14,15,27]. The end-systolic pressure-volume relation (ESPVR) and end-diastolic pressure-volume relation (EDPVR) are characterized by the slope of end-systolic and end-diastolic pressure-volume point, which are load-independent indexes of myocardial contractility and ventricular compliance, respectively.

4. Determination of Cardiac Histology

After the hemodynamic data recordings, the heart was rapidly removed and perfused on a modified Langendorff perfusion apparatus with Krebs buffer (containing: 110 mM NaCl, 2.6 mM KCl, 1.2 mM KH_2PO_4, 1.2 mM $MgSO_4$, 25 mM $NaHCO_3$ and 11 mM glucose (pH 7.4)). After blood was washed out, the weight of heart was measured, and calculated the heart to body weight ratio (HW/BW). The hearts were fixed in 4% paraformaldehyde, embedded in paraffin, and sectioned horizontally in 4 μm slice. HE and masson's trichrome stain were performed for histological and fibrosis analysis.

5. Determination of Plasma DPP4 Activity, GLP-1, and Angiotensin II Concentration

Blood sample was collected and centrifuged at 13,000 rpm for 5 min, and plasma was stored at −80°C until used. Gly-Pro-AMC (St. Louis, MO, USA), a synthesis substrate of DPP4, was used for the determination of DPP4 activity as described previously. The amount of AMC liberation is positive related with DPP4 activity. Plasma in 50 mM HEPES solution containing 100 uM Gly-Pro-AMC was incubated at 37°C for 60 min, and the liberation of free AMC was monitored at 370-nm excitation and 440-nm emission. Plasma GLP-1 levels were measured by ELISA kit (Millipore Corporation, Billerica, MA, USA). Plasma angiotensin II levels were measured by ELISA kit (BioRay, Laguna Hills, CA, USA).

6. Statistical Analysis

All values were presented as means ± SE. The results were analyzed using ANOVA followed by Bonferroni's post hoc tests. $P<0.05$ was considered as significant difference.

Results

1. AAB-induced Cardiac Dysfunction was Recovered in DPP-4 Deficient Rats

The cardiac functions were measured by PV loops in the sham and AAB conditions after 6 and 10 weeks of surgery. The representative PV loops were shown (Fig. 1), and the hemodynamic data were all quantified (Table 1). In the 6-week sham condition, similar cardiac performance was found in two kinds of rats. After 6 weeks of AAB surgery, cardiac function was compensated against pressure overload. AAB caused an enhancement of LVSP in wild-type rats, being 1.17-fold of sham. LVESV also increased after AAB surgery, being 2.2-fold of sham, while LVEDV remained unchanged, so both SV and EF were reduced in wild-type rats. The compensatory cardiac response to AAB was also found in DPP4-deficient rats. LVSP was elevated to 1.21 fold of sham in DPP4-deficient rats after 6 weeks of AAB. However, the increase of LVESV was less in DPP4-deficient rats; therefore, SV and EF remained. ESPVR and EDPVR, load-independent cardiac function indexes, elevated more after AAB in wild-type rats, suggesting that AAB resulted in the increase of myocardial contractility and stiffness, whereas EDPVR in DPP4-deficient rats remained unchanged after 6 week of AAB.

Without AAB, aging worsen ventricular contractility and relaxation performance, especially the rate of relaxation was suppressed more in wild-type than in DPP4-deficient rats in the 10-week sham condition. Cardiac function was deteriorated after 10 weeks of AAB surgery. In wild-type rats, the reduction of LVSP and the elevation of LVEDP were found, being 0.87- and 1.45-fold of sham group, respectively. Furthermore, the suppression of +dP/dt and −dP/dt were also observed, indicating the deterioration of systolic and diastolic functions. AAB caused a more

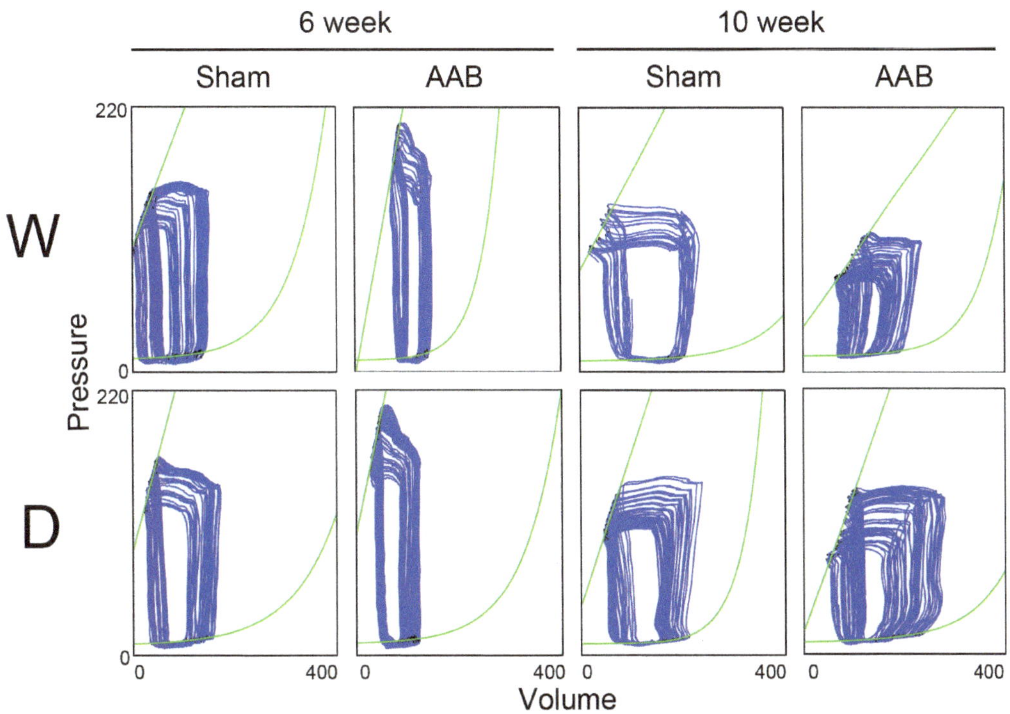

Figure 1. Pressure-volume loops of wild-type and DPP4-deficient rat hearts in sham and AAB conditions after 6 or 10 weeks of surgery. W indicates wild-type, and D indicates DPP4 deficiency.

significant elevation of LVESV in wild-type rats than that in DPP4-deficient rats, being 1.94- and 1.32-fold of sham group, respectively, which resulted in a more prominent reduction of EF in wild-type rats. AAB induced cardiac dysfunction was preserved in DPP4-deficient rats.

2. Cardiac Hypertrophy after AAB Surgery

The heart weight to body weight ratio were measured (Fig. 2B), and the representative heart from each condition were shown (Fig. 2A). DPP4-deficient rats gained weight slower than wild-type rats [28]. Male wild-type rats gained about 40 g weight during 4 weeks, but DPP4-deficient rats gained less than 20 g weight. Heart weight and body weight were both less in DPP4-deficient rats than that in wild-type rats. In the 6-week sham group, HW/BW is similar in two rats. Without AAB, aging can increase HW/BW slightly. After 6 week of AAB, HW/BW increased in wild-type rats, being 1.2-fold of sham, and remained higher after 10 week of AAB. There was no significant difference in HW/BW between two kinds of rats.

Histological analysis of hematoxylin and eosin-stained cardiac sections was shown (Fig. 2C and F). After 10 weeks of AAB surgery, the dimension of left ventricle posterior wall and interventricular septum increased in wild-type rats, along with a smaller ventricular cavity. Cardiomyocyte cross-sectional diameter and cell number were calculated (Fig. 2D and E). In the sham condition, cardiomyocyte was little bigger in wild-type rats than that in DPP4-deficient rats. Morphology analyses showed that AAB induced a marked time-dependent increase in diameter of cardiomyocyte associated with the loss of cell number in wild-type hearts. These changes were less in DDP4-deficient hearts.

3. GLP-1 Signaling Upregulated in DPP4-deficient Rats

In the 6-week sham groups, DPP4 activity was 5.5-fold higher in wild-type rats than that in DPP4-deficient rats (Fig. 3A). AAB did not alter DPP4 activity in two kinds of rats after 6 weeks of surgery. However, DPP4 activity increased in wild-type rats after 10 weeks of AAB surgery, but it remained unchanged in DPP4-deficient rats.

In the 6-week sham condition, plasma GLP-1 concentration was 3.0-fold higher in DPP4-deficient rats than that in wild-type rats (Fig. 3B). AAB surgery did not alter GLP-1 level in wild-type rats. However, in DPP4-deficient rats, GLP-1 concentration was elevated by AAB after 6 weeks of surgery, and it remained higher than sham group after10 weeks.

4. AAB Induced Ang II Level Elevation was Alleviated in DPP4-deficient Rats

Angiotensin II (Ang II) plays an important role in the progression of pressure overload-induced heart failure. Without AAB, aging cause Ang II concentration increased in both rats (Fig. 3C). Ang II level markedly increased after 6 weeks of AAB surgery, and became much higher after 10 weeks. This elevation was alleviated in DPP4-deficient rats.

5. AAB Induced Massive Collagen Deposits was Less in DPP4-deficient Rats

Masson's trichrome staining revealed fibrosis in blue (Fig. 4A), and the percentage of fibrosis was also quantified (Fig. 4B). Collagen deposits was not found in myocardium at 6 weeks after AAB surgery (data not shown), but prominent collagen deposits appeared in myocardium of wild-type rats at 10 weeks after AAB surgery. The collagen deposit was less in DPP4-deficient rats.

Table 1. Invasive hemodynamic analysis of 6 and 10 weeks after AAB in wild-type and DPP4-deficient rats.

	Wild-type		DPP4-deficiency	
	Sham	AAB	Sham	AAB
		6 week		
LVSP (mmHg)	147.9±6.0	172.8±4.9*	153.9±2.9	185.8±6.5†
LVEDP (mmHg)	4.5±0.3	5.6±0.5	4.9±0.5	5.3±1.0
LVDP (mmHg)	143.4±5.8	167.2±5.2*	148.3±3.2	180.5±7.5†
HR (bpm)	396.1±10.8	389.9±21.5	405.4±8.5	395.4±24.0
+max dPdt (mmHg/s)	14243.5±1069.4	13778.4±695.6	14445.6±114.6	13974.1±253.7
−min dPdt (mmHg/s)	12700.2±1129.1	12229.8±753.9	12525.8±193.9	12539.9±168.9
LVESV (μL)	36.2±5.7	81.4±15.9*	36.2±2.0	43.3±9.5#
LVEDV (μL)	143.7±7.2	161.1±7.7	155.9±18.8	141.8±3.4
SV (μL)	107.5±3.1	79.7±11.9*	119.7±17.6	98.4±7.9#
%EF	75.3±2.9	50.5±10.7*	75.9±1.9	69.3±7.4#
ESPVR slope (mmHg/μL)	1.21±0.05	2.30±0.45*	1.31±0.11	1.97±0.05†
EDPVR slope (mmHg/μL)	0.0220±0.0026	0.0416±0.0089*	0.0230±0.0031	0.0194±0.0083
		10 week		
LVSP (mmHg)	136.2±7.2	119.6±8.9@	141.7±3.5	140.3±7.2&
LVEDP (mmHg)	4.7±0.5	6.8±1.9@	4.5±0.1	4.4±0.9&
LVDP (mmHg)	131.4±7.1	112.8±8.1@	137.2±4.2	135.9±8.1&
HR (bpm)	378.1±35.0	366.3±13.3	393.7±15.2	385.3±17. 3
+max dPdt (mmHg/s)	11353.8±652.5	8568.2±483.9@	10924.9±945.5	10039.7±246.4&
−min dPdt (mmHg/s)	8301.3±683.9*	6994.9±519.9@	10696.8±1028.7	8644.8±850.2&
LVESV (μL)	64.6±11.4	125.3±2.8@	67.5±7.9	89.2±14.7&
LVEDV (μL)	194.9±26.7	231.1±19.7	213.2±38.7	249.2±37.3
SV (μL)	130.2±16.2	105.9±18.1@	145.7±31.2	159.9±22.7&
%EF	67.5±2.5	44.7±4.6@	67.6±2.2	64.4±0.9&
ESPVR slope (mmHg/μL)	1.10±0.35	0.59±0.20@	1.12±0.23	0.82±0.20&
EDPVR slope (mmHg/μL)	0.0145±0.0033	0.0126±0.0023	0.0172±0.0069	0.0168±0.0027

Values are means ± SE (n = 8) MAP : mean arterial pressure, HR : heart rate, LVSP : left ventricular systolic pressure, LVEDP : left ventricular end-diastolic pressure, DP : development pressure, +dP/dt and −dP/dt : maximal slope of the systolic pressure increment and diastolic pressure decrement, respectively, Ves : end-systolic volume, Ved : end-diastolic volume, SV : stroke volume, EF : ejection fraction, ESPVR : end-systolic pressure–volume relation, and EDPVR : end-diastolic pressure–volume relation.
*p<0.05 vs. wild-type 6 week sham,
#p<0.05 vs. wild-type 6 week AAB,
@p<0.05 vs. wild-type 10 week sham,
&p<0.05 vs. wild-type 10 week AAB,
†p<0.05 vs. DPP4 deficiency 6 week sham.

Discussion

Sustained pressure overload stimulates pathological cardiac hypertrophy, which changes calcium sensitivity and associates with the consequent defects in calcium homeostatsis [29]. Exitation-contraction coupling is a physiological process of converting an electrical stimulus to a mechanical response, which involves calcium release and uptake in association with the contraction and relaxation of heart [30]. The alteration of calcium handling related genes may result in heart failure [30,31]. These changes include the decrease of the sarcoplasmic reticulum calcium pump (SERCA) activity during heart failure [31]. SERCA is inhibited by phospholamban. When phospholamban is phosphorylated, its ability to inhibit SERCA is lost [32]. Gene transfer of antisense phospholamban can improve the contractile function in failing human myocardium [33]. The modulation of calcium handling is a strategy to improve function of falling heart. In our study, aging decreased the rate of relaxation in wild-type rat hearts, indicating an impairment of calcium reuptake. In chronic pressure overload model, the decreased of ventricular systolic and diastolic function were found, suggesting a deterioration of calcium homeostasis. DPP4-deficient rats had better cardiac performance than wild-type rats in response to aging and pressure overload. These results coincide with other studies. DPP4 inhibitor exhibited significant attenuation of heart failure related dysfunction and remarkably mitigated pulmonary congestion [34,35]. Indeed, sepsis induced decrease of phospholamban phosphorylation level was alleviated by DPP4 deficiency, and associated with the cardiac function preservation [14]. The capability of regulating calcium homeostasis by DPP4 deficiency may be partly contributed to the cardiac function improvement during heart failure.

Hypertrophic cardiomyopathy is an example of myocardial energy depletion [36]. Change of fatty acid to glucose utilization is increased early in heart failure. The switch in substrate utilization is a protective mechanism, which allows the heart to produce more ATP per molecule of oxygen consumed [1,37]. In advanced

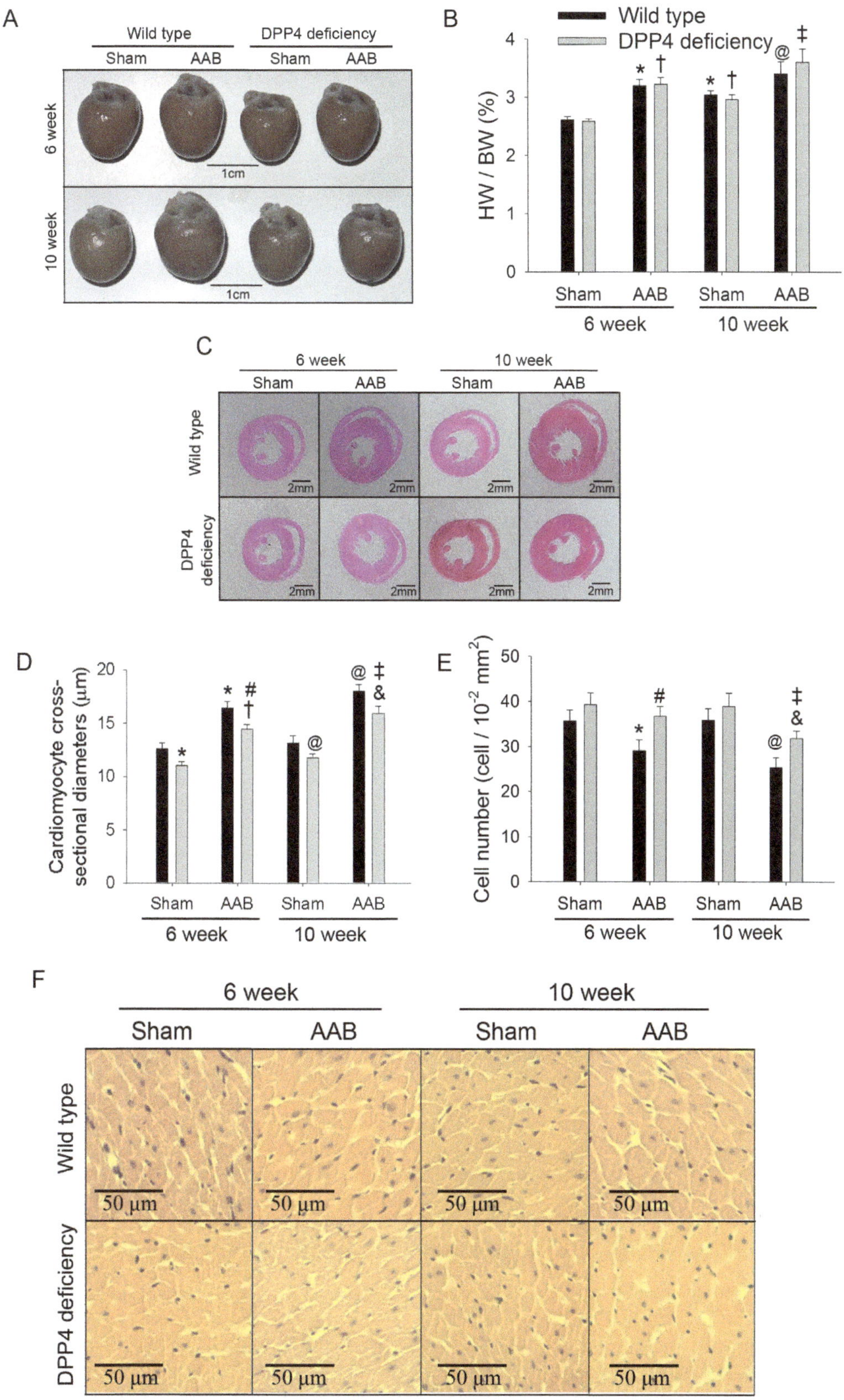

Figure 2. Effects of AAB on heart morphology. Representative heart (A), heart weight to body weight ratio (n = 8) (B), and histological sections (C and F) of HE staining were shown after 6 and 10 weeks of AAB surgery. Cardiomyocyte cross-sectional diameter (D) and cell number (E) were also measured. (300 cell in 5 rats each). *$p<0.05$ vs. wild-type 6 week sham, $^{\#}p<0.05$ vs. wild-type 6 week AAB, $^{@}p<0.05$ vs. wild-type 10 week sham, $^{\&}p<0.05$ vs. wild-type 10 week AAB, $^{\dagger}p<0.05$ vs. DPP4 deficiency 6 week sham, $^{\ddagger}p<0.05$ vs. DPP4 deficiency 10 week sham.

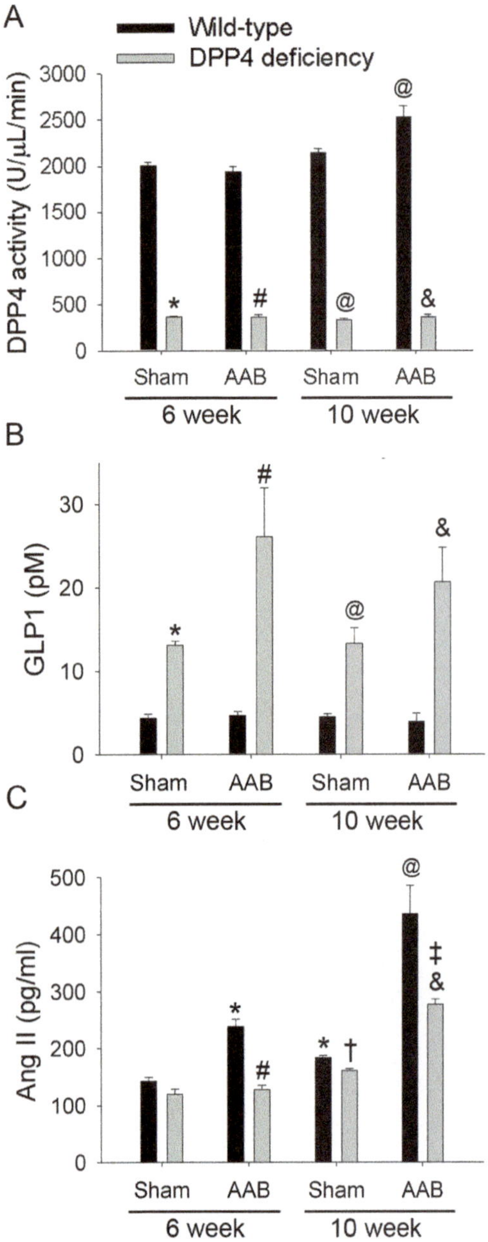

Figure 3. Effects of AAB on plasma GLP-1, DPP4 activity, and Ang II level. Plasma DPP4 activity (A), plasma GLP-1 concentration (B), and plasma Ang II level (C) were measured in wild-type and DPP4-deficient rats after 6 and 10 weeks of AAB surgery. (n = 8) $^{*}p<0.05$ vs. wild-type 6 week sham, $^{\#}p<0.05$ vs. wild-type 6 week AAB, $^{@}p<0.05$ vs. wild-type 10 week sham, $^{\&}p<0.05$ vs. wild-type 10 week AAB, $^{\dagger}p<0.05$ vs. DPP4 deficiency 6 week sham, $^{\ddagger}p<0.05$ vs. DPP4 deficiency 10 week sham.

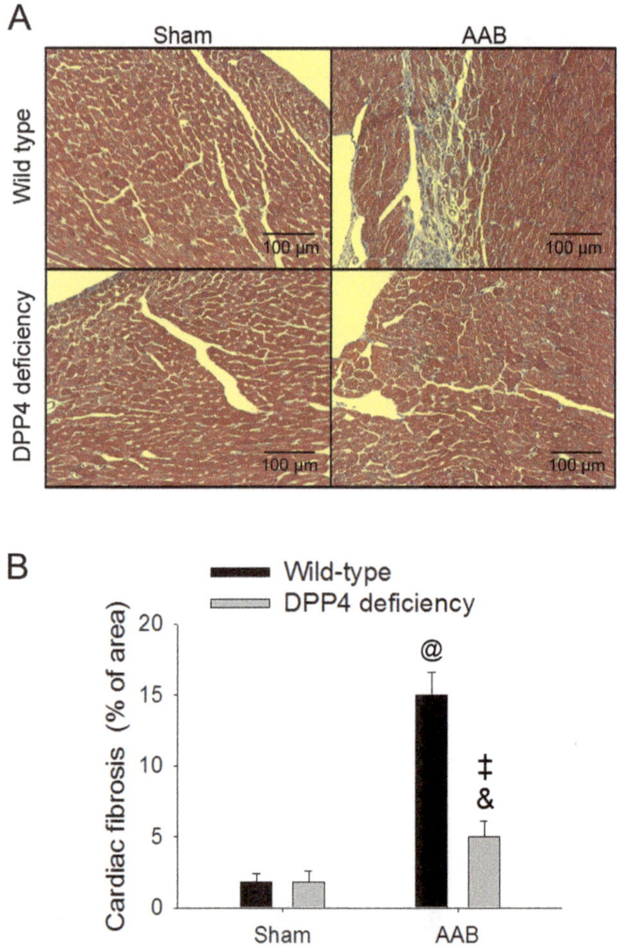

Figure 4. Effects of AAB on cardiac fibrosis. Masson's trichrome staining revealed collagen deposits in blue, which was measured in wild-type and DPP4-deficient rats after 10 weeks of surgery. Representative masson's trichrome sections (A), and quantification of the collagen accumulation (B). $^{@}p<0.05$ vs. wild-type sham, $^{\&}p<0.05$ vs. wild-type AAB, $^{\ddagger}p<0.05$ vs. DPP4 deficiency sham.

hypertrophy heart failure, insulin resistance develops in heart failure result in decreasing glucose metabolism, which reduces the overall ability of the heart to generate sufficient ATP [1,36]. A strategy for metabolic intervention in heart failure is to modulate substrate utilization [36]. Cardiac specific overexpression of GLUT1 prevents the development of heart failure in pressure overload mice [38]. Increase of myocardial glucose uptake contributes to the improvement of left ventricular performance in dogs with pace-inducing dilated cardiomyopathy [39]. In our previous study, the cardiac GLUT4 expression and glucose uptake elevated in normal DPP4-difficient rats [15,21], and remain higher than that in wild-type rats during stress [21]. The improvement of cardiac function during pressure overload in DPP4-deficient rats may be partly attributed to the modulation of energy utilization.

Lack of DPP4 activity resulted in accumulation of plasma GLP-1 in DPP4-deficient rats. Evidence from several studies suggested that GLP-1 receptor agonist improved cardiac function in heart failure in both animal and clinical studies [39–42]. GLP-1 or its receptor agonist improved glucose uptake in heart by increasing GLUT4 translocation, and led to cardiac performance preservation in dilated cardiomyopathy or myocardial infarction induced failing heart [16,43,44]. Besides glucose homeostasis controlling, GLP-1 increased AKT, ERK, and eNOS signaling to activate survival pathway, which were associated with the outcome improvement after myocardial infarction [13,41]. Liraglutide, a GLP-1 receptor agonist, ameliorated high fat diet induced cardiac endoplasmic reticulum stress, lipid accumulation, collagen deposition, and preserved cardiac function via an AMPK dependent pathway [45]. In our study, GLP-1 concentration increased after AAB in DPP4-deficient rats but not in wild-type rats, while DPP4 activity increased only in wild-type rats. Similar result was found in clinical study, enhanced serum DPP4 activity has been observed in patients with cardiovascular disease [34,46]. The plasma DPP4 activity was correlated negatively with cardiac function in heart failure patients [34]. Higher DPP4 activity involve in degradation of GLP-1, so the elevation of GLP-1 concentration was only found in DPP4-deficient rats. GLP-1 is mainly secreted by L cells in distal small intestine, and also secreted by microglial cell in areas of the brain distinct from the brainstem and the NTS neurons [47]. However, the precise mechanism that regulating GLP-1 secretion during pressure overload remains unclear. We speculated that a great quantity of GLP-1 accumulation during pressure overload in DPP4-deficient rats can lead to cardioprotection.

Ventricular hypertrophy is associated with cell death. Several factors implicated as triggers of cardiomyocyte apoptosis, including a great quantity cytosolic calcium concentration, the formation of oxygen free radicals, excess levels of Ang II and cytokines [48]. Cell number in ventricle after abdominal aortic banding was less in wild-type rats than that in DPP4-deficient rats. DPP4 deficiency increased the capability against stress in cardiomyocyte during abdominal aortic banding. In our previous study, DPP4-deficient cardiomyocytes preserved catalase activity after H_2O_2 exposure. Bax/Bcl2 ratio and caspase3 activity were also found to be lower in DPP4-deficient cardiomyocytes than in wild-type cardiomyocytes after exposure to H_2O_2. The improvement of cell viability during ROS stress was partly contributed to GLP-1 accumulation in DPP4-deficient cardiomyocytes. [21]. Progression of cell death result in the replacement of fibrosis in the ventricle is the major determinant of increased myocardial stiffness and myocardial dysfunction [49]. Ang II promote the development of myocardial fibrosis in hypertensive heart disease and chronic heart failure [50]. Administration of Ang II in cardiac fibroblasts stimulate cell proliferation and collagen synthesis [51], while suppression of Ang II signaling is a well-known treatment for cardiovascular diseases. GLP-1 was reported to decrease the circulating concentration of Ang II in healthy men [52], and also had an anti-hypertensive effect through the attenuation of Ang II -induced high-salt sensitivity [53]. In our study, whether higher GLP-1 level contribute to the lower Ang II concentration in DPP4-deficient rat, or cardiac function improvement lead to the decrease of Ang II secretion after ABB remain unclear. In addition, DPP4 inhibitor decreased cardiac mRNA expression of fibrosis markers, such as TGF-β1, TIMP-1, collagen type I α1 and III α1 in uremic rats [54], and was also shown to alleviate collagen deposits in ventricle during heart failure [34,35]. It can be conceivable that DPP4 deficiency alleviated left ventricular fibrosis after AAB in rat, which associated with the amelioration of heart dysfunction.

Besides GLP-1, Stromal cell derived factor-1α (SDF-1α), another DPP4 substrate, was also shown to exert cardioprotective effect [7]. G-CSF administration in combination with DPP4 inhibitor leads to the stabilization of active SDF-1α, which attracted stem cells to the injury sites and improved outcome after myocardial infarction [55]. In addition to the peptidase activity, DPP4 involve in immune system. DPP4 known as CD26 is a homodimeric type II transmembrane glycoprotein [7,56], which is one of the accessory molecules of helper T cells, mediates as a co-stimulatory effect on T cell activation [7]. In this study, the cardioprotective effect of DPP4 deficiency during heart failure may be involved in more than GLP-1 signaling.

In conclusion, we have demonstrated that DPP4-deficient rats had better cardiac performance than wild-type rats after AAB surgery, which was associated with an alleviation of Ang II level and collagen deposits. The cardioprotective effect of DPP4 deficiency may be partly contributed to a great quantity of GLP-1 accumulation. These results prove that DPP4 deficiency improved cardiovascular outcome after pressure overload. The safety of long term loss of DPP4 activity should be further concerned.

Author Contributions

Conceived and designed the experiments: HCK MJS. Performed the experiments: HCK MJS. Analyzed the data: HCK MJS. Contributed reagents/materials/analysis tools: HCK MJS. Wrote the paper: HCK MJS.

References

1. Bernardo BC, Weeks KL, Pretorius L, McMullen JR (2010) Molecular distinction between physiological and pathological cardiac hypertrophy: experimental findings and therapeutic strategies. Pharmacol Ther 128: 191–227.
2. Stefan N, Fritsche A, Haring HU (2005) Insulin resistance and congestive heart failure. Jama 294: 2578; author reply 2578.
3. Boudina S, Abel ED (2010) Diabetic cardiomyopathy, causes and effects. Rev Endocr Metab Disord 11: 31–39.
4. Carley AN, Severson DL (2005) Fatty acid metabolism is enhanced in type 2 diabetic hearts. Biochim Biophys Acta 1734: 112–126.
5. Rijzewijk LJ, van der Meer RW, Lamb HJ, de Jong HW, Lubberink M, et al. (2009) Altered myocardial substrate metabolism and decreased diastolic function in nonischemic human diabetic cardiomyopathy: studies with cardiac positron emission tomography and magnetic resonance imaging. J Am Coll Cardiol 54: 1524–1532.
6. Fehmann HC, Goke R, Goke B (1995) Cell and molecular biology of the incretin hormones glucagon-like peptide-I and glucose-dependent insulin releasing polypeptide. Endocr Rev 16: 390–410.
7. Morimoto C, Schlossman SF (1998) The structure and function of CD26 in the T-cell immune response. Immunol Rev 161: 55–70.
8. Doupis J, Veves A (2008) DPP4 inhibitors: a new approach in diabetes treatment. Adv Ther 25: 627–643.
9. Michel MC, Fliers E, Van Noorden CJ (2008) Dipeptidyl peptidase IV inhibitors in diabetes: more than inhibition of glucagon-like peptide-1 metabolism? Naunyn Schmiedebergs Arch Pharmacol 377: 205–207.
10. Capuano A, Sportiello L, Maiorino MI, Rossi F, Giugliano D, et al. (2013) Dipeptidyl peptidase-4 inhibitors in type 2 diabetes therapy-focus on alogliptin. Drug Des Devel Ther 7: 989–1001.
11. Ban K, Noyan-Ashraf MH, Hoefer J, Bolz SS, Drucker DJ, et al. (2008) Cardioprotective and vasodilatory actions of glucagon-like peptide 1 receptor are mediated through both glucagon-like peptide 1 receptor-dependent and -independent pathways. Circulation 117: 2340–2350.
12. Gros R, You X, Baggio LL, Kabir MG, Sadi AM, et al. (2003) Cardiac function in mice lacking the glucagon-like peptide-1 receptor. Endocrinology 144: 2242–2252.
13. Noyan-Ashraf MH, Momen MA, Ban K, Sadi AM, Zhou YQ, et al. (2009) GLP-1R agonist liraglutide activates cytoprotective pathways and improves outcomes after experimental myocardial infarction in mice. Diabetes 58: 975–983.
14. Ku HC, Chen WP, Su MJ (2010) GLP-1 signaling preserves cardiac function in endotoxemic Fischer 344 and DPP4-deficient rats. Naunyn Schmiedebergs Arch Pharmacol 382: 463–474.
15. Ku HC, Chen WP, Su MJ (2011) DPP4 deficiency preserves cardiac function via GLP-1 signaling in rats subjected to myocardial ischemia/reperfusion. Naunyn Schmiedebergs Arch Pharmacol 384: 197–207.
16. Vyas AK, Yang KC, Woo D, Tzekov A, Kovacs A, et al. (2011) Exenatide improves glucose homeostasis and prolongs survival in a murine model of dilated cardiomyopathy. PLoS One 6: e17178.
17. Monji A, Mitsui T, Bando YK, Aoyama M, Shigeta T, et al. (2013) Glucagon-like peptide-1 receptor activation reverses cardiac remodeling via normalizing cardiac steatosis and oxidative stress in type 2 diabetes. Am J Physiol Heart Circ Physiol 305: H295–304.
18. Liu J, Liu Y, Chen L, Wang Y, Li J (2013) Glucagon-Like Peptide-1 Analog Liraglutide Protects against Diabetic Cardiomyopathy by the Inhibition of the Endoplasmic Reticulum Stress Pathway. J Diabetes Res 2013: 630537.
19. Wang XM, Yang YJ, Wu YJ (2013) The emerging role of dipeptidyl peptidase-4 inhibitors in cardiovascular protection: current position and perspectives. Cardiovasc Drugs Ther 27: 297–307.
20. Sauve M, Ban K, Momen MA, Zhou YQ, Henkelman RM, et al. (2010) Genetic deletion or pharmacological inhibition of dipeptidyl peptidase-4 improves cardiovascular outcomes after myocardial infarction in mice. Diabetes 59: 1063–1073.

21. Ku HC, Chen WP, Su MJ (2013) DPP4 deficiency exerts protective effect against H2O2 induced oxidative stress in isolated cardiomyocytes. PLoS One 8: e54518.
22. Scheen AJ (2013) Cardiovascular effects of gliptins. Nat Rev Cardiol 10: 73–84.
23. Monami M, Ahren B, Dicembrini I, Mannucci E (2013) Dipeptidyl peptidase-4 inhibitors and cardiovascular risk: a meta-analysis of randomized clinical trials. Diabetes Obes Metab 15: 112–120.
24. Scirica BM, Bhatt DL, Braunwald E, Steg PG, Davidson J, et al. (2013) Saxagliptin and Cardiovascular Outcomes in Patients with Type 2 Diabetes Mellitus. N Engl J Med.
25. Patten RD, Hall-Porter MR (2009) Small animal models of heart failure: development of novel therapies, past and present. Circ Heart Fail 2: 138–144.
26. Ni L, Zhou C, Duan Q, Lv J, Fu X, et al. (2011) beta-AR blockers suppresses ER stress in cardiac hypertrophy and heart failure. PLoS One 6: e27294.
27. Chen WP, Tzeng HJ, Ku HC, Ho YJ, Lee SS, et al. (2010) Thaliporphine ameliorates cardiac depression in endotoxemic rats through attenuating TLR4 signaling in the downstream of TAK-1 phosphorylation and NF-kappaB signaling. Naunyn Schmiedebergs Arch Pharmacol 382: 441–453.
28. Klemann C, Schade J, Pabst R, Leitner S, Stiller J, et al. (2009) CD26/dipeptidyl peptidase 4-deficiency alters thymic emigration patterns and leukcocyte subsets in F344-rats age-dependently. Clin Exp Immunol 155: 357–365.
29. Guinto PJ, Haim TE, Dowell-Martino CC, Sibinga N, Tardiff JC (2009) Temporal and mutation-specific alterations in Ca2+ homeostasis differentially determine the progression of cTnT-related cardiomyopathies in murine models. Am J Physiol Heart Circ Physiol 297: H614–626.
30. Hasenfuss G (1998) Alterations of calcium-regulatory proteins in heart failure. Cardiovasc Res 37: 279–289.
31. Ashrafian H, McKenna WJ, Watkins H (2011) Disease pathways and novel therapeutic targets in hypertrophic cardiomyopathy. Circ Res 109: 86–96.
32. East JM (2000) Sarco(endo)plasmic reticulum calcium pumps: recent advances in our understanding of structure/function and biology (review). Mol Membr Biol 17: 189–200.
33. Monte del F, Harding SE, Dec GW, Gwathmey JK, Hajjar RJ (2002) Targeting phospholamban by gene transfer in human heart failure. Circulation 105: 904–907.
34. Dos Santos L, Salles TA, Arruda-Junior DF, Campos LC, Pereira AC, et al. (2013) Circulating Dipeptidyl Peptidase IV Activity Correlates with Cardiac Dysfunction in Human and Experimental Heart Failure. Circ Heart Fail.
35. Takahashi A, Asakura M, Ito S, Min KD, Shindo K, et al. (2013) Dipeptidyl-peptidase IV inhibition improves pathophysiology of heart failure and increases survival rate in pressure-overloaded mice. Am J Physiol Heart Circ Physiol 304: H1361–1369.
36. Neubauer S (2007) The failing heart–an engine out of fuel. N Engl J Med 356: 1140–1151.
37. van Bilsen M, van Nieuwenhoven FA, van der Vusse GJ (2009) Metabolic remodelling of the failing heart: beneficial or detrimental? Cardiovasc Res 81: 420–428.
38. Liao R, Jain M, Cui L, D'Agostino J, Aiello F, et al. (2002) Cardiac-specific overexpression of GLUT1 prevents the development of heart failure attributable to pressure overload in mice. Circulation 106: 2125–2131.
39. Nikolaidis LA, Elahi D, Hentosz T, Doverspike A, Huerbin R, et al. (2004) Recombinant glucagon-like peptide-1 increases myocardial glucose uptake and improves left ventricular performance in conscious dogs with pacing-induced dilated cardiomyopathy. Circulation 110: 955–961.
40. Khan MA, Deaton C, Rutter MK, Neyses L, Mamas MA (2013) Incretins as a novel therapeutic strategy in patients with diabetes and heart failure. Heart Fail Rev 18: 141–148.
41. Zhao TC (2013) Glucagon-like peptide-1 (GLP-1) and protective effects in cardiovascular disease: a new therapeutic approach for myocardial protection. Cardiovasc Diabetol 12: 90.
42. Nikolaidis LA, Elahi D, Shen YT, Shannon RP (2005) Active metabolite of GLP-1 mediates myocardial glucose uptake and improves left ventricular performance in conscious dogs with dilated cardiomyopathy. Am J Physiol Heart Circ Physiol 289: H2401–2408.
43. Liu Q, Anderson C, Broyde A, Polizzi C, Fernandez R, et al. (2010) Glucagon-like peptide-1 and the exenatide analogue AC3174 improve cardiac function, cardiac remodeling, and survival in rats with chronic heart failure. Cardiovasc Diabetol 9: 76.
44. Fisman EZ, Tenenbaum A (2009) A cardiologic approach to non-insulin antidiabetic pharmacotherapy in patients with heart disease. Cardiovasc Diabetol 8: 38.
45. Noyan-Ashraf MH, Shikatani EA, Schuiki I, Mukovozov I, Wu J, et al. (2013) A glucagon-like peptide-1 analog reverses the molecular pathology and cardiac dysfunction of a mouse model of obesity. Circulation 127: 74–85.
46. Lee SA, Kim YR, Yang EJ, Kwon EJ, Kim SH, et al. (2011) CD26/DPP4 levels in peripheral blood and T cells in patients with type 2 diabetes mellitus. J Clin Endocrinol Metab 98: 2553–2561.
47. Kappe C, Tracy LM, Patrone C, Iverfeldt K, Sjoholm A (2012) GLP-1 secretion by microglial cells and decreased CNS expression in obesity. J Neuroinflammation 9: 276.
48. Sabbah HN (2000) Apoptotic cell death in heart failure. Cardiovasc Res 45: 704–712.
49. Biernacka A, Frangogiannis NG (2011) Aging and Cardiac Fibrosis. Aging Dis 2: 158–173.
50. Brilla CG (2000) Renin-angiotensin-aldosterone system and myocardial fibrosis. Cardiovasc Res 47: 1–3.
51. Bouzegrhane F, Thibault G (2002) Is angiotensin II a proliferative factor of cardiac fibroblasts? Cardiovasc Res 53: 304–312.
52. Skov J, Dejgaard A, Frokiaer J, Holst JJ, Jonassen T, et al. (2013) Glucagon-like peptide-1 (GLP-1): effect on kidney hemodynamics and renin-angiotensin-aldosterone system in healthy men. J Clin Endocrinol Metab 98: E664–671.
53. Hirata K, Kume S, Araki S, Sakaguchi M, Chin-Kanasaki M, et al. (2009) Exendin-4 has an anti-hypertensive effect in salt-sensitive mice model. Biochem Biophys Res Commun 380: 44–49.
54. Chaykovska L, von Websky K, Rahnenfuhrer J, Alter M, Heiden S, et al. (2011) Effects of DPP-4 inhibitors on the heart in a rat model of uremic cardiomyopathy. PLoS One 6: e27861.
55. Zaruba MM, Theiss HD, Vallaster M, Mehl U, Brunner S, et al. (2009) Synergy between CD26/DPP-IV inhibition and G-CSF improves cardiac function after acute myocardial infarction. Cell Stem Cell 4: 313–323.
56. Thompson MA, Ohnuma K, Abe M, Morimoto C, Dang NH (2007) CD26/dipeptidyl peptidase IV as a novel therapeutic target for cancer and immune disorders. Mini Rev Med Chem 7: 253–273.

Silencing Mutant Ataxin-3 Rescues Motor Deficits and Neuropathology in Machado-Joseph Disease Transgenic Mice

Clévio Nóbrega[1,◑], Isabel Nascimento-Ferreira[1,2,◑], Isabel Onofre[1,2], David Albuquerque[3], Hirokazu Hirai[4], Nicole Déglon[5], Luís Pereira de Almeida[1,2]*

1 CNC - Center for Neurosciences & Cell Biology, University of Coimbra, Coimbra, Portugal, **2** Faculty of Pharmacy, University of Coimbra, Coimbra, Portugal, **3** Faculty of Sciences and Technology, University of Coimbra, Coimbra, Portugal, **4** Department of Neurophysiology, Gunma University Graduate School of Medicine, Maebashi, Gunma, Japan, **5** Lausanne University Hospital, Department of Clinical Neurosciences, Laboratory of Cellular and Molecular Neurotherapies, Lausanne, Switzerland

Abstract

Machado-Joseph disease (MJD) or spinocerebellar ataxia type 3 (SCA3) is an autosomal dominantly-inherited neurodegenerative disorder caused by the over-repetition of a CAG codon in the *MJD1* gene. This expansion translates into a polyglutamine tract that confers a toxic gain-of-function to the mutant protein – ataxin-3, leading to neurodegeneration in specific brain regions, with particular severity in the cerebellum. No treatment able to modify the disease progression is available. However, gene silencing by RNA interference has shown promising results. Therefore, in this study we investigated whether lentiviral-mediated allele-specific silencing of the mutant ataxin-3 gene, after disease onset, would rescue the motor behavior deficits and neuropathological features in a severely impaired transgenic mouse model of MJD. For this purpose, we injected lentiviral vectors encoding allele-specific silencing-sequences (shAtx3) into the cerebellum of diseased transgenic mice expressing the targeted C-variant of mutant ataxin-3 present in 70% of MJD patients. This variation permits to discriminate between the wild-type and mutant forms, maintaining the normal function of the wild-type allele and silencing only the mutant form. Quantitative analysis of rotarod performance, footprint and activity patterns revealed significant and robust alleviation of gait, balance (average 3-fold increase of rotarod test time), locomotor and exploratory activity impairments in shAtx3-injected mice, as compared to control ones injected with shGFP. An important improvement of neuropathology was also observed, regarding the number of intranuclear inclusions, calbindin and DARPP-32 immunoreactivity, fluorojade B and Golgi staining and molecular and granular layers thickness. These data demonstrate for the first time the efficacy of gene silencing in blocking the MJD-associated motor-behavior and neuropathological abnormalities after the onset of the disease, supporting the use of this strategy for therapy of MJD.

Editor: Roberto Chiesa, Dulbecco Telethon Institute and Mario Negri Institute for Pharmacological Research, Italy

Funding: This work was supported by the Portuguese Foundation for Science and Technology (grants PTDC/SAU-NEU/099307/2008 and PTDC/SAU-FAR/116535/2010 and fellowships to CN, INF, IO and DA), the Center for Science and Technology of Madeira (CN), the National Ataxia Foundation and the Richard Chin and Lily Lock Machado-Joseph disease Research Fund. The funders had no role in study design, data collection and analysis, decision to publish, or preparation of the manuscript.

Competing Interests: The authors have declared that no competing interests exist.

* E-mail: luispa@ci.uc.pt

◑ These authors contributed equally to this work.

Introduction

Machado-Joseph disease (MJD), also designated spinocerebellar ataxia type 3 (SCA3), is the most common dominantly-inherited cerebellar ataxia worldwide [1–4]. It is part of a group of nine known polyglutamine (polyQ) disorders which share expanded CAG repeat mutations that translate into polyQ tracts [5,6]. The signs and symptoms of MJD include progressive postural instability, gait and limb ataxia, weight loss and, in severe cases, premature death [2,7,8]. The pathology of MJD includes severe neuronal loss in the spinal cord and selective brain regions such as dentate nuclei (cerebellum), pontine nuclei (brainstem), substantia nigra, and striatum [2,7–10]. MJD is caused by ataxin-3 carrying a stretch of 54–84 consecutive glutamines (mutant ataxin-3) in opposition to normal ataxin-3 whose glutaminic stretch has 14–37 repetitions [11,12]. The polyQ expansion confers a toxic gain-of-function to the mutant protein, leading to the formation of neuronal intranuclear inclusions, neuronal dysfunction and degeneration [13].

Several therapeutic strategies are under study for MJD treatment such as modulation of Ca2+ signalling, inhibition of calpain-mediated proteolysis of mutant ataxin-3 in the brain or promotion of degradation of mutant ataxin-3 species, either by activation of the proteasome or of the beclin-1 autophagy pathway [14–17]. Although such approaches offer promise, the most direct solution to block the pathogenesis of MJD would be to prevent the translation of the mutant ataxin-3 protein in the brain. Accordingly, in conditional transgenic MJD, HD and SCA1 mice shutting off expression of the mutant transgene dramatically slows disease progression and, for selected features, even reverses severe diseases pathology [18,19].

Decreasing the expression of mutant protein can be achieved by using the RNA interference (RNAi) mechanism to inhibit the expression of the target gene. RNAi-based strategies have been used to supress the expression of toxic polyQ proteins in transgenic mouse models of the polyQ disorders SCA1 and Huntington's disease [20–23]. This approach has been refined to discriminate between the wild-type and mutant forms of the messenger RNA by developing silencing sequences targeting single nucleotide polymorphisms (SNP) [24]. In MJD patients, an intragenic single nucleotide polymorphism (SNP) at the 3′ end of the CAG tract of the ataxin-3 gene is present in more than 70% of the cases [25]. This SNP can be used to selectively inactivate mutant ataxin-3, significantly decreasing the severity of the neuropathological abnormalities associated with concomitant induction of MJD, as we previously showed in a (LV)-based rat model of MJD [26]. Nevertheless, no study has ever evaluated the efficacy of gene silencing in a) a transgenic mouse model of MJD, b) exihibiting a severe phenotype, c) regarding motor behavior impairments, d) when initiated after disease onset.

Therefore, in the present study we used lentiviral vectors (LV) encoding short-hairpin RNAs (shRNAs) targeting this SNP, to downregulate mutant ataxin-3 in the cerebellum of a transgenic MJD mouse model that exhibits an early and very severe motor and neuropathological phenotype [15]. This model was generated in a C57/BL6 background, by the introduction of a truncated form of human ataxin-3 with 69 repeats [15,27] and encoding the C variant of the human ataxin-3 transgene sequence that is present in 70% of the MJD patients [15,25,28], making this model particularly adequate for the allele-specific silencing here performed. Transgenic mice expressing a truncated form of the mutant protein have also been used for investigation of gene silencing approaches in Huntington's disease [21,29]. Moreover, this MJD model displays robust ataxic behavior and neuropathological features from 21 days of age, which makes it time and cost-effective, and mimics a late stage of disease.

In this work we show for the first time that allele-specific silencing of mutant ataxin-3, even when initiated at a late stage of disease dramatically improves motor coordination and reduces neuropathological abnormalities related with MJD.

Results

Allele-specific silencing reduces mutant ataxin-3 protein levels and the number of intranuclear inclusions

The polyQ expansion in mutant ataxin-3 confers aggregation properties to the mutant protein, leading to the formation of neuronal intranuclear inclusions, which despite its unclear role in the disease process are a hallmark of MJD [13]. We investigated if specific silencing would reduce the number of aggregates in the cerebellum, a region that is particularly involved in MJD, in a transgenic mouse model of MJD with particular expression of mutant ataxin-3 in Purkinje cells [15]. These transgenic mice were generated using the human ataxin-3 sequence [15] with a polymorphism that is present in 70% of MJD patients [25,28], which permitted the design of a shRNA that specifically targets mutant ataxin-3 [26]. Therefore, we injected transgenic mice at 3 weeks of age (P21–25) with LV encoding the allele-specific silencing sequence for the human ataxin-3 (shAtx3) or a control sequence (shGFP), see Figure S1. Preliminary experiments in this transgenic model showed that a single injection of 6 µl of lentiviral vectors encoding GFP ($n=8$) mediated an extensive transduction (around 60%) of the cerebellar cortex (Figure S2 and Figure S3).

In the shGFP-injected mice cerebella, we found increased ataxin-3 immunoreactivity in aggregated inclusions (Fig. 1A–F), as compared to shAtx3-injected mice which displayed a diffuse subcellular distribution of ataxin-3 in the cytoplasm of the Purkinje cells (Fig. 1G–L). Importantly, the number of inclusions was robustly and significantly decreased in mice injected with the silencing vectors (shAtx3) as compared to the control ones ($n=6$; 4.30±0.07 versus 19.5±0.5 nuclear aggregates per 100 Purkinje cells, $P=0.0018$; Fig. 1M). Further analysis revealed that mice injected with the control vectors encoding the short-hairpin targeting GFP (shGFP) exhibit ataxin-3 aggregates (HA tag) that co-localize with the co-expressed LacZ reporter gene; whereas aggregates from mice injected with vectors encoding the short-hairpin against mutant ataxin-3 (shAtx3) do not co-localize with LacZ and therefore correspond to non-transduced cells (Figure S4). This indicates that expression of the shRNAs against mutant ataxin-3 but not the control sequences prevented formation of ataxin-3 aggregates in the transduced cells. We further investigated the effects of mutant ataxin-3 silencing by western blot analysis of protein levels in the transgenic mice cerebella (Fig. 2A). Probing the membrane with the 1C2 antibody, a marker of the polyglutamine-expanded proteins, revealed a significant decrease in the levels of oligomerized and aggregated ataxin-3 in transgenic mice injected with shAtx3 vectors compared to the control mice injected with shGFP vectors ($n=3$; aggregates at the upper level of the running and stacking gels Fig. 2B, 2D: 0.44±0.03 versus 0.57±0.03 in control, $P=0.0023$, normalizing with actin; 0.40±0.02 versus 0.51±0.01 in control, $P=0.0028$, normalizing with lacZ; and oligomers Fig. 2C, 2E: 3.66±0.17 versus 4.96±0.29 in control, $P=0.018$ normalizing with actin; 2.32±0.21 versus 3.06±0.14 in control, $P=0.04$, normalizing with lacZ).

Silencing of mutant ataxin-3 preserves Purkinje cells integrity and molecular and granular layer thickness in the cerebellar cortex

Purkinje cells are key and vulnerable elements regulating the cerebellar function. Therefore we analyzed if silencing mutant ataxin-3 would prevent loss of Purkinje cell markers immunoreactivity and degeneration as well as neurons from molecular and granular layers. Control mice (shGFP) exhibited reduced calbindin immunoreactivity of Purkinje cells (Fig. 3A, C), which was robustly and significantly preserved in mice injected with shAtx3 (Fig. 3B, D, G: $n=6$; 3.060±0.546 versus 0.852±0.079 in control, $P=0.0026$). Similarly, silencing mutant ataxin-3 prevented loss of DARPP-32 immunoreactivity (Figure S5).

The cerebellar cortex is characterized by strong cellular interconnectivity, which may propagate the degeneration of one cell to the other. Cresyl violet staining (Fig. 3E, F) revealed that the thickness of the molecular and granular layers were significantly larger in mice injected with shAtx3 as compared to control mice, which suggests a prevention of neurodegeneration (Fig. 3H; molecular layer: $n=6$; 111.1±7.14 µm versus 67.32±3.08 µm in control, $P=0.0005$; Fig. 3I; granular layer: $n=6$; 107.2±2.43 µm versus 77.6±4.42 µm in control, $P=0.0004$). Furthermore, FluoroJade-B staining, revealed a consistently increased staining suggestive of increased neuronal degeneration in control mice (Fig. 4J), as compared to mice injected with shAtx3 (Fig. 4K). Finnally, Golgi staining revealed reduced arborization in the molecular layer of shGFP-injected mice (Fig. 4L), as compared to mice injected with shAtx3 silencing sequences (Fig. 4M).

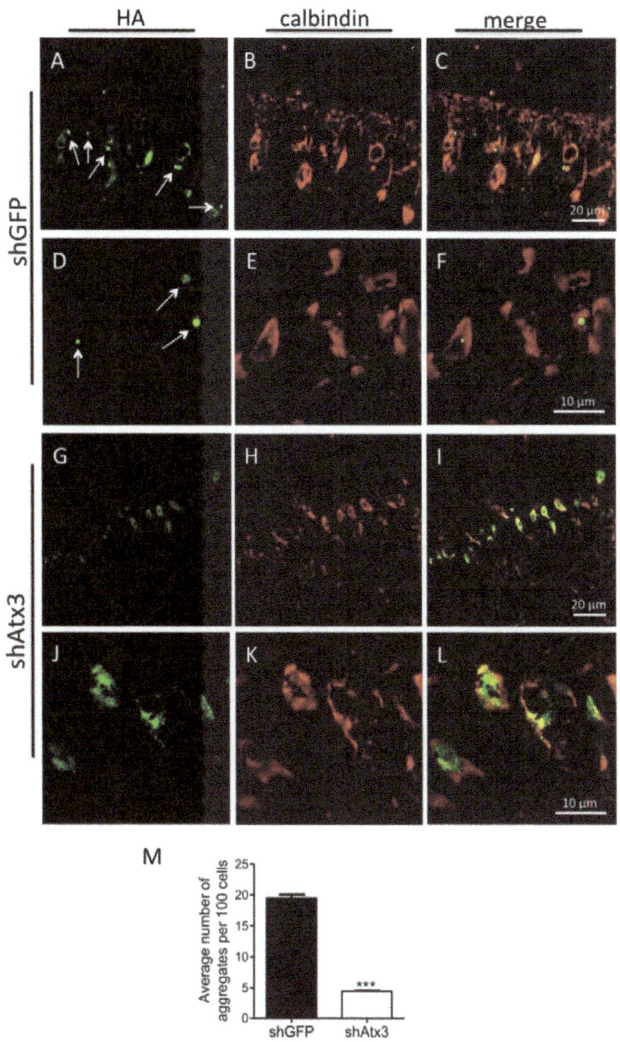

Figure 1. Allele-specific silencing of mutant ataxin-3 reduces the number of intranuclear inclusions. A–L) Confocal analysis of mutant ataxin-3 aggregates in the Purkinje cells at 10 weeks post-injection. Transgenic MJD mice injected at 21–25 days of age with shGFP (control) display intranuclear inclusions of mutant ataxin-3 (revealed by immunohistochemistry with an HA antibody) in Purkinje cells (revealed by calbindin immunohistochemistry) versus a diffuse expression of mutant ataxin-3 in mice injected with the silencing shAtx3. M) The number of intranuclear inclusions is significantly reduced in transgenic mice injected with shAtx3 compared to control animals injected with shGFP ($n = 6$, ***$P < 0.001$; Unpaired Student's *t*-test).

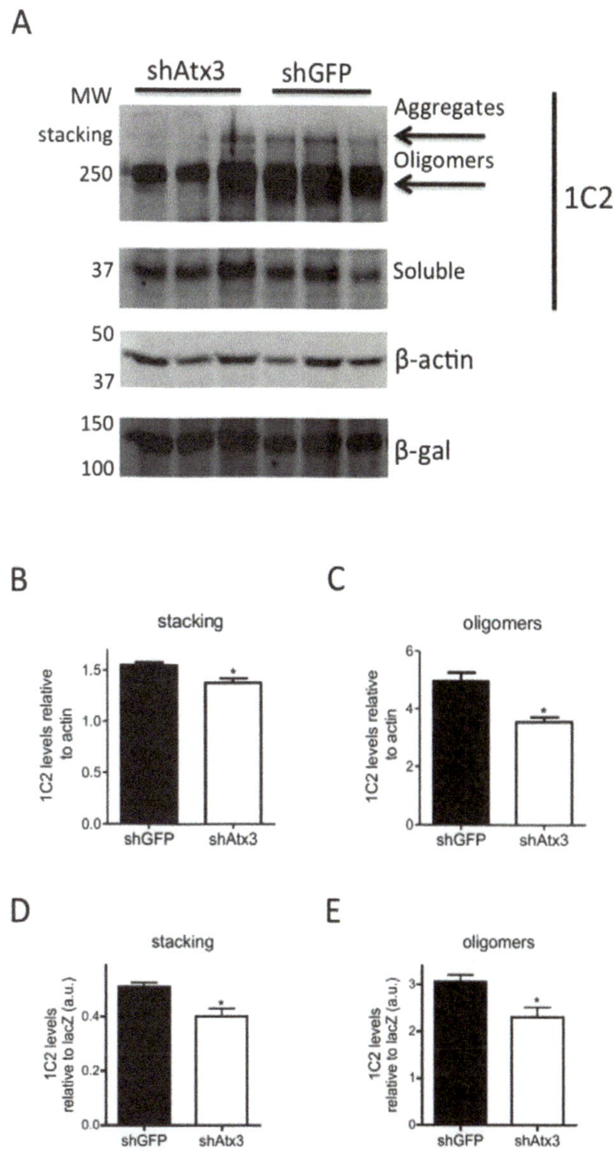

Figure 2. Allele-specific silencing of mutant ataxin-3 reduces the levels of mutant aggregated protein. A) Western blot analysis of cerebellar lysates stained with 1C2 antibody (for the polyglutamine expansion). Differences in levels of high molecular weight protein species were detected between transgenic mice injected with shAtx3 ($n = 3$) and animals injected with shGFP ($n = 3$). Normalization of protein levels was made with β-actin protein endogenous levels, and with β-gal marker (reporter gene of shRNAs lentiviral vector). B, D). The quantification of aggregated protein levels (in the stacking gel) revealed significant differences between transgenic mice injected with shAtx3 and control animals injected with shGFP ($n = 3$; *$P < 0.05$; Unpaired Student's *t*-test). C,E). Oligomerized species were also significantly reduced in animals injected with shAtx3 compared to control animals ($n = 3$; *$P < 0.05$; Unpaired Student's *t*-test).

Silencing of mutant ataxin-3 alleviates balance and motor coordination impairments in the MJD transgenic mouse

In order to evaluate whether gene silencing would alleviate gait and limb ataxia the MJD transgenic mice used in this study [15], stereotaxicaly injected with the silencing shAtax3 vectors, were submitted to behavior testing every two weeks until ten weeks post-injection (Figure S6 A). Already at the time of injection, at 21–25 days old, the animals displayed a marked phenotype characterized by an ataxic movement and difficulty to walk and to equilibrate. At the first test performed before the stereotaxic injection, animals barely stayed in the rotarod for 5 seconds, and, throughout the study time course, control mice (injected with shGFP) exhibited drastic impairment in the rotarod task. Silencing of the mutant ataxin-3 robustly enhanced the rotarod performance with mice consistently exhibiting better performances than mice injected with shGFP. These differences were significant and robust from 4 weeks after the injection at constant rotarod velocity (Fig. 4A; $n = 8$; shAtx3 55.5±9.4 s versus shGFP 11.92±2.1 s; $P < 0.001$) and in the accelerated test (Fig. 4B; $n = 8$; shAtx3 39.93±3.8 s versus shGFP 17.52±4.59 s; $P < 0.001$), and were maintained

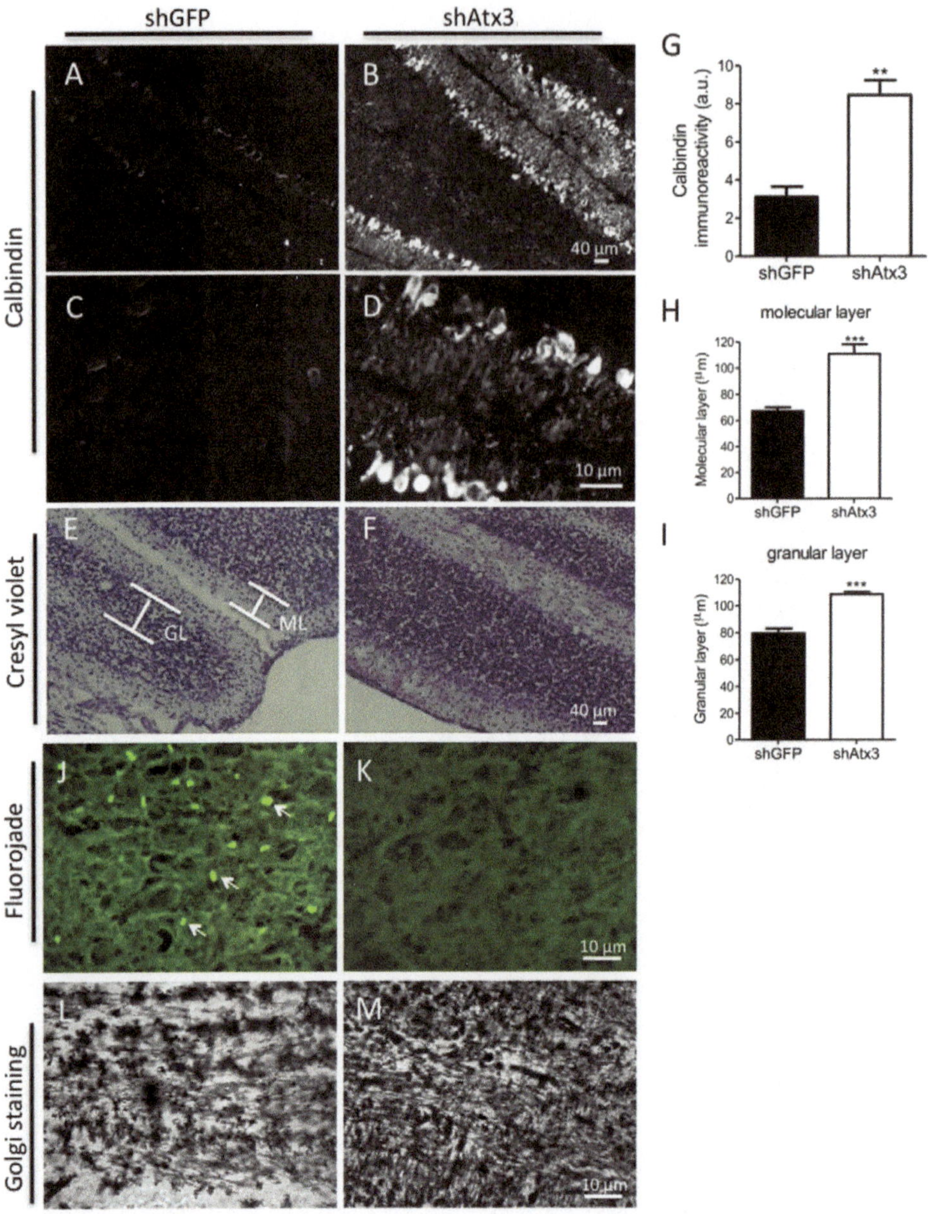

Figure 3. Silencing mutant ataxin-3 reduces Purkinje cell pathology, and prevents neurodegeneration and shrinkage of cellular layers within the cerebellar cortex. A–D) Fluorescence microscopy analysis of calbindin revealed a loss of dentritic arborizations and a shrinkage of Purkinje cells in the mice injected with shGFP ($n=8$) compared to control mice injected with shAtx3 ($n=8$). G) Quantification of optical densitometry of calbindin immunoreactivity. Silencing of mutant ataxin-3 preserved immunoreactivity for calbindin in the shAtx3-injected mice as compared to control mice (n = 6; **$P<0.01$; Unpaired Student's *t*-test). E–F) Cresyl violet staining. A more severe loss of neurons was observed in the control mice (shGFP, $n=8$) as compared to mice injected with shAtx3. H–I) paralleled by a significant decrease in the size of the molecular (ML) and granular layers (GL) in control mice as compared to shAtx3-injected mice. *Statistical significance ($n=6$; ***$P<0.01$; Unpaired Student's *t*-test). J–K) Fluorojade-B staining. Degenerating fluorescently-labelled neurons were only observed in control mice (shGFP, $n=8$), while no degeneration was observed in mice injected with shAtx3 ($n=8$). L–M) Golgi staining. A reduction in the number of dentritic arborizations in the molecular layer was observed in mice injected with shGFP ($n=8$) compared to a higher density of arborizations detected in mice injected with shAtx ($n=8$). The figure shows representative images that were reproducible among the different groups of animals.

along the experiment time course (Fig. 4A, B; $n=8$; 10 weeks; constant: shAtx3 61.95±11.9 s, and shGFP 19.25±4.1 s; accelerated: shAtx3 44.92±5.72 s, and shGFP 24.2±4.02 s). This improvement in motor functions of mice injected with shAtx3 was further investigated by analysis of the footprint patterns. From 4 weeks post-injection mice injected with shAtx3 showed a consistently larger stride length (Fig. 4C) than mice injected with shGFP, a difference that was significant from 6 weeks on ($n=8$; $P<0.01$), until the end of the experiment at 10 weeks post-injection ($n=8$; $P<0.01$). A similar pattern and an even more robust effect was registered upon analysis of the footprint overlap (Fig. 4D), where mice injected with shAtx3 revealed a significantly better overlap measure since 6 weeks after the injection compared to control mice ($n=8$; $P<0.001$). Another quantitative parameter, the hindbase width of footprint patterns confirmed the improved motor performance of shAtx3-injected mice (Figure S6 B).

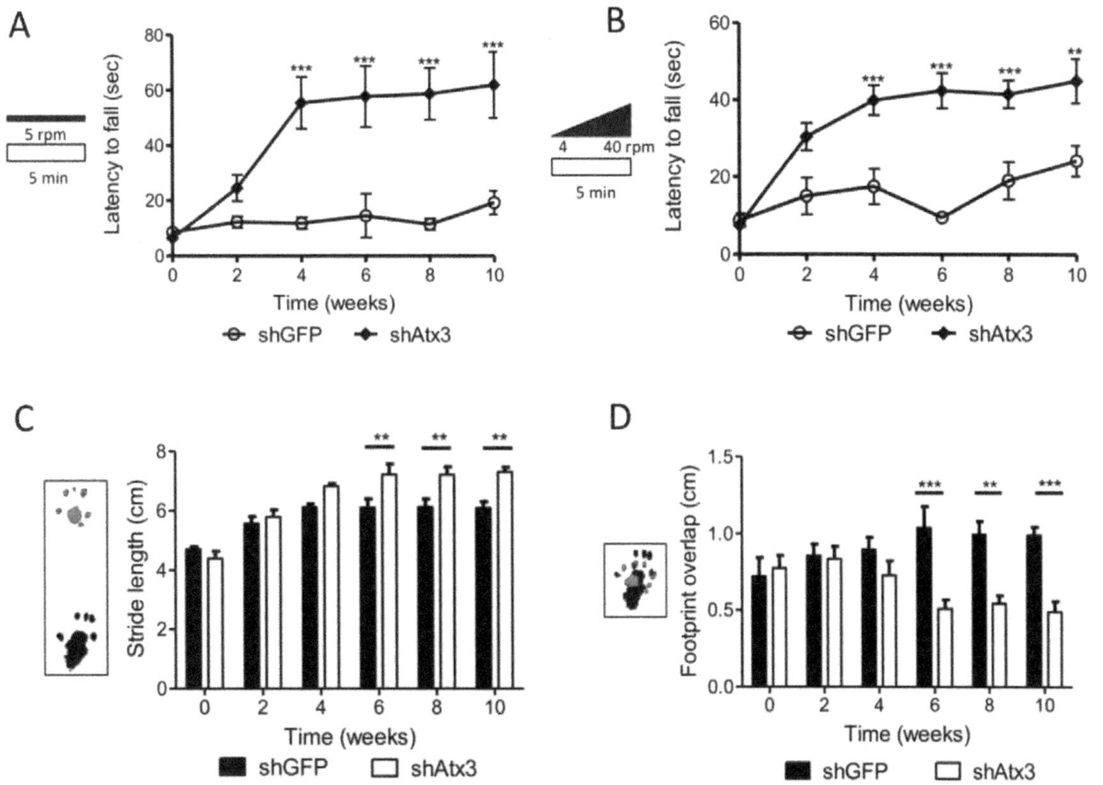

Figure 4. Silencing mutant ataxin-3 improves balance and motor coordination in transgenic MJD mice. A–B) Rotarod performance. Mice injected at 21–25 days of age with shAtx3 ($n = 8$) exhibited a significant better performance in constant velocity (5 r.p.m.) and accelerated rotarod, compared to control mice (shGFP, $n = 8$) from 4 weeks after the injection. C–D) Footprints patterns quantitative analysis, revealed a significantly better motor coordination in mice injected with shAtx3 ($n = 8$) from 6 weeks post-injection compared to control ones (shGFP, $n = 8$), with a higher stride length and a better footprint overlap. *Statistical significance (**$P<0.01$; ***$P<0.001$; 2-way ANOVA, Bonferonni post-hoc test).

Silencing of mutant ataxin-3 enhances locomotor and exploratory activities and reduces anxiety in the MJD transgenic mice

We further analyzed the locomotor activity every two weeks after transduction of mice cerebella by monitoring behavior in an activity box for 30 minutes, after a 10 minutes period of habituation. Animals injected with shAtx3 ($n = 8$) exhibited increased movement as compared to shGFP-injected mice ($n = 8$; Fig. 5A). These differences revealed that animals injected with shAtx3 consistently traveled longer distances than control mice (Fig. 5B; $n = 8$; 6 weeks, $P<0.05$; and 8 weeks, $P<0.01$), exhibited an increase in medium velocity (Fig. 5C), and maximum velocity of movement (Figure S6 C). Analysis of the first 10 minutes in the activity box at 10 weeks post-injection also revealed that mice injected with shAtx3 traveled a longer distance in the first 10 minutes as compared to control mice, revealing an increase in the exploratory behavior (Figure S6 D). Furthermore, when the arena was divided in two zones, it became clear that shAtx3-injected mice traveled a longer distance in zone 1 (Fig. 5D). Also, the control animals (shGFP-injected) traveled to the center of the cage significantly less often than shAtx3-injected mice, suggesting that these latter animals were less anxious (Fig. 5E). In fact, when the total distance or the number of entries in zone 2 was normalized with the total distance, the animals injected with shAtx3 exhibited a less anxious behavior, comparing to control mice shGFP-injected (Fig. 5GF, G).

Discussion

In this study we show for the first time that allele-specific silencing of mutant ataxin-3 with lentiviral vectors robustly rescues Machado-Joseph disease behavioral and neuropathological phenotype in a severely affected transgenic mouse model.

We previously showed that allele-specific silencing of mutant ataxin-3 expression could mitigate MJD neuropathology induced by viral-mediated expression of mutant ataxin-3 in the rat striatum [26]. In the present work we addressed the question whether this strategy would rescue behavioral and neuropathological impairments, when applied in the cerebellum of a transgenic mouse model already presenting a severe MJD-associated phenotype at the time of viral injection [15]. The transgenic mouse model here used was particularly suited for this purpose as it encoded the human C variant of the human ataxin-3 transgene sequence that is present in 70% of the MJD patients, which allowed allele-specific gene silencing. Moreover, this model encodes a truncated mutant ataxin-3 that induces the strong phenotype that would typically be found in a clinical setting at the time of intervention. Transgenic mouse models expressing truncated forms of the protein have been extensively used to model polyglutamine disorders such as Huntington's disease [21,29]. In contrast to some of the MJD transgenic mice expressing full-length ataxin-3 developed in the last years [30–36], whose behavioral and neuropathological deficits are mild and develop after several months of age, the use of a truncated form of ataxin-3 leads to an early and severe MJD phenotype with robust abnormal neuropathological and

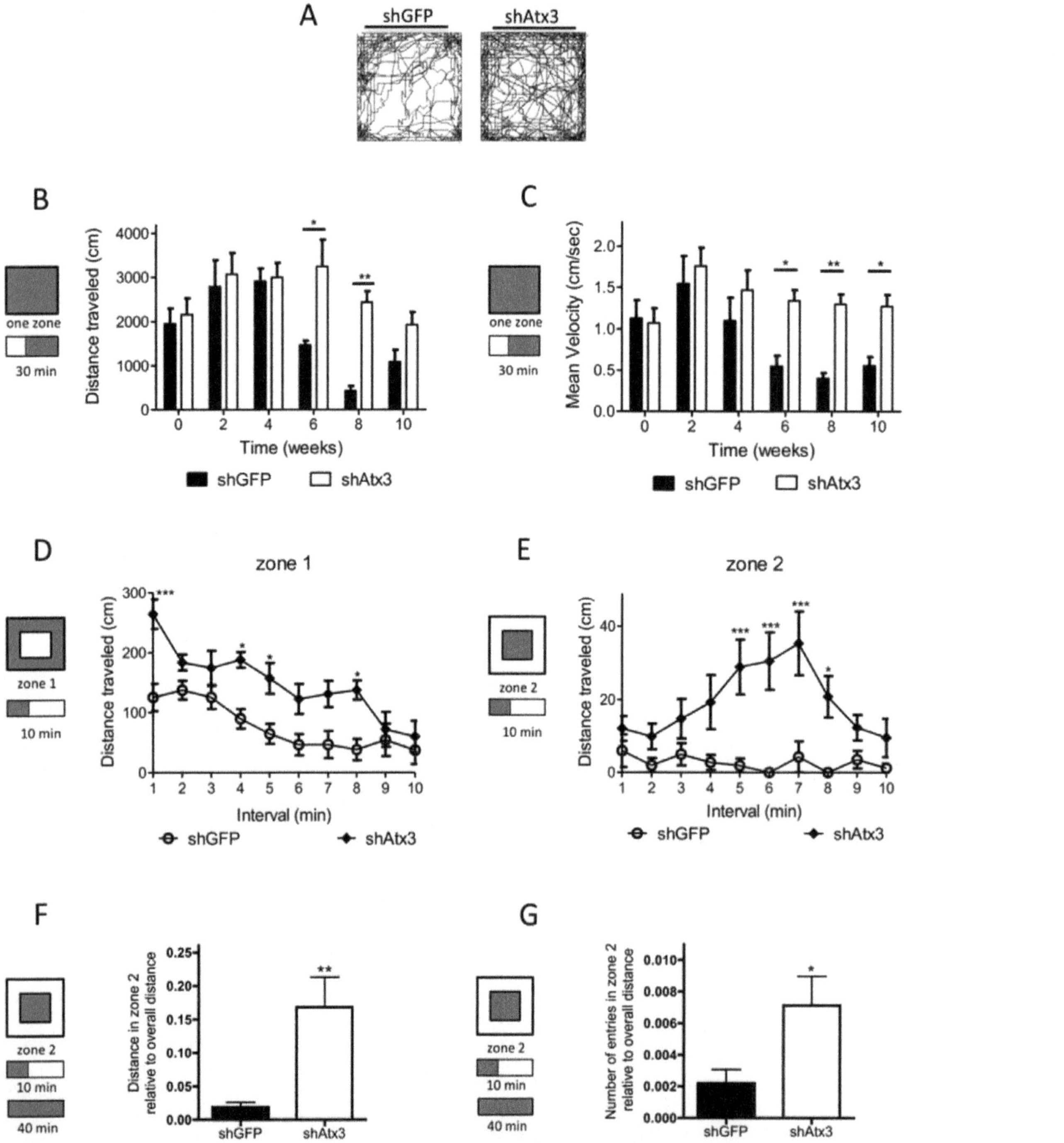

Figure 5. Allele specific silencing of mutant ataxin-3 improves exploratory and locomotor activity and reduces anxiety in MJD transgenic mice. A) Representative plots of moved track at 10 weeks post-injection during 40 minutes in the activity box. Animals injected with shAtx3 ($n=8$) traveled more that shGFP-injected mice ($n=8$). The figure shows representative images that were reproducible among the different groups of animals. B–C) Locomotor horizontal activity of mice was tracked for 30 minutes (after a 10 minutes habituation period), and revealed a significantly better locomotor activity of mice injected with shAtx3 ($n=8$) as compared to control mice (shGFP, $n=8$), by traveling longer distances with a higher medium velocity. D–E) Two zones analysis of exploratory behavior, revealed that shAtx3-injected mice ($n=8$) had increased exploratory behavior and were significantly less anxious than controls (shGFP; $n=8$). F–G) This reduced anxiety of shAtx3-injected mice was confirmed through analyzis of the distance in zone two normalized with the overall distance in the arena, and also the number of entries in zone 2 normalized with the overall distance in the arena. *Statistical significance (*$P<0.05$; **$P<0.01$; ***$P<0.001$; 2-way ANOVA, Bonferonni post-hoc test).

behavior deficits while maintaining physiological relevance for this particular question [15,37].

Gene silencing has been successfully used to downregulate the expression of mutant genes and to rescue phenotypes in various neurodegenerative diseases [21–23,38–40]. However, a potential risk of gene silencing lacking discrimination between normal and mutant forms of the causative protein is the loss of the normal protein function. While for some diseases such silencing may not present any drawback [29], for other disorders it is unlikely that the loss of normal protein function would be tolerated. Even though no degenerative phenotype was observed either in a MJD knockout animal [41] or upon non-specific gene silencing in a lentiviral rat model [10], absence of ataxin-3 has been reported to mediate cytoskeletal disorganization and increase cell death in cellular lines [42]. Therefore it may be prudent to either avoid complete silencing of the targeted protein or to employ allele-specific approaches able to preserve the wild-type protein as used in this study. We previously reported that a shRNA allele-specific lentiviral vector permitted selective silencing of mutant ataxin-3 in the striatum, while preserving wild-type ataxin-3, by targeting a C variant of the human ataxin-3 sequence that is present in 70% of the MJD patients [26]. Specific silencing in cellular models has also been reported to SNPs targeting ataxin-7 in SCA7 [43] and huntingtin in Huntington's disease [44,45]. In the present work, we knocked-down the mutant ataxin-3 transgene carrying the C-polymorphism present in the transgenic animal here used with the previously generated lentiviral vectors encoding an allele-specific shRNA. As there is no homology between mouse ataxin-3 and human ataxin-3 in the region targeted by the silencing sequence (See Fig. S1C), this means that the shRNA used in this study was specific only to human ataxin-3 and therefore does not allow to further infer about the selectivity of the approach, which had been previously demonstrated. Nevertheless, it shows that this allele-specific sequence is highly effective, mediating strong knock-down of the mutant ataxin-3 transgene in the cerebellum and rescuing motor impairments and neuropathology in a severe MJD model, even when initiated at late stage of disease. To our knowledge this is the first demonstration of alleviation of neuropathology and ataxia upon gene transfer of an allele-specific silencing sequence to the mouse cerebellum.

The cerebellum plays an important role in motor coordination and motor learning [46]. Damage of the cerebellum causes motor coordination problems such as gait and limb ataxia that partially explain the disabling clinical features of MJD, as this is one of the most affected regions in the disorder [2,8]. The MJD transgenic mouse used in this study [15], exhibits mutant ataxin-3 preferential expression in Purkinje cells of the cerebellar cortex, a marked atrophy of the cerebellum already at 3 weeks of age, and pronounced ataxic motor behavior. Similarly, motor incoordination and ataxia development related with progressive Purkinje neuron dendritic and somatic atrophy have been reported in SCA1 transgenic mice [47]. In this study we show that allele-specific silencing of mutant ataxin-3 mediates preservation of dendritic arborizations and Purkinje cells that project into the molecular layer. This may presumably underlie the improved motor performance of animals transduced with shAtx3 as compared to control ones.

The different cell types in the cerebellum are strongly interconnected to each other, and degeneration of one type can lead to atrophy of the interconnected cells. Purkinje cells account for a large fraction of the molecular layer thickness and its atrophy in aging has been associated with a reduction in dendritic arborizations of Purkinje cells [48,49]. The preservation of Purkinje cells in mice injected with shAtx3 may therefore explain the larger molecular layer thickness in these mice, while Purkinje cell degeneration in control animals probably induces shrinkage of the molecular layer.

Interestingly, in MJD patients the cerebellar cortex and the Purkinje layer are initially relatively preserved, in contrast to other spinocerebellar ataxia such as SCA1 and SCA2 [50]. Instead, neuronal degeneration affects initially dentate nucleus neurons, displaying grumose degeneration [50]. However, as pathology progresses Purkinje cells dysfunction and degeneration develops in MJD mice and patients [51,52], which may contribute to the retrograde atrophy of cells from molecular and granular layer. Degeneration of Purkinje cells may reduce directly the information input, reduce afferent efficacy, and affect information integration and signal transmission, consequently leading to a reduction in overall cerebellar function. Thus, mutant ataxin-3 silencing in Purkinje cells and other cells in the cerebellar cortex, may prevent the neurodegeneration and cell death in MJD affected regions of the cerebellum. However, translation of this approach to the clinic may have higher chances of success if also targeting the deep cerebellar nuclei, the region with primary degeneration in MJD.

The number of mutant ataxin-3 nuclear inclusions was also significantly reduced in mice injected with shAtx3, as previously observed in the lentiviral-based model [26], in parallel with a subcellular diffuse distribution of the protein in the cytoplasm. The presence of ataxin-3 in the nucleus has been shown to drastically aggravate the pathology in Machado-Joseph disease [32]. Therefore, silencing mutant ataxin-3 may prevent nuclear accumulation and degeneration contributing to preservation of neuronal function and motor behavior of mice injected with shAtx3.

The rotarod test for motor performance is a valid indicator of progressive ataxia; evidentiating proper foot placement and coordination in response to the rotating rod challenges the cerebellum. The allele-specific silencing of mutant ataxin-3 greatly improved the rotarod performance from 4 weeks after the injection, allowing animals to equilibrate in the rotarod for more than 60 seconds, on average 3-fold the time found for control mice. Even though this performance is far from the one that is commonly found in wild-type mice (around 120 s), it corresponds to a very robust preservation of performance. Footprints analysis and activity measures also revealed that mutant ataxin-3 silencing strongly alleviates the phenotype of the MJD transgenic mice. In other polyglutamine disorders silencing of mutant protein also reduced the neurological abnormalities [20–22]. Nevertheless, improvement in the rotarod performance upon mutant protein silencing did exceed 0.5 to 1-fold compared to the controls [20,21].

Our data suggest that shRNA specific silencing reduces the neuropathology and the motor behavior deficits of a severely impaired transgenic MJD mouse model. These results suggest that this approach is a promising therapeutic strategy to MJD patients. Importantly, several questions should be addressed, from the shRNAs long-term safety, to the precise brain regions to be targeted.

Materials and Methods

Lentiviral vectors

Viral vectors encoding a short-hairpin targeting the mutant ataxin-3 and the GFP, with the LacZ reporter gene (26) were produced described previously [53].

In vivo experiments

Animals. MJD transgenic mice (C57BL/6 background) expressing the N-terminal-truncated human ataxin-3 with 69 glutamine repeats and an N-terminal hemagglutinin (HA) epitope driven specifically in cerebellar Purkinje cells by the L7 promoter were maintained at CNC by backcrossing heterozygous males with C57BL/6 females [15,54]. The present study used 8 heterozygous females and 8 heterozygous males injected with lentiviral vectors (in two separate cohorts, with 4 animals per group in each one) encoding the short-hairpin targeting the mutant ataxin-3 (shAtx3; $n = 8$), and as control the short hairpin targeting the GFP (shGFP;

$n = 8$). Behavioral testing began at 21–25 days of age (P21–25), and one day after the first behavioral testing animals were submitted to stereotaxic surgery. The experiments were carried out in accordance with the European Community Council directive (86/609/EEC) for the care and use of laboratory animals.

Stereotaxic injection of lentiviral vectors

The animals (P21–25) received a single 6 µl injection of LV (200 000 ng of p24/ml), at 0.25 µl/min at the coordinates: −1.6 mm rostral to lambda, 0 mm midline, and 1 mm ventral to the skull surface, with the mouth bar set at −3.3.

Behavioral testing

Mice were trained on a battery of motor tests starting at 21–25 days of age (P21–25) and performed every 2 weeks until 10 weeks by an experienced operator in a blind fashion way. All tests were performed in the same dark room after 30 minutes of acclimatization. Mean values for each measure were calculated and statistical analysis was performed by two-way ANOVA with GraphPad (La Jolla, USA). Data are represented as mean ± SEM.

Rotarod

Motor coordination and balance were evaluated in a rotarod apparatus (Letica Scientific Instruments, Panlab, Barcelona, Spain). Mice were placed on the rotarod at a constant speed (5 rpm) for a maximum of 5 min, and at accelerated speed (4 to 40 rpm in 5 min) and the latency to fall was recorded. Mice were allowed to perform four trials for each test and time point, with 15 min rest between trials. For analysis, the mean latency to fall off the rotarod of 3–4 trials was used.

Footprint patterns analysis

The footprint test was used to compare the gait of mice injected with vectors encoding shAtx3 with that of injected with shGFP (control). Hind- and forefeet were coated with black and green nontoxic paints, and the animals allowed walking along a 100-cm-long, 10-cm-wide runaway (with 15-cm high walls) over a fresh sheet of white paper. The footprint patterns were analyzed for (1) stride length, corresponding to the average distance of forward movement between each stride; (2) hind-base width and (3) front-base width measured as the average distance between left and right hind footprints, respectively, and (4) distance from left or right front footprint/hind footprint overlap to measure uniformity of step alternation. The distance between the center of the hind footprint and the center of the preceding front footprint was recorded over a sequence of six consecutive steps, excluding footprints made at the beginning and end of the run. The same operator made all footprints measurements blindly.

Open field analysis

For the assessment of mice explorative behavior, and locomotor horizontal activity, mice were placed in a 50×50 cm arena with 50 cm high walls and movement activity was recorded for 40 min using Acti-Track System (Panlab, Barcelona, Spain). The collected data was analyzed for the first 10 min, and for the last 30 minutes.

Histological processing

Tissue preparation. Tissue was prepared as previously described (16). Slices throughout the entire cerebellum were collected in superfrost plus microscope slides (Thermo Fisher Scientific, U.S.A) and stored at −20°C before immunohistochemical processing. For each animal the entire cerebellum was collected into 12 slides with 8 coronal sections each, distant 240 µm from each other.

Immunohistochemical procedure

The immunohistochemical procedure was initiated with a 30 min dehydration at 37°C followed by a 30 min hydration in 0.1 PBS and 1 h blocking in a 0.3% triton in 0.1 PBS with 10% normal goat serum both at room temperature (RT). The following primary antibodies diluted in blocking solution with 0.1% triton were used: mouse monoclonal anti-HA (InvivoGen, San Diego, CA, USA; 1:1000; O/N, 4°C), and rabbit polyclonal anti-Calbindin D-28K (Chemicon, Temecula, CA, USA; 1:1000, O/N, 4°C). Sections were then incubated in secondary antibody, goat-anti rabbit and/or mouse conjugated to alexa 488 or 594 (Invitrogen) for 2 h/RT and then mounted in Fluorsave (Calbiochem, Germany) with 4′,6′-diamidino-2-phenylindole (DAPI). Fluorescence images were acquired with a Zeiss Axiovert 200 imaging microscope or LSM Zeiss microscope for double staining experiments.

Fluorojade B staining

Cerebellar sections were stained with FluoroJade-B (Chemicon, Temecula, CA), an anionic fluorescein derivative that stains neurons undergoind degeneration. The sections were first washed in water and then mounted on silanecoated glass slides, dehydrated, and stained according to the supplier's manual. All photographs for comparison were taken under identical image acquisition conditions and uniform adjustments of brightness and contrast were made to all images.

Golgi staining

Golgi neurohistological staining was performed following a modification of classical Golgi procedure described previously [55].

Western blot

Mice cerebella were removed after a sodium pentobarbital overdose, and incubated on ice in a radioimmunoprecipitation assay-buffer solution (50 mM Tris HCl, pH 8, 150 nM NaCl, 1% NP-40, 0.5% sodium deoxycholate, 0.1% sodium dodecyl sulphate) containing proteases inhibitors (Roche diagnostics GmbH) followed by a 4 sec ultra-sound pulse (1 pulse/sec). Total protein lysates were stored at −80°C, protein concentration was determined with the Bradford protein assay (BioRad), and 20 µg of protein extract was resolved in sodium dodecyl sulphate-polyacrylamide gels (4% stacking and 8% running). The proteins were transferred onto polyvinylidene difluoride membranes (GE Healthcare) according to standard protocols. The immunobloting procedure was performed as described previously [10] with the respective primary antibody (1C2, 1:1000, Millipore), followed by incubation wit the corresponding alkaline phosphatase-linked secondary antibody. Bands were visualized with Enhanced Chemifluorescence substrate (ECF, GE Healthcare) and chemifluorescence imaging (VersaDoc Imaging System Model 3000, Bio-Rad). Membranes were stripped using 0.1 M glycine pH 2.3 (30 min, room temperature) and reprobed with mouse monoclonal anti-β-actin antibody (1:5000, Sigma), and anti-β-gal antibody (1:5000, Cell Signaling). Densitometric analysis was carried out in the same gel using Image J software (NIH, USA).

Quantitative analysis of HA aggregates

Quantification of HA positive inclusions was performed blindly by scanning 4 coronal sections spread over the anterior-posterior extent of the cerebellum of each animal (inter-section distance:

240 μm), using a 20× objective on a Zeiss Axiovert 200 imaging microscope and the image J acquisition and analysis software (NIH, USA). For each coronal section, 8 fields covering the entire cerebellar cortex were digitalized. The total number of HA inclusions, and the number of Purkinje cells were counted, and the average number of inclusions per 100 Purkinje cells was plotted.

Quantitative analysis of calbindin neuronal expression

Quantification of calbindin immunoreactivity was performed blindly by scanning 4 coronal sections spread over the anterior-posterior extent of the cerebellum (inter-section distance: 240 μm), using a 40× objective on a Zeiss Axiovert 200 imaging microscope. For each coronal section, 8 fields covering the entire cerebellar cortex were digitalized. Optical densitometry analysis was performed with Image J (NIH, USA). Values are represented as the mean value of calbindin optical density per section ± SEM.

Cresyl violet staining

Cerebellar sections were stained with cresyl violet for 2 minutes, differentiated in acetate buffer pH 3.8 to 4 (2.72% sodium acetate and 1.2% acetic acid; 1:4 v/v), dehydrated by passing twice through ethanol and toluol solutions, and mounted with Eukitt® (O. Kindler GmbH & CO. Freiburg, Germany).

Quantification of granular and molecular layers size

Quantification was made over 4 cresyl violet staining coronal sections spread over the anterior-posterior extent of the cerebellum in a blind fashion (inter-section distance: 200–300 μm), using a 20× objective. For each coronal section, 8 fields covering the entire cerebellar cortex were digitalized. For each acquired field at least 6 measurements were made blindly in the same region for all animals, and results converted to μm using Image J software (NIH).

Supporting Information

Figure S1 A) The transgene used to generate the Q69 transgenic mouse model (Torashima et al., 2008) was isolated from the human ataxin-3 gene with the polymorphism (G→C transition, highlighted in blue) that is present in 70% of MJD patients (Stevanin et al., 1995; Gaspar et al., 1996). B) The presence of this polymorphism permitted the design of an allele-specific silencing of mutant ataxin-3 (Alves et al., 2008), using shRNAs in a LV backbone (with a separate cassette containing the lacZ reporter gene). As control a shRNA targeting GFP in a LV was used. C) There is no homology between mouse ataxin-3 and human ataxin-3 in the region targeted by the silencing sequence, which means that the shRNA used in this study was specific only to human ataxin-3.

Figure S2 Intracerebellar injection of 6 μl (200.000 ng/ml) of lentiviral vectors (LV) encoding for Green Fluorescent Protein (GFP) in the cerebellar vermis of transgenic mice (P21–25) mediates an extensive antero-posterior transduction of the cerebellar cortex ($n = 8$). This extensive transduction was observed from the place of the injection and covered almost 60% of the area of the cerebellar cortex, mainly in the molecular layer and in Purkinje cells.

Figure S3 The intracerebellar injection of LV encoding for GFP in the cerebellar vermis mediates an extensive transduction of the cells in the cerebellar cortex, mainly the molecular layer cells (ML), and the Purkinje layer cells (PCL).

Figure S4 Transgenic mice injected with LV encoding a short-hairpin against GFP (shGFP) exhibit ataxin-3 aggregates (HA tag, green) co-localizing with LacZ (red, white arrows); whereas aggregates from mice injected with short-hairpins against mutant ataxin-3 (shAtx3) do not co-localize. This indicates that the shRNAs targeting mutant ataxin-3, but not the control shRNAs targeting GFP, prevent the formation of ataxin-3 aggregates.

Figure S5 DARPP-32 staining revealed a preservation of immunoreactivity in transgenic mice injected with LV encoding shAtx3 ($n = 8$) as compared to mice injected with shGFP (arrows, $n = 8$). The figure shows representative images that were reproducible among the different groups of animals. Scale bar: 40 μm.

Figure S6 Behavior analysis of studied transgenic mice. A) Time course of mice behavior tests: rotarod performance test, footprints patterns analysis and activity box monitoring were assessed every 2 weeks post-injection until 10 weeks. B) Footprints patterns quantitative analysis. Hindbase width measures of shGFP injected mice ($n = 8$) show a significantly greater distance between left and right limb compared to shAtx3 injected mice ($n = 8$), indicating a higher deficit of coordination in control mice. *Statistical significance (*$P<0.05$; ***$P<0.001$; 2-way ANOVA, Bonferonni *post*-test). C) Locomotor horizontal activity of mice was tracked for 30 minutes (after a 10 minutes habituation period) and analyzed for maximum velocity (cm/sec). Mice injected with shAtx3 ($n = 8$) revealed a significantly better locomotor activity than control mice (shGFP, $n = 8$), as shown by faster movement from 4 weeks post-injection. *Statistical significance (*$P<0.05$; 2-way ANOVA, Bonferonni *post*-test). D) Allele specific silencing of mutant ataxin-3 improves exploratory activity. Analysis of the first 10 minutes in the cage at 10 weeks post-injection for one zone (arena not divided) revealed significantly-increased traveled distance in mice injected with shAtx3 ($n = 8$) compared to mice injected with shGFP ($n = 8$).

Acknowledgments

Authors would like to thank to Juliana Bortolleto for the scientific drawing of the mouse brain.

Author Contributions

Financial support and administrative support: LPdA.. Conceived and designed the experiments: CN INF LPdA. Performed the experiments: CN INF IO DA. Analyzed the data: CN INF HH ND LPdA. Contributed reagents/materials/analysis tools: HH ND LPdA. Wrote the paper: CN LPdA.

References

1. Schols L, Vieira-Saecker AM, Schols S, Przuntek H, Epplen JT, et al. (1995) Trinucleotide expansion within the MJD1 gene presents clinically as spinocerebellar ataxia and occurs most frequently in German SCA patients. Human molecular genetics 4: 1001–1005.
2. Durr A, Stevanin G, Cancel G, Duyckaerts C, Abbas N, et al. (1996) Spinocerebellar ataxia 3 and Machado-Joseph disease: clinical, molecular, and neuropathological features. Annals of neurology 39: 490–499.

3. Silveira I, Coutinho P, Maciel P, Gaspar C, Hayes S, et al. (1998) Analysis of SCA1, DRPLA, MJD, SCA2, and SCA6 CAG repeats in 48 Portuguese ataxia families. American journal of medical genetics 81: 134–138.
4. Jardim LB, Silveira I, Pereira ML, Ferro A, Alonso I, et al. (2001) A survey of spinocerebellar ataxia in South Brazil - 66 new cases with Machado-Joseph disease, SCA7, SCA8, or unidentified disease-causing mutations. Journal of neurology 248: 870–876.
5. Zoghbi HY, Orr HT (2000) Glutamine repeats and neurodegeneration. Annual review of neuroscience 23: 217–247.
6. Nakamura K, Jeong SY, Uchihara T, Anno M, Nagashima K, et al. (2001) SCA17, a novel autosomal dominant cerebellar ataxia caused by an expanded polyglutamine in TATA-binding protein. Human molecular genetics 10: 1441–1448.
7. Fowler HL (1984) Machado-Joseph-Azorean disease. A ten-year study. Archives of neurology 41: 921–925.
8. Sudarsky L, Coutinho P (1995) Machado-Joseph disease. Clinical neuroscience 3: 17–22.
9. Wullner U, Reimold M, Abele M, Burk K, Minnerop M, et al. (2005) Dopamine transporter positron emission tomography in spinocerebellar ataxias type 1, 2, 3, and 6. Archives of neurology 62: 1280–1285.
10. Alves S, Regulier E, Nascimento-Ferreira I, Hassig R, Dufour N, et al. (2008) Striatal and nigral pathology in a lentiviral rat model of Machado-Joseph disease. Human molecular genetics 17: 2071–2083.
11. Kawaguchi Y, Okamoto T, Taniwaki M, Aizawa M, Inoue M, et al. (1994) CAG expansions in a novel gene for Machado-Joseph disease at chromosome 14q32.1. Nature genetics 8: 221–228.
12. van Alfen N, Sinke RJ, Zwarts MJ, Gabreels-Festen A, Praamstra P, et al. (2001) Intermediate CAG repeat lengths (53,54) for MJD/SCA3 are associated with an abnormal phenotype. Annals of neurology 49: 805–807.
13. Schmidt T, Landwehrmeyer GB, Schmitt I, Trottier Y, Auburger G, et al. (1998) An isoform of ataxin-3 accumulates in the nucleus of neuronal cells in affected brain regions of SCA3 patients. Brain pathology 8: 669–679.
14. Chen X, Tang TS, Tu H, Nelson O, Pook M, et al. (2008) Deranged calcium signaling and neurodegeneration in spinocerebellar ataxia type 3. The Journal of neuroscience : the official journal of the Society for Neuroscience 28: 12713–12724.
15. Torashima T, Koyama C, Iizuka A, Mitsumura K, Takayama K, et al. (2008) Lentivector-mediated rescue from cerebellar ataxia in a mouse model of spinocerebellar ataxia. EMBO reports 9: 393–399.
16. Nascimento-Ferreira I, Santos-Ferreira T, Sousa-Ferreira L, Auregan G, Onofre I, et al. (2011) Overexpression of the autophagic beclin-1 protein clears mutant ataxin-3 and alleviates Machado-Joseph disease. Brain : a journal of neurology 134: 1400–1415.
17. Simoes AT, Goncalves N, Koeppen A, Deglon N, Kugler S, Duarte CB and Pereira de Almeida L. (2012) Calpastatin-mediated inhibition of calpains in the mouse brain prevents mutant ataxin 3 proteolysis, nuclear localization and aggregation, relieving Machado-Joseph disease. Brain: 135:2428–39.
18. Yamamoto A, Lucas JJ, Hen R (2000) Reversal of neuropathology and motor dysfunction in a conditional model of Huntington's disease. Cell 101: 57–66.
19. Zu T, Duvick LA, Kaytor MD, Berlinger MS, Zoghbi HY, et al. (2004) Recovery from polyglutamine-induced neurodegeneration in conditional SCA1 transgenic mice. The Journal of neuroscience : the official journal of the Society for Neuroscience 24: 8853–8861.
20. Xia H, Mao Q, Eliason SL, Harper SQ, Martins IH, et al. (2004) RNAi suppresses polyglutamine-induced neurodegeneration in a model of spinocerebellar ataxia. Nature medicine 10: 816–820.
21. Harper SQ, Staber PD, He X, Eliason SL, Martins IH, et al. (2005) RNA interference improves motor and neuropathological abnormalities in a Huntington's disease mouse model. Proceedings of the National Academy of Sciences of the United States of America 102: 5820–5825.
22. Rodriguez-Lebron E, Denovan-Wright EM, Nash K, Lewin AS, Mandel RJ (2005) Intrastriatal rAAV-mediated delivery of anti-huntingtin shRNAs induces partial reversal of disease progression in R6/1 Huntington's disease transgenic mice. Molecular therapy : the journal of the American Society of Gene Therapy 12: 618–633.
23. DiFiglia M, Sena-Esteves M, Chase K, Sapp E, Pfister E, et al. (2007) Therapeutic silencing of mutant huntingtin with siRNA attenuates striatal and cortical neuropathology and behavioral deficits. Proceedings of the National Academy of Sciences of the United States of America 104: 17204–17209.
24. Miller VM, Xia H, Marrs GL, Gouvion CM, Lee G, et al. (2003) Allele-specific silencing of dominant disease genes. Proceedings of the National Academy of Sciences of the United States of America 100: 7195–7200.
25. Gaspar C, Lopes-Cendes I, Hayes S, Goto J, Arvidsson K, et al. (2001) Ancestral origins of the Machado-Joseph disease mutation: a worldwide haplotype study. American journal of human genetics 68: 523–528.
26. Alves S, Nascimento-Ferreira I, Auregan G, Hassig R, Dufour N, et al. (2008) Allele-specific RNA silencing of mutant ataxin-3 mediates neuroprotection in a rat model of Machado-Joseph disease. PloS one 3: e3341.
27. Yoshizawa T, Yamagishi Y, Koseki N, Goto J, Yoshida H, et al. (2000) Cell cycle arrest enhances the in vitro cellular toxicity of the truncated Machado-Joseph disease gene product with an expanded polyglutamine stretch. Human molecular genetics 9: 69–78.
28. Gaspar C, Lopes-Cendes I, DeStefano AL, Maciel P, Silveira I, et al. (1996) Linkage disequilibrium analysis in Machado-Joseph disease patients of different ethnic origins. Human genetics 98: 620–624.
29. Boudreau RL, McBride JL, Martins I, Shen S, Xing Y, et al. (2009) Nonallele-specific silencing of mutant and wild-type huntingtin demonstrates therapeutic efficacy in Huntington's disease mice. Molecular therapy : the journal of the American Society of Gene Therapy 17: 1053–1063.
30. Cemal CK, Carroll CJ, Lawrence L, Lowrie MB, Ruddle P, et al. (2002) YAC transgenic mice carrying pathological alleles of the MJD1 locus exhibit a mild and slowly progressive cerebellar deficit. Human molecular genetics 11: 1075–1094.
31. Goti D, Katzen SM, Mez J, Kurtis N, Kiluk J, et al. (2004) A mutant ataxin-3 putative-cleavage fragment in brains of Machado-Joseph disease patients and transgenic mice is cytotoxic above a critical concentration. The Journal of neuroscience : the official journal of the Society for Neuroscience 24: 10266–10279.
32. Bichelmeier U, Schmidt T, Hubener J, Boy J, Ruttiger L, et al. (2007) Nuclear localization of ataxin-3 is required for the manifestation of symptoms in SCA3: in vivo evidence. The Journal of neuroscience : the official journal of the Society for Neuroscience 27: 7418–7428.
33. Chou AH, Yeh TH, Ouyang P, Chen YL, Chen SY, et al. (2008) Polyglutamine-expanded ataxin-3 causes cerebellar dysfunction of SCA3 transgenic mice by inducing transcriptional dysregulation. Neurobiology of disease 31: 89–101.
34. Boy J, Schmidt T, Wolburg H, Mack A, Nuber S, et al. (2009) Reversibility of symptoms in a conditional mouse model of spinocerebellar ataxia type 3. Human molecular genetics 18: 4282–4295.
35. Boy J, Schmidt T, Schumann U, Grasshoff U, Unser S, et al. (2010) A transgenic mouse model of spinocerebellar ataxia type 3 resembling late disease onset and gender-specific instability of CAG repeats. Neurobiology of disease 37: 284–293.
36. Silva-Fernandes A, Costa Mdo C, Duarte-Silva S, Oliveira P, Botelho CM, et al. (2010) Motor uncoordination and neuropathology in a transgenic mouse model of Machado-Joseph disease lacking intranuclear inclusions and ataxin-3 cleavage products. Neurobiology of disease 40: 163–176.
37. Ikeda H, Yamaguchi M, Sugai S, Aze Y, Narumiya S, et al. (1996) Expanded polyglutamine in the Machado-Joseph disease protein induces cell death in vitro and in vivo. Nature genetics 13: 196–202.
38. van Bilsen PH, Jaspers L, Lombardi MS, Odekerken JC, Burright EN, et al. (2008) Identification and allele-specific silencing of the mutant huntingtin allele in Huntington's disease patient-derived fibroblasts. Human gene therapy 19: 710–719.
39. Lombardi MS, Jaspers L, Spronkmans C, Gellera C, Taroni F, et al. (2009) A majority of Huntington's disease patients may be treatable by individualized allele-specific RNA interference. Experimental neurology 217: 312–319.
40. Pfister EL, Kennington L, Straubhaar J, Wagh S, Liu W, et al. (2009) Five siRNAs targeting three SNPs may provide therapy for three-quarters of Huntington's disease patients. Current biology : CB 19: 774–778.
41. Schmitt I, Linden M, Khazneh H, Evert BO, Breuer P, et al. (2007) Inactivation of the mouse Atxn3 (ataxin-3) gene increases protein ubiquitination. Biochemical and biophysical research communications 362: 734–739.
42. Rodrigues AJ, do Carmo Costa M, Silva TL, Ferreira D, Bajanca F, et al. (2010) Absence of ataxin-3 leads to cytoskeletal disorganization and increased cell death. Biochimica et biophysica acta 1803: 1154–1163.
43. Scholefield J, Greenberg LJ, Weinberg MS, Arbuthnot PB, Abdelgany A, et al. (2009) Design of RNAi hairpins for mutation-specific silencing of ataxin-7 and correction of a SCA7 phenotype. PloS one 4: e7232.
44. Hu J, Matsui M, Gagnon KT, Schwartz JC, Gabillet S, et al. (2009) Allele-specific silencing of mutant huntingtin and ataxin-3 genes by targeting expanded CAG repeats in mRNAs. Nature biotechnology 27: 478–484.
45. Zhang S, Feany MB, Saraswati S, Littleton JT, Perrimon N (2009) Inactivation of Drosophila Huntingtin affects long-term adult functioning and the pathogenesis of a Huntington's disease model. Disease models & mechanisms 2: 247–266.
46. Ito M (2002) Historical review of the significance of the cerebellum and the role of Purkinje cells in motor learning. Annals of the New York Academy of Sciences 978: 273–288.
47. Clark HB, Burright EN, Yunis WS, Larson S, Wilcox C, et al. (1997) Purkinje cell expression of a mutant allele of SCA1 in transgenic mice leads to disparate effects on motor behaviors, followed by a progressive cerebellar dysfunction and histological alterations. The Journal of neuroscience : the official journal of the Society for Neuroscience 17: 7385–7395.
48. Hadj-Sahraoui N, Frederic F, Zanjani H, Delhaye-Bouchaud N, Herrup K, et al. (2001) Progressive atrophy of cerebellar Purkinje cell dendrites during aging of the heterozygous staggerer mouse (Rora(+/sg)). Brain research Developmental brain research 126: 201–209.
49. Zhang C, Hua T, Zhu Z, Luo X (2006) Age-related changes of structures in cerebellar cortex of cat. Journal of biosciences 31: 55–60.
50. Koeppen AH (2005) The pathogenesis of spinocerebellar ataxia. Cerebellum 4: 62–73.
51. Munoz E, Rey MJ, Mila M, Cardozo A, Ribalta T, et al. (2002) Intranuclear inclusions, neuronal loss and CAG mosaicism in two patients with Machado-Joseph disease. Journal of the neurological sciences 200: 19–25.
52. Shakkottai VG, do Carmo Costa M, Dell'Orco JM, Sankaranarayanan A, Wulff H, et al. (2011) Early changes in cerebellar physiology accompany motor dysfunction in the polyglutamine disease spinocerebellar ataxia type 3. The

Journal of neuroscience : the official journal of the Society for Neuroscience 31: 13002–13014.

53. de Almeida LP, Zala D, Aebischer P, Deglon N (2001) Neuroprotective effect of a CNTF-expressing lentiviral vector in the quinolinic acid rat model of Huntington's disease. Neurobiology of disease 8: 433–446.

54. Oue M, Mitsumura K, Torashima T, Koyama C, Yamaguchi H, et al. (2009) Characterization of mutant mice that express polyglutamine in cerebellar Purkinje cells. Brain research 1255: 9–17.

55. Friedland DR, Los JG, Ryugo DK (2006) A modified Golgi staining protocol for use in the human brain stem and cerebellum. Journal of neuroscience methods 150: 90–95.

4

Effect of Feeding Selenium-Fertilized Alfalfa Hay on Performance of Weaned Beef Calves

Jean A. Hall[1]*, Gerd Bobe[2,3], Janice K. Hunter[2¤a], William R. Vorachek[1], Whitney C. Stewart[2¤b], Jorge A. Vanegas[4], Charles T. Estill[2,4], Wayne D. Mosher[2], Gene J. Pirelli[2]

1 Department of Biomedical Sciences, College of Veterinary Medicine, Oregon State University, Corvallis, Oregon, United States of America, **2** Department of Animal and Rangeland Sciences, College of Agricultural Sciences, Oregon State University, Corvallis, Oregon, United States of America, **3** Linus Pauling Institute, Oregon State University, Corvallis, Oregon, United States of America, **4** Department of Clinical Sciences, College of Veterinary Medicine, Oregon State University, Corvallis, Oregon, United States of America

Abstract

Selenium (Se) is an essential micronutrient in cattle, and Se-deficiency can affect morbidity and mortality. Calves may have greater Se requirements during periods of stress, such as during the transitional period between weaning and movement to a feedlot. Previously, we showed that feeding Se-fertilized forage increases whole-blood (WB) Se concentrations in mature beef cows. Our current objective was to test whether feeding Se-fertilized forage increases WB-Se concentrations and performance in weaned beef calves. Recently weaned beef calves (n = 60) were blocked by body weight, randomly assigned to 4 groups, and fed an alfalfa hay based diet for 7 wk, which was harvested from fields fertilized with sodium-selenate at a rate of 0, 22.5, 45.0, or 89.9 g Se/ha. Blood samples were collected weekly and analyzed for WB-Se concentrations. Body weight and health status of calves were monitored during the 7-wk feeding trial. Increasing application rates of Se fertilizer resulted in increased alfalfa hay Se content for that cutting of alfalfa (0.07, 0.95, 1.55, 3.26 mg Se/kg dry matter for Se application rates of 0, 22.5, 45.0, or 89.9 g Se/ha, respectively). Feeding Se-fertilized alfalfa hay during the 7-wk preconditioning period increased WB-Se concentrations ($P_{Linear}<0.001$) and body weights ($P_{Linear}=0.002$) depending upon the Se-application rate. Based upon our results we suggest that soil-Se fertilization is a potential management tool to improve Se-status and performance in weaned calves in areas with low soil-Se concentrations.

Editor: Pascale Chavatte-Palmer, INRA, France

Funding: Funded in part by Animal Health and Disease Project Formula Funds, Oregon State University, Corvallis, Oregon 97331-4802, United States of America (JAH, Principal Investigator). The funders had no role in study design, data collection and analysis, decision to publish, or preparation of the manuscript. No additional external funding received for this study.

Competing Interests: The authors have declared that no competing interests exist.

* E-mail: Jean.Hall@oregonstate.edu

¤a Current address: Deep Springs College, Dyer, Nevada, United States of America
¤b Current address: Texas AgriLife Research, San Angelo, Texas, United States of America

Introduction

Selenium (Se) is an essential micronutrient of cattle. Provision of adequate Se is important to prevent Se-responsive diseases in growing cattle such as nutritional myodegeneration and Se-responsive unthriftiness [1]. Many parts of the world, including Oregon, USA, are known to have soil conditions conducive to deficient forage-Se content, potentially leading to clinical signs of Se deficiency in livestock grazing or fed crops raised on them [2]. Soils are the major source of Se for plants and soil-Se exits in various forms including selenides, elemental Se, selenites, selenates and organic Se compounds [3]. Soil-Se content varies considerably depending upon geographic location. Low soil pH and high concentrations of sulfur and phosphorus from fertilization decrease Se availability for plants. Leaching from the topsoil in areas of high rainfall or irrigation also lowers forage-Se content. Plant species also differ in their ability to incorporate Se from soil. Most forage plants are categorized as non-Se accumulator plants.

The bioavailability of Se is not straightforward because of wide variation in Se content of foods (determined by a combination of geographical and environmental factors) and chemical forms in which Se may be absorbed and metabolized [4]. In general, organic forms are absorbed and retained more efficiently than inorganic forms [5,6]. Selenium is normally present in the diet in organic forms, e.g., as selenomethionine (SeMet) or selenocysteine (SeCys) [7]. Inorganic Na-selenite and Na-selenate are present in the diet in very small amounts.

Although the essentiality of Se has been known for five decades, the most effective method of Se delivery to cattle for optimum performance is still being investigated. Several means of administering Se to deficient ruminants are available [8]. For example, there are a number of injectable preparations, which often include vitamin E. Selenium can also be added to feed, mineral, and protein supplements. Sustained-release boluses with a life of several months may be used. Because of their weight, these boluses stay in the rumen whereby they gradually release Se. Selenium supplemented by these methods is usually inorganic Na-selenite or Na-selenate. One limitation of supplementing with inorganic Se in salt or feed is the apparent short duration of Se storage in the animal [8,9]. Other limitations to these methods of Se delivery include individual variation or sporadic intake, extra labor requirements, added expenses, and seasonal grazing practices that result in limited access to Se for extended periods of time. Therefore, animals may be Se deficient by the end of the grazing season.

Agronomic biofortification is defined as increasing the bioavailable concentrations of essential elements in edible portions of crop plants through the use of fertilizers. The potential for using Se-containing fertilizers to increase forage Se concentrations and, thus, dietary Se intake has been demonstrated in Finland, New Zealand, and Australia where it has proven to be both effective and safe [10–14]. The predominant chemical form of Se in Se-fortified grains and hays is SeMet [7]. We previously reported that Se-replete beef cows fed Se-fertilized forage for 6 wk had elevated WB-Se concentrations for 20 wk, which ensured adequate WB-Se while grazing forage on Se-deficient soils [15]. We have also shown that the FDA-approved supplementation rate for sheep (0.3 mg of Se/kg of diet as fed, which is equivalent to 0.7 mg of Se/d or 4.9 mg of Se/wk per sheep) for organic Se supplementation was equally effective as supranutritional rates of Na-selenite supplementation (14.7 and 24.5 mg of Se/wk) in increasing whole-blood (WB) Se concentrations, demonstrating the greater oral bioavailability of organic Se in sheep. In addition, short-term exposure to Se-fertilized forage results in whole-body Se status sufficient to maintain adequate WB-Se concentrations throughout grazing periods when there is limited access to Se supplements [16]. In another study Stewart et al. [17] showed that growth and survival was better in lambs from ewes receiving Se-yeast at 5 times the FDA-allowed supplementation rate compared to lambs from ewes receiving the FDA-allowed supplementation rate or no Se.

The transition period between weaning and movement to a feedlot is one of the most stressful times for beef calves. Because Se plays an important role in the immune response in cattle [18], calves may have greater Se requirements during the transitional period. Performance in weaned beef calves is enhanced if a preconditioning program is utilized before calves enter the feedlot. Several weeks in a preconditioning program are recommended to reduce the stress associated with weaning, dehorning, castration, and vaccination with the goal of reducing morbidity and mortality after arrival at the feedlot. Preconditioning also reduces the number of calves pulled to sick pens, and improves weight gain and feed efficiency of calves after arrival at a feedlot [19–21].

The objectives of this study were to evaluate WB-Se status and performance in weaned beef calves fed alfalfa hay fertilized with Se at increasing rates for 7 wk in a preconditioning program prior to entering the feedlot. We hypothesized that feeding weaned beef calves forage fertilized with increasing amounts of Na-selenate would improve both WB-Se status and growth rate.

Materials and Methods

Animal Ethics Statement and Study Design

The experimental protocol was reviewed and approved by the Oregon State University Animal Care and Use Committee (ACUP Number: 4051). This was a prospective clinical trial of 7-wk duration (August 29 through October 14, 2010) involving 60 weaned beef calves, primarily of Angus breeding. The calves ranged in age from 4.5 to 6 mo (166±2 d; mean ± SEM) and originated from the Oregon State University Beef Ranch, Corvallis, OR, USA. Body weights at weaning ranged from 181 to 310 kg (239±3.6 kg, mean ± SEM), and body condition scores ranged from 6 to 7 (1 to 9 scale). There were 27 heifers and 33 steer calves in the study.

Corvallis is located at an elevation of 72 m, midway in the Willamette Valley, 74 km east of the Oregon Coast, and 137 km south of Portland. Like the rest of the Willamette Valley, Corvallis falls within the Marine West Coast climate zone with some Mediterranean characteristics. Temperatures are mild year round, with warm, dry, sunny summers and mild, wet winters with persistent overcast skies. Spring and fall are also moist seasons with persistent cloudiness, and light rain falling for extended periods. Winter snow is rare, but occasionally does fall, usually in the form of heavy wet snow, ranging between a dusting to several cm that does not persist on the ground for more than a day. During the mid-winter months after extended periods of rain, thick persistent fogs can form, sometimes lasting the entire day. Rainfall total is surprisingly variable, ranging from an average of 168.7 cm per year in the far northwest hills to 110.9 cm per year at Oregon State University, which is located in the center of Corvallis. Typical distribution of precipitation includes about 50 percent of the annual total from December through February, lesser amounts in the spring and fall, and very little during summer. Rainfall tends to vary inversely with temperatures, with the cooler months being the wettest, and the warmer summer months being the driest. Because of its close proximity to the coast range, Corvallis can experience slightly cooler temperatures, particularly in the hills, compared with the rest of the Willamette Valley. Despite this, temperatures dropping below freezing are a rare event. Average monthly temperatures for September are 25.1°C (high) and 9.0°C (low).

Using a randomized complete block design, calves were blocked at the time of weaning by body weight (BW) and then assigned to one of 4 treatment groups of 15 calves each. Ear tags were used to identify the calves. All calves were put together in a large dry field and fed non Se-fortified grass hay for 4 d. Calves were then placed by treatment group into dry barn lots (11–15 m^2/calf; concrete flooring in open lots that were strip cleaned once weekly; dirt flooring in loafing sheds with 5–6 m^2/calf; concrete bunks with 64–97 cm of feeder space/calf; all measurements exceeded requirements [22]) with continuous access to water, feed bunks, and shelter. Calves were fed alfalfa hay once daily. The alfalfa hay was grown in fields fertilized with sodium-selenate at an application rate of 0, 22.5, 45.0, or 89.9 g Se/ha. Calves were transitioned to their respective alfalfa hay sources over a 10-d period. Alfalfa hay was fed as follows: 0.64 kg/head/d 1; 0.79 kg/head/d 2; 1.59 kg/head/d 3; 3.2 kg/head/d 4 to 7; 4.77 kg/head/d 8 to 11; and 6.4 kg/head/d 12. In addition, grass hay was offered for the first 10 d, and then discontinued. By the beginning of the third wk, calves were consuming on average 6.4 kg (as fed) of alfalfa hay per head, which was approximately 2.6% of their BW. A specified quantity of hay was offered each day and intake was uniform among groups. Hay was available all day, but by the next morning bunkers were empty and less than 5% of hay (visual estimate) was wasted on the ground.

In addition, calves were fed grain-based concentrate (0.23 kg as fed/head/d for 5 wk and then 0.46 kg as fed/head/d for 2 wk; **Table 1**) containing a coccidiostat (Rumensin® 80; Elanco Animal Health Co, Indianapolis, IN; 0.0195%) and added Se (0.200 mg/kg). Grain concentrate was offered once a day beginning the day calves were placed into the barn lots; grain was placed into the bunkers and consumed before hay was fed. The grain concentrate consisted of 34% steam flaked corn, USDA grade 2; 28% rolled barley; 19% wheat middlings; 10% dried distiller's grains with solubles from an ethanol plant; and 4% dried distiller's grains containing mainly wheat. The ration was formulated for growing beef calves in the 200 to 300 kg weight range to achieve a target average daily gain of 0.5 kg/d.

Table 1. Alfalfa hay and grain concentrate nutrient composition (dry matter basis).

Nutrient	Alfalfa Hay	Grain Concentrate
Dry matter, g/kg	906	944
Crude protein, g/kg	183	158
Acid detergent fiber, g/kg	351	84
Neutral detergent fiber, g/kg	406	148
Nonfiber carbohydrates, g/kg[1]	323	608
Fat, g/kg	10	34
Ash, g/kg	78	52
Calcium, g/kg	15.3	7.4
Phosphorus, g/kg	2.7	5.4
Magnesium, g/kg	4.5	3.1
Potassium, g/kg	14.9	7.5
Sodium, g/kg	1.5	3.1
Copper, mg/kg	12	13
Iron, mg/kg	357	115
Manganese, mg/kg	43	61
Zinc, mg/kg	22	67

[1]Nonfiber carbohydrates calculated by difference.

Prior to this study, dams and calves had free-choice access to a mineral supplement containing 120 mg/kg Se from sodium-selenite. After weaning and during this study, all calves had free-choice access to the same type of mineral supplement, however Se was not added to the mixture. The mineral supplement (dry matter basis) was in loose granular format and contained 57.0 to 64.0 g/kg calcium; 30.0 g/kg phosphorus; 503 to 553 g/kg salt (NaCl); 50.0 g/kg magnesium; 50 mg/kg cobalt; 2,500 mg/kg copper; 200 mg/kg manganese; 200 mg/kg iodine; 6,500 mg/kg zinc (Wilbur-Ellis Company, Clackamas, OR). During the first 10 d, one bloat block containing 13 mg/kg Se (Bloat Guard® POL6.6 Pressed, SWEETLIX® Livestock Supplement System; Mankato, MN) was also offered to each group of calves. Routine farm management practices, including vaccinations and deworming, were the same for all treatment groups with the exception that one calf in the 45.0 g Se/ha group was castrated at the beginning of the trial.

Selenium Fortified-Alfalfa Hay and Other Selenium Analyses

The soil was enriched with Se by mixing sodium-selenate (RETORTE Ulrich Scharrer GmbH, Röthenbach, Germany) with water and spraying it onto the soil surface of an alfalfa field at an application rate of 0, 22.5, 45.0, or 89.9 g Se/ha immediately after the first cutting of hay in June 2010. Fields were approximately 1.2 ha each. The application rates were chosen based on work with Selcote Ultra® (10 g Se/kg as 1:3 Na_2SeO_4:$BaSeO_4$; Terralink, Vancouver, British Columbia, Canada) in previous studies [15,16]. Second-cutting alfalfa hay was harvested 40 d after Se application and then analyzed for nutrient and Se content. Alfalfa yield was approximately 4.9 ton/ha. To determine whether Se remained in the soil after harvesting second-cutting alfalfa hay, third-cutting alfalfa hay was also harvested from the respective field plots after another 50 d and analyzed for Se content. A Penn State forage sampler was used to take 25 cores from random bales in each alfalfa hay source (0, 22.5, 45.0, or 89.9 g Se/ha). This sampling regime was repeated 3 times (all samples collected mid way through the feeding trial) for each alfalfa hay source. Core samples were mixed well and representative samples selected for analysis. Alfalfa hay samples were submitted to commercial laboratories for routine nutrient analysis (**Table 1**; Cumberland Analytical Services, Maugansville, MD) and Se analysis (Utah Veterinary Diagnostic Laboratory, Logan, UT). Alfalfa hay dry matter determination was completed at a temperature of 105°C for 12 to 14 h in a forced draught oven. Methods for crude protein (CP), acid detergent fiber (ADF), ash, and minerals were performed according to the Association of Official Analytical Chemists [23]. The neutral detergent fiber (NDF) was determined according to Van Soest et al. [24]. Soluble protein was determined according to Krishnamoorthy et al. [25]. Plant samples were prepared for Se analysis as previously described [26], and Se was analyzed using inductively coupled argon plasma emission spectroscopy (ICP-MS; ELAN 6000, Perkin Elmer, Shelton, CT). Quantification of Se was performed by the standard addition method, using a 4-point standard curve. A quality-control sample (in similar matrix) was analyzed after every 5 samples, and analysis was considered acceptable if the Se concentration of the quality-control sample fell within ±5% of the standard/reference value for the quality control.

Grain samples were prepared for Se analysis the same manner as alfalfa hay. Salt samples were ground using a mortar and pestle. The ground salt material (0.50 g) was placed into a labeled 30-ml digestion tube (Oak Ridge Teflon digestion tube, Nalge Nunc International, Rochester, NY). Trace metal-grade nitric acid (4.0 mL; Thermo Fisher Scientific Inc., Waltham, MA) was added to the digestion tubes. The tubes were then heated at 90°C for 1 h with the caps loose on the tubes. After digestion, tubes were allowed to cool and 5.0 mL of ultrapure water was added and the samples were again digested at 90°C for 1 h. Contents were increased to 10 mL by adding trace metal-grade nitric acid. One milliliter of the digest was transferred into another trace metal-free tube containing 9.0 mL of ultrapure water to make up a 5% (v/v) nitric acid matrix and was centrifuged at 520×g for 10 min. The supernatant was removed and the samples were analyzed to quantify Se using ICP-MS in the same manner as for plant and grain samples.

Performance and Whole-blood Selenium Assay

Health was monitored daily during the 7-wk feeding trial. Body weights were measured at the beginning of the treatment period (baseline), and at 3 wk, 6 wk, and 7 wk (end of the Se supplementation period). To assess the effect of Se supplementation on WB-Se status, all calves were bled at 0 time (baseline) and once each week for 7 wk until study termination. Jugular venous blood was collected into evacuated ethylenediaminetetraacetic acid (EDTA) tubes (2 mL; final EDTA concentration 2 g/L; Becton Dickinson, Franklin Lakes, NJ) and stored on ice until they were frozen at −20°C. Whole-blood Se concentrations were determined by a commercial laboratory (Center for Nutrition, Diagnostic Center for Population and Animal Health, Michigan State University, E. Lansing, MI) using an ICP-MS method with modifications as previously described [15].

Statistical Analyses

Statistical analyses were performed using SAS version 9.2 [27]. Whole-blood Se concentrations and BW were analyzed as repeated-measures-in-time using PROC MIXED. Fixed effects in the model were Se application rate (0, 22.5, 45.0, and 89.9 g Se/ha), sex of calf, (male, female), BW block, time (wk 0, 1, 2, 3, 4,

5, 6, and 7 of feeding experiment for WB-Se concentrations and wk 0, 3, 6, and 7 of feeding experiment for BW), and the interaction between Se application rate and time. Fixed effects in the model for BW were Se application rate, sex of calf, BW block, time (wk 0, 3, 6, and 7 of feeding experiment), and the interaction between Se application rate and time. An unstructured variance-covariance matrix was used to account for variation of measures within calves. The unstructured variance-covariance matrix provided the most parsimonious variance-covariance matrix based on the lowest value by the Aikaike Information Criterion.To evaluate the effect of Se application rate, linear, quadratic, and cubic contrasts were constructed. In addition, the linear response of the dependent variables Se forage content or WB-Se concentrations of beef calves to the independent variable Se fertilization rate were evaluated using univariate regression in PROC REG. Data are reported as least square means ± SEM. Statistical significance was declared at $P \leq 0.05$ and a tendency at $0.05 < P \leq 0.10$.

Results

Effect of Soil-Se Fertilization on Se Concentrations in Alfalfa Hay

Fertilizing fields with increasing amounts of sodium-selenate increased in a dose-dependent manner the Se-content of second-cutting alfalfa hay from 0.07 to 0.95, 1.55, and 3.26 mg Se/kg dry matter for sodium-selenate application rates of 0 (non-fertilized control), 22.5, 45.0, or 89.9 g Se/ha, respectively (**Figure 1**). Subsequent third-cutting alfalfa hay had carry-over Se concentrations of 0.16, 0.28, and 0.60 mg Se/kg dry matter for sodium-selenate application rates of 22.5, 45.0, or 89.9 g Se/ha, which is equivalent to 16.8%, 18.1%, and 18.4% carry over, respectively.

Effect of Supranutritional Se-supplementation to Se-replete Weaned Beef Calves on Whole- blood Se Concentrations and Performance during the 7-wk Feeding Period

Based on the total amount of alfalfa hay and grain concentrate fed to each group of calves, average dry matter intake per head was calculated at 5.59 kg/head/d for alfalfa hay starting on day 12, and 0.20 kg/head/d for grain concentrate for the first 5 wk and 0.40 kg/head/d for the last 2 wk. Using measured alfalfa hay and grain concentrate values for CP, net energy for gain (NE_g), net energy for maintenance (NE_m), and total digestible nutrients (TDN), we calculated CP (1.09 kg/head/d), NE_g (4.83 Mcal/head/d), NE_m (8.29 Mcal/head/d), and TDN consumption (3.7 kg/head/d) and compared them to National Research Council (NRC) [28] requirements {CP (0.44 kg/head/d), NE_g (0.62 Mcal/head/d), NE_m (4.50 Mcal/head/d), and TDN (2.9 kg/head/d)} for growing beef calves in this weight range to verify adequate nutrient intake for growth.

Calculated Se intake from alfalfa hay was 0.4, 5.3, 8.7, and 18.2 mg Se/head/d for calves consuming hay with Se concentrations of 0.07, 0.95, 1.55, and 3.26 mg Se/kg dry matter. The measured Se concentration of the grain concentrate was 1.41 mg Se/kg dry matter. Calculated Se intake from grain concentrate was 0.28 mg Se/head/d (first 5 wk) and 0.56 mg Se/head/d (last 2 wk). The average intake of mineral supplement was 17.5 mg/head/d. The measured Se concentration of the mineral supplement without added Se was 0.10 mg Se/kg dry matter. Calculated Se intake from the mineral supplement was 0.002 mg Se/head/d. The average intake of bloat block was 120 g/head/d during the first 10 d. The reported Se concentration of the bloat block offered during the first 10 d was 13 mg Se/kg. Calculated Se intake from the bloat block fed during the first 10 d was 1.55 mg Se/head/d.

Feeding Se-fertilized alfalfa hay was effective at increasing WB-Se concentrations in weaned beef calves ($P_{Treatment}$, P_{Time}, and $P_{Treatment \times Time}$: *all* $P<0.001$; **Figure 2**). The normal reference interval for WB-Se concentrations of adult cows is 120–300 ng/mL [15]. Heifer calves had greater WB-Se concentrations than male calves ($P=0.03$). The WB-Se response increased with greater amounts of sodium-selenate applied to the soil ($P_{Linear}<0.001$; **Figure 3**). No significant quadratic ($P=0.11$) or cubic ($P=0.30$) Se-dose response was detected. The WB-Se concentrations continued to increase throughout the 7-wk feeding period.

Feeding Se-fertilized alfalfa hay was effective at increasing BW in weaned beef calves ($P_{Treatment}=0.002$; $P_{Time}<0.001$; and $P_{Treatment \times Time}=0.03$; **Figure 4**). Calf sex did not affect BW ($P=0.98$). The BW response increased with greater amounts of sodium-selenate applied to the soil ($P_{Linear}<0.001$; **Figure 4**). No significant quadratic ($P=0.84$) or cubic ($P=0.11$) Se-dose response was detected. Weight data were not confounded by adverse health events in the preconditioning period because only two calves were retreated for pink eye during the first week of the feeding period (existent before the start of the experiment). None of the calves died during the 7-wk feeding period or showed signs of Se-deficiency or Se-toxicosis.

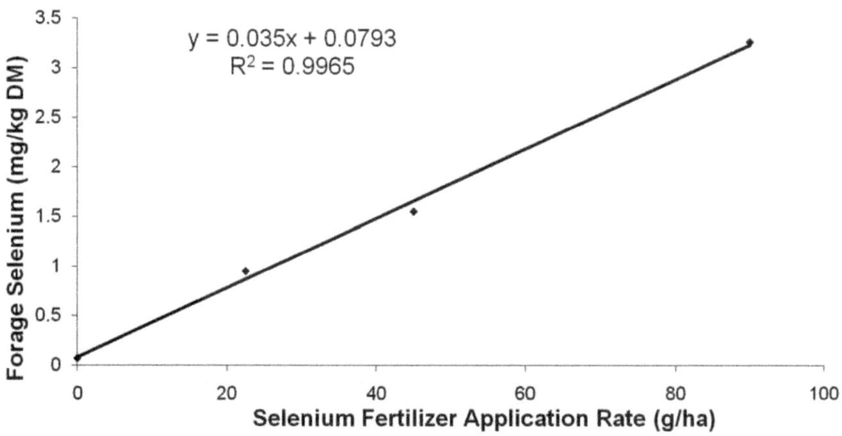

Figure 1. Relationship between amount of Se applied by fertilization (g Se/ha) and observed forage Se content (g Se/kg DM).

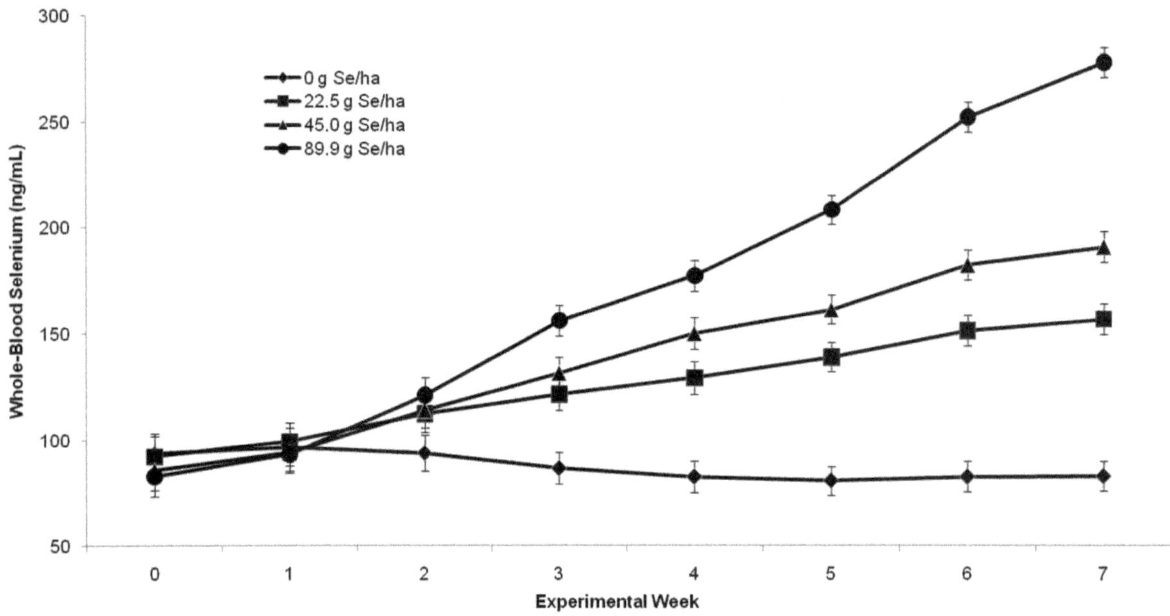

Figure 2. Comparison of whole-blood Se concentrations (mean ± SEM) in weaned beef calves consuming alfalfa hay grown in fields not fertilized with Se (0 g Se/ha), or harvested from fields fertilized with sodium-selenate at an application rate of 22.5, 45.0, or 89.9 g Se/ha for 7 wk (n = 15 calves per group). The normal reference interval for whole-blood Se concentrations of beef cattle is 120 to 300 ng/mL.

Discussion

The objectives of this study were to evaluate whether fertilizing the soil of alfalfa hay fields with increasing amounts of sodium-selenate, and subsequent feeding of Se-fertilized alfalfa hay to recently weaned beef calves would improve in a dose-dependent manner WB-Se status and, consequently, increase growth rate in the preconditioning period prior to entering the feedlot. Fertilizing alfalfa hay fields with sodium-selenate increased Se content of alfalfa hay in a dose-dependent manner. Feeding Se-fertilized alfalfa hay during the 7-wk preconditioning program increased WB-Se concentrations and BW in Se-replete calves in a dose-dependent manner. Our results suggest that fertilization of alfalfa fields with sodium-selenate is a potential management tool to improve Se status and performance in weaned beef calves.

Effect of Fertilizing Soil with Increasing Application Rates of Sodium-selenate on Se Concentrations in Alfalfa Hay

Agronomic Se-biofortification has been used in several countries with regions of low soil-Se concentrations including Finland, Denmark, New Zealand, and the United Kingdom to increase Se concentrations in the food chain [10,29,30]. In the United States, the use of feedstuffs that are naturally high in Se content is not regulated; Se fertilization, however, is not allowed in any state except for Oregon, where the Department of Agriculture does not control the use of Se as a plant fertilizer. Therefore, in Oregon it is

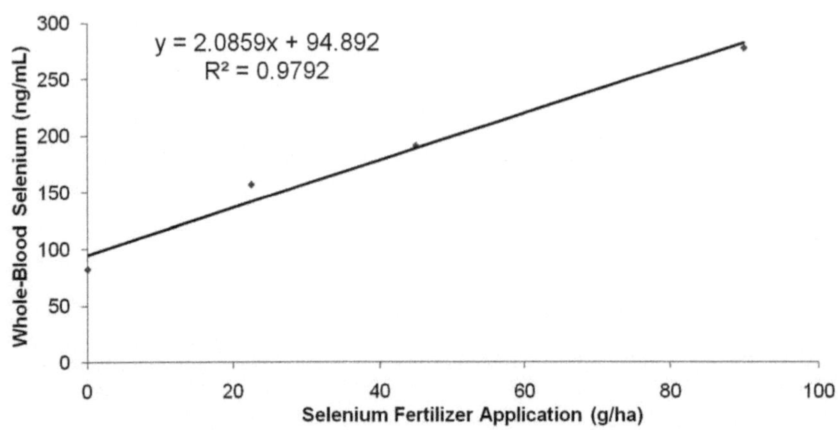

Figure 3. Relationship between amount of Se applied by fertilization (g Se/ha) and observed WB-Se content (ng/mL) in weaned beef calves consuming alfalfa hay grown in fields not fertilized with Se (0 g Se/ha), or harvested from fields fertilized with sodium-selenate at an application rate of 22.5, 45.0, or 89.9 g Se/ha for 7 wk (n = 15 calves per group). The normal reference interval for whole-blood Se concentrations of beef cattle is 120 to 300 ng/mL.

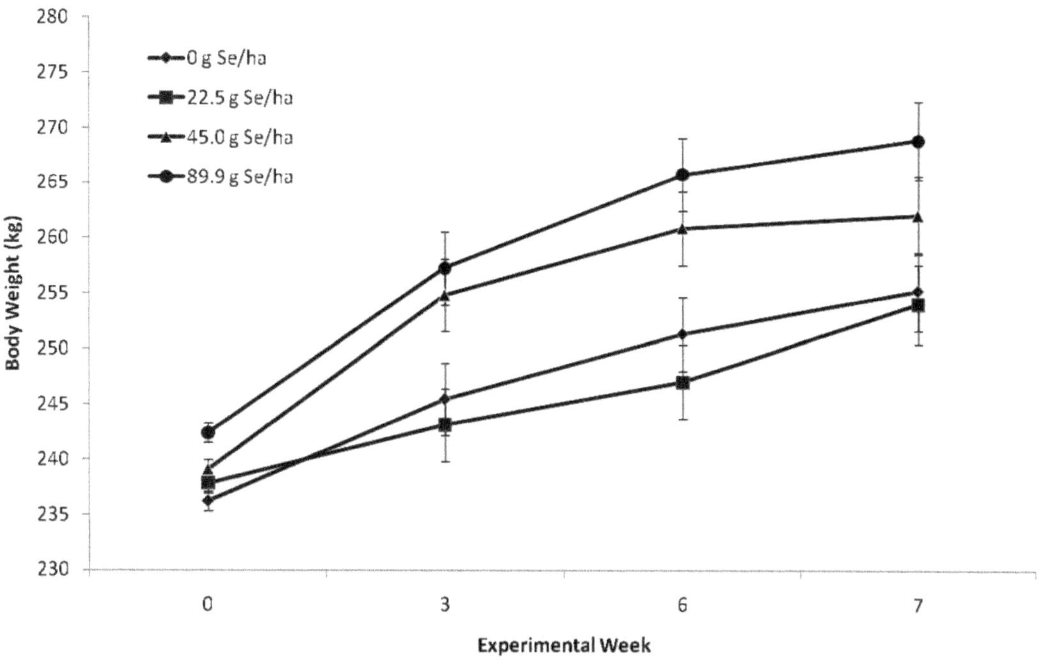

Figure 4. Comparison of block-adjusted BW (kg; mean ± SEM) of weaned beef calves (primarily of Angus breeding and ranging in age from 4.5-to-6-mo) after consuming alfalfa hay grown in fields not fertilized with Se (0 g Se/ha), or harvested from fields fertilized with sodium-selenate at an application rate of 22.5, 45.0, or 89.9 g Se/ha for 7 wk (n = 15 calves per group). Initial BW (baseline) ranged from 181 to 310 kg (239±3.6 kg, mean ± SEM). Final BW (7-wk) ranged from 183 to 346 kg (260±4.1 kg, mean ± SEM).

possible to produce feedstuffs with increased Se concentrations by applying Se as a fertilizer. Plants absorb Se from the soil in the form of selenate and synthesize selenoamino acids with SeMet being the major selenocompound in grassland legumes [7].

We have shown that sodium-selenate can be solubilized in water and sprayed onto soil surfaces of established alfalfa hay fields after the first cutting of alfalfa hay at three application rates in a 1×, 2×, and 4× ratio (0, 22.5, 45.0, or 89.9 g Se/ha). Hay harvested from respective field plots has a similar dose-dependent Se content (0.07, 0.95, 1.55, and 3.26 mg Se/kg dry matter) (**Figure 1**).The Se fertilizer application rates in the current study were chosen based upon WB-Se concentrations attained and absence of clinical signs of toxicosis in previous Se-fertilization studies [15,16].

In Hall et al. [15], Selcote Ultra® (10 g Se/kg as sodium selenate; Terralink, Vancouver, British Columbia, Canada) was mixed with urea-sol fertilizer and applied to pasture at a rate of 3.4 kg Selcote/ha (34 g Se/ha; 1.5× our lowest Se application rate in the current study). This resulted in forage Se of 0.11, 1.52, and 1.06 mg/kg (dry matter basis) at pre-fertilization, day 1, and day 42 of the grazing period, respectively. Beef cows grazing this forage had WB-Se concentrations of 186±5 ng/mL immediately post-grazing (day 42), which was within the normal reference interval for WB-Se concentrations of adult cows at the Michigan State University diagnostic laboratory (120–300 ng/mL). In a sheep study [16], Se applied to subclover-fescue sward type pasture at 3.4 kg Selcote/ha resulted in Se concentrations of up to 2.02 mg/kg (dry matter basis) in green forage grazed by the sheep, and WB-Se concentrations in ewes after grazing for 6 weeks of 573±20 ng/mL, with no clinical signs of Se toxicosis. When comparing the results of the current study with results from these two studies and one other that was conducted by our group that used Selcote Ultra® for Se fertilization (all at different locations in Oregon), a linear relationship between sodium-selenate application rate and Se content of forage was observed (forage selenium concencentrations in mg/kg DM = 0.0299×selenium fertilizer application rate in g/ha +0.3297; $r^2 = 0.9967$) [15,16].

The linear relationship between Se fertilizer application rate and forage Se concentration is surprising given the fact that plant species, field location, and soil differed among the studies [15,16]. It is well documented that plant species, chemical species of Se, and soil pH, sulfur, and iron content alter Se availability for plant uptake [31,32]. Based upon our results, we suggest that Se content of soil primarily determines Se content of common forage species. This is supported by the observation that, regardless of initial Se-application rate, subsequent third-cutting alfalfa hay had approximately 18% of the Se content of hay harvested after initial application of Se. The Se concentrations of third-cutting alfalfa hay were 0.16, 0.28, and 0.60 mg Se/kg (dry matter basis) for sodium-selenate application rates of 22.5, 45.0, or 89.9 g Se/ha, respectively. To our knowledge, this is the first report of Se carry-over in the soil to the next cutting of alfalfa hay.

Effects of Feeding Alfalfa Hay Fertilized with Increasing Application Rates of Se for 7 wk on Whole-blood Se Concentrations and Performance of Weaned Beef Calves

Feeding Se-fertilized alfalfa hay was effective at increasing WB-Se concentrations in Se-replete weaned beef calves. The WB-Se concentrations increased with greater amounts of sodium-selenate applied to soil (**Figures 2 and 3**). For example, the increase in WB-Se concentrations for calves consuming alfalfa hay from fields fertilized with 89.9 vs. 45.0 g Se/ha was double the increase for calves consuming alfalfa hay from fields fertilized with 45.0 vs. 22.5 g Se/ha. Combining our results with a previous study in mature beef cows [15] indicates that there is a linear relationship between sodium-selenate application rate and WB-Se concentrations in beef cattle (WB-Se concentration in ng/mL = 1.9688×selenium fertilizer application rate in g/ha +105.67; $r^2 = 0.9594$).

The WB-Se concentrations continued to increase throughout the 7-wk feeding period.

The majority of dietary Se was supplied by the alfalfa hay, except in calves consuming non-Se fertilized alfalfa hay. In the first 5 wk, 0.28 mg Se/head/d was provided by the grain concentrate with essentially none provided by the mineral supplement. In the last 2 wk, 0.56 mg Se/head/d was provided by the grain concentrate. Thus, in the last 2 wk, total dietary Se intake was 0.96, 5.86, 9.26, and 18.76 mg Se/d for calves consuming alfalfa hay with sodium-selenite application rates of 0, 22.5, 45.0, or 89.9 g Se/ha (alfalfa hay provided 41, 90, 94, and 97% of dietary Se intake, respectively). During the first 10 d of the feeding trial, calculated Se intake from the bloat block was 1.55 mg Se/head/d. This represented the greatest source of Se for this short-term period in those calves receiving non-Se fertilized alfalfa hay.

In the United States, the FDA [33] allows Se to be added to cattle diets as sodium-selenite, sodium-selenate, or Se-yeast in complete feeds not to exceed 0.3 mg of Se/kg of diet (as fed basis), or in supplements for limit feeding not to exceed 3 mg of Se/head/day. The non-Se fortified alfalfa hay (0.07 mg Se/kg dry matter) was below this level. We were able to provide additional dietary Se using Se-fortified alfalfa hay. Our results for WB-Se concentrations in weaned beef calves fed Se-biofortified hay are similar to what we reported [15] when grazing adult beef cattle on Se-fertilized forage. No clinical signs of Se deficiency or toxicosis were observed.

Our results are consistent with SeMet from the grassland legumes being absorbed in the duodenum and incorporated into general body proteins in place of methionine. The concentration of SeMet is not regulated and ultimately reflects dietary intake [7,32]. Selenomethionine acts as a storage form of Se in body proteins, including hemoglobin and albumin in WB, from which it is slowly released by protein catabolism. The SeMet can also be trans-selenated into selenocysteine and subsequently used to provide Se requirements for selenoprotein synthesis [34].

Production benefits of agronomic Se biofortification were assessed by comparing BW gains at the end of the 7-wk Se-supplementation period. Feeding Se-fertilized alfalfa hay increased BW in a linear manner (**Figure 4**). Calves receiving Se-fertilized alfalfa hay with the highest concentration of Se had the highest BW. There results were not confounded by morbidly because none of the calves were sick during the 7-wk preconditioning period. None experienced bloating, went off feed, or showed any clinical signs of illness. These results are consistent with our previous results for lambs [17] whereby ewes raising multiples that received the highest Se-yeast dose (24.5 mg Se/wk) had heavier lambs at weaning (120 d of age) in yr 1 of supplementation and at 60 d of age in yr 2 of supplementation than ewes receiving Se-yeast at the maximum FDA-allowed level (4.9 mg Se/wk).

Selenium's role in animal performance is based upon the functions of selenoproteins, many of which have antioxidant activities [4]. Although reactive oxygen species and free radicals are a natural result of the body's normal metabolic activity, excessive stress as a result of weaning, dietary and environmental changes, comingling with other animals, and disease can lead to the over production of free radicals or accumulation of free radicals because of a lack of antioxidants. Therefore, it is important that micronutrients involved in antioxidant functions be present in tissues to provide oxidant-antioxidant balance. Although the level of Se supplementation needed for adequate performance may be less under optimum conditions, in periods of transition such as weaning, preconditioning, and shipping, where stress is a confounding factor for optimum performance, Se requirements may be increased.

Because all calves were visually healthy, it is unlikely that the observed BW response to Se supplementation is explained solely by the antioxidant activity of selenoproteins. Two selenoprotein families, the iodothyronine deiodinases, responsible for metabolism of thyroid hormones, and the thioredoxin reductases, responsible for reducing thioredoxin, are directly or indirectly through regulation of transcription factors, involved in cell growth and control of apoptosis, as well as maintenance of cellular redox status [35]. Our results suggest that supranutritional Se supplementation may have growth-promoting properties in beef cattle. Future studies are warranted to examine the effects of supranutritional Se supplementation on gene and protein expression of the iodothyronine deiodinases, thioredoxin reductases, and transcription factors regulating cell growth.

In summary, Se fertilization of alfalfa fields in a region with Se deficient soils increased in a dose-dependent manner the Se content of alfalfa hay. Supranutritional Se supplementation of recently weaned beef calves with Se-fortified alfalfa hay resulted in increased WB-Se concentrations and improved growth rates. Our results suggest that building Se-body reserves by feeding supranutritional Se levels from sodium-selenate fertilized alfalfa hay during the preconditioning program is an effective management strategy to optimize growth and health in weaned beef calves.

Acknowledgments

Appreciation is expressed to KC Bare and Opal Springs Farms, LLC, Culver, OR for precise Se application rates to alfalfa fields and growing the alfalfa hay for the conduct of these experiments.

Author Contributions

Conceived and designed the experiments: JAH WDM GJP. Performed the experiments: JAH JKH WRV WCS JAV CTE WDM GJP. Analyzed the data: JAH GB. Contributed reagents/materials/analysis tools: JAH GB. Wrote the paper: JAH GB CTE GJP.

References

1. Koller LD, South PJ, Exon JH, Whitbeck GA (1983) Selenium deficiency of beef cattle in Idaho and Washington and a practical means of prevention. Cornell Vet 73: 323–332.
2. Stevens JB, Olson WG, Kraemer R, Archambeau J (1985) Serum selenium concentrations and glutathione peroxidase activities in cattle grazing forages of various selenium concentrations. Am J Vet Res 46: 1556–1560.
3. Surai PF, Fisinin VI, Papazyan TT (2008) Selenium deficiency in Europe: causes and consequences. In: Surai PF, Taylor-Pickard JA, editors. Current advances in selenium research and applications. The Netherlands: Wageningen Academic Publishers. 13–44.
4. Fairweather-Tait SJ, Collings R, Hurst R (2010) Selenium bioavailability: current knowledge and future research requirements. Am J Clin Nutr 91: 1484S–1491S.
5. Qin S, Gao J, Huang K (2007) Effects of different selenium sources on tissue selenium concentrations, blood GSH-Px activities and plasma interleukin levels in finishing lambs. Biol Trace Elem Res 116: 91–102.
6. Hall JA, Van Saun RJ, Bobe G, Stewart WC, Vorachek WR, et al. (2012) Organic and inorganic selenium: I. Oral bioavailability in ewes. J Ani Sci 90: 568–576.
7. Whanger PD (2002) Selenocompounds in plants and animals and their biological significance. J Am Col Nutr 21: 223–232.
8. Surai PF (2006) Selenium in ruminant nutrition. In: Surai PF, editor. Selenium in nutrition and health. Nottingham: Nottingham University Press. 487–587.
9. Surai PF (2006) Selenium in food and feed: selenomethionine and beyond. In: Surai PF, editor. Selenium in nutrition and health. Nottingham: Nottingham University Press. 151–212.
10. Broadley MR, White PJ, Bryson RJ, Meacham MC, Bowen HC, et al. (2006) Biofortification of UK food crops with selenium. Proc Nutr Soc 65: 169–181.
11. Makela AL, Nanto V, Makela P, Wang W (1993) The effect of nationwide selenium enrichment of fertilizers on selenium status of healthy Finnish medical students living in south western Finland. Biol Trace Elem Res 36: 151–157.

12. Whelan BR (1989) Uptake of selenite fertilizer by subterranean clover pasture in Western Australia. Australian J Experimental Agriculture 29: 517–522.
13. Whelan BR, Barrow NJ, Peter DW (1994a) Selenium fertilizers for pastures grazed by sheep. 1. Selenium concentrations in whole-blood and plasma. Australian J Agricultural Research 45: 863–875.
14. Whelan BR, Barrow NJ, Peter DW (1994b) Selenium fertilizers for pastures grazed by sheep. 2. Wool and liveweight responses to selenium. Australian J Agricultural Research 45: 877–887.
15. Hall JA, Harwell AM, Van Saun RJ, Vorachek WR, Stewart WC, et al. (2011) Agronomic biofortification with selenium: Effects on whole blood selenium and humoral immunity in beef cattle. Anim Feed Sci Technol 164: 184–190.
16. Hall JA, Van Saun RJ, Nichols T, Mosher W, Pirelli G (2009) Comparison of Se status in sheep after short-term exposure to high-Se-fertilized forage or mineral supplement. Small Ruminant Res 82: 40–45.
17. Stewart WC, Bobe G, Pirelli GJ, Mosher WD, Hall JA (2012) Organic and inorganic selenium: III. Ewe and progeny performance. J Anim Sci 90: 4536–4543.
18. Finch JM, Turner RJ (1996) Effects of selenium and vitamin E on the immune responses of domestic animals. Res Vet Sci 60: 97–106.
19. Cole NA (1985) Preconditioning calves for the feedlot. Vet Clin North Am Food Anim Pract 1: 401–411.
20. Pritchard RH, Mendez JK (1990) Effects of preconditioning on pre- and post-shipment performance of feeder calves. J Anim Sci 68: 28–34.
21. Duff GC, Galyean ML (2007) Board-invited review: recent advances in management of highly stressed, newly received feedlot cattle. J Anim Sci 85: 823–840.
22. MWPS-6 (1987) Beef Housing and Equipment Handbook, 4th ed. Midwest Plan Service, Iowa State University, Ames, IA, USA.
23. AOAC (2000) Official Methods of Analysis, 17th ed. Association of Official Analytical Chemists, Arlington, VA, USA.
24. Van Soest PJ, Robertson JB, Lewis BA (1991) Methods for dietary fiber, neutral detergent fiber and nonstarch polysaccharides in relation to animal nutrition. J Dairy Sci 74: 3583–3597.
25. Krishnamoorthy U, Muscato TV, Sniffen CJ, Van Soest PJ (1982) Nitrogen fractions in selected feedstuffs. J Dairy Sci 65: 217–225.
26. Davis TZ, Stegelmeier BL, Panter KE, Cook D, Gardner DR, et al. (2012) Toxicokinetics and pathology of plant-associated acute selenium toxicosis in steers. J Vet Diagn Invest 24: 319–327.
27. SAS Institute (2009) SAS User's Guide. Statistics, Version 9.2. SAS Inst Inc, Cary, NC.
28. NRC (1996) Nutrient Requirements of Beef Cattle, seventh revised edition. Natl Acad Press, Washington, DC.
29. Wang WC, Mäkelä AL, Näntö V, Mäkelä P, Lagström H (1998) The serum selenium concentrations in children and young adults: a long-term study during the Finnish selenium fertilization programme. Eur J Clin Nutr 52: 529–535.
30. Gupta UC, Gupta SC (2002) Quality of animal and human life as affected by selenium management of soils and crops. Comm Soil Sci Plant Anal 33: 2537–2555.
31. NRC (1983) Selenium in Nutrition, revised edition. Natl Acad Press, Washington, DC.
32. NRC (2007) Nutrient Requirements of Small Ruminants. Natl Acad Press, Washington, DC.
33. FDA (2009) Title 21. Food and Drugs: Food additives permitted in feed and drinking water of animals. http://www.accessdata.fda.gov/scripts/cdrh/cfdocs/cfcfr/CFRSearch.cfm?fr = 573.920 Accessed 2012 July 24.
34. Rayman MP (2008) Food-chain selenium and human health: emphasis on intake. Br J Nutr 100: 254–268.
35. Rooke JA, Robinson JJ, Arthur JR (2004) Effects of vitamin E and selenium on the performance and immune status of ewes and lambs. J Agric Sci 142: 253–262.

Loss of Hippocampal Neurons after Kainate Treatment Correlates with Behavioral Deficits

Gisela H. Maia[1,2,3], José L. Quesado[1], Joana I. Soares[2,3], Joana M. do Carmo[3], Pedro A. Andrade[3], José P. Andrade[1], Nikolai V. Lukoyanov[1,2,3]*

1 Departamento de Anatomia, Faculdade de Medicina da Universidade do Porto, Porto, Portugal, **2** Neural Networks Group, Instituto de Biologia Molecular e Celular da Universidade do Porto, Porto, Portugal, **3** Programa Doutoral em Neurociências, Universidade do Porto, Porto, Portugal

Abstract

Treating rats with kainic acid induces status epilepticus (SE) and leads to the development of behavioral deficits and spontaneous recurrent seizures later in life. However, in a subset of rats, kainic acid treatment does not induce overt behaviorally obvious acute SE. The goal of this study was to compare the neuroanatomical and behavioral changes induced by kainate in rats that developed convulsive SE to those who did not. Adult male Wistar rats were treated with kainic acid and tested behaviorally 5 months later. Rats that had experienced convulsive SE showed impaired performance on the spatial water maze and passive avoidance tasks, and on the context and tone retention tests following fear conditioning. In addition, they exhibited less anxiety-like behaviors than controls on the open-field and elevated plus-maze tests. Histologically, convulsive SE was associated with marked neuron loss in the hippocampal CA3 and CA1 fields, and in the dentate hilus. Rats that had not experienced convulsive SE after kainate treatment showed less severe, but significant impairments on the spatial water maze and passive avoidance tasks. These rats had fewer neurons than control rats in the dentate hilus, but not in the hippocampal CA3 and CA1 fields. Correlational analyses revealed significant relationships between spatial memory indices of rats and neuronal numbers in the dentate hilus and CA3 pyramidal field. These results show that a part of the animals that do not display intense behavioral seizures (convulsive SE) immediately after an epileptogenic treatment, later in life, they may still have noticeable structural and functional changes in the brain.

Editor: Liset Menendez de la Prida, Consejo Superior de Investigaciones Cientificas - Instituto Cajal, Spain

Funding: This work was supported by FEDER Funds through the Programa Operacional Factores de Competitividade – COMPETE and National Funds through FCT - Fundação para a Ciência e a Tecnologia within the scope of the Project FCOMP-01-0124-FEDER-015919– PTDC/SAU-NSC/115506/2009, Strategic Project Centro de Morfologia Experimental (CME/FM/UP) – 2011–2012, Project PEst-OE/SAU/UI0121/2011. The funders had no role in study design, data collection and analysis, decision to publish, or preparation of the manuscript.

Competing Interests: The authors have declared that no competing interests exist.

* E-mail: luk@med.up.pt

Introduction

Temporal lobe epilepsy (TLE) can be associated with hippocampal cell loss (hippocampal sclerosis) and other structural changes in the medial temporal lobe region, such as fiber sprouting and synaptic reorganization [1,2,3]. It has also been repeatedly reported that TLE patients with hippocampal sclerosis may show learning and memory deficits, increased anxiety, depression and mood disorders [4,5,6]. Thus, it seems reasonable to assume that these behavioral changes are secondary to the TLE-related hippocampal neurodegeneration. However, it is also possible that they are not so straightforwardly related to classical hippocampal sclerosis, but may rather reflect some subtle functional or structural alterations in neuronal circuits, either associated with or even unrelated to epileptic seizures.

Most of the neuropathological aspects of TLE can be reproduced using appropriate animal models. For example, many studies have shown that the treatment of rats with kainic acid (KA), a specific agonist for glutamate receptors, induces acute status epilepticus (SE), accompanied by severe convulsive (motor) seizures, and that later in life these animals show spontaneous recurrent seizures, hippocampal cell loss and memory deficits [7,8,9]. However, in this epilepsy model some rats are considerably more resistant to KA-induced convulsive SE than the others and, for that reason, are usually excluded from the studies [9,10,11,12,13]. Thus, given that the research on the neuronal effects of KA in this subset of animals has been scarce [14], which represents an important gap in the understanding of epilepsy-related neuropathology, the present study was designed to directly compare the behavioral and neuroanatomical outcomes in rats that developed convulsive SE after kainate treatment to those that did not. We thought that the presence of behavioral deficits in KA-treated rats that had not experienced convulsive SE would support the hypothesis that at least some of the behavioral impairments seen in epilepsy can be explained by relatively mild structural or functional changes in the hippocampal region. To address this issue, we assessed the behavioral profile of KA-treated rats of both phenotypes using the Morris water maze [15], passive avoidance and fear conditioning tests, as well as in the open field and elevated plus maze tests. Additionally, absolute numbers of neurons in the dentate gyrus hilus and pyramidal layers of the hippocampus proper were estimated using stereological approaches in order to compare the hippocampal cell loss between the two groups.

Materials and Methods

The handling and care of the animals were conducted according to the "Principles of laboratory animal care" (NIH publication No. 86-23, revised 1985) and Directive 2010/63/EU of the European Parliament and of the Council of 22 September 2010 on the protection of animals used for scientific purposes. The experimental protocol has been approved by the Ethics Committee of the Faculty of Medicine of Porto and the General Veterinary Direction (03.04.2012) for the FCT application grant PTDC/SAU-NSC/115506/2009. All efforts were made to minimize the number of animals used and their suffering.

Animals and Treatments

Male Wistar rats, maintained under standard laboratory conditions, were used in this study. At 10 weeks of age (mean body weight 323±69 g), they were randomly divided into two groups: KA group (n = 28) and control group (n = 8). In the first group, rats were treated with KA according to the protocol developed by Dudek and coworkers and described in detail elsewhere [16,17]. Briefly, the rats were injected every hour with 5 mg/kg of KA (i.p., Sigma) until the onset of convulsive SE, which was defined as the appearance of behavioral symptoms corresponding to stage 3, 4 or 5 seizures on the Racine scale [18], i.e. forelimb clonus, rearing, and rearing with falling. When necessary, treatment was continued in order to assure that SE lasted for at least 3 hours. All rats in this group demonstrated numerous wet-dog-shake seizures after either the second or third injection of KA. The SE, lasting 3–6 hours, was detected in 17 rats after 3–5 injections, with the mean cumulative dose of KA equal to 21±3 mg/kg. These rats were ascribed to the "behavioral status epilepticus" group (BSE; n = 17). Remaining 11 rats after 4 to 6 KA injections either became motionless (unchanged position in the observation cage for tens of minutes, weak reactivity to touching by experimenter's hand), or showed excessive activity (long runs, jumping). Because we and others [17] have previously observed that continuation of the treatment of animals in these states often leads to their death, the treatment was terminated. The mean cumulative dose of KA received by these rats, which were ascribed to the "no behavioral status epilepticus" group (no-BSE; n = 11) was 24±4 mg/kg. It was previously reported that the treatment protocol employed in our study is associated with a relatively low mortality rate of the animals when compared to the single high-dose kainate treatments [16,17]. However, because animal mortality is a prominent cause of bias in quantitative evaluation of neuronal loss [19], special efforts were made in order to further improve the survival rate in the kainate-treated rats. In particular, the animals were periodically injected with saline (s.c.) during the first 48 hours of the recovery period. On the following days, the rat diet was supplemented with apples that were sliced and left at the bottom of the cage. The rats that refused to eat or drink were hand-fed using a plastic syringe. Rats in the control group were injected with saline.

Following the treatments, all animals were given a 5-month recovery period. In the last week of each month, the behavior of rats in their home cages was videorecorded during the light phase of the 12-h light/dark cycle (between 08:00 h and 20:00 h) over a period of 5 consecutive days. The recording was performed using the multichannel video-tracking system EthoVision 8.5 (Noldus, The Netherlands). The video files generated by the system and stored on the hard disk of the computer were then analyzed in a fast-motion mode by a person blind to treatment groups using the VLC media player (VideoLAN, France). In addition, the rats were daily (except weekends) observed for spontaneous behavioral seizures during 1-h intervals between 09:00 h and 10:00 h, as well as during all the procedures associated with handling and behavioral testing.

Open-field and Elevated Plus-maze Tests

Behavioral testing began when animals were 7 months old. Animals were handled 3 minutes per day during 5 days and subjected to the open-field test and plus-maze as previously described in detail [20]. The rats were counterbalanced so that, in each group, half received the open-field test first and the other half received the plus-maze test first. There was a 3-day interval between the two tests.

The behavior of rats in the open field (100×100×40 cm) was tested during 5-minute sessions. Distances traveled in the outer zone of the open field, defined as 20 cm from any wall, and in its inner zone were measured using the computerized EthoVision video-tracking system.

The elevated plus maze was arranged as a cross with two opposite open and two opposite closed arms (50×12 cm), connected by a common central square (12×12 cm). The rats were allowed to explore the apparatus for 5 minutes. Behavior of rats was recorded and analyzed using the video-tracking system. The percentages of time spent by rats in the open arms, closed arms, and in the central square were measured. The number of entries into the arms of the apparatus was also registered.

Step-through Passive Avoidance and Fear Conditioning

Following a 3-day interval, all rats were submitted to the passive avoidance and fear conditioning tests. The rats were counterbalanced so that, in each group, half received the passive avoidance test first and the other half received the fear conditioning first. There was a 3-day interval between the two tests.

The apparatus used in the passive avoidance task was comprised of two adjacent compartments separated by a guillotine door [20]. The larger compartment (45×45×45 cm) was brightly lit and the smaller compartment (30×16×16 cm) was dark. The floor of both compartments was composed of stainless steel bars, 0.5 cm in diameter and spaced 1.2 cm apart, but only the floor of the dark compartment was wired to the stimulus generator (Hugo-Sachs Elektronik, Germany). On the first day, the rats were allowed to explore the apparatus with the guillotine door open for 5 minutes. The following day, each rat was placed into the brightly lit compartment and the latency to enter into the dark compartment was recorded. Upon entry into the dark compartment, the door was lowered and a 1-mA, 1-s footshock was delivered 3 times at 5-s intervals. Twenty four hours later, this procedure was repeated, with the exception that no footshock was delivered, and the latency to enter the dark compartment was again recorded.

All rats were given a single session of fear conditioning as previously described in detail [20]. The conditioning chamber (San Diego Instruments, USA) was equipped with a metal grid floor, wired to the stimulus generator and acoustic stimulus unit. The grid floor was composed of stainless steel bars, 0.6 cm in diameter, spaced 1.4 cm apart. The chamber was scented with 1% acetic acid solution (placed into base pan underneath the grid floor) in order to provide a distinct context odor. Rats were placed inside the apparatus and left undisturbed for 3 minutes. During next 3-minute period they received 5 tone-footshock conditioning trials, with 30-s intervals. Each conditioning trial consisted of a 10-s tone conditioned stimulus (80 dB, 2.8 kHz), which coterminated with a 1-s footshock unconditioned stimulus (0.8 mA) delivered via the grid floor.

One day later, all rats were submitted to retention tests. The order of testing was counterbalanced so that half of the rats in each group were tested for retention of the conditioned context by repeating the procedure employed during training (in the absence of the tone or footshock), whereas remaining rats were tested for retention of the conditioned tone in a different context. In the later test, the rats were placed into a novel chamber which was located in a novel behavioral room. The floor of the new chamber was composed of a piece of a black carpet. The chamber was scented with lemon instead of acetic acid. The animals remained in the chamber for a period of 6 minutes and the conditioned stimulus (10-s tone) was presented 5 times during the last 3 minutes of this period. Four hours later, these rats were tested for retention of the context as described above, whereas the animals from the context test were now subjected to the tone retention test.

All training and testing trials were recorded with a video camera connected to a SVHS recorder for subsequent analysis. The rat's behavior was analyzed by an observer blinded with respect to experimental group. Freezing (defined as the absence of all movement other than that required for breathing) was scored if the rat remained inactive for at least 3 seconds. The percentage of accumulated time spent freezing was calculated.

Morris Water Maze Test

The procedure of the water maze training, previously described in detail [21], initiated one week following the termination of the passive avoidance and fear conditioning tests. The maze consisted of a black circular tank, 180 cm in diameter and 50 cm deep, and was located in a corner of a room containing extramaze cues. The apparatus was filled with water at room temperature (21±1°C) to a depth of approximately 35 cm. The water was made opaque by adding a non-toxic paint. The maze was divided, by imaginary lines, into four equal-size quadrants. An escape platform, 10 cm in diameter, was placed in the center of one of the quadrants. It was located 2 cm below the surface of the water. The animals were trained to find the escape platform during 14 consecutive days with two 60-s daily trials separated by a 30-s inter-trial interval. Each rat was placed in the water facing the pool wall at one of the four starting points that were used in a pseudo-random order so that each position was used once in each block of four trials. However, in each trial the starting positions were the same for all animals. If the rats did not find the escape platform within 60 s, the experimenter guided them to the platform where they were allowed to remain for 15 s. The platform location was not changed during the 14-day training period. The swim path length and swim speed were registered using the EthoVision video-tracking system. One day after completion of the acquisition, animals were submitted to a single 60-s probe trial in which the platform was removed from the pool. The number of times the rats swam through the zone where the platform had been located (platform crossings) and the percentage of time spent by rats swimming in the training quadrant were recorded.

Performance of animals on the visible platform task was assessed during a 2-day period. The rats were given 1 block of 4 trials per day separated by 30-s inter-trial intervals. The platform, painted in white, was exposed 3 cm above the water surface. The position of the platform was different in each trial. The distances swum to locate the platform were recorded and averaged across 8 trials.

Tissue Preparation

Following the completion of the behavioral experiments, six animals in each group, selected at random, were deeply anesthetized with pentobarbital (90 mg/kg) and injected intracardially with 0.1 mL of a heparin solution, followed by 1 mL of 1% sodium nitrite in saline. Then, they were perfused transcardially with 150 mL of 0.1 M phosphate buffer (pH 7.4) for vascular rinse, followed by 250 mL of a fixative solution containing 4% paraformaldehyde in phosphate buffer. The brains were removed from the skulls, immersed for 2 hours in the fixative, and infiltrated during 36 h in 10% sucrose solution at 4°C. After the frontal poles were trimmed away, the blocks were mounted on a vibratome and sectioned in the coronal plane at 40 μm. Every sixth section containing the hippocampal formation was collected using a systematic random sampling procedure [22], which provided an average of 14 sections per animal, mounted serially on gelatin-coated slides and air-dried. The sections were stained with a Giemsa, dehydrated in a series of ethanol solutions (50%, 70%, 90% and 100%) and coverslipped using Histomount (National Diagnostics, Atlanta, GA, USA).

Morphological Analysis

The Giemsa-stained tissue sections were visualized and imaged using an Axio Scope.A1 microscope equipped with a AxioCam MRc5 digital camera (Zeiss, Germany). The total numbers of neurons were estimated by applying the optical fractionator method [23]. The boundaries of the hilus of the dentate gyrus and of the CA3 and CA1 hippocampal fields were consistently defined at all levels along the rostrocaudal axis of the brain on the basis of cell morphology and cytoarchitectonic criteria [24]. Neurons belonging to the CA2 hippocampal field were included in the CA3 region. Estimations were carried out using the C.A.S.T.-Grid System (Olympus, Denmark). Beginning at a random starting position, visual fields were systematically sampled along the x and y axes, using a raster pattern procedure. Neurons were counted in every frame using the optical disector at a final magnification of ×2,000. Tissue thickness was estimated at each counting frame and guard zones of 2 μm were implemented. The nucleus of the neurons was used as the counting unit. The coefficient of error (CE) of the individual estimates was calculated according to Gundersen et al. [25] and it ranged between 0.06 and 0.09.

Statistical Analysis

Before conducting statistical comparisons, data were tested for normality using the Shapiro-Wilk's W test, which is known to be the most powerful normality test, especially in the case of small sample sizes [26]. Because all data samples passed the normality test ($p>0.05$), they were analyzed for statistical significance using parametric tests. Data derived from the passive avoidance test, acquisition trials of the water maze task, and training trials and tone retention trials of the fear conditioning test were analyzed using repeated measures ANOVA. The remaining data were analyzed using one-way ANOVA. Newman-Keuls post hoc test was used where appropriate. Because some of the behavioral tests were presented in a counterbalanced manner, the data derived from these tests were collapsed across order of presentation. Pearson's product moment regression model was used to analyze correlations between the neuron numbers and the two behavioral indices derived from the water maze test: the percentage of savings assessed by the difference in the mean distance swum between the first four trials and the last four trials (index of learning), and the percentage of time spent by rats swimming in the quadrant where the platform had been located during training (index of retention). All behavioral data are presented as the mean±SEM, while morphological results are expressed as the mean±SD. Differences were considered as significant at the $p<0.05$ level.

Results

Monitoring during the Recovery Period

Despite all the efforts made to reduce the mortality rate in KA-treated rats (see above), a significant proportion of them have died during the recovery period. In particular, from the 17 rats that had experienced convulsive SE 3 rats have died within 48 h following the treatment and 3 rats have died during the first month of the recovery period. From the 11 rats that did not show motor manifestations of SE 2 rats have died within 48 h following the treatment. Repeated behavioral seizures of stage 3 to 5 on the Racine scale were observed in a total of 9 rats: by the end of the first month of recovery – in just 1 rat; by the end of the second and third months – additionally in 3 and 5 rats, respectively. Interestingly, during the treatment with KA, all of these rats showed a typical pattern of behavioral SE, that is, continuous stage 3 to 5 seizures on the Racine scale. Three of the rats in this group had motor seizure attacks during the behavioral tests: 1 rat – during the water maze training (twice) and 2 rats during the tone test of the fear conditioning procedure. The respective trials were repeated for these animals on the next day. None of the rats had motor seizures during the 1-h intervals preceding or following the behavioral tests. The remaining 2 rats, which experienced convulsive SE during the treatment with KA, showed no spontaneous stage 3, 4 or 5 seizures during the entire 6-month observation period (including the period of behavioral tests). However, one of the experimenters detected in these 2 rats repeated mouth and facial movements (stage 1 seizures on the Racine scale), but this observation was not confirmed by other experimenters. Thus, the final group size for BSE rats was n = 11. No signs of behavioral seizures (stages 1–5 on the Racine scale) were detected in the remaining 9 KA-treated rats. These rats, in contrast, did not show clearly expressed motor signs of SE during the treatment with KA (no-BSE group; final n = 9). No behavioral abnormalities were observed in the control group.

Behavioral Effects of KA Treatment in the Open-field and Elevated Plus-maze

The results obtained in the open-field tests are shown in Fig. 1A. ANOVA revealed a significant effect of the treatment on the locomotion activity of rats both in the outer zone of the apparatus ($F_{2,25} = 4.70$, $p<0.05$) and in its inner zone ($F_{2,25} = 4.07$, $p<0.05$). Post-hoc comparisons showed that animals from BSE group traveled significantly longer distances in the peripheral zone of the open field than did control rats and no-BSE rats ($p<0.05$). Similarly, rats from BSE group showed increased locomotion in the inner part of the open field. However, in this zone, the difference was significant only when BSE group was compared to no-BSE group ($p<0.05$), but it did not reach the level of significance when compared to control group ($p = 0.08$).

Rats in all groups showed clear preference to protected areas of the plus maze, as they spent more time in the closed arms and central zone of the apparatus in comparison to its open arms (Fig. 1B). However, although the percentages of time spent by rats in the closed arms and central zone did not differ between the groups ($Fs<1$, $p>0.05$), ANOVA revealed a significant effect of the treatment on the time spent by rats in the open arms ($F_{2,25} = 3.31$, $p<0.05$). Particularly, rats in the BSE group spent significantly more time in the open arms than those from control group and no-BSE group ($p<0.05$). No significant differences were detected between the groups regarding the total number of entries into the arms of the apparatus.

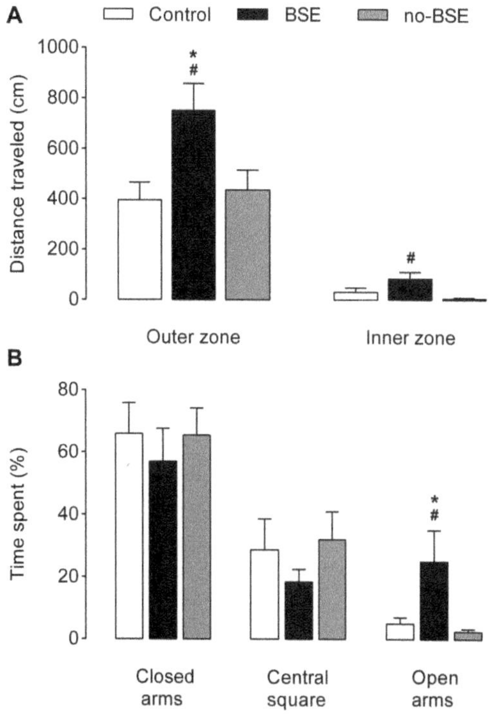

Figure 1. Behavior of kainate-treated rats in the open-field (A) and elevated plus-maze (B) tests. The y-axis represents the distances travelled by rats during the 5-min testing sessions in the outer and inner zones of the open-field apparatus, and in the central zone and open and closed arms of the plus-maze apparatus. Note that rats who had experienced behavioral status epilepticus (BSE group) traveled longer distances in the outer and inner zones of the open field when compared to control rats and rats that had not experienced behavioral status epilepticus (no-BSE group), respectively. In the plus-maze test, rats in the BSE group spent more time in the open arms of the apparatus than rats in the other two groups. *$p<0.05$ vs. control group and #$p<0.05$ vs. no-BSE group. Data are presented as the mean±SEM.

Effects of KA Treatment on Passive Avoidance and Conditioned Freezing Behavior

The results of the passive avoidance testing are shown in Fig. 2. Repeated measures ANOVA did not reveal a significant main effect of treatment on the performance of rats on this task ($F_{2,25} = 1.47$, $P = 0.25$). However, there was a significant effect of training ($F_{1,25} = 14.30$, $p<0.001$) and a significant interaction between these two effects ($F_{2,25} = 5.09$, $p<0.01$). Post-hoc comparisons showed that, although rats in no-BSE group did not differ from control rats in the baseline latency to enter the dark compartment, on the retention trial, the entry latencies were significantly reduced in this group relative to controls ($p<0.05$). The latencies to enter the dark compartment on the retention test were not significantly different between control group and BSE group. However, the baseline latency was significantly increased in BSE rats relative to control rats ($p<0.05$), suggesting that the high entry latencies registered in this group on the retention test might be related to a decreased motivation to enter the dark compartment of the apparatus, for example, due to reduced levels of anxiety.

Conditioning produced an increase in the amount of freezing time in either group ($F_{1,25} = 156.11$, $p<0.0001$; Fig. 3A) and examination of the data revealed no significant differences between the groups. Once placed into the training chamber 1

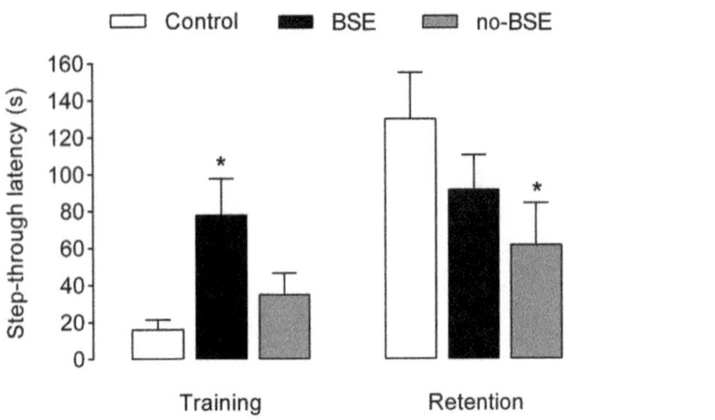

Figure 2. Performance of kainate-treated rats on the passive avoidance test. The y-axis represents the mean (±SEM) step-through latencies during training trial as well as on the retention trial given 24 h later. Note that rats which did not show behavioral signs of status epilepticus following the treatment with kainate (no-BSE group) showed lower retention scores than did control rats. In contrast, rats that had experienced behavioral status epilepticus (BSE group) showed increased baseline entry latency, but appeared normal on the retention test. *$p<0.05$ vs. control group.

day later, the animals from all groups displayed increased levels of freezing behavior (Fig. 3B). However, the amount of freezing differed between the groups ($F_{2,25}=4.68$, $p<0.05$), so that rats in the BSE group froze to the familiar context significantly less than control rats and no-BSE rats ($p<0.05$). However, no-BSE and control groups did not differ significantly in this test. In the novel context, the levels of freezing displayed by all groups in the absence of the tone were inferior to those observed during the tone presentation ($F_{1,25}=99.42$, $p<0.0001$; Fig. 3C). Repeated-measures ANOVA of these data did not reveal a significant main effect of treatment, but yielded a significant treatment by tone presentation interaction ($F_{2,25}=3.29$, $p<0.05$). Post-hoc tests showed that, although in the absence of the tone the 3 groups did not differ with respect to the amount of freezing, during the tone presentation, the rats in BSE group showed relatively low levels of fear, freezing significantly less than control rats ($p<0.05$).

Water Maze Navigation

The mean distances swum by rats in the control, BSE and no-BSE groups to locate the hidden platform in the place learning task are presented in Fig. 4A. The statistical analysis by repeated measures ANOVA showed that the rats included in this experiment progressively improved their ability to find the platform over the 14 days of acquisition ($F_{6,150}=25.980$ $p<0.00001$). However, the overall performance of the rats on this task differed among the groups studied, as indicated by a significant main effect of treatment ($F_{2,25}=12.89$, $p<0.0001$) and significant treatment × trial block interaction ($F_{12,150}=1.86$, $p<0.05$). Post-hoc tests for the main effect of treatment revealed that animals from the BSE group and no-BSE group performed the task less well than did control animals ($p<0.001$ and $p<0.01$, respectively), but animals in the BSE group were more severely impaired than those in the no-BSE group ($p<0.05$). Post-hoc tests for the treatment × trial block interaction showed that BSE group performed significantly worse than control group on the 2nd to 7th trial blocks ($p<0.0001$). In turn, rats in the no-BSE group also traveled greater distances than did control rats on the 2nd to 7th trial blocks ($p<0.001$). Moreover, rats from the BSE group were less efficient in finding the escape platform than rats from the no-BSE group on the 4th, 5th, and 7th trial blocks ($p<0.05$).

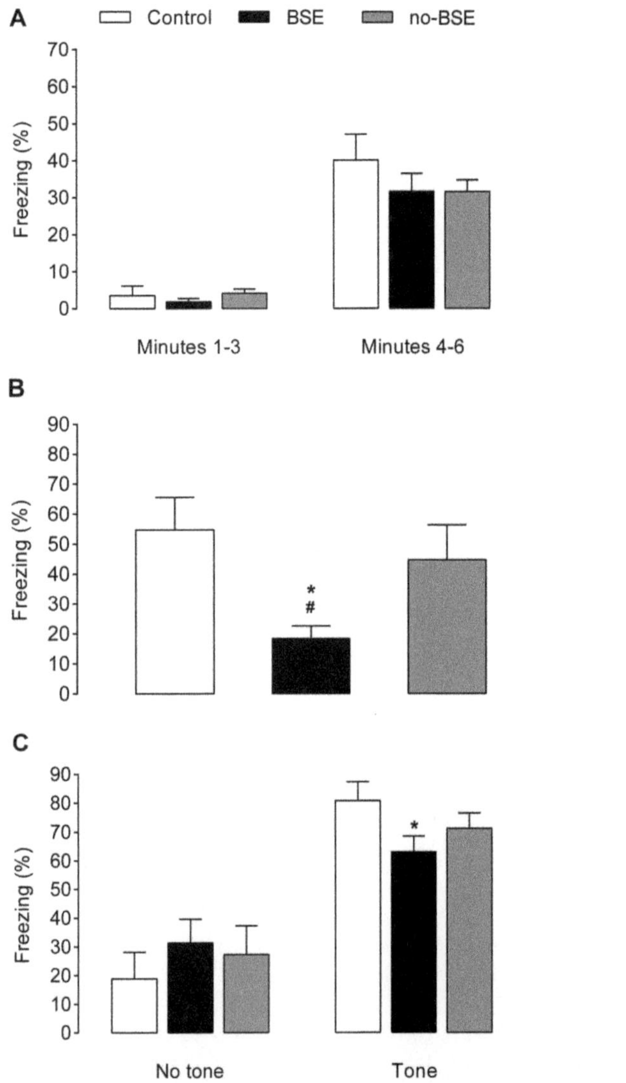

Figure 3. Effects of kainate treatment on the acquisition and retention of conditioned fear. The y-axes represent percentage of freezing time for each of the 3-minute periods of the acquisition session (**A**), during the context retention test (**B**), and during each of the 3-minute periods of the tone retention test which was performed in a novel context (**C**). No tone or footshock was delivered during the first 3-minute periods of the acquisition session (**A**) and of the tone retention test (**C**). Note that rats that had experienced behavioral status epilepticus (BSE group) showed reduced levels of fear, freezing significantly less than control rats on both retention tests. However, no-BSE group (composed of rats that had not experienced behavioral status epilepticus after kainate treatment) and control group did not differ significantly on these tests. *$p<0.05$ vs. control group and #$p<0.05$ vs. no-SE group. Data are presented as the mean±SEM.

Behavioral measures obtained on the probe trial are shown in Fig. 4B. ANOVA revealed a significant effect of treatment on the total percentages of time spent by rats swimming in the training quadrant ($F_{2,25}=7.01$, $p<0.01$) and a significant effect of treatment on the number of times they crossed the former location of the escape platform (platform crossings; $F_{2,25}=6.40$, $p<0.01$). Post-hoc tests showed that both control and no-BSE rats spent more time in the training quadrant than did rats in the BSE group ($p<0.01$ and $p<0.05$, respectively). However, the no-BSE group did not differ from the control group on this measure, suggesting

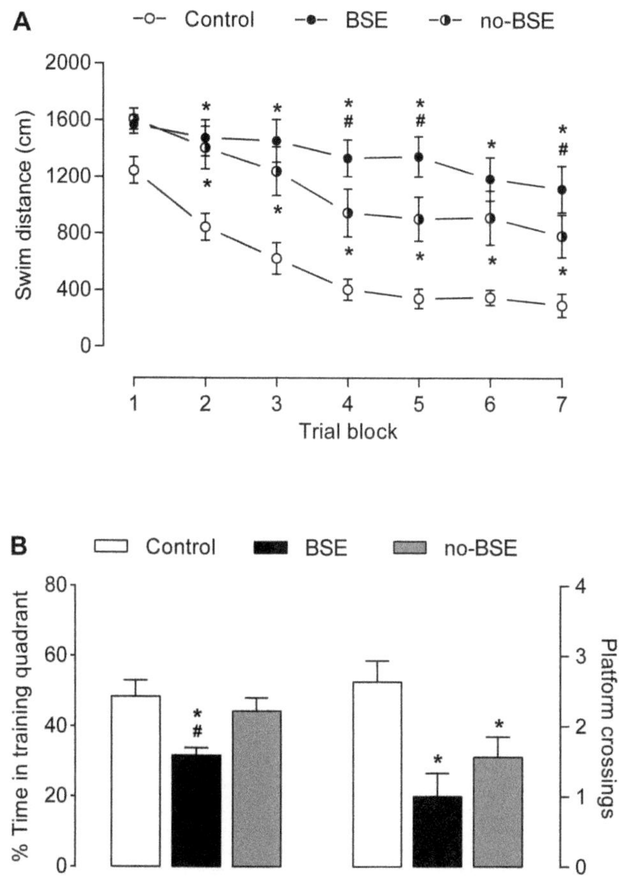

Figure 4. Effects of kainate treatment on the performance of rats on the water maze. (**A**) The y-axis represents the mean swim distance (±SEM) in centimeters to reach a hidden platform throughout the acquisition of the water maze task. The data were averaged across seven blocks of four consecutive trials each (the x-axis). Note that the performance of rats that had experienced behavioral status epilepticus (BSE group) on this task was extremely poor. However, spatial learning of the rats that showed no convulsive status epilepticus (no-BSE group) was also considerably impaired. *$p<0.0001$ for BSE group vs. control group and $p<0.001$ for no-BSE group vs. control group; #$p<0.05$ vs. no-BSE group. (**B**) The y-axes represent the percentage of time spent by rats in the training quadrant of the water maze during the probe trial and the number of times they swam through the zone where the platform had been located (platform crossings). Unlike control rats, rats in the BSE group did not show a preference for the training quadrant of the water maze and were impaired in their ability to localize the former platform position. However, the acuity of spatial search was also impaired in the no-BSE rats, as indicated by the fact that they crossed the former platform position less frequently than did control rats. *$p<0.05$ for no-BSE group vs. control group and $p<0.01$ for BSE group vs. control group; #$p<0.05$ vs. no-BSE group. Data are presented as the mean±SEM.

that rats in both these groups, unlike those in the BSE group, were able to learn a spatial strategy when searching for the escape platform. Nevertheless, the acuity of spatial search was impaired in both KA-treated groups, as indicated by the fact that control rats crossed the former platform position more frequently when compared to both BSE group ($p<0.01$) and no-BSE group ($p<0.05$).

ANOVA revealed that there was a significant main effect of treatment on the speed of swimming averaged across all trials of the water maze acquisition ($F_{2,25}=3.29$, $p<0.05$). The animals from the BSE and no-BSE groups swam somewhat faster than control rats (24.7±1.2 and 22.3±1.5 cm/s vs. 20.1±1.1 cm/s, respectively). However, only BSE rats differed significantly from control rats on this measure ($p<0.05$). Animals in all groups were able to learn the visible platform task. The distances swum to locate the platform, averaged over eight trials, were 398±102 cm in the control group, 513±115 cm in the BSE group and 466±95 cm in the no-BSE group. ANOVA failed to reveal a significant effect of treatment on this parameter. These data indicate that, with respect to sensorimotor functions, KA-treated rats were not impaired when compared to control rats.

Effects of KA Treatment on Hippocampal Morphology

Representative photographs of brain sections illustrating treatment-related morphological changes in the hippocampal formation of rats in the BSE and no-BSE groups are shown in Fig. 5 and Fig 6. The qualitative observation of the sections inferred that the neuronal density of the CA3 and CA1 pyramidal cells was considerably smaller in BSE-rats than in control rats and no-BSE rats. In some hippocampal regions, CA1 pyramids were practically absent in BSE rats (Fig. 5B). In addition, the dentate gyrus was deformed and its granule cell layer was dispersed in animals that experienced behavioral SE. Consistent with prior anatomical studies [27,28], the density of hilar neurons in control rats was higher in the sections obtained from the temporal (ventral) pole of the dentate gyrus when compared to those obtained from its septal (dorsal) pole (Fig. 5A and Fig. 6A). The density of hilar neurons was reduced in KA-treated rats that experienced convulsive SE throughout the entire dentate gyrus, but neuron loss was particularly extensive in the ventral hilus (Fig. 6B). However, in rats from the no-BSE group hilar neuron loss was less extensive and most of the neurons of the ventral hilus were preserved (Fig. 6C).

The estimates of the total numbers of neurons in the hilus of the dentate gyrus, and in the CA3 and CA1 fields of the hippocampus proper are shown in Fig. 7. Analysis of these data revealed that there was a significant main effect of seizures on neuron numbers in all three hippocampal subdivisions: dentate hilus ($F_{2,15}=6.77$, $p<0.01$), pyramidal CA3 field ($F_{2,15}=31.13$, $p<0.0001$) and pyramidal CA1 field ($F_{2,15}=30.72$, $p<0.0001$). Post-hoc comparisons showed that in rats in which KA treatment resulted in the development of behavioral SE there was a significant neuron loss in the dentate hilus (40%; $p<0.01$), and hippocampal CA3 (50%; $p<0.001$) and CA1 (80%; $p<0.001$) fields. In sharp contrast, in rats from the no-BSE group there was a significant neuron loss only in the dentate hilus (30%; $p<0.05$), but not in the hippocampal pyramidal fields.

Relationship between KA-induced Neuron Loss and Spatial Learning and Memory

Correlational analyses were conducted to determine whether individual variations in learning and memory abilities of rats are related to KA-induced loss of hippocampal neurons. The analyses were conducted separately for hilar neurons, and CA3 and CA1 hippocampal pyramidal neurons. Data obtained from 6 animals from each group, whose brains were randomly selected for histological assessment, were included in the analysis (final n = 18). The results of the correlational analyses, shown in Fig. 8, revealed significant correlations between the total number of hilar neurons and two behavioral indexes of spatial learning and memory, i.e. the percentage of savings between the first four trials and the last four trials of the water maze acquisition ($r=0.57$, $p<0.01$; Fig. 8A) and the percentage of time spent in the training quadrant during the probe trial ($r=0.49$, $p<0.05$; Fig. 8B). Interestingly, both behavioral measures tend to exhibit even stronger correlations

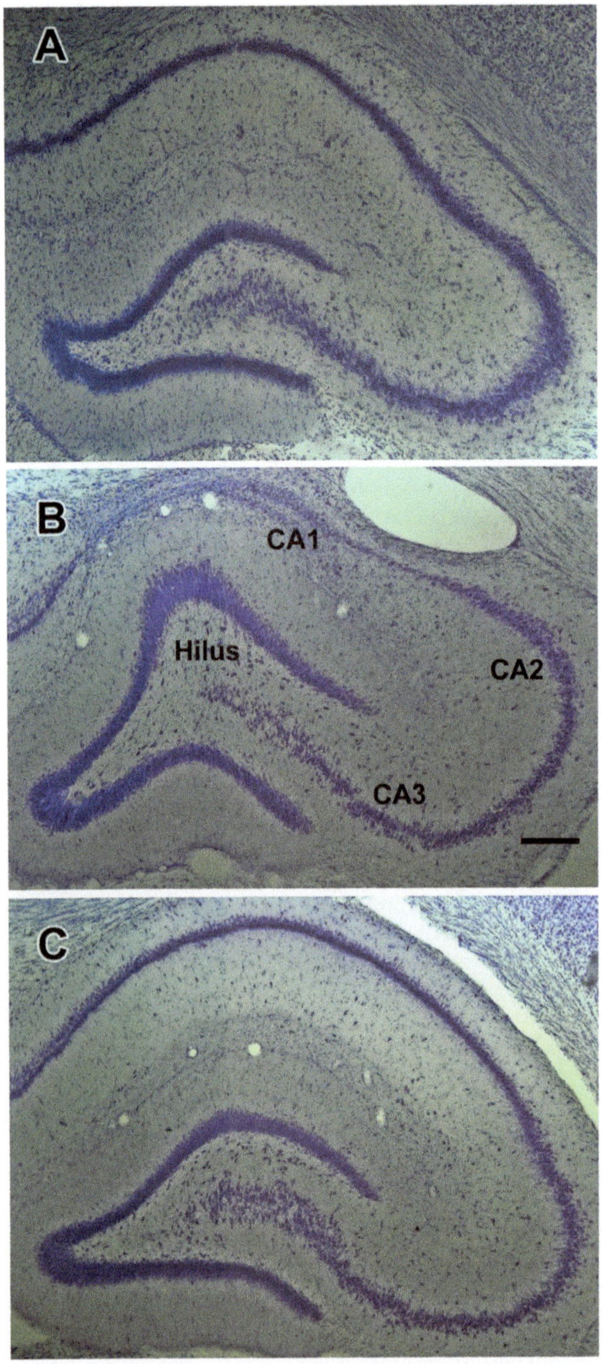

Figure 5. Effects of kainate treatment on hippocampal morphology. Photomicrographs of representative Nissl-stained coronal sections containing the dentate gyrus, CA3 and CA1 hippocampal fields taken from a control rat (**A**), from a rat in the behavioral status epilepticus (BSE) group (**B**) and from a kainate-treated rat that had not experienced behavioral status epilepticus (no-BSE; **C**). Note that the density of neurons in the dentate hilus is considerably reduced in the rats with (**B**) and without (**C**) behavioral SE when compared to the control rat (**A**). Note also dramatic loss of the CA3 and especially CA1 neurons in the rat from the BSE group; neurons in the CA2 field are relatively preserved (**B**). Scale bar shown in (**B**) = 200 μm.

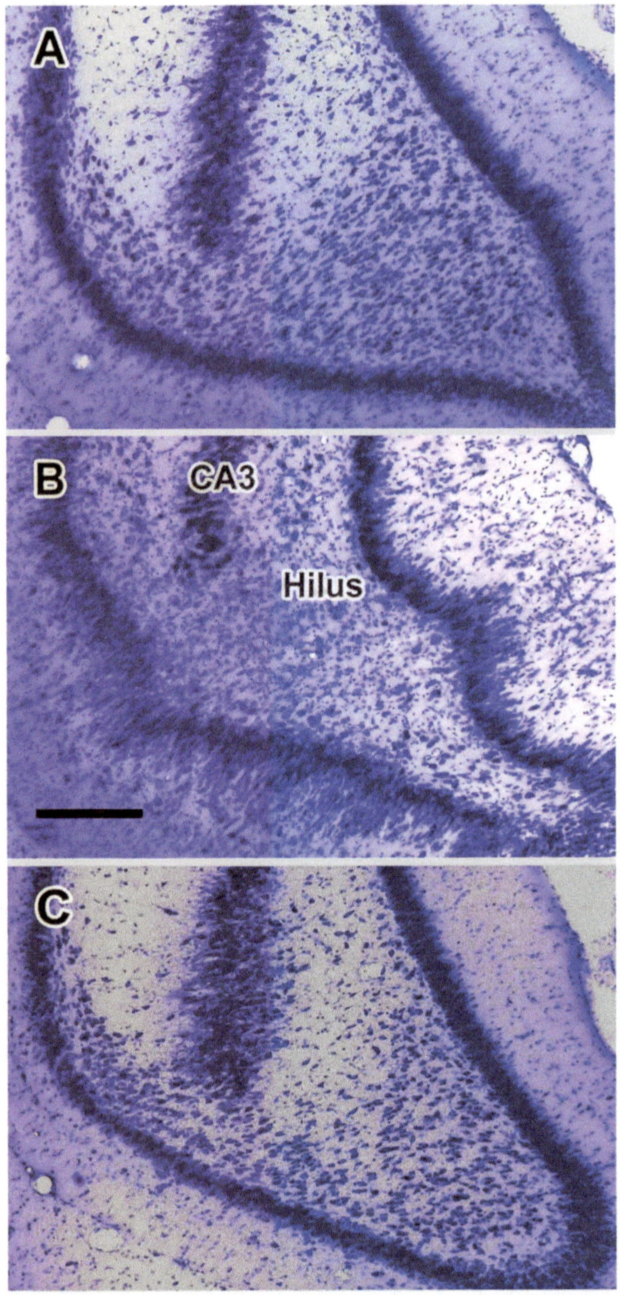

Figure 6. Kainate-induced loss of neurons in the ventral dentate hilus. Photomicrographs of representative Nissl-stained coronal sections containing the ventral portion of the dentate gyrus taken from a control rat (**A**), from a rat in the behavioral status epilepticus (BSE) group (**B**) and from a kainate-treated rat that had not experienced behavioral status epilepticus (no-BSE; **C**). Note that the density of neurons in the dentate hilus is dramatically reduced in the rat that experienced behavioral SE (**B**). In striking contrast, most of the neurons of the ventral hilus are preserved in the rat from the no-BSE group (**C**). Scale bar shown in (**B**) = 250 μm.

with the total number of neurons in the CA3 hippocampal field ($r = 0.74$, $p < 0.0005$ and $r = 0.6$, $p < 0.01$ for the percentage of savings and percentage of time in the training quadrant, respectively; Fig. 8C, D). In contrast, we found no significant correlations between the two behavioral indexes and the total number of the CA1 pyramidal neurons (Fig. 8E, F).

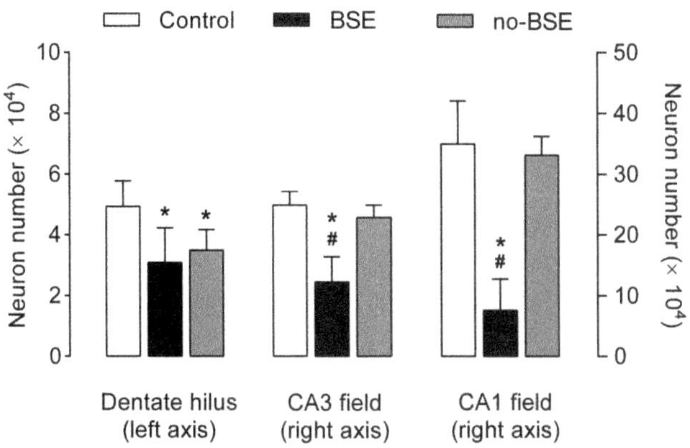

Figure 7. Loss of hippocampal neurons in kainate-treated rats. The y-axes represent the total numbers of neurons (mean±SD) in the dentate hilus and hippocampal CA3 and CA1 pyramidal fields in control rats and in kainate-treated rats with and without behavioral status epilepticus (BSE group and no-BSE group, respectively). The number of hilar neurons is reduced in both groups of kainate-treated rats when compared to control group. Note that the number of cells in the CA3 and CA1 pyramidal regions was strikingly reduced in rats from the BSE group, but not in those from the no-BSE group. Dentate hilus: *$p<0.05$ for no-BSE group vs. control group and $p<0.01$ for BSE group vs. control group; CA3 and CA1 fields: *$p<0.001$ vs. control group and #$p<0.001$ vs. no-BSE group.

Discussion

The results of this experiment show that rats treated with similar doses of KA may have two quite distinct sets of brain changes. In particular, the neuropathological sequelae of the treatment were different in the following ways: (1) in one group of rats KA induced a clear pattern of convulsive SE including repeated stage 4 and 5 seizures on the Racine scale, but this was not observed in remaining rats; (2) majority of the rats that had experienced behavioral SE developed spontaneous motor seizures 1 to 3 months later, whereas the rats in the other group did not; (3) histological analysis revealed massive loss of hippocampal pyramidal neurons in rats that had experienced behavioral SE, whereas no pyramidal cell loss was detected in rats who did not show motor signs of SE; in either group of KA-treated rats, the total number of neurons in the hilus of the dentate gyrus was reduced; (4) although rats in both KA-treated groups were impaired in behavioral tests, the impairments were significantly more profound and wide-ranging in the BSE group. These data are consistent with previous studies showing that animals of the same strain may have different seizure susceptibility [14,29,30,31], which is believed to have both genetic [29,30] and environmental [32] underpinnings.

Rats in the BSE group were not only impaired in learning the precise platform position in the water maze, but were, in fact, unable to adopt a spatial strategy to solve the task. In addition, they showed severely impaired performance on the passive avoidance task and on the fear conditioning tests. Rats in this group were hyperactive in the open-field test and spent more time walking on the open arms of the elevated plus maze, suggesting a deficient emotional responsiveness. Our results are consistent with many prior studies in which behavioral changes were detected in animals following SE induced by either electrical stimulation [11,33] or injections of chemoconvulsants (pilocarpine [34,35], lithium-pilocarpine [9,13,36], and kainate [8,9,11,12,13,37,38] models). However, the behavioral deficits observed in the present study appear to be somewhat more pronounced than those reported in prior studies using the same model of SE [9,39,40]. This discrepancy is likely due to differences in experimental conditions between different studies. For example, it has been reported that the induction of SE in adult rats is followed by more robust behavioral deficits than in premature rats [8,37,38] and that old rats, in turn, are more susceptible to SE when compared to adult rats [12]. In addition, it has been shown that the seizure-related behavioral impairments are strain-specific [9,13,31,41]. Furthermore, it is also well documented that different treatment protocols, e.g. duration of SE, may be associated with very different neuropathological outcomes in experimental animals [9,16,42]. It may be also noted that we performed behavioral testing of rats approximately 5 months after the induction of SE, that is, at a time when seizure frequency is maximal [43]. For comparison, studies on animals just 2 or 3 months post-SE, when seizures are relatively less frequent and neuroanatomical changes are likely to be less dramatic, generally reveal less severe behavioral deficits [9,40]. However, it remains to be seen whether SE-related behavioral deficits actually worsen with increased duration of epileptic state. In fact, it has been recently reported that animals tested two months and twenty months after the induction of SE showed similar cognitive impairments [44].

There are a few prior studies in which animals that did not develop convulsive SE after the respective treatments were evaluated behaviorally in parallel to those who did. In particular, Szyndler et al. [36] reported that lithium-pilocarpine-treated rats resistant to develop SE did not differ behaviorally from control rats, whereas rats that developed SE were impaired in the fear conditioning and open-field tests. Müller et al. [34] have shown that pilocarpine-treated mice without motor signs of SE differed significantly from control mice in several measures derived from the open-field, hole-board and light-dark transition tests (mainly related to their increased activity), but were intact, unlike mice which had behavioral SE, on the elevated plus-maze, novel object exploration and spatial water maze tests. However, histological analysis carried out in the latter study revealed significant loss of hippocampal neurons only in mice that developed convulsive SE [34]. Interestingly, it has been also reported that treating rats with KA during early postnatal period induces electrographic and mild behavioral seizures, but does not evoke a typical pattern of SE, i.e. stage 4–5 seizures on the Racine scale [37,38]. Yet, later in life, these rats show significant cognitive deficits accompanied by long-term impairments in hippocampal plasticity [37,38]. In this respect, it is worth noting the results of a recent study by Suárez et al. [14] who reported that treating rats with KA, even if it induces neither motor SE, nor spontaneous seizures later in life, may lead to enduring changes in synaptic plasticity. In the present study, KA-treated rats that did not develop motor SE had cell loss in only one of the hippocampal subdivisions, the dentate hilus, but were impaired on the passive avoidance and spatial water maze tasks. Thus, taken together, these findings indicate that a part of the animals that do not manifest intense behavioral seizures immediately after an epileptogenic treatment, later in life, they may still have conspicuous structural and functional changes in the brain.

It is well documented that epilepsy models based on the induction of SE are typically associated with marked neuronal loss in various brain regions, such as the hippocampal formation, amygdala, thalamus, piriform, entorhinal, perirhinal, retrosplenial and olfactory cortices [7,9,10,11,33,35,45,46,47], many of which are implicated in cognitive and emotional processes [15,48,49,50,51,52]. Therefore, it is likely that brain damage induced by SE and subsequent spontaneous seizures is directly responsible for the behavioral deficits observed in these models. In

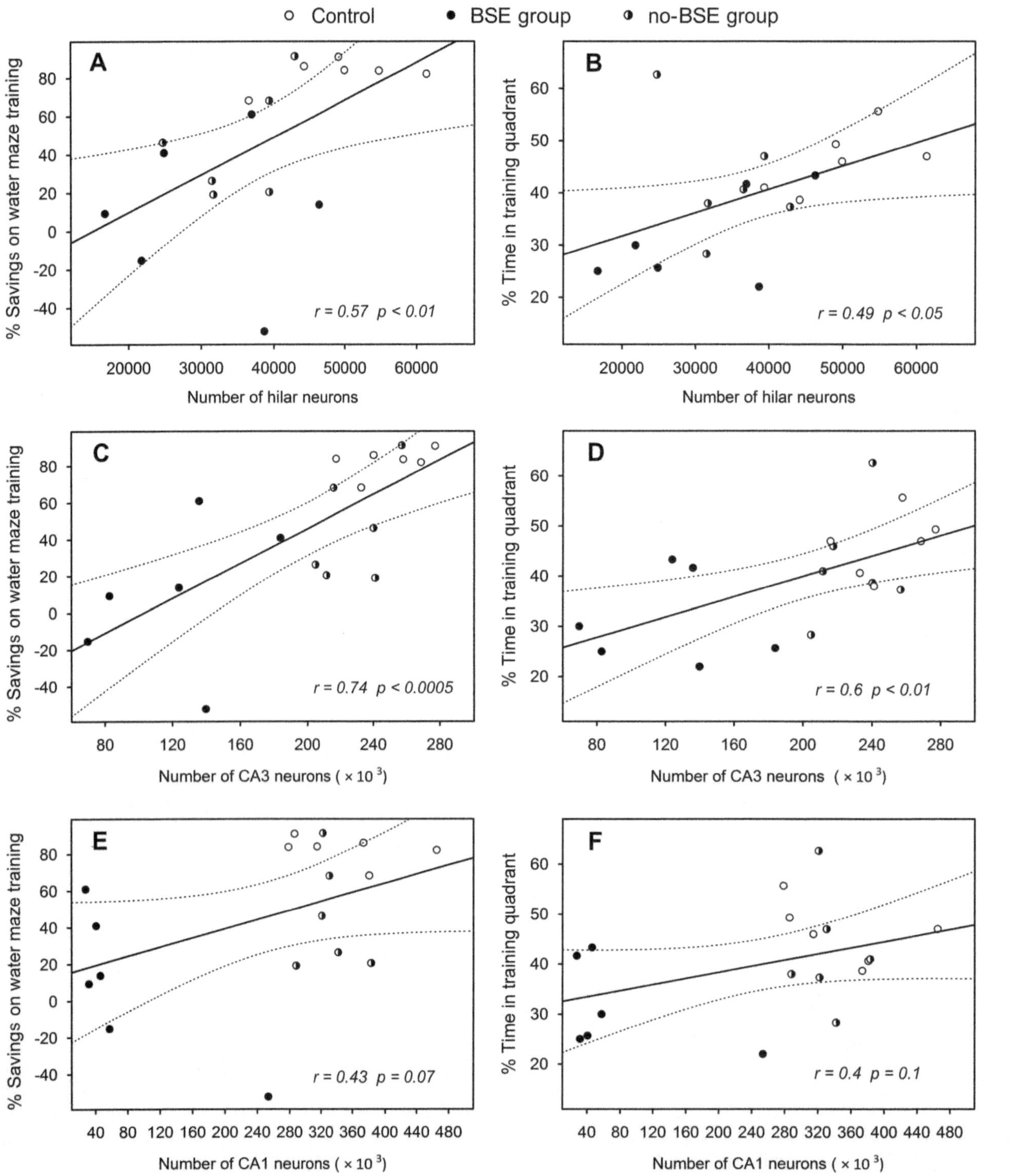

Figure 8. Correlations between hippocampal neuron numbers and spatial learning and memory indices. The x-axes represent the total numbers of neurons in the dentate hilus (**A**, **B**), and hippocampal CA3 (**C**, **D**) and CA1 (**E**, **F**) pyramidal fields in control and kainate-treated rats. The y-axes represent the percentages of savings during water maze training, i.e. the differences, in %, between the mean distances swum by rats on the first four trials and on the last four trials (**A**, **C**, **E**), and the percentages of time spent by rats in the training quadrant of the water maze during the probe trial (**B**, **D**, **F**). Data obtained from 6 animals randomly selected from each group (final n = 18) were used in the correlational analyses. Correlation coefficients are shown at the bottom of each plot and dashed lines represent 95% confidence intervals.

effect, it has been recently shown that hippocampal volume of epileptic rats significantly correlates with their learning and memory abilities [9,53]. Similar relationships were found between the volumetric estimates of the amygdala and the indices of emotionality [9,53]. In this study, we detected marked loss of hippocampal CA3 and CA1 pyramidal neurons in rats that had experienced behavioral SE, whereas in rats without behavioral manifestations of SE these neuronal populations were preserved. However, the loss of hilar neurons was detected in both KA-treated groups, even though it was predictably greater in BSE group than in no-BSE group. Thus, our data suggest that hilar cell loss alone may be sufficient to cause functional deficits in hippocampal circuits as indicated by the impaired performance of no-BSE rats on the water maze and passive avoidance tests. This suggestion is consistent with the evidence that hilar neurons may influence mnemonic processes by synchronizing the activity of the dentate gyrus granule cells and, more generally, by modulating the transmission of information from the entorhinal cortex to the hippocampal pyramidal fields [54,55]. This possibility is also supported by the results of the correlational analyses which revealed significant correlations between individual spatial memory indices of rats used in this experiment and the total number of hilar neurons, but not of CA1 pyramidal neurons. The strongest correlation, however, was found between the cognitive indices and the total number of CA3 neurons. This result may appear surprising given that significant neuronal loss in this region was observed in only one of the KA-treated group, namely in rats that had experienced convulsive SE. However, it is possible that even small loss of CA3 neurons, perhaps, undetectable in small-size samples (such as those used in the present study), may have a substantial impact on hippocampal function. This possibility is consistent with accepted theoretical models of hippocampal circuits [56,57,58] and experimental studies [59,60,61] implicating the CA3 region of the hippocampus in key cognitive processes underlying episodic memory, such as pattern separation (reducing the overlap between similar neuronal representations) and pattern completion (converging neuronal representations into a single pattern, thus, facilitating their retrieval using only partial cues). Therefore, it is plausible that the memory deficits observed in KA-treated rats in this study can be accounted for, partly at least, by the disruption of cognitive processes in CA3 neuronal networks, which may be associated with the loss of neurons in this region.

Although in this study we have not estimated neuronal numbers separately for the dorsal and ventral parts of the hippocampal formation, visual inspection of the brain sections clearly showed that neuronal loss associated with behavioral SE was more marked in the ventral division of the dentate hilus than in its dorsal part (Fig. 6). This finding is in line with a previous study demonstrating that kainate-induced SE produces more severe loss at the temporal (ventral) pole of the dentate hilus when compared to the septal (dorsal) pole [28]. However, in the no-BSE group, the neurons of the ventral hilus were less severely affected. This observation is important for the interpretation of the finding that rats in the no-BSE group, unlike those that had experienced motor SE, showed normal behavior in the open-field, elevated plus-maze and fear-conditioning tests. Indeed, considerable evidence suggests that the ventral hippocampus has a preferential role in the regulation of emotional behaviors in rodents [49]. Therefore, the different extents of cell loss in the ventral hilus of rats from the BSE and no-BSE groups can be related to the differences in their performance on the anxiety-related tests. This explanation, however, should be viewed with caution given that brain regions other than hippocampus, e.g. the amygdala, play important roles in emotional behaviors and were not analyzed histologically in the present study. Another implication of this morphological observation is that the greater loss of neurons in the ventral dentate hilus of BSE rats may be associated with the increased susceptibility to the development of spontaneous motor seizures. Relevant to this hypothesis is the finding by Inostrosa et al [9], who showed that kainate-induced convulsive SE produces significant mossy fiber sprouting only in the ventral part of the dentate gyrus, suggesting an enhanced susceptibility of this hippocampal subdivision to epileptogenesis [28]. Furthermore, it has been previously reported that KA injection into the ventral hippocampus induces local electrographic seizures, which rapidly propagate to the dorsal hippocampus and cingulate cortex and are accompanied by severe motor manifestations including stage 4 and 5 seizures on the Racine scale [62]. In contrast, injection of KA into the dorsal hippocampus induces only local electrographic seizures, which do not propagate to the cingulate cortex and do not evoke behavioral seizures [62]. Therefore, it is possible that the absence of spontaneous motor seizures in no-BSE rats may be related to the fact that the ventral hippocampus was better preserved in this group when compared to the BSE group, in which all rats showed spontaneous motor seizures. This, of course, does not mean that the absence of convulsive SE after the treatment with KA precludes the development of epilepsy later in life. Indeed, because in the present study we did not perform EEG monitoring, spontaneous subconvulsive seizures might have occurred in the no-BSE group, but were not registered. In addition, the latency between the kainate treatment and the onset of the first spontaneous seizures may exceed the 5-month period of observation implemented in the current study [16].

In conclusion, this study was designed to compare the behavioral and neuroanatomical outcomes in rats that developed convulsive SE after kainate treatment to those that did not. It was found that those rats in which KA did not induce motor signs of SE were, however, impaired in hippocampal-dependent memory tasks later in life, which can be explained by long-term functional abnormalities in hippocampal circuits associated with the loss of hilar cells and, probably, CA3 pyramidal neurons. However, the extent of neuronal loss and the severity of cognitive deficits in this group were far less striking than those observed in the rats that had experienced motor SE. Thus, our findings add further weight to the evidence from former studies in animal models [9,53] and humans [4,6,63] showing that behavioral deficits in TLE correlate with the extent of hippocampal damage.

Author Contributions

Conceived and designed the experiments: GHM NVL. Performed the experiments: GHM JIS JMC. Analyzed the data: GHM JLQ PAA JAP NVL. Contributed reagents/materials/analysis tools: JPA NVL. Wrote the paper: GHM NVL.

References

1. Margerison JH, Corsellis JA (1966) Epilepsy and the temporal lobes. A clinical, electroencephalographic and neuropathological study of the brain in epilepsy, with particular reference to the temporal lobes. Brain 89: 499–530.
2. Sutula T, Cascino G, Cavazos J, Parada I, Ramirez L (1989) Mossy fiber synaptic reorganization in the epileptic human temporal lobe. Ann Neurol 26: 321–330.
3. Mathern GW, Babb TL, Pretorius JK, Leite JP (1995) Reactive synaptogenesis and neuron densities for neuropeptide Y, somatostatin, and glutamate decarboxylase immunoreactivity in the epileptogenic human fascia dentata. J Neurosci 15: 3990–4004.
4. Aldenkamp AP, Overweg J, Gutter T, Beun AM, Diepman L, et al. (1996) Effect of epilepsy, seizures and epileptiform EEG discharges on cognitive function. Acta Neurol Scand 93: 253–259.
5. Piazzini A, Canevini MP, Maggiori G, Canger R (2001) Depression and anxiety in patients with epilepsy. Epilepsy Behav 2: 481–489.
6. Alessio A, Damasceno BP, Camargo CH, Kobayashi E, Guerreiro CA, et al. (2004) Differences in memory performance and other clinical characteristics in patients with mesial temporal lobe epilepsy with and without hippocampal atrophy. Epilepsy Behav 5: 22–27.

7. Ben-Ari Y, Tremblay E, Ottersen OP, Meldrum BS (1980) The role of epileptic activity in hippocampal and "remote" cerebral lesions induced by kainic acid. Brain Res 191: 79–97.
8. Stafstrom CE, Chronopoulos A, Thurber S, Thompson JL, Holmes GL (1993) Age-dependent cognitive and behavioral deficits after kainic acid seizures. Epilepsia 34: 420–432.
9. Inostroza M, Cid E, Brotons-Mas J, Gal B, Aivar P, et al. (2011) Hippocampal-dependent spatial memory in the water maze is preserved in an experimental model of temporal lobe epilepsy in rats. PLoS ONE 6: e22372.
10. Covolan L, Mello LE (2000) Temporal profile of neuronal injury following pilocarpine or kainic acid-induced status epilepticus. Epilepsy Res 39: 133–152.
11. Kemppainen EJ, Nissinen J, Pitkänen A (2006) Fear conditioning is impaired in systemic kainic acid and amygdala-stimulation models of epilepsy. Epilepsia 47: 820–829.
12. Hattiangady B, Kuruba R, Shetty AK (2011) Acute seizures in old age leads to a greater loss of CA1 pyramidal neurons, an increased propensity for developing chronic TLE and a severe cognitive dysfunction. Aging Dis 2: 1–18.
13. Inostroza M, Cid E, Menendez de la Prida L, Sandi C (2012) Different emotional disturbances in two experimental models of temporal lobe epilepsy in rats. PLoS ONE 7: e38959.
14. Suárez LM, Cid E, Gal B, Inostroza M, Brotons-Mas JR, et al. (2012) Systemic injection of kainic acid differently affects LTP magnitude depending on its epileptogenic efficiency. PLoS ONE 7: e48128.
15. Morris R (1984) Developments of a water-maze procedure for studying spatial learning in the rat. J Neurosci Methods 11: 47–60.
16. Hellier JL, Patrylo PR, Buckmaster PS, Dudek FE (1998) Recurrent spontaneous motor seizures after repeated low-dose systemic treatment with kainate: assessment of a rat model of temporal lobe epilepsy. Epilepsy Res 31: 73–84.
17. Hellier JL, Dudek FE (1999) Spontaneous motor seizures of rats with kainate-induced epilepsy: effect of time of day and activity state. Epilepsy Res 35: 47–57.
18. Racine RJ (1972) Modification of seizure activity by electrical stimulation. II. Motor seizure. Electroencephalogr Clin Neurophysiol 32: 281–294.
19. Herguido MJ, Carceller F, Roda JM, Avendaño C (1999) Hippocampal cell loss in transient global cerebral ischemia in rats: a critical assessment. Neuroscience 93: 71–80.
20. Lukoyanov NV, Lukoyanova EA (2006) Retrosplenial cortex lesions impair acquisition of active avoidance while sparing fear-based emotional memory. Behav Brain Res 173: 229–236.
21. Lukoyanov NV, Lukoyanova EA, Andrade JP, Paula-Barbosa MM (2005) Impaired water maze navigation of Wistar rats with retrosplenial cortex lesions: effect of nonspatial pretraining. Behav Brain Res 158: 175–182.
22. Gundersen HJ, Jensen EB (1987) The efficiency of systematic sampling in stereology and its prediction. J Microsc 147: 229–263.
23. West MJ, Slomianka L, Gundersen HJ (1991) Unbiased stereological estimation of the total number of neurons in the subdivisions of the rat hippocampus using the optical fractionator. Anat Rec 231: 482–497.
24. Amaral D, Witter M (1995) Hippocampal formation. In: Paxinos G, editor. The rat nervous system. San Diego: Academic Press. 443–493.
25. Gundersen HJ, Jensen EB, Kiêu K, Nielsen J (1999) The efficiency of systematic sampling in stereology-reconsidered. J Microsc 193: 199–211.
26. Shapiro SS, Wilk MB, Chen HJ (1968) A Comparative Study of Various Tests for Normality. J Am Stat Assoc 63: 1343–1372.
27. Gaarskjaer FB (1978) Organization of the mossy fiber system of the rat studied in extended hippocampi. I. Terminal area related to number of granule and pyramidal cells. J Comp Neurol 178: 49–72.
28. Buckmaster PS, Dudek FE (1997) Neuron loss, granule cell axon reorganization, and functional changes in the dentate gyrus of epileptic kainate-treated rats. J Comp Neurol 385: 385–404.
29. Gilby KL, Da Silva AG, McIntyre DC (2005) Differential GABA(A) subunit expression following status epilepticus in seizure-prone and seizure-resistant rats: a putative mechanism for refractory drug response. Epilepsia 46 Suppl 5: 3–9.
30. McIntyre DC, Gilby KL (2007) Genetically seizure-prone or seizure-resistant phenotypes and their associated behavioral comorbidities. Epilepsia 48 Suppl 9: 30–32.
31. Langer M, Brandt C, Löscher W (2011) Marked strain and substrain differences in induction of status epilepticus and subsequent development of neurodegeneration, epilepsy, and behavioral alterations in rats. Epilepsy Res 96: 207–224.
32. Gilby KL, Sydserff S, Patey AM, Thorne V, St-Onge V, et al. (2009) Postnatal epigenetic influences on seizure susceptibility in seizure-prone versus seizure-resistant rat strains. Behav Neurosci 123: 337–346.
33. Brandt C, Gastens AM, Sun M, Hausknecht M, Löscher W (2006) Treatment with valproate after status epilepticus: effect on neuronal damage, epileptogenesis, and behavioral alterations in rats. Neuropharmacology 51: 789–804.
34. Müller CJ, Gröticke I, Bankstahl M, Löscher W (2009) Behavioral and cognitive alterations, spontaneous seizures, and neuropathology developing after a pilocarpine-induced status epilepticus in C57BL/6 mice. Exp Neurol 219: 284–297.
35. Cardoso A, Lukoyanova EA, Madeira MD, Lukoyanov NV (2011) Seizure-induced structural and functional changes in the rat hippocampal formation: comparison between brief seizures and status epilepticus. Behav Brain Res 225: 538–546.
36. Szyndler J, Wierzba-Bobrowicz T, Skórzewska A, Maciejak P, Walkowiak J, et al. (2005) Behavioral, biochemical and histological studies in a model of pilocarpine-induced spontaneous recurrent seizures. Pharmacol Biochem Behav 81: 15–23.
37. Sayin U, Sutula TP, Stafstrom CE (2004) Seizures in the developing brain cause adverse long-term effects on spatial learning and anxiety. Epilepsia 45: 1539–1548.
38. Lynch M, Sayin U, Bownds J, Janumpalli S, Sutula T (2000) Long-term consequences of early postnatal seizures on hippocampal learning and plasticity. Eur J Neurosci 12: 2252–2264.
39. Gayoso MJ, Primo C, al-Majdalawi A, Fernandez JM, Garrosa M, et al. (1994) Brain lesions and water-maze learning deficits after systemic administration of kainic acid to adult rats. Brain Res 653: 92–100.
40. Barkas L, Redhead E, Taylor M, Shtaya A, Hamilton DA, et al. (2012) Fluoxetine restores spatial learning but not accelerated forgetting in mesial temporal lobe epilepsy. Brain 135: 2358–2374.
41. Hort J, Brozek G, Komárek V, Langmeier M, Mares P (2000) Interstrain differences in cognitive functions in rats in relation to status epilepticus. Behav Brain Res 112: 77–83.
42. Sloviter RS (1987) Decreased hippocampal inhibition and a selective loss of interneurons in experimental epilepsy. Science 235: 73–76.
43. Williams PA, White AM, Clark S, Ferraro DJ, Swiercz W, et al. (2009) Development of spontaneous recurrent seizures after kainate-induced status epilepticus. J Neurosci 29: 2103–2112.
44. Cavarsan CF, Queiroz CM, Dos Santos JG Jr, Xavier GF, Mello LE, et al. (2013) Reduced hippocampal dentate cell proliferation and impaired spatial memory performance in aged-epileptic rats. Front Neurol doi: 10.3389/fneur.2013.00106.
45. Tuunanen J, Lukasiuk K, Halonen T, Pitkänen A (1999) Status epilepticus-induced neuronal damage in the rat amygdaloid complex: distribution, time-course and mechanisms. Neuroscience 94: 473–495.
46. Chen S, Buckmaster PS (2005) Stereological analysis of forebrain regions in kainate-treated epileptic rats. Brain Res 1057: 141–152.
47. Cardoso A, Madeira MD, Paula-Barbosa MM, Lukoyanov NV (2008) Retrosplenial granular b cortex in normal and epileptic rats: a stereological study. Brain Res 1218: 206–214.
48. Olton DS (1983) Memory functions and the hippocampus. In: Seifert W, editor. Neurobiology of the hippocampus. New York: Academic Press. 335–374.
49. LeDoux JE (2000) Emotion circuits in the brain. Annu Rev Neurosci 23: 155–184.
50. Bannerman DM, Rawlins JN, McHugh SB, Deacon RM, Yee BK, et al. (2004) Regional dissociations within the hippocampus - memory and anxiety. Neurosci Biobehav Rev 28: 273–283.
51. Furtak SC, Wei SM, Agster KL, Burwell RD (2007) Functional neuroanatomy of the parahippocampal region in the rat: the perirhinal and postrhinal cortices. Hippocampus 17: 709–722.
52. Aggleton JP, O'Mara SM, Vann SD, Wright NF, Tsanov M, et al. (2010) Hippocampal-anterior thalamic pathways for memory: uncovering a network of direct and indirect actions. Eur J Neurosci 31: 2292–2307.
53. Kubová H, Mareš P (2013) Are morphologic and functional consequences of status epilepticus in infant rats progressive? Neuroscience 235: 232–249.
54. Paulsen O, Moser EI (1998) A model of hippocampal memory encoding and retrieval: GABAergic control of synaptic plasticity. Trends Neurosci 21: 273–278.
55. Vogt KE, Nicoll RA (1999) Glutamate and gamma-aminobutyric acid mediate a heterosynaptic depression at mossy fiber synapses in the hippocampus. Proc Natl Acad Sci U S A 96: 1118–1122.
56. Marr D (1971) Simple memory: a theory for archicortex. Philos Trans R Soc Lond B Biol Sci 262: 23–81.
57. Treves A, Rolls ET (1994) Computational analysis of the role of the hippocampus in memory. Hippocampus 4: 374–391.
58. O'Reilly RC, Bhattacharyya R, Howard MD, Ketz N (2011) Complementary Learning Systems. Cogn Sci doi: 10.1111/j.1551-6709.2011.01214.x.
59. Vazdarjanova A, Guzowski JF (2004) Differences in hippocampal neuronal population responses to modifications of an environmental context: evidence for distinct, yet complementary, functions of CA3 and CA1 ensembles. J Neurosci 24: 6489–6496.
60. Leutgeb S, Leutgeb JK, Treves A, Moser MB, Moser EI (2004) Distinct ensemble codes in hippocampal areas CA3 and CA1. Science 305: 1295–1298.
61. Bakker A, Kirwan CB, Miller M, Stark CE (2008) Pattern separation in the human hippocampal CA3 and dentate gyrus. Science 319: 1640–1642.
62. Akaike K, Tanaka S, Tojo H, Fukumoto S, Imamura S, et al. (2001) Kainic acid-induced dorsal and ventral hippocampal seizures in rats. Brain Res 900: 65–71.
63. Bonilha L, Alessio A, Rorden C, Baylis G, Damasceno BP, et al. (2007) Extrahippocampal gray matter atrophy and memory impairment in patients with medial temporal lobe epilepsy. Hum Brain Mapp 28: 1376–1390.

6

Genetic Knock-Down of *Hdac3* Does Not Modify Disease-Related Phenotypes in a Mouse Model of Huntington's Disease

Lara Moumné[1], Ken Campbell[2], David Howland[3], Yingbin Ouyang[2], Gillian P. Bates[1]*

1 Department of Medical and Molecular Genetics, King's College London, London, United Kingdom, **2** Taconic, Cranbury, New Jersey, United States of America, **3** CHDI Management/CHDI Foundation, Princeton, New Jersey, United States of America

Abstract

Huntington's disease (HD) is an autosomal dominant progressive neurodegenerative disorder caused by an expansion of a CAG/polyglutamine repeat for which there are no disease modifying treatments. In recent years, transcriptional dysregulation has emerged as a pathogenic process that appears early in disease progression and has been recapitulated across multiple HD models. Altered histone acetylation has been proposed to underlie this transcriptional dysregulation and histone deacetylase (HDAC) inhibitors, such as suberoylanilide hydroxamic acid (SAHA), have been shown to improve polyglutamine-dependent phenotypes in numerous HD models. However potent pan-HDAC inhibitors such as SAHA display toxic side-effects. To better understand the mechanism underlying this potential therapeutic benefit and to dissociate the beneficial and toxic effects of SAHA, we set out to identify the specific HDAC(s) involved in this process. For this purpose, we are exploring the effect of the genetic reduction of specific HDACs on HD-related phenotypes in the R6/2 mouse model of HD. The study presented here focuses on HDAC3, which, as a class I HDAC, is one of the preferred targets of SAHA and is directly involved in histone deacetylation. To evaluate a potential benefit of *Hdac3* genetic reduction in R6/2, we generated a mouse carrying a critical deletion in the *Hdac3* gene. We confirmed that the complete knock-out of *Hdac3* is embryonic lethal. To test the effects of HDAC3 inhibition, we used *Hdac3*$^{+/-}$ heterozygotes to reduce nuclear HDAC3 levels in R6/2 mice. We found that *Hdac3* knock-down does not ameliorate physiological or behavioural phenotypes and has no effect on molecular changes including dysregulated transcripts. We conclude that HDAC3 should not be considered as the major mediator of the beneficial effect induced by SAHA and other HDAC inhibitors in HD.

Editor: Xiao-Jiang Li, Emory University, United States of America

Funding: This work was funded by CHDI Foundation, a privately-funded non-profit biomedical research organization exclusively dedicated to discovering and developing therapeutics that slow the progression of Huntington's disease. DH was involved in the design and generation of the Hdac3 KO mouse. With the exception of this, the funders had no role in study design, data collection and analysis, decision to publish, or preparation of the manuscript.

Competing Interests: DH is employed by CHDI Management, Inc. and YO and KC are employees of Taconic. There are no patents, products in development or marketed products to declare.

* E-mail: gillian.bates@kcl.ac.uk

Introduction

Huntington's disease (HD) is an autosomal dominant progressive neurodegenerative disorder with a mean age of onset of 40 years. The main clinical manifestations are chorea, cognitive impairment, psychiatric disorders and weight loss. The disease duration is 15–20 years and in the absence of disease modifying treatments, the disease progresses inexorably until death [1]. The mutation responsible for HD is an unstable expansion of a CAG repeat in the *HTT* gene that leads to a polyglutamine expansion in the N-terminus of the huntingtin (HTT) protein [2]. Neuropathologically, HD is characterized by neuronal loss in several brain regions including the striatum and the cortex as well as the deposition of nuclear and cytoplasmic HTT-containing aggregates [3].

A variety of mouse models have been used to study the pathogenic pathways involved in HD [4]. These include the R6/2 model, which is transgenic for a single-copy of exon 1 of human *HTT*, which in our colony carries approximately 200 CAG repeats [5,6] and the *Hdh*Q150 knock-in model in which 150 CAGs have been inserted into the mouse *Htt* gene [7,8]. The R6/2 mouse has an early onset progressive phenotype that recapitulates many features of the human disease. Motor and cognitive impairment appear before 6 weeks, HTT aggregation can clearly be detected from 3 weeks, whereas neuronal cell loss in the striatum occurs at later stages [9,10,11]. Mice with an average 200 CAG repeats are not usually kept beyond 15 weeks. The early and reproducible phenotype of this mouse line has made it an ideal model screening compounds and performing genetic crosses. At late-stage disease, the R6/2 and *Hdh*Q150 models have developed remarkably similar phenotypes [8,12,13,14] supporting the hypothesis that the generation of a toxic fragment is a critical event in HD pathogenesis [15].

In recent years, transcriptional dysregulation has emerged as a pathogenic process that appears early in disease progression and has been recapitulated across multiple HD model systems including the R6/2 and *Hdh*Q150 mouse models [12,16]. Histone acetylation/deacetylation is an important mechanism involved in gene transcriptional regulation and altered histone acetylation has been proposed to underlie transcriptional dysregulation in HD [17]. In general, acetylation of lysines by histone acetyl transferases (HAT) leads to activation of transcription whereas removal of acetyl groups by histone deacetylases (HDAC) leads to gene silencing [18]. There are two evolutionary divergent types of HDACs: the zinc-dependent HDACs (human HDAC1-11) and the NAD^+ dependent HDACs also called sirtuins (human SIRT1-7). Inhibitors of the zinc-dependant HDACs include suberoylanilide hydroxamic acid (SAHA) and trichostatin A (TSA) [19], whereas sirtuins are inhibited by nicotinamide [20]. The human zinc-dependent HDAC family comprises eleven members divided into four classes: I (HDAC1, 2, 3 and 8), IIa (HDAC4, 5, 7 and 9), IIb (HDAC6 and 10) and IV (HDAC11). HDAC inhibition has been shown to improve polyglutamine-dependent phenotypes in several models of HD [21,22,23,24,25,26]. Although administration of SAHA improves HD-related phenotypes in the R6/2 mouse model, therapeutic doses have toxic effects and induce profound weight loss [24].

To better understand the mechanism underlying this potential therapeutic benefit and to dissociate the beneficial and toxic effects of SAHA, we set out to identify the specific HDAC(s) involved in this process. For this purpose, we are exploring the effect of the genetic reduction of specific HDACs on HD-related phenotypes in R6/2 mice. Genetic reduction (knock-down or complete knock-out) of *Hdac5, 6, 7 and 9* failed to induce a phenotypic improvement ([27,28] and unpublished data) whereas knock-down of *Hdac4* induces a significant beneficial effect (unpublished data).

The study presented here focuses on HDAC3, which is the most highly expressed class I HDAC in the brain [29]. This HDAC is of particular interest for several reasons. Class I HDACs are directly involved in histone deacetylation and as a class I HDAC, HDAC3 is one of the main cellular targets of SAHA [30]. A recent study showed that the class I inhibitor HDACi **4b**, which is reported to be more specific for HDAC3 than the other class I HDACs, ameliorated the disease phenotype and reversed many of the transcriptional abnormalities found in the brain of R6/2 mice [26]. Moreover, studies involving genetic reduction of specific HDACs in invertebrate models of HD have implicated class I HDACs in the reduction of polyglutamine-dependent toxicity. In *Drosophila melanosgaster*, knock-down of the class I HDAC *rpd3*, that is homologous to human *HDAC1/2* but also partially homologous to *HDAC3*, suppresses pathogenesis [31]. Similarly, in *C. Elegans*, only the knock-down of the *hda-3*, that shares homology with human class I HDACs, suppressed polyglutamine-dependant toxicity [21]. Finally, the catalytic domain of HDAC4 interacts with HDAC3 via the transcriptional corepressor complex N-CoR/SMRT, and suppression of HDAC4 binding to SMRT/N-CoR and to HDAC3 results in the loss of enzymatic activity associated with HDAC4 [32]. Given the beneficial effect of *Hdac4* knock-down on HD-related phenotypes in R6/2 mice, we might expect that a reduction of *Hdac3* expression would lead a reduced HDAC4 activity and an improvement in R6/2 phenotypes.

To evaluate a potential benefit of *Hdac3* genetic reduction in R6/2, we generated a genetically engineered mouse in which part of the *Hdac3* gene is deleted. We observed that a complete knock-out of *Hdac3* is embryonic lethal. *Hdac3* mRNA levels were reduced to 50% of wild type (WT) in the brains of *Hdac3*$^{+/-}$ heterozygotes (knock-down) with a corresponding reduction in HDAC3 protein in the nucleus but not in the cytoplasm. We found that *Hdac3*$^{+/-}$ heterozygotes are viable and fertile, with no overt phenotype. We performed a genetic cross between R6/2 mice and *Hdac3*$^{+/-}$ heterozygotes and found that *Hdac3* knock-down does not ameliorate physiological or behavioural phenotypes in R6/2 mice, does not modulate HTT aggregation and has no effect on transcriptional dysregulation. We conclude that HDAC3 should not be considered as the major mediator of the beneficial effect induced by SAHA and other HDAC inhibitors in HD.

Results

Conventional heterozygous deletion of *Hdac3*

We generated a conventional null allele of *Hdac3* in order to evaluate whether a reduction in HDAC3 level has beneficial effects in the R6/2 mice. For this purpose, loxP sites were introduced upstream of exon 11 and within exon 15 by homologous recombination inducing a deletion covering exon 11 to 14 and the 5′ end of exon 15 (Fig. 1A). This mutation removes a part of the nuclear localization signal and a C-terminal region necessary for both deacetylase activity and transcriptional repression [33,34]. Heterozygous F1 mice were generated and deletion of *Hdac3* was confirmed at the genomic level by PCR and sequencing (Fig. 1B). We intercrossed *Hdac3*$^{+/-}$ mice in order to generate nullizygotes, but did not obtain any *Hdac3* knock-out offspring at birth or between embryonic stages E8.5 and E15.5. This result reveals that a complete lack of *Hdac3* induces an embryonic lethality prior to E8.5 and confirms previous studies showing that *Hdac3*$^{-/-}$ mice die before E9.5 [35,36]. Given this embryonic lethality we used *Hdac3*$^{+/-}$ heterozygotes for our study. These mice are viable and fertile, with no overt phenotype.

It was important first to confirm that the *Hdac3* expression level was reduced in *Hdac3*$^{+/-}$ mice as it has been previously shown that *Hdac1*, another class I HDAC, autoregulates its expression to wild-type (WT) levels in *Hdac1*$^{+/-}$ heterozygous mice [37]. It was also important to determine whether *Hdac3* levels might be regulated by the R6/2 transprotein. Therefore, we crossed R6/2 males to *Hdac3*$^{+/-}$ heterozygous females to generate WT, *Hdac3*$^{+/-}$, R6/2 and R6/2::*Hdac3*$^{+/-}$ double mutant (Dbl) mice and performed RT-qPCR on cDNA prepared from cortex, cerebellum and striatum of 15-week old mice. The RT-qPCR primer sequences were located within the deletion (at the junction between exon 14 and 15) so that only the WT allele can be detected. We observed a significant reduction of *Hdac3* mRNA to 50% of the WT level in all the brain regions irrespective of the presence of the R6/2 transprotein (Fig. 2A). In order to verify the specificity of *Hdac3* reduction we analyzed the expression level of the other *Hdacs* by RT-qPCR and showed that these were not changed in *Hdac3*$^{+/-}$ mice compared to WT (Fig. S1).

In order to see whether this mRNA reduction translates into a reduction at the protein level, we performed western blotting on whole brain lysates using an HDAC3 specific antibody. Surprisingly, the protein level was only reduced to 80% of the WT level (Fig. S2). As HDAC3 is present in both the nucleus and the cytoplasm in mammalian cells [33], we investigated whether HDAC3 levels might differ between these two intracellular compartments. For this purpose, we performed nuclear and cytoplasmic fractionation of whole brains from 4 week old mice. We first confirmed that HDAC3 is present in both the nucleus and cytoplasm in WT mouse brain (Fig. 2B). Analysis of brains from WT, *Hdac3*$^{+/-}$, R6/2 and Dbl mice revealed that cytoplasmic HDAC3 is not reduced in *Hdac3*$^{+/-}$ and Dbl brains as compared to WT and R6/2. In contrast, its expression is reduced to 60% of the WT level in *Hdac3*$^{+/-}$ heterozygotes irrespective of the presence of the R6/2 transgene (Fig. 2C and D). As we are primarily interested in the involvement of HDAC3 in transcriptional repression through histone deacetylation, a specific reduction of HDAC3 in the nucleus would be expected to be

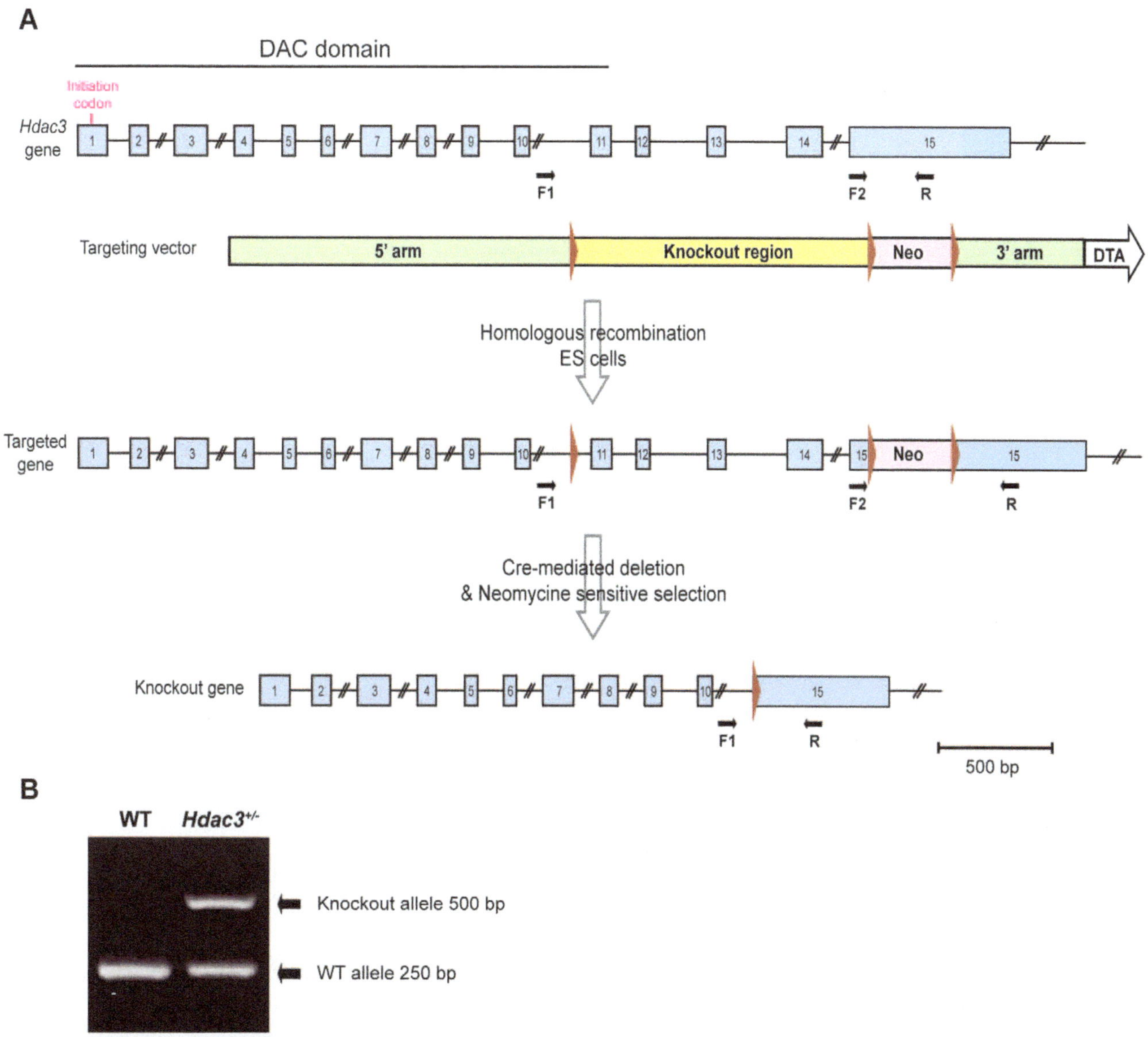

Figure 1. Generation of an *Hdac3* convention knock-out allele. (A) Strategy to generate an *Hdac3* conventional knock-out allele. The genomic structure and the targeting vector are shown. The *Hdac*3 gene contains 15 exons (blue rectangles). LoxP sites (red triangles) were introduced upstream exon 11 and within exon 15. The vector contains a 5′ homology arm covering the exonic and intronic region from intron 3–4 to intron 10–11 and a 3′ homology arm covering a part of exon 15 and the 3′UTR (green rectangle). The conditional knock-out region (yellow rectangle) covers exon 11 to 14 and 5′ end of exon 15. This conditional allele was introduced by homologous recombination in ES cells. The neomycine cassette (pink rectangle) flanked by 2 LoxP sites was removed by electroporation of Cre recombinase in ES cells and the cells containing the allele corresponding to a complete deletion of exon 11 to 14 were selected. Primers used for genotyping are represented as black arrows. F1 = forward 1; F2 = forward 2; R = Reverse. (B) Representative genotyping PCR on mouse genomic DNA. Duplex PCR with F1, F2 and R primers detects both the WT (250 bp band with primers F2/R) and the knock-out (500 bp with primers F1/R) allele.

sufficient to mimic inhibition of HDAC3 histone deacetylase activity and explore its potential benefit in HD.

Genetic reduction of *Hdac3* does not modify the R6/2 phenotype

A set of previously established quantitative tests was used to evaluate whether a genetic reduction of *Hdac3* had an effect on HD-related phenotypes in R6/2 mice. In order to generate mice for this analysis, R6/2 males were bred with *Hdac3*$^{+/-}$ females to generate at least 12 mice per genotype (males and females, WT n = 16, *Hdac3*$^{+/-}$ n = 14, R6/2 n = 12, Dbl n = 15). The mice were born over a period of 5 days and the 4 genotypes were recovered in a Mendelian ratio. The average CAG repeat size did not differ between the R6/2 group and the Dbl group (204±3.16 vs 205±0.91; p = 0.11). Body weight, RotaRod performance, grip strength and exploratory activity were monitored from 4 to 15 weeks of age, and in each case, a specific test was performed by the same operator, on the same day and at the same time during the weeks in which measurements were taken.

Mice were weighed weekly from 4 to 15 weeks of age. As expected, R6/2 mice weighed less overall than WT mice [$F_{(1,49)} = 32.657$, $p<0.001$] and gained weight at a slower rate [$F_{(4,539)} = 38.086$, $p<0.001$] (Fig. 3A). No significant differences in the overall weight [$F_{(1,49)} = 3.267$, $p = 0.077$] or in weight gain

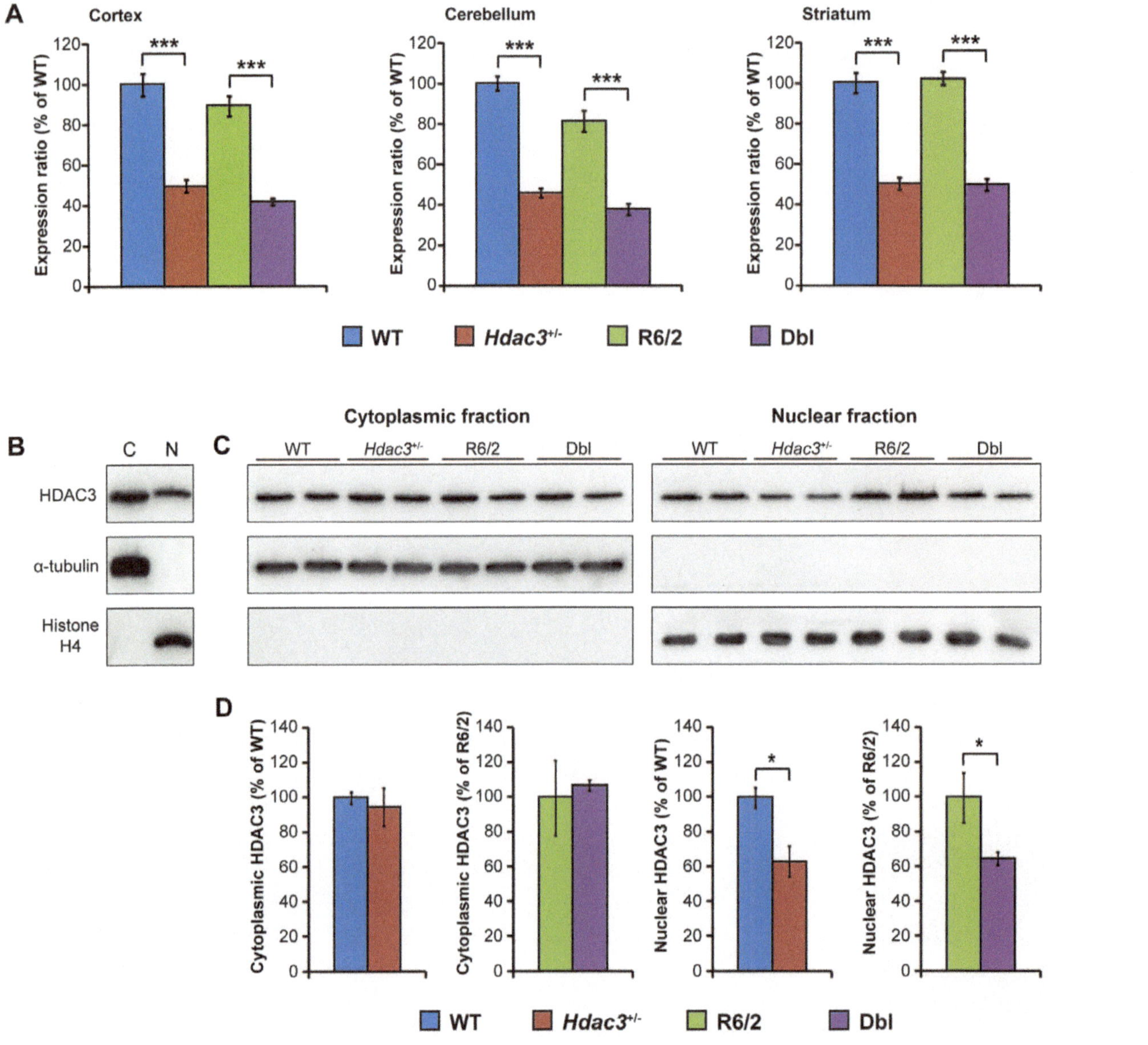

Figure 2. *Hdac3* mRNA and protein expression in *Hdac3+/−* heterozygous mouse brain. (A) *Hdac3* mRNA expression levels in 15 week old mouse cortex, cerebellum and striatum are shown as a relative expression ratio to the WT level. *Hdac3*$^{+/-}$ (red) and Dbl (purple) mice express the mRNA at 50% of the WT (blue) and R6/2 (green) levels in all brain regions. Error bars correspond to S.E.M. (n = 8) ***p<0.001. (B) Western blot showing the expression of HDAC3 protein in the cytoplasmic (C) and the nuclear (N) fraction of WT mouse whole brain. Antibodies to α-tubulin (cytoplasmic) and histone H4 (nuclear) were used to control for the purity of the fractions. (C) Representative western blot and (D) quantification of cytoplasmic and nuclear fraction prepared from WT (blue), *Hdac3*$^{+/-}$ (red), R6/2 (green) and Dbl (purple) 15 week old-mouse whole brains. Antibodies to α-tubulin (cytoplasmic) and histone H4 (nuclear) were used as both purity and loading controls. Cytoplasmic HDAC3 was not affected by *Hdac3* deletion whereas nuclear HDAC3 was reduced to 60% of the WT level. Error bars correspond S.E.M. (n = 3) *p<0.05.

[$F_{(4,539)} = 0.254$, $p = 0.887$] between WT and *Hdac3*$^{+/-}$ mice were observed. However, there was a trend for *Hdac3*$^{+/-}$ mice to weigh more than WT ($p = 0.077$). Genetic reduction of *Hdac3* had no effect on R6/2 weight [$F_{(1,49)} = 0.218$, $p = 0.643$] and did not attenuate the rate of R6/2 weight loss [$F_{(4,539)} = 1.458$, $p = 0.222$]. Therefore, genetic reduction of *Hdac3*$^{+/-}$ does not improve the weight loss phenotype in the R6/2 mice.

RotaRod performance is a sensitive indicator of balance and motor coordination, which has been reliably shown to decline in R6/2 mice. RotaRod was assessed at 4, 8, 10, 12 and 14 weeks of age. Consistent with previous data, R6/2 mice displayed a significant decline in overall RotaRod performance [$F_{(1,49)} = 75.523$, $p < 0.001$] and this worsened with age [$F_{(3,980)} = 23.764$, $p < 0.001$] (Fig. 3B). *Hdac3* knock-down had no effect on overall RotaRod performance of *Hdac3*$^{+/-}$ mice [$F_{(1,49)} = 0.386$, $p = 0.537$] and did not significantly change the overall performance of R6/2 mice [$F_{(1,49)} = 0.478$, $p = 0.493$]. Furthermore, *Hdac3* knock-down had no effect on the deterioration of R6/2 RotaRod performance with age [$F_{(3,980)} = 0.344$, $p = 0.792$].

Forelimb grip strength was assessed at 4 weeks of age and then weekly when the mice were 11 to 14 weeks old. Consistent with previous data, R6/2 mice performance was significantly decreased compared to WT [$F_{(1,49)} = 105.705$, $p < 0.001$] and deteriorated with age [$F_{(3,392)} = 25.804$, $p < 0.001$] (Fig. 3C). The grip strength

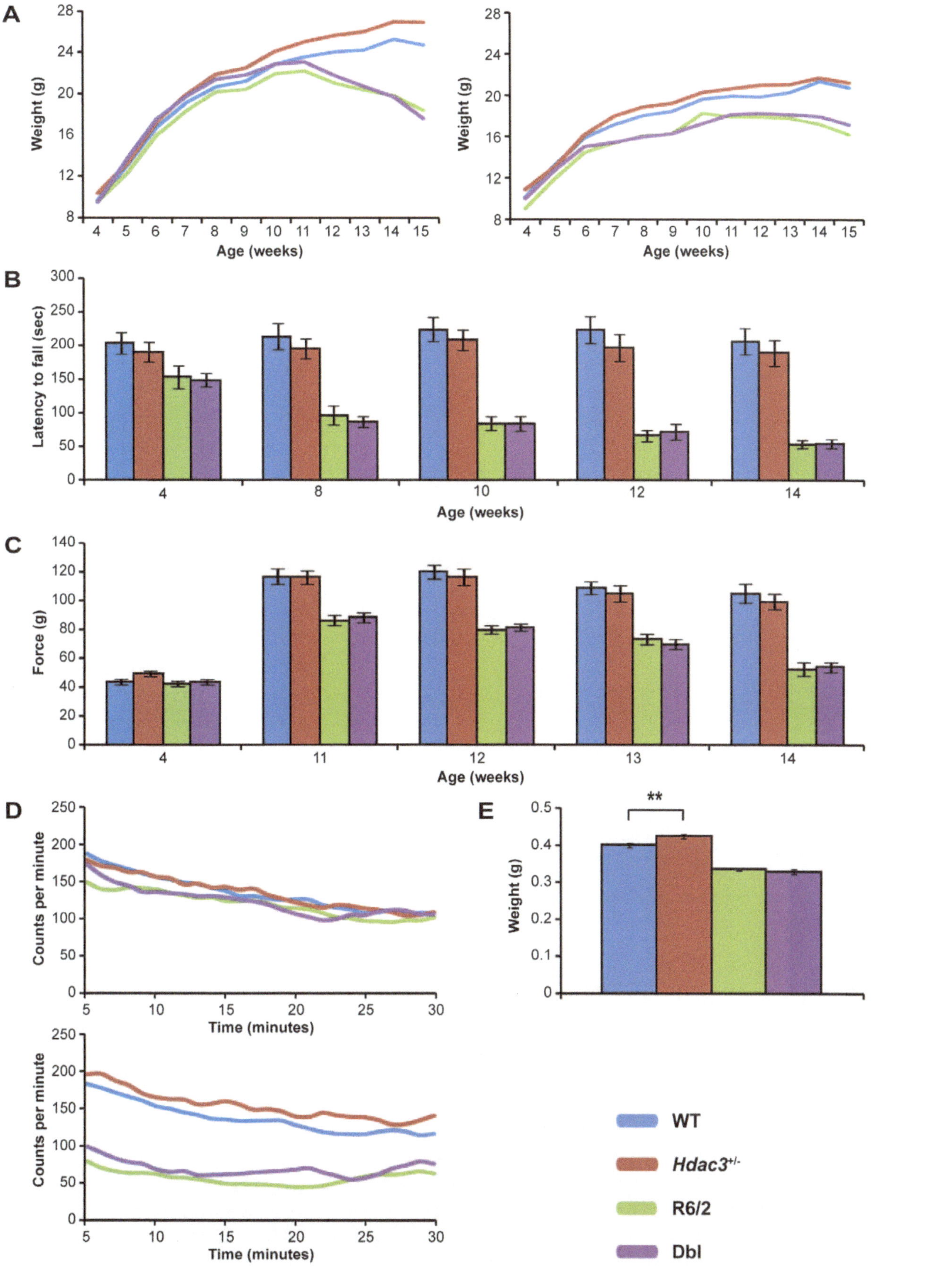

Figure 3. *Hdac3* genetic reduction does not modify R6/2 phenotypes. (A) Weight loss in males (left panel) and females (right panel) are shown between 4 and 15 weeks of age. *Hdac3* genetic reduction did not induce a significant increase of the body weight in R6/2 (B) RotaRod performance is represented as the average latency to fall in each group at 4, 8, 10, 12 and 14 weeks. *Hdac3* genetic reduction did not ameliorate the

impairment in RotaRod performance in R6/2 (C) Average grip strength in each group is represented at 4, 11, 12, 13 and 14 weeks. *Hdac3* genetic reduction did not induce a significant improvement in the grip strength in R6/2 mice (D) Average activity for each genotype is shown at 5 (upper panel) and 13 (lower panel) weeks of age. *Hdac3* genetic reduction did not reverse the hypoactivity observed in R6/2 mice (E) Average brain weight for each group was measured at 15 weeks of age. *Hdac3* genetic reduction did not modify the brain weight loss in R6/2 but a slight increase in brain weight was observed in WT animals **$p<0.01$. Error bars correspond to SEM ($n>12$). The same color code (blue = WT; red = Hdac3; green = R6/2 and purple = Dbl) was used for all measured parameters.

of *Hdac3*$^{+/-}$ mice was comparable to WT mice, both overall [$F_{(1,49)} = 0.230$, $p = 0.634$] and over the course of the experiment [$F_{(3,392)} = 0.675$, $p = 0.586$]. Furthermore, genetic reduction of *Hdac3* did not improve R6/2 overall grip strength performance [$F_{(1,49)} = 0.603$, $p = 0.441$] or deterioration with age [$F_{(3,392)} = 0.846$, $p = 0.483$].

Exploratory activity was assessed fortnightly from 5 to 13 weeks of age as described previously and analyzed by repeated measures general linear model (GLM) ANOVA. Mice were assessed for a period of 30 min for total activity, rearing, centre rearing and mobility. The *p*-values obtained for each parameter are displayed in Table S1. For all genotypes, the mice are highly active and mobile during the first part of the assessment period and this decreased over the course of the 30 min (Fig. 3D and Fig. S3). R6/2 mice show an overall hypoactivity and decreased mobility relative to WT mice from 7 weeks onwards. Rearing and centre rearing are also significantly decreased in R6/2 mice from 9 and 11 weeks of age respectively. *Hdac3*$^{+/-}$ mice were indistinguishable from WT mice for all of the parameters. Furthermore, *Hdac3* genetic reduction does not improve hypoactivity in the R6/2 mice.

Finally, we evaluated whether *Hdac3* genetic reduction had an effect on brain weight. Loss of brain weight has been observed in R6/2 mice in previous studies [6]. Brains were harvested from mice at 15 weeks of age, at the end of the trial. As expected, R6/2 mice displayed a significant decrease in brain weight [$F_{(1,49)} = 182.333$, $p<0.001$] (Fig. 3E). ANOVA analysis failed to detect any significant change in brain weight as a result of *Hdac3* genetic reduction [$F_{(1,49)} = 1.865$, $p = 0.178$]. Analysis of the results by Student's *t*-test revealed that *Hdac3* knock-down induces a significant increase in brain weight in the WT context ($p = 0.0074$) but no significant change was observed in the transgenic context (Fig. 3E). The brain weight of the Dbl mice is not significantly different from R6/2 ($p = 0.3857$). This result suggests that *Hdac3* knock-down has no effect on R6/2 mice brain weight loss.

Hdac3 genetic reduction does not ameliorate the dysregulated expression of genes of interest in R6/2 mouse brains

HDAC inhibitors have been shown to reverse transcriptional dysregulation in wide variety of HD models [17,26,38]. In order to evaluate whether a genetic reduction of *Hdac3* might modulate the transcriptional changes observed in HD, we used RT-qPCR to analyze the expression level of a set of 'genes of interest' in different brain regions. This included cortical *Bdnf* as well as cerebellar and striatal genes that are dysregulated in HD mouse models [8,12]. This analysis was performed on brain regions taken from 15 week old WT, *Hdac3*$^{+/-}$, R6/2 and Dbl mice. Changes in gene expression induced by the transgene were consistent with previous results for all the genes that were tested (Fig. 4A, B, C). Genetic reduction of *Hdac3* had no effect on the levels of the dysregulated transcripts in R6/2 brain regions except for *Bdnf* V (decreased expression, $p = 0.037$) and *Cnr1* (increased expression, $p = 0.028$) for which very mild but statistically significant changes were observed in the cortex and in striatum respectively (Fig. 4A and C). Similarly *Hdac3* genetic reduction had no effect on the expression of these genes in non-transgenic animals except for *Cnr1* for which a significant decrease in expression was observed in *Hdac3*$^{+/-}$ compared to WT striata ($p = 0.012$; Fig. 4C). We also found that *Hdac3* reduction does not affect the expression of the R6/2 transgene in the cortex (Fig. 4D), cerebellum (Fig. 4E) and striatum (Fig. 4F).

Hdac3 genetic reduction does not reduce Huntingtin aggregation in R6/2 mouse brains

HTT aggregation is a hallmark of brain pathology in HD and has been consistently observed in all HD models including the R6/2 mouse. Aggregation can be detected in several R6/2 brain regions from 3–4 weeks of age and increases with age [11]. In order to evaluate whether *Hdac3* reduction has an effect on this aggregation, we used a SEPRION ligand-based ELISA that provides a highly quantitative assay for measuring HTT aggregation in mouse brains [14]. Aggregates were captured from proteins extracted from the cortex, hippocampus and brain stem of 4, 9 and 15 week-old mice and detected with the MW8 antibody [39]. As expected, HTT aggregation was detectable at 4 weeks in all three brain regions, increased with age and this was not modified by *Hdac3* reduction. Levels of soluble HTT in these lysates can be quantified by western blot using the HTT specific antibody S830 [40]. The soluble trans-protein migrates at around 95 kDa whereas the aggregated protein remains in the stacking gel. Fig. 5B shows the results that we obtained for hippocampi from mice of 4, 9 and 15 weeks of age. The soluble fraction decreased with age in both R6/2 and Dbl mice and did not differ significantly between these genotypes at any time point (Fig. 5C). Similarly, a qualitative difference of the aggregated fraction was not detected. Taken together, the data obtained from the ELISA and western blotting revealed that *Hdac3* genetic reduction has no effect on HTT aggregation.

Discussion

HDAC inhibitors were first investigated as therapeutics in the context of HD to counteract transcriptional dysregulation, thought to contribute to the molecular pathogenesis of HD. They were first shown to be protective in a *Drosophila melanogaster* model [25] and following on from the demonstration that SAHA, a potent HDAC inhibitor, improves HD-related phenotypes in the R6/2 mouse, (Hockly et al., 2003), protective effects of the butyrates in HD mouse models were also reported [22,23]. However, therapeutic doses of SAHA were not well tolerated, leading to a marked weight loss in both WT and R6/2 animal. Therefore, if HDAC inhibitors are to be further developed as therapeutics for HD, it is essential that efficacy can be uncoupled from toxicity as far as is possible. As a pan-HDAC inhibitor, SAHA targets the eleven zinc-dependant HDACs with a preference for class I HDACs (1, 2, 3 and 8) and the class IIb HDAC, HDAC6 [30]. Identifying more specific inhibitors that target the HDAC(s) responsible for the beneficial effects observed in HD mouse models might be one mechanism by which the toxic effects could be diminished. Toward this aim, we are systematically investigating the effects of genetically reducing the levels of individual HDAC enzymes in the R6/2 mouse. In this study, we focus on HDAC3 and show that the

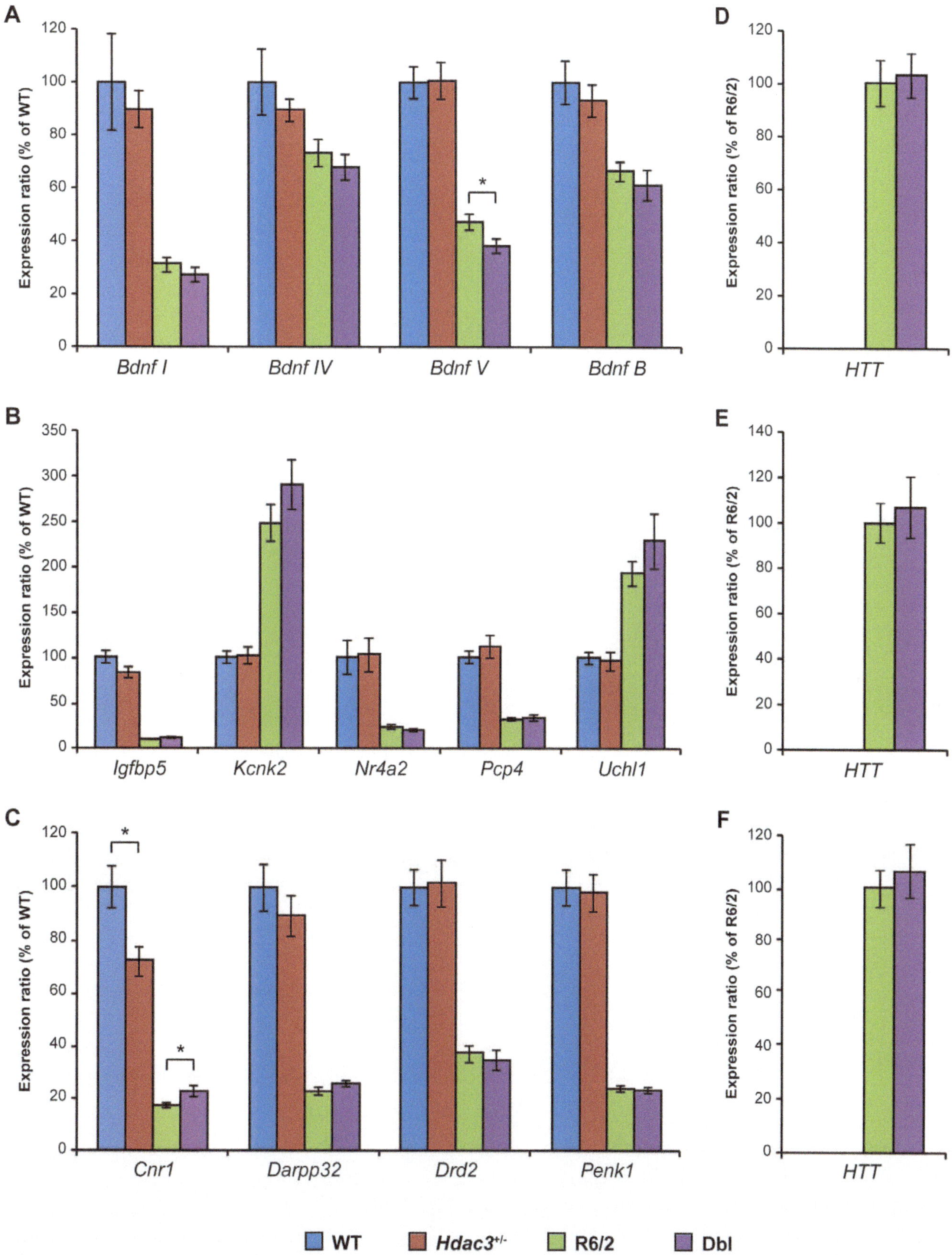

Figure 4. ***Hdac3* genetic reduction does not reverse transcriptional dysregulation in R6/2.** (A) Expression of *Bdnf* transcripts from different promoters (*Bdnf I, IV* and *V*) and the coding region (*Bdnf B*) in the cortex are represented as a percent of WT expression levels. With the exception of a slight decrease in *Bdnf* V, *Hdac3* reduction did not affect *Bdnf* expression. (B) Expression levels of genes specifically altered in the cerebellum of R6/2

mice are represented as a percent of WT expression. No significant difference was induced by *Hdac3* genetic reduction. (C) Expression levels of genes specifically altered in the striatum of R6/2 mice are represented as a percent of WT expression. A significant decrease in the expression of *Cnr1* in non-transgenic animals was observed as well as a slightly significant increase in *Cnr1* expression in R6/2 striata. Expression of the R6/2 transgene in Dbl brains is represented as a percent of that in R6/2 brains for cortex (D), cerebellum (E) and striatum (F). *Hdac3* reduction did not induce a significant change in transgene expression. Error bars correspond to S.E.M. (n = 8) *$p<0.05$. The same color code (blue = WT; red = Hdac3; green = R6/2 and purple = Dbl) was used for all the graphs. *Bdnf I, IV V*, brain derived neurotrophic factor promoter I, IV, V; *Bdnf B*, brain derived neurotrophic factor coding exon B; *Igfbp5*, insulin-like growth factor binding protein 5; *Kcnk2*, potassium channel subfamily K, member 2; *Nr4a2*, nuclear receptor subfamily 4, group A, member 2; *Pcp4*, Purkinje cell protein 4; *Uchl1*, ubiquitin C-terminal hydrolase L1; *Cnr1*, cannabinoid receptor 1; *Darpp32*, dopamine and cAMP regulated neuronal phosphoprotein; *Drd2*, dopamine D2 receptor; *Penk1*, proenkephalin.

reduction of nuclear HDAC3 to approximately 60% of WT levels has no effect on HD-related behavioral and physiological phenotypes or HTT aggregation. Importantly, the expression levels of a specific set of dysregulated gene transcripts were not restored.

HDAC3 was an interesting candidate for a number of reasons. The specific HDACs involved in the therapeutic benefits observed in invertebrate models of HD have also been explored by systematic genetic invalidation-based studies. Although a direct comparison with the mammalian enzymes cannot be made, as there are only four HDACs in *Drosophila melanogaster* and eight in *Caenorhabditis elegans* [21,31] both studies pointed to the class I HDACs as being the critical mediators of the beneficial effects. In *Drosophila melanogaster*, *rpd3* knock-down suppressed neurodegeneration and extended the survival of transgenic flies [31]. *Rpd3* is a class I HDAC that shares amino-acid sequence similarities with human *HDAC1*, *2* and *3* at 82%, 76% and 57% respectively. In *Caenorhabditis elegans*, of the eight HDAC enzymes, it was only knock-down of *Hda-3* that induced a neuroprotective effect, whereas knock-down of other HDACs, including the other class 1 orthologues *Hda-1* and *Hda-2*, exacerbated toxicity [21]. However, it was a pharmacological study in the R6/2 mouse model of HD that indicated that HDAC3 rather than the other class I HDACs might be the most relevant in a mammalian system. HDACi **4b** is one of a series of inhibitors for which the preferred cellular target has been reported to be HDAC3 [41]. HDACi **4b** was shown to improve motor performance and body weight of R6/2 mice, associated with a reversal of histone H3 hypoacetylation and the correction of specific brain mRNA levels [26]. Given that we have found that the genetic reduction of HDAC4 has beneficial effects in the R6/2 mouse (unpublished data) and that the enzyme activity of HDAC4 is dependent on its presence in the multi-protein complex that contains HDAC3 and the SMRT/N-CoR transcriptional repressors [32], HDAC3 was a therapeutic target that merited further investigation.

Although HDAC3 is the most highly expressed HDAC in the brain [29] little is known about its function in the central nervous system. Specific silencing of *Hdac3* in the hippocampus has been shown to lead to long-term memory enhancement [42]. The overexpression of *Hdac3* in rat cerebellar neurons induces neuronal death whereas suppression of its expression by shRNA protects against low-potassium-induced neuronal death [43]. Unlike HDAC1 and 2, which are predominantly nuclear proteins, HDAC3 contains a nuclear export signal and is present in both the nucleus and the cytoplasm in mammalian cells [33,44]. Investigations into the functions of the class I HDACs have been confined to their roles in the nucleus: HDACs 1 and 2 are part of the CoREST/mSin3a/NuRD repressor complex whereas HDAC3 interacts with NCoR/SMRT to repress gene expression [44], and in cortical neurons, HDAC3 is likely to be important for SMRT nuclear retention [45]. We were surprised to find that the cytoplasmic levels of HDAC3 in the brains of $Hdac3^{+/-}$ heterozygous mice were equivalent to that in WT brains, whereas nuclear levels were reduced to approximately 60% of WT. The fact that cytoplasmic HDAC3 levels are maintained in the $Hdac3^{+/-}$ heterozygous state suggests that cytoplasmic HDAC3 may be tightly bound in a protein complex that sequesters HDAC3 and prevents it from migrating to the nucleus.

Our inability to detect any beneficial consequences of reducing nuclear HDAC3 levels in R6/2 mice was disappointing. The fact that $Hdac3^{-/-}$ nullizygous mice die early in embryogenesis indicates that another protein cannot compensate for HDAC3 function. We might expect that reduction of nuclear HDAC3 to approximately 60% of WT levels could be similar to the level of inhibition achieved by pharmacological approaches. Therefore, the $Hdac3^{+/-}$ heterozygous mice present a suitable model with which to further investigate the therapeutic potential of reducing HDAC3 nuclear function. Our negative results would appear to contradict the report that the administration of HDACi **4b** to R6/2 mice improves transcriptional dysregulation as well as behavioural phenotypes [26]. However these previously reported beneficial effects are surprising in the light of the physical stability and metabolic issues concerning HDACi **4b** that have recently been highlighted (Beconi et al. unpublished data). Taken together, the current data raise doubts as to the validation of HDAC3 inhibition as a therapeutic target for HD.

Materials and Methods

Generation of an *Hdac3* null allele

The *Hdac3*-targeting vector was generated by PCR and standard cloning. The 5′ homology arm (4.7 kb), the 3′ homology arm (4.9 kb) and the knock-out region (4.5 kb) were generated by PCR using high fidelity Prime Star HS DNA polymerase (Takara) and RP23-24H23 BAC clone as a template. Aside from the homologous arms, the final vector (HighQ1B-mHdac3) also contains loxP sequences flanking the knock-out region (4.5 kb), a neomycine resistance cassette (Neo, for positive selection of the ES cells), and a diphtheria toxin gene cassette (DTA, for negative selection of the ES cells). The targeting vector was linearized with NotI and electroporated into C57BL/6NTac ES cells (an ES cell line derived from C57BL/6 Taconic mice). The cells were selected with 200 ug/ml G418 and 192 neo resistant ES clones were screened for homologous recombination by Southern blot analysis. Targeted ES cells were electroporated with the Cre-recombinase and G418 sensitive clones were picked and analyzed by Southern blot. Clones carrying a deletion of Exon 11 to 14 were identified and microinjected into the blastocysts of C57BL/6 females to generate chimeric mice. Chimeras were bred to C57BL/6 females to achieve germ-line transmission and generate heterozygotes. The deletion was confirmed at the genomic level by PCR and sequencing.

Mouse maintenance and breeding

All animal work was approved by the King's College London Ethical Review Panel and experimental procedures were performed in accordance with the UK Home Office regulations.

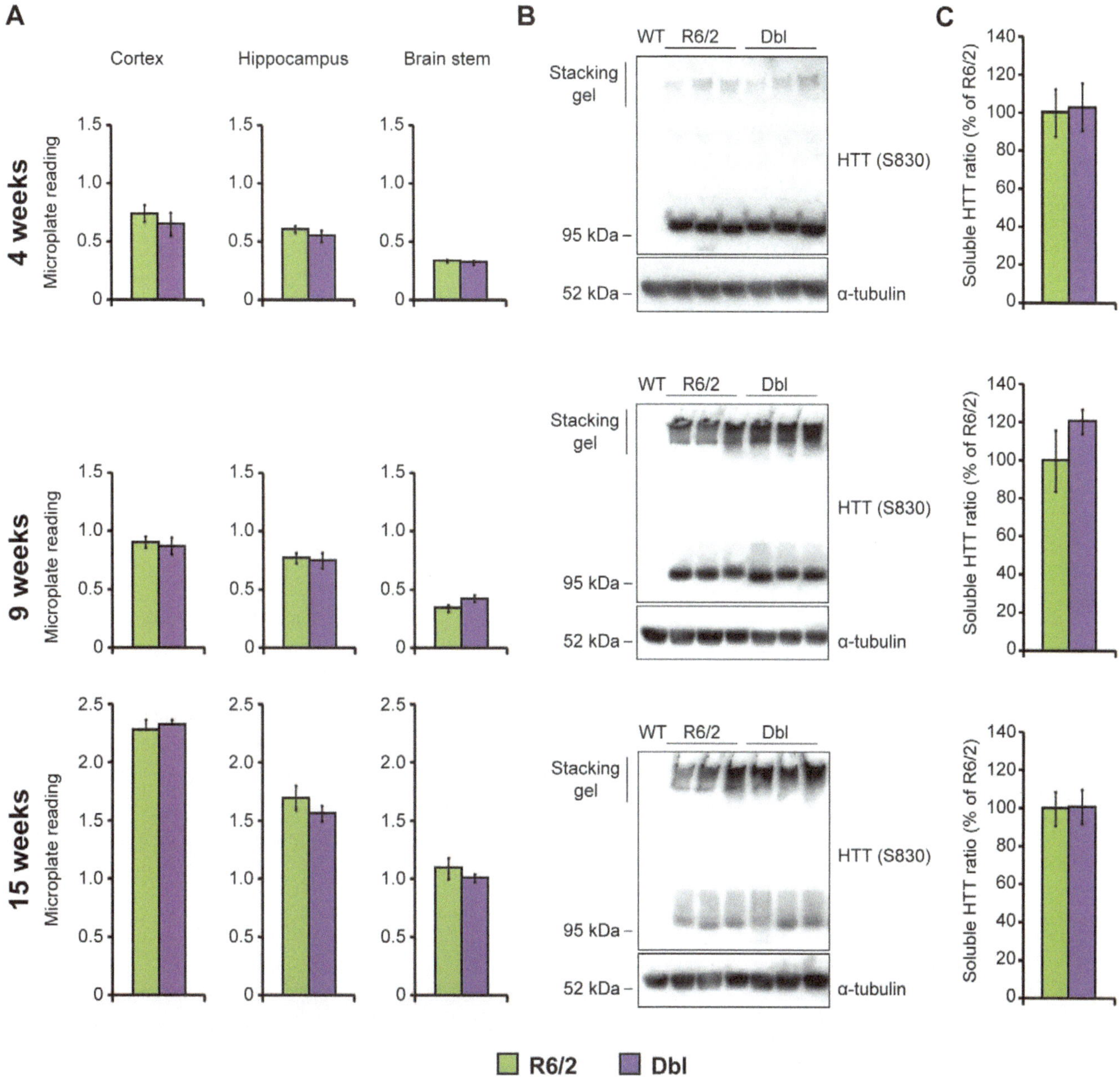

Figure 5. Hdac3 genetic reduction does not reduce HTT aggregation in R6/2 mouse brain. (A) The SEPRION ligand based ELISA assay was used to quantify HTT aggregation in the cortex, hippocampus and brain stem of 4, 9 and 15 week-old mice. The graphs represent the microtitre-plate reading of R6/2 (green) and Dbl (purple) lysates. Background readings obtained with WT and *Hdac3* lysates were comparable to water. Aggregation levels augment with age but are not modified by *Hdac3* reduction. Error bars correspond to S.E.M. (n>6). (B) Representative western blot of hippocampal lysates at 4, 9 and 15 weeks of age. The aggregated HTT fraction (stacking gel) augments with age whereas the soluble fraction decreases with age. α-tubulin was used as a loading control. (C) Quantification of (B). Soluble HTT is represented as a percentage of the soluble fraction in R6/2. Error bars correspond to S.E.M. (n = 6). The same color code (R6/2 = green; Dbl = purple) is used in (A) and (C).

Hemizygous R6/2 mice were bred and reared in our colony by backcrossing R6/2 males to (CBA×C57Bl/6) F1 (CBF) females (B6CBAF1/OlaHsd, Harlan Olac). *Hdac3*$^{+/-}$ heterozygotes originated in a C57Bl/6 background and were backcrossed once to CBF females before breeding to R6/2 males. All animals had unlimited access to water and breeding chow (Special Diet Services, Witham, UK). Housing conditions and environmental enrichment were used as previously described [46]. *Hdac3*$^{+/-}$ females were crossed to R6/2 males to generate WT, *Hdac3*$^{+/-}$, R6/2 and Dbl mice. All cages contained at least one mouse of each genotype and mice were additionally given mash that consist of powdered chow mixed with water from 4 to 15 weeks. Mice were subject to a 12 h light/12 h dark cycle.

Genotyping of R6/2 and *Hdac3*$^{+/-}$ mice and CAG repeat sizing

R6/2 mice were genotyped by PCR of tail-tip genomic DNA. PCR was performed using 1× Thermo-start master mix (Thermo scientific), 10% DMSO, 10 ng/ul forward primer 33727 (5′-CGCAGGCTAGGGCTGTCAATCATGCT-3′), 10 ng/ul reverse primer 32252 (5′-TCATCAGCTTTTCCAGGGTCGC-CAT-3′) and 100 ng genomic DNA. Amplification conditions were: 15 min at 94°C, 35× (30 s at 94°C; 30 s at 60°C; 60 s at 72°C), 10 min at 72°C. The amplicon for the R6/2 transgene is 272 bp. The CAG repeat size was determined as follow: amplification of the CAG repeat from R6/2 mouse DNA was performed with a FAM labelled forward primer (5′-GA-GTCCCTCAAGTCCTTCCAGCA-3′) and reverse primer (5′-GCCCAAACTCACGGTCGGT-3′) with 0.2 mM dNTPs; 10% DMSO; AM buffer (67 mM TrisHCL pH 8.8; 16.6 mM $(NH_4)SO_4$; 2 mM $MgCl_2$; 0.17 mg/ml BSA) and 0.5 U AmpliTaq DNA polymerase (Applied Biosystems). Amplification conditions were: 90 s at 94°C, 24× (30 s at 94°C; 30 s at 65°C; 90 s at 72°C), 10 min at 72°C. All instruments and materials were obtained from Applied Biosystems unless indicated. The FAM-tagged PCR product (1 ul) together with MegaBACETM ET900 (AmershamBioscience) internal size standard (0.04 ul) were denatured for 5 min at 94°C in 9 ul of HiDi-formamide and analysed using an ABI3730 sequencer. Data analysis was performed using platemanager application GeneMapper v5.2- 3730XL. *Hdac3*$^{+/-}$ mice were genotyped by duplex PCR in a 10 ul reaction containing 100 ug gDNA, 2 ul of 5× Promega buffer, 1.2 ul of 25 mM MgCl2, 1 ul of 10× PCR Enhancer (Invitrogen), 1 ul of 2 mM dNTP, 1 ul of 10 uM forward 1 (F1) primer (5′- GCTTAGCCTACTTGGCAAGTGCCAG-3′), 1 ul of 10 uM forward 2 (F2) primer (5′-GGCCAGAAGCACCCAAT-GAGTTCTA-3′), 1 ul of 10 uM reverse (R) primer (5′- ACAAT-CATCAGGCCGTGAGAGTTTG-3′), 0.2 ul of Promega Taq DNA polymerase and 4.4 ul H_2O. Amplification conditions were: 10 min at 94°C, 40× (30 s at 94°C; 30 s at 58°C; 60 s at 72°C) 10 min at 72°C. The WT allele product is 250 bp and the *Hdac3* knock-out allele is 500 bp.

Phenotypic analysis

The phenotypic analysis was performed on mice from 4 weeks to 15 weeks of age (WT males n = 9, females n = 7; *Hdac3*$^{+/-}$ males n = 7, females n = 7; R6/2 males n = 5, females n = 7; Dbl males n = 10, females n = 5). Mice were weaned 4 days before the beginning of the analysis. Mice were weighed weekly to the nearest 0.1 g. Motor coordination was assessed using an Ugo Basile 7650 accelerating RotaRod with an acceleration set to run from 4 to 40 RPM in 300 seconds (Linton Instrumentation, UK). At 4 weeks of age, mice were tested on four consecutive days, with three trials per day. At 8, 10, 12 and 14 weeks of age, mice were tested on three consecutive days with three trials per day. Forelimb grip strength was measured once a week at 4 and from 11 to 14 weeks using a San Diego Instruments Grip Strength Meter (San Diego, CA, USA) as described in [24]. Exploratory spontaneous motor activity was recorded and assessed fortnightly at 5, 7, 9, 11 and 13 weeks of age for 30 min during the day using infra-red activity monitoring cages (Linton Instruments, AM1053) as described in [47]. Briefly, activity (total number of beam breaks in the lower level), mobility (at least two consecutive beam breaks in the lower level) and rearing (number of rearing beam breaks) were measured. The data were collected and analyzed as described in [47]. At 15 weeks of age mice were sacrificed, brains were extracted and weighed to the nearest 0.01 g.

RNA extraction and gene expression analysis

Brain regions were dissected, snap-frozen in liquid nitrogen and stored at −80°C until required. Cortices, cerebella and striata were homogenized in Qiazol reagent and extracted using Qiagen RNeasy Mini kit (Qiagen). After quantification with a Nanodrop 1000 spectrophotometer (Labtech international), 1 ug of total RNA was reverse transcribed in 20 ul RT reaction using MMLV-RT kit (Invitrogen) and P21 random hexamers as described in Benn et al, 2008. RT reaction was diluted 10 times in nuclease-free Sigma water and 5 ul was used in 25 ul reaction containing Precision MasterMix (PrimerDesign), 300 nM gene-specific primers and 200 nM fluorescent reporting probe (FAM labelled and TAMRA quenched) specific to each target gene of interest using the Opticon2 Real Time PCR detection system (MJ research). Housekeeping genes previously identified [48] were used as ready-mixed PerfectProbe sets containing primers and probe (PrimerDesign). The geometric mean of three housekeeping genes for each brain region (*Atp5b*, *Canx*, and *Ubc* for cortex; *Atp5b*, *Eif4a2*, and *Ubc* for cerebellum, *Atp5b*, *Ubc* and *Ywhaz* for striatum) was used as a reference in order to determine relative expression ratio of the genes of interest using $2^{-\Delta\Delta Ct}$ method. For *Hdac3* expression analysis, primers and probe were designed using Primer3 software at the junction between exon 14 and 15 so only the WT allele can be detected (forward primer 5′- CGACGCTGAAGAGAGA-GGTC -3′, reverse primer 5′- TTTCCTTGTCGTTGTC-ATGG-3′; probe 5′- CCGAGGAGAACTACAGCAGG-3′). The amplicon was sequence-verified. For the other target genes, primers and probes were used as in previous studies [28,48,49] and their sequences are available in Table S2.

Isolation of nuclear and cytoplasmic fractions from mouse brains

Brains were extracted from mice, snap-frozen in −50°C isopenthane and stored at −80°C until required. All procedures were performed on ice. Each brain was homogenized in 2 ml ice-cold buffer 1 (575 mM sucrose; 25 mM KCl; 50 mM Triethanolamine; 5 mM MgCl2; 5 mM DTT; 0.5 mM PMSF; 1 tablet complete protease inhibitor cocktail from Roche) with 5–10 strokes in a 5 ml ice-cold homogenizer. The homogenate was then centrifuged at 800 g for 15 min at 4°C. The supernatant corresponds to the cytoplasmic fraction. The pellet was then homogenized in 3 ml buffer 1 and 2 volumes (6 ml) of Buffer 2 (2.3 M sucrose; 24.8 mM KCl; 50 mM Triethanolamine; 5 mM MgCl2; 1 mM DTT; 0.5 mM PMSF; 1 tablet complete protease inhibitor cocktail from Roche) were added and mixed to the homogenate. The mix was then centrifuged at 124000 g for 1 hour at 4°C (SW40 rotor in Beckman ultracentrifuge). The pellet (nuclear fraction) was washed twice in Buffer 1 and finally resuspended in 100 ul of Buffer 1.

Seprion ligand ELISA assay

Brain regions were dissected, snap-frozen in liquid nitrogen and stored at −80°C until required. A 2.5% lysate was prepared by ribolysing tissues for 3×30 sec in Lysing matrix tubes (Lysing matrix D; MP Biomedicals) in ice-cold RIPA buffer (50 mM Tris-HCl pH 8.0; 120 mM NaCl; 1% Igepal; 3.125% sodium deoxycholate; 0.01% SDS; 1 mM β-mercaptoethanol; 1 uM PMSF; 1 mM DTT; 1 tablet complete protease inhibitor cocktail from Roche). Lysates were then stored at −80°C and used within 12 h. Fifteen microliters of homogenate was mixed to 3 ul 10% SDS, 62 ul H_2O and 20 ul 5× capture buffer (Microsens Biotechnologies). A total of 100 ul was transferred to the well of a Seprion ligand-coated ELISA plate, and incubated with shaking for 1 h at room temperature (RT). The well was then washed 5× in PBS-T (PBS; 0.1% Tween) and incubated shaking for 1 h at RT with 100 ul MW8 primary antibody (mouse monoclonal antibody [39]) diluted 1:3000 in conjugate buffer (150 mM NaCl; 4% BSA; 1% non-fat dried milk; 0.1% Tween 20 in PBS). After 5 washes with PBS-T, the well was incubated shaking for 45 min at RT with 100 ul horse radish peroxidase (HRP)-conjugated anti-mouse secondary antibody (DAKO) diluted 1:2000 in conjugate buffer. The well was then washed 5× in PBS-T and

100 ul of RT TMB substrate (SerTec) was added and incubated in the dark at RT for 5 min. Reactions were terminated by adding 100 ul 0.5 M HCl and the absorption at 450 nm was measured using a plate reader (Biorad).

Western blotting and antibodies

The protein concentration of the whole brain cytoplasmic fraction and RIPA extracted brain regions (ELISA) were determined using the BCA assay kit (Thermo scientific). Twenty micrograms of proteins were denatured in one volume of 2× sample buffer (125 mM Tris-HCl pH 6.8, 4% SDS, 20% Glycerol, 10% β-mercaptoethanol, 0.004% bromophenol blue) for 10 min at 95°C. Four microliters of the total 100 ul nuclear fraction was also denatured in one volume of 2× sample buffer for 10 min at 95°C. Lysates were then fractionated on 12% (nuclear and cytoplasmic fraction) or 10% (RIPA-extracted protein lysates) SDS-PAGE gels and electro-transferred into nitrocellulose membranes (Whatman) by submerged transfer apparatus (Bio-Rad) in (25 mM Tris, 192 mM glycine, 20% v/v methanol). Membranes were blocked for 1 h at RT in 5% non-fat dried milk in PBS-T (PBS; 0.2% Tween). Membranes containing nuclear and cytoplasmic fractions were then incubated with gentle agitation over night at 4°C with an HDAC3 antibody (rabbit polyclonal Abcam, 1/1000) or 20 min RT with a mouse monoclonal α-tubulin antibody (1/30000, Sigma) or rabbit polyclonal histone H4 antibody (1/5000, Millipore) in PBS-T, 5% non-fat dried milk. Membranes containing RIPA-extracted protein lysates were incubated 1 h RT with the sheep polyclonal S830 huntingtin antibody [40] diluted 1:2000 in PBS-T, 5% non-fat dried milk or α-tubulin antibody (20 min). For chemiluminescent detection, membranes were washed 3× in PBS-T probed with HRP-linked secondary antibodies (HRP conjugated anti-rabbit mouse or sheep antibody (1:10000, Dako) in PBS-T, 5% non-fat dried milk for 45 min at RT and washed 3× in PBS-T. Protein was detected by chemiluminescense (ECL reagent GE healthcare) according to the manufacturer's instructions. The signals were quantified using a GS-800 calibrated densitometer (Bio-Rad).

Statistical analysis

Statistical analysis was performed by Student's *t*-test (Excel or SPSS), one-way ANOVA, two-way ANOVA and repeated measures GLM ANOVA, with the Greenhouse–Geisser correction for non-sphericity using SPSS.

Supporting Information

Figure S1 *Hdac3* genetic reduction does not affect the expression of the other *Hdacs*. Expression of *Hdac1-11* transcripts are represented as a percent of WT expression levels in the cortex (A), the cerebellum (B) and the striatum (C) of 6 week old WT and $Hdac3^{+/-}$ mice. The level of *Hdac3* is the only significant difference between WT and $Hdac3^{+/-}$ brain regions. Error bars correspond to S.E.M. (n = 6) ***p<0.001. The same color code (blue = WT; red = Hdac3) was used for all the graphs.

Figure S2 HDAC3 protein expression in *Hdac3+/−* heterozygous mouse brain. (A) Representative western blot showing the expression of the HDAC3 protein in 4 week old mouse whole brains extracted with RIPA buffer. α-tubulin was used as a loading control (B) Quantification of (A). A slight significant decrease (≈20%) was induced by *Hdac3* genetic reduction. Error bars correspond to S.E.M. (n = 3) *p<0.05. Blue = WT; red = Hdac3.

Figure S3 *Hdac3* genetic reduction does not modify R6/2 exploratory activity. Average activity (left) rearing (middle) and mobility (right) for each genotype is shown at 5, 7, 9, 11 and 13 weeks of age. The same color code (blue = WT; red = $Hdac3^{+/-}$; green = R6/2 and purple = Dbl) was used for all the graphs. R6/2 mice show an overall hypoactivity and decreased mobility relative to WT mice from 7 weeks onwards and rearing is significantly decreased in R6/2 mice from 9 weeks of age. $Hdac3^{+/-}$ mice were indistinguishable from WT mice for all of the parameters. Genetic reduction of *Hdac3* failed to induce any improvement for these parameters in R6/2 mice.

Table S1 *Hdac3* genetic reduction does not modify R6/2 exploratory activity. The numbers displayed in Table S1 indicate the *p*-values for each of the parameters analysed (R6/2 genotype, *Hdac3* genotype, time in the activity cages) at 5, 7, 9, 11 and 13 weeks of age. Significant *p*-values are highlighted in yellow for $p<0.05$, orange for $p<0.01$ and pink for $p<0.001$. Mice in activity cages are shown to behave differently with respect to measures of exploratory behaviour (as determined by measuring activity, mobility and rearing) over a period of 30 min (Time). R6/2 mice exhibit an overall hypoactivity and a reduced mobility from 7 weeks and rearing is significantly reduced from 9 weeks (R6/2 genotype) and (Time*R6/2 genotype). *Hdac3* genetic reduction does not influence the behaviour of mice for any parameter assessed (*Hdac3* genotype) and (Time**Hdac3* genotype) and furthermore, there appears to be no interaction between *Hdac3* and R6/2 genotypes (R6/2**Hdac3*) and (R6/2**Hdac3**time).

Table S2 Sequences of primers and Taqman probes used in real-time PCR assays. *Bdnf I, IV V*, brain derived neurotrophic factor promoter I, IV, V; *Bdnf B*, brain derived neurotrophic factor coding exon B; *Cnr1*, cannabinoid receptor 1; *Darpp32*, dopamine and cAMP regulated neuronal phosphoprotein; *Drd2*, dopamine D2 receptor; *Hdac1-11*, histone deacetylase 1–11; *Htt*, huntingtin gene; *Igfbp5*, insulin-like growth factor binding protein 5; *Kcnk2*, potassium channel subfamily K, member 2; *Nr4a2*, nuclear receptor subfamily 4, group A, member 2; *Pcp4*, Purkinje cell protein 4; *Penk1*, proenkephalin; *Uchl1*, ubiquitin C-terminal hydrolase L1.

Acknowledgments

We thank Donna Groom and members of the Neurogenetics Laboratory for the provision of R6/2 males for breeding.

Author Contributions

Conceived and designed the experiments: LM KC DH YO GB. Performed the experiments: LM KC YO. Analyzed the data: LM KC YO GB. Contributed reagents/materials/analysis tools: DH. Wrote the paper: LM GB.

References

1. Bates G, Harper PS, Jones L (2002) Huntington's disease. Oxford: Oxford University Press.
2. The Huntington's Disease Collaborative Research Group (1993) A novel gene containing a trinucleotide repeat that is expanded and unstable on Huntington's

disease chromosomes. The Huntington's Disease Collaborative Research Group. Cell 72: 971–983.
3. DiFiglia M, Sapp E, Chase KO, Davies SW, Bates GP, et al. (1997) Aggregation of huntingtin in neuronal intranuclear inclusions and dystrophic neurites in brain. Science 277: 1990–1993.
4. Crook ZR, Housman D. Huntington's disease: can mice lead the way to treatment? Neuron 69: 423–435.
5. Mangiarini L, Sathasivam K, Seller M, Cozens B, Harper A, et al. (1996) Exon 1 of the HD gene with an expanded CAG repeat is sufficient to cause a progressive neurological phenotype in transgenic mice. Cell 87: 493–506.
6. Davies SW, Turmaine M, Cozens BA, DiFiglia M, Sharp AH, et al. (1997) Formation of neuronal intranuclear inclusions underlies the neurological dysfunction in mice transgenic for the HD mutation. Cell 90: 537–548.
7. Lin CH, Tallaksen-Greene S, Chien WM, Cearley JA, Jackson WS, et al. (2001) Neurological abnormalities in a knock-in mouse model of Huntington's disease. Hum Mol Genet 10: 137–144.
8. Woodman B, Butler R, Landles C, Lupton MK, Tse J, et al. (2007) The Hdh(Q150/Q150) knock-in mouse model of HD and the R6/2 exon 1 model develop comparable and widespread molecular phenotypes. Brain Res Bull 72: 83–97.
9. Carter RJ, Lione LA, Humby T, Mangiarini L, Mahal A, et al. (1999) Characterization of progressive motor deficits in mice transgenic for the human Huntington's disease mutation. J Neurosci 19: 3248–3257.
10. Lione LA, Carter RJ, Hunt MJ, Bates GP, Morton AJ, et al. (1999) Selective discrimination learning impairments in mice expressing the human Huntington's disease mutation. J Neurosci 19: 10428–10437.
11. Stack EC, Kubilus JK, Smith K, Cormier K, Del Signore SJ, et al. (2005) Chronology of behavioral symptoms and neuropathological sequela in R6/2 Huntington's disease transgenic mice. J Comp Neurol 490: 354–370.
12. Kuhn A, Goldstein DR, Hodges A, Strand AD, Sengstag T, et al. (2007) Mutant huntingtin's effects on striatal gene expression in mice recapitulate changes observed in human Huntington's disease brain and do not differ with mutant huntingtin length or wild-type huntingtin dosage. Hum Mol Genet 16: 1845–1861.
13. Moffitt H, McPhail GD, Woodman B, Hobbs C, Bates GP (2009) Formation of polyglutamine inclusions in a wide range of non-CNS tissues in the HdhQ150 knock-in mouse model of Huntington's disease. PLoS One 4: e8025.
14. Sathasivam K, Lane A, Legleiter J, Warley A, Woodman B, et al. (2010) Identical oligomeric and fibrillar structures captured from the brains of R6/2 and knock-in mouse models of Huntington's disease. Hum Mol Genet 19: 65–78.
15. Landles C, Sathasivam K, Weiss A, Woodman B, Moffitt H, et al. Proteolysis of mutant huntingtin produces an exon 1 fragment that accumulates as an aggregated protein in neuronal nuclei in Huntington disease. J Biol Chem 285: 8808–8823.
16. Cha JH (2007) Transcriptional signatures in Huntington's disease. Prog Neurobiol 83: 228–248.
17. Sadri-Vakili G, Bouzou B, Benn CL, Kim MO, Chawla P, et al. (2007) Histones associated with downregulated genes are hypo-acetylated in Huntington's disease models. Hum Mol Genet 16: 1293–1306.
18. Kuo MH, Allis CD (1998) Roles of histone acetyltransferases and deacetylases in gene regulation. Bioessays 20: 615–626.
19. Codd R, Braich N, Liu J, Soe CZ, Pakchung AA (2009) Zn(II)-dependent histone deacetylase inhibitors: suberoylanilide hydroxamic acid and trichostatin A. Int J Biochem Cell Biol 41: 736–739.
20. Avalos JL, Bever KM, Wolberger C (2005) Mechanism of sirtuin inhibition by nicotinamide: altering the NAD(+) cosubstrate specificity of a Sir2 enzyme. Mol Cell 17: 855–868.
21. Bates EA, Victor M, Jones AK, Shi Y, Hart AC (2006) Differential contributions of Caenorhabditis elegans histone deacetylases to Huntingtin polyglutamine toxicity. J Neurosci 26: 2830–2838.
22. Ferrante RJ, Kubilus JK, Lee J, Ryu H, Beesen A, et al. (2003) Histone deacetylase inhibition by sodium butyrate chemotherapy ameliorates the neurodegenerative phenotype in Huntington's disease mice. J Neurosci 23: 9418–9427.
23. Gardian G, Browne SE, Choi DK, Klivenyi P, Gregorio J, et al. (2005) Neuroprotective effects of phenylbutyrate in the N171-82Q transgenic mouse model of Huntington's disease. J Biol Chem 280: 556–563.
24. Hockly E, Richon VM, Woodman B, Smith DL, Zhou X, et al. (2003) Suberoylanilide hydroxamic acid, a histone deacetylase inhibitor, ameliorates motor deficits in a mouse model of Huntington's disease. Proc Natl Acad Sci U S A 100: 2041–2046.
25. Steffan JS, Bodai L, Pallos J, Poelman M, McCampbell A, et al. (2001) Histone deacetylase inhibitors arrest polyglutamine-dependent neurodegeneration in Drosophila. Nature 413: 739–743.
26. Thomas EA, Coppola G, Desplats PA, Tang B, Soragni E, et al. (2008) The HDAC inhibitor 4b ameliorates the disease phenotype and transcriptional abnormalities in Huntington's disease transgenic mice. Proc Natl Acad Sci U S A.
27. Bobrowska A, Paganetti P, Matthias P, Bates GP (2011) Hdac6 knock-out increases tubulin acetylation but does not modify disease progression in the R6/2 mouse model of Huntington's disease. PLoS One 6: e20696.
28. Benn CL, Butler R, Mariner L, Nixon J, Moffitt H, et al. (2009) Genetic knock-down of HDAC7 does not ameliorate disease pathogenesis in the R6/2 mouse model of Huntington's disease. PLoS One 4: e5747.
29. Broide RS, Redwine JM, Aftahi N, Young W, Bloom FE, et al. (2007) Distribution of histone deacetylases 1–11 in the rat brain. J Mol Neurosci 31: 47–58.
30. Salisbury CM, Cravatt BF (2007) Activity-based probes for proteomic profiling of histone deacetylase complexes. Proc Natl Acad Sci U S A 104: 1171–1176.
31. Pallos J, Bodai L, Lukacsovich T, Purcell JM, Steffan JS, et al. (2008) Inhibition of specific HDACs and sirtuins suppresses pathogenesis in a Drosophila model of Huntington's disease. Hum Mol Genet 17: 3767–3775.
32. Fischle W, Dequiedt F, Hendzel MJ, Guenther MG, Lazar MA, et al. (2002) Enzymatic activity associated with class II HDACs is dependent on a multiprotein complex containing HDAC3 and SMRT/N-CoR. Mol Cell 9: 45–57.
33. Yang WM, Tsai SC, Wen YD, Fejer G, Seto E (2002) Functional domains of histone deacetylase-3. J Biol Chem 277: 9447–9454.
34. Guenther MG, Barak O, Lazar MA (2001) The SMRT and N-CoR corepressors are activating cofactors for histone deacetylase 3. Mol Cell Biol 21: 6091–6101.
35. Bhaskara S, Knutson SK, Jiang G, Chandrasekharan MB, Wilson AJ, et al. Hdac3 is essential for the maintenance of chromatin structure and genome stability. Cancer Cell 18: 436–447.
36. Montgomery RL, Potthoff MJ, Haberland M, Qi X, Matsuzaki S, et al. (2008) Maintenance of cardiac energy metabolism by histone deacetylase 3 in mice. J Clin Invest 118: 3588–3597.
37. Schuettengruber B, Simboeck E, Khier H, Seiser C (2003) Autoregulation of mouse histone deacetylase 1 expression. Mol Cell Biol 23: 6993–7004.
38. Ryu H, Lee J, Olofsson BA, Mwidau A, Dedeoglu A, et al. (2003) Histone deacetylase inhibitors prevent oxidative neuronal death independent of expanded polyglutamine repeats via an Sp1-dependent pathway. Proc Natl Acad Sci U S A 100: 4281–4286.
39. Ko J, Ou S, Patterson PH (2001) New anti-huntingtin monoclonal antibodies: implications for huntingtin conformation and its binding proteins. Brain Res Bull 56: 319–329.
40. Sathasivam K, Woodman B, Mahal A, Bertaux F, Wanker EE, et al. (2001) Centrosome disorganization in fibroblast cultures derived from R6/2 Huntington's disease (HD) transgenic mice and HD patients. Hum Mol Genet 10: 2425–2435.
41. Xu C, Soragni E, Chou CJ, Herman D, Plasterer HL, et al. (2009) Chemical probes identify a role for histone deacetylase 3 in Friedreich's ataxia gene silencing. Chem Biol 16: 980–989.
42. McQuown SC, Barrett RM, Matheos DP, Post RJ, Rogge GA, et al. HDAC3 is a critical negative regulator of long-term memory formation. J Neurosci 31: 764–774.
43. Bardai FH, D'Mello SR. Selective toxicity by HDAC3 in neurons: regulation by Akt and GSK3beta. J Neurosci 31: 1746–1751.
44. Yang XJ, Seto E (2008) The Rpd3/Hda1 family of lysine deacetylases: from bacteria and yeast to mice and men. Nat Rev Mol Cell Biol 9: 206–218.
45. Soriano FX, Hardingham GE (2011) In cortical neurons HDAC3 activity suppresses RD4-dependent SMRT export. PLoS One 6: e21056.
46. Hockly E, Woodman B, Mahal A, Lewis CM, Bates G (2003) Standardization and statistical approaches to therapeutic trials in the R6/2 mouse. Brain Res Bull 61: 469–479.
47. Hockly E, Tse J, Barker AL, Moolman DL, Beunard JL, et al. (2006) Evaluation of the benzothiazole aggregation inhibitors riluzole and PGL-135 as therapeutics for Huntington's disease. Neurobiol Dis 21: 228–236.
48. Benn CL, Fox H, Bates GP (2008) Optimisation of region-specific reference gene selection and relative gene expression analysis methods for pre-clinical trials of Huntington's disease. Mol Neurodegener 3: 17.
49. Mielcarek M, Benn CL, Franklin SA, Smith DL, Woodman B, et al. (2011) SAHA decreases HDAC 2 and 4 levels in vivo and improves molecular phenotypes in the R6/2 mouse model of Huntington's disease. PLoS One in press.

7

Including Dominance Effects in the Genomic BLUP Method for Genomic Evaluation

Motohide Nishio*, Masahiro Satoh

NARO Institute of Livestock and Grassland Science, Tsukuba, Japan

Abstract

We evaluated the performance of GBLUP including dominance genetic effect (GBLUP-D) by estimating variances and predicting genetic merits in a computer simulation and 2 actual traits (T4 and T5) in pigs. In simulation data, GBLUP-D explained more than 50% of dominance genetic variance. Moreover, GBLUP-D yielded estimated total genetic effects over 1.2% more accurate than those yielded by GBLUP. In particular, when the dominance genetic variance was large, the accuracy could be substantially improved by increasing the number of markers. The dominance genetic variances in T4 and T5 accounted for 9.6% and 6.3% of the phenotypic variances, respectively. Estimates of such small dominance genetic variances contributed little to the improvement of the accuracies of estimated total genetic effects. In both simulation and pig data, there were nearly no differences in the estimates of additive genetic effects or their variance between GBLUP-D and GBLUP. Therefore, we conclude GBLUP-D is a feasible approach to improve genetic performance in crossbred populations with large dominance genetic variation and identify mating systems with good combining ability.

Editor: Xinping Cui, University of California, Riverside, United States of America

Funding: The authors have no support or funding to report.

Competing Interests: The authors have declared that no competing interests exist.

* E-mail: mtnishio@affrc.go.jp

Introduction

Genomic selection refers to the use of genome-wide dense single nucleotide polymorphism (SNP) markers to predict breeding values and subsequently select individuals [1]. Several approaches of genomic prediction have been presented. One of them is the genomic best linear unbiased prediction (GBLUP), which uses genomic information in the form of a genomic relationship matrix that defines the additive genetic covariance between individuals [2,3]. The genomic relationship coefficients are estimated with higher accuracy than when using pedigree information because genomic information can capture of Mendelian sampling across the genome. GBLUP has become popular approach in genomic selection of dairy cattle [4,5] because it is simple and has low computational requirements [6,7].

Most published models only include additive genetic effects [8], and little research has been performed to expand these models to predict genetic merits to account for dominance genetic effects. It can be argued that such expansion is difficult because calculation becomes complicated and de-regressed estimated breeding values are used as phenotypes in most applications of genomic selection [9]. However, dominance genetic effect is of theoretical and practical important because it is heavily used in crosses of animal breeds. In fact, assortative mating and mate allocation boost the field performances of livestock [10]. Genomic selection has, therefore, renewed the interest in the prediction of dominance genetic effects. For example, the dominance genetic variance accounted for 5.6% of the phenotypic variance by GBLUP including dominance genetic effect [11]. More recently, GBLUP method including dominance genetic effect was suggested and the software (GVCBLUP) are already available online (http://animalgene.umn.edu/) [12–14].

The present study aimed to evaluate the performance of GBLUP including dominance genetic effect by estimating variance components and predicting genetic effects for both simulation and actual pig data.

Materials and Methods

Stochastic Simulation

A historical population was simulated to establish mutation drift equilibrium. The simulated genome comprised one chromosome 1 Morgan long that contained 6,000 SNP markers and 300 randomly spaced biallelic quantitative trait loci (QTL). In the first generation of the historical population, the initial allele frequencies of all markers and QTL were assumed to be 0.5. The recurrent mutation process was applied, and the mutation rate of markers and QTL was 5.0×10^{-4} per locus per generation. Recombinations were sampled from a Poisson distribution with a mean of 1 per Morgan and were then randomly placed along the chromosome. The historical population evolved over 2,000 generations of random mating and random selection with a population size of 100 (50 males and 50 females) to reach mutation–drift balance.

After 2,000 historical generations, a base population (G0) and the subsequent 6 generations (G1 to G6) were generated as a recent population. The population size of G0 increased to 300 (150 males and 150 females). In G1 to G6, 30 sires were randomly selected and mated to 150 dams in each generation. Each dam had 1 son and 1 daughter; thus, each sire had 5 sons and 5 daughters.

In G0, 1,000 markers and 50 QTL were randomly selected among the segregating markers and QTL with minor allele frequencies >0.05. Let Q_{1j} and Q_{2j} be 2 alleles at the *j*th QTL. The genetic values are then given by α_j, δ_j, and $-\alpha_j$ for genotypes $Q_{1j}Q_{1j}$, $Q_{1j}Q_{2j}$, and $Q_{2j}Q_{2j}$, respectively. The value of α_j was drawn from a gamma distribution with a shape parameter of 0.42; its sign was drawn at random with equal chance. According to the previous simulation study including dominance genetic effects [15], the value of δ_j was determined as the product of the absolute of α_j and the degree of dominance, which was drawn from a normal distribution $N(0,\tau^2)$. The total genetic effect (g_i) of the *j*th animal was calculated by summing all QTL genotypic values, and its variance (σ_g^2) was calculated as the sum of additive and dominance genetic variances (σ_a^2 and σ_d^2) [16], which were calculated as follows:

$$\sigma_a^2 = \sum_{j=1}^{Nq} 2q_j(1-q_j)\{\alpha_j + (1-2q_j)\delta_j\}^2,$$

$$\sigma_d^2 = \sum_{j=1}^{Nq} \{2q_j(1-q_j)\delta_j\}^2,$$

where q_j is the frequency of Q_{1j}. The broad-sense heritability (h^2) of the trait was 0.3. To obtain phenotypic values, an environmental effect was added to the total genetic effect, which was sampled from a normal distribution $N(0,(1-h^2)\sigma_g^2/h^2)$.

The phenotypes and genotypes of SNP markers were available for 1,500 and 1,800 individuals from G1 to G5 and G1 to G6, respectively. Thus, the reference population with both phenotypes and genotypes comprised 1,500 individuals from G1 to G5, and the test population with only genotypes comprised 300 individuals in G6.

In a standard simulation scenario, τ and the number of markers in G0 (N_m) were set to 0.5 and 1,000, respectively. To investigate the effects of τ and N_m on the performance of the present method, 2 alternative scenarios were simulated in addition to the standard scenario. Three different values of τ (0.25, 0.5, and 1.0) and N_m (200, 1,000, and 5,000) were simulated in the first and seconds groups, respectively. For all of these alternatives, only the intended parameter differed from the standard scenario. Twenty replicates were simulated for each scenario.

PIC Pig Data

Publicly available data including pedigree, genotypic, and phenotypic information on a single Pig Improvement Company (PIC) nucleus pig line were used (http://www.g3journal.org/content/2/4/429/Suppl/DC1). The total number of individuals was 3,512, and all have phenotypes for 2 traits (T4 and T5) and genotypes available from the PorcineSNP60 chip (N=64,233). These phenotypes were already adjusted for environmental fixed effects such as sex, farm, and year of birth [17]. Then, 1,800 individuals were randomly selected from all individuals whose accuracies from the full PIC dataset exceeded 0.8. From these 1,800 individuals, 1,500 were selected from old generations and defined as the reference population while the other 300 individuals were defined as the test population. Genotypes were filtered for minor allele frequencies less than 0.05. The pig data are summarized in Table 1.

Genomic BLUP Model

The GBLUP including additive and dominance genetic effects termed GBLUP-D. The statistical model of GBLUP-D can be expressed as:

$$\mathbf{y} = \mathbf{Xb} + \mathbf{Za} + \mathbf{Zd} + \mathbf{e},$$

where **y** is the vector of phenotypes; **b** is the vector of fixed effects; **a** and **d** are the vector of additive and dominance genetic effects of animals; **X** and **Z** are incident matrices for the fixed effects, additive, and dominance genetic effect, respectively; and **e** is the vector of residuals. Additive and dominance genetic effects were assumed to follow normal distributions: $\mathbf{a} \sim N(0,\mathbf{G}\sigma_a^2)$ and $\mathbf{d} \sim N(0,\mathbf{D}\sigma_d^2)$, where **G** and **D** are additive and dominance genomic relationship matrices, respectively. These matrices describe the relationships between genotyped individuals and can be constructed from the information on genome-wide SNP markers. Let A_{1j} and A_{2j} be 2 alleles at the *j*th marker locus and p_j be the frequency of A_{2j}. The **G** matrix is created as follows [3]:

$$\mathbf{G} = \frac{\mathbf{M_a M_a}'}{\sum_{j=1}^{Nm} 2p_j(1-p_j)},$$

where $\mathbf{M_a}$ is the $n \times N_m$ matrix (n is a number of individuals) and the element of $\mathbf{M_a}$ for the *i*th individual at the *j*th marker is calculated as follows:

$$\mathbf{M}_{\mathbf{a}i,j} = \begin{cases} -2p_j(A_1A_1) \\ 1-2p_j(A_1A_2). \\ 2-2p_j(A_2A_2) \end{cases}$$

Similarly, $\mathbf{M_d}$ is assumed to be the $n \times N_m$ matrix and the element of $\mathbf{M_d}$ for the *i*th individual at the *j*th marker can be calculated as:

Table 1. Summary of PIC pig dataset.

		Number of animals		Phenotype		Accuracy of estimated breeding value from all PIC datasets	
Trait	No. SNPs	Reference	Test	Mean	SD	Mean	SD
T4	27,391	1,500	300	−1.125	2.417	0.875	0.048
T5	27,287	1,500	300	44.107	60.315	0.880	0.048

Table 2. Variance component estimates (±standard errors) and heritabilities for simulation data with 3 dominance degrees (0.25, 0.5, and 1.0).

		Genetic variance components				
Condition	Method	Additive	Dominance	Residual variance	Narrow-sense heritability	Broad-sense heritability
τ = 0.25	True value	0.259	0.032	0.736	0.267	0.299
N_m = 1,000	GBLUP	0.277±0.036	–	0.793±0.022	0.256	–
	GBLUP-D	0.277±0.036	0.029±0.012	0.764±0.024	0.259	0.286
τ = 0.5	True value	0.191	0.108	0.736	0.185	0.289
N_m = 1,000	GBLUP	0.205±0.029	–	0.865±0.024	0.192	–
	GBLUP-D	0.208±0.030	0.066±0.016	0.797±0.026	0.194	0.256
τ = 1.0	True value	0.138	0.147	0.706	0.139	0.290
N_m = 1,000	GBLUP	0.152±0.024	–	0.857±0.024	0.151	–
	GBLUP-D	0.152±0.024	0.080±0.016	0.773±0.025	0.151	0.231

$$\mathbf{M}_{\mathbf{d}i,j} = \begin{cases} -2p_j^2(A_1A_1) \\ 2p_j(1-p_j)(A_1A_2) \\ -2(1-p_j)^2(A_2A_2) \end{cases}.$$

This element describes the coefficients of δ_j in dominance deviations. Therefore, **d** and its variance can be derived as follows:

$$\mathbf{d} = \mathbf{M_d}\boldsymbol{\delta},$$

where $\boldsymbol{\delta}$ is the N_m dimensional vector of the *j*th element, which is δ_j. This dominance formula was also used in GVCBLUP [14]. Assuming the dominance genetic effects at different marker loci are identically and independently distributed normal variables, the variance of the genome-wide dominance effect is calculated as follows:

$$\mathrm{V}(\mathbf{d}) = \mathbf{M_d}\mathbf{M_d}'\sigma_\delta^2 = \mathbf{D}\sigma_d^2,$$

$$\sigma_d^2 = \sum_{j=1}^{Nm} \{2p_j(1-p_j)\}^2 \sigma_\delta^2.$$

Consequently, **D** can be calculated using $\mathbf{M_d}$:

$$\mathbf{D} = \frac{\mathbf{M_d}\mathbf{M_d}'}{\sum_{j=1}^{Nm} \{2p_j(1-p_j)\}^2}.$$

Variance components were estimated with average information restricted maximum likelihood (REML) [18]. The dataset of reference population were used to predict genetic effects of the

Table 3. Variance component estimates (±standard errors) and heritabilities for simulation data with 200, 1,000, and 5,000 markers.

		Genetic variance components				
Condition	Method	Additive	Dominance	Residual variance	Narrow-sense heritability	Broad-sense heritability
τ = 0.5	True value	0.195	0.103	0.739	0.188	0.287
N_m = 200	GBLUP	0.158±0.026	–	0.878±0.024	0.153	–
	GBLUP-D	0.158±0.026	0.048±0.012	0.831±0.024	0.152	0.199
τ = 0.5	True value	0.191	0.108	0.736	0.185	0.289
N_m = 1,000	GBLUP	0.205±0.029	–	0.865±0.024	0.192	–
	GBLUP-D	0.208±0.030	0.066±0.016	0.797±0.026	0.194	0.256
τ = 0.5	True value	0.211	0.075	0.685	0.217	0.295
N_m = 5,000	GBLUP	0.206±0.027	–	0.761±0.021	0.213	–
	GBLUP-D	0.207±0.028	0.063±0.016	0.698±0.024	0.214	0.279

Table 4. Accuracies of estimates ($r_{\hat{a},a}$, $r_{\hat{d},d}$, and $r_{\hat{g},g}$) and regression coefficients of estimates on their true values ($b_{\hat{a},a}$, $b_{\hat{d},d}$, and $b_{\hat{g},g}$) in the test population for simulation data with 3 dominance degrees (0.25, 0.5, and 1.0).

Condition	Method	$r_{\hat{a},a}$	$r_{\hat{d},d}$	$r_{\hat{g},g}$	$b_{\hat{a},a}$	$b_{\hat{d},d}$	$b_{\hat{g},g}$
τ = 0.25	GBLUP	0.803	-	0.760	0.898	-	0.939
N_m = 1,000	GBLUP-D	0.804	0.212	0.769	0.902	0.609	0.942
τ = 0.5	GBLUP	0.743	-	0.616	0.891	-	0.976
N_m = 1,000	GBLUP-D	0.745	0.339	0.664	0.900	0.893	0.994
τ = 1.0	GBLUP	0.711	-	0.466	1.001	-	0.937
N_m = 1,000	GBLUP-D	0.712	0.478	0.581	1.006	1.189	1.035

Table 5. Accuracies of estimates ($r_{\hat{a},a}$, $r_{\hat{d},d}$, and $r_{\hat{g},g}$) and regression coefficients of estimates on their true values ($b_{\hat{a},a}$, $b_{\hat{d},d}$, and $b_{\hat{g},g}$) in the test population for simulation data with 200, 1,000, and 5,000 markers.

Condition	Method	$r_{\hat{a},a}$	$r_{\hat{d},d}$	$r_{\hat{g},g}$	$b_{\hat{a},a}$	$b_{\hat{d},d}$	$b_{\hat{g},g}$
τ = 0.5	GBLUP	0.689	-	0.552	1.049	-	1.027
N_m = 200	GBLUP-D	0.696	0.246	0.563	1.062	0.547	0.938
τ = 0.5	GBLUP	0.743	-	0.616	0.891	-	0.976
N_m = 1,000	GBLUP-D	0.745	0.339	0.664	0.900	0.893	0.994
τ = 0.5	GBLUP	0.799	-	0.694	1.004	-	1.035
N_m = 5,000	GBLUP-D	0.801	0.374	0.723	1.010	0.961	1.032

genotyped individuals in test population. The GBLUP-D model solutions yielded the estimates of additive ($\hat{\mathbf{a}}$) and dominance genetic effects ($\hat{\mathbf{d}}$). The estimates of the total genetic effect ($\hat{\mathbf{g}}$) were calculated by the sum of $\hat{\mathbf{a}}$ and $\hat{\mathbf{d}}$. In GBLUP, $\hat{\mathbf{g}}$ equals $\hat{\mathbf{a}}$, because the dominance genetic effect is not considered. The predictive ability of the model was evaluated from accuracy and unbiasedness of estimates in the test population. The accuracies of estimated additive, dominance, and total genetic effects ($r_{\hat{a},a}$, $r_{\hat{d},d}$, and $r_{\hat{g},g}$, respectively) were measured as the correlations between the estimates and true values. Unbiasedness ($b_{\hat{a},a}$, $b_{\hat{d},d}$, and $b_{\hat{g},g}$) was measured using the regressions of estimates on true values. A regression coefficient of one denotes unbiasedness. Since the true values are unknown in pig data, estimated breeding values from the full PIC dataset (a^*) and phenotypes (y) were used instead of true additive and total genetic effects, respectively. In real data, the predictive ability of dominance genetic effect cannot be calculated and was inferred from that of total genetic effect.

Results

Stochastic Simulation

Tables 2 and 3 show the estimates of variance components and heritability in simulation data with various values of τ and N_m. For all values of τ and N_m, there were nearly no differences in estimates of additive genetic variance and narrow-sense heritability between GBLUP and GBLUP-D. The ratios of dominant genetic variance estimated by GBLUP-D were 90.6%, 61.1%, and 54.4% with a τ of 0.25, 0.5, and 1.0 and 46.6%, 61.1%, and 84.0% with an N_m of 200, 1,000, and 5,000, respectively.

Tables 4 and 5 show the accuracies and unbiasedness of estimated genetic values calculated by GBLUP and GBLUP-D in simulation data. For all values of τ and N_m, $r_{\hat{a},a}$ and $b_{\hat{a},a}$ were almost equal between GBLUP-D and GBLUP. In GBLUP-D, $r_{\hat{d},d}$ and $b_{\hat{d},d}$ increased with increasing τ and N_m. Meanwhile, $r_{\hat{g},g}$ values in GBLUP-D exceeded those in GBLUP by 1.2%, 7.8%, and 24.7% with a τ of 0.25, 0.5, and 1.0 and by 1.9%, 7.8%, and 4.2% with an N_m of 200, 1,000, and 5,000, respectively. In GBLUP-D, larger values of τ resulted in $b_{\hat{g},g}$ being closer to 1.

PIC Pig Data

In T4 and T5 from the pig data, dominance genetic variance accounted for 9.6% and 6.3% of the phenotypic variance, respectively (Table 6). The estimated additive genetic variances and residual variances calculated by GBLUP-D were smaller than those calculated by GBLUP. Thus, GBLUP-D consequently yielded lower narrow-sense heritability and higher broad-sense heritability than GBLUP.

In T4 and T5 from the pig data, there were nearly no differences between GBLUP and GBLUP-D with respect to $r_{\hat{a},a^*}$ (Table 7). In T5, $r_{\hat{g},y}$ and $b_{\hat{g},y}$ were slightly higher in GBLUP-D than GBLUP, whereas $r_{\hat{g},y}$ and $b_{\hat{g},y}$ were not higher in GBLUP-D in T4.

Discussion

Stochastic Simulation

The GBLUP-D method captured the substantial ratios of the dominance genetic variances and estimated the individual dominance genetic effects although there were nearly no differences in estimates of additive genetic variance and narrow-sense heritability between GBLUP and GBLUP-D. Our result indicates that GBLUP-D is expected to improve performance of the crossbreds, in particular when degree of dominance is large.

In the present study, the simulated genome comprised one chromosome 1 Morgan long. However, the whole genome sizes of livestock are larger. Here, another simulation data were constructed to evaluate the effect of the genome size on predictive ability. This simulation data comprised five chromosomes of 1 Morgan. The numbers of markers and QTL set to be 5,000 and 250 to obtain the same distances between markers and QTL as the initial simulation. Table 8 shows the accuracies and unbiasedness of estimated genetic values in this simulation data. The dominance genetic effects could be captured in this data, but the accuracies of additive, dominance and total genetic effects decreased in comparison with the genome of 1 chromosome. Hence, large size of reference population would be required when genome size is large.

PIC Pig Data

The degrees of estimated dominance genetic variance in T4 and T5 were nearly equal to those in simulation data with a τ of 0.25. When τ was 0.25, $r_{\hat{g},g}$ and $b_{\hat{g},g}$ were 1.2% and 0.3% higher in GBLUP-D than GBLUP. These results indicate the predictive ability of GBLUP-D in T4 and T5 only improved slightly because the degrees of dominance genetic variance were too small. In fact, $r_{\hat{g},y}$ and $b_{\hat{g},y}$ in GBLUP-D were little improved in comparison with GBLUP. In general, the degree of dominance genetic variance is expected to be much larger in crossbred populations than purebred ones. Since the present study was based on data from purebred PIC pig data, the degrees of dominance genetic variance in T4 and T5 might have been small.

Table 6. Variance components estimates (±standard errors) and heritabilities for PIC pig data.

		Genetic variance components				
Trait	Method	Additive	Dominance	Residual variance	Narrow-sense heritability	Broad-sense heritability
T4	GBLUP	1.909±0.189	-	3.678±0.138	0.342	-
	GBLUP-D	1.735±0.201	0.537±0.219	3.331±0.185	0.310	0.405
T5	GBLUP	1298.1±123.2	-	2198.2±84.3	0.371	-
	GBLUP-D	1239.0±129.3	220.8±125.9	2049.3±113.1	0.353	0.416

Practical Use of GBLUP-D

GBLUP-D has two practical uses. First, selection on the basis of GBLUP-D in crossbreds is useful for commercial production. In swine and poultry, crossbreds are the end product. The marker information from a purebred and its crossbred relatives enables the selection of candidate purebreds for the performance of their crossbred offspring [11]. Second, GBLUP-D could allow mating allocation to exploit dominance. An extra response will be obtained when an appropriate design of future matings using mating allocation techniques is implemented [10,19].

Additive and Dominance Relationship Matrix

In the presence of dominance genetic effects, the breeding values of $A_{1j}A_{1j}$, $A_{1j}A_{2j}$, and $A_{2j}A_{2j}$ at the jth locus are $-2p_j\{\alpha_j+(1-2p_j)\delta_j\}$, $(1-2p_j)\{\alpha_j+(1-2p_j)\delta_j\}$,and$(2-2p_j)$ $\{\alpha_j+(1-2p_j)\delta_j\}$, respectively. The additive genomic relationship matrix in GBLUP-D should be constructed considering δ_j. If all QTL are of complete dominance, then δ_j is 1. However, in practice, the value of δ_j cannot be determined because the degree of dominance is unknown. Therefore, in the present study, the additive genomic relationship matrix was the same in GBLUP-D and GBLUP. Although there are nearly no differences in the estimates of additive genetic effects or their variance between GBLUP and GBLUP-D, the additive genomic relationship matrix including δ may yield good estimates of them.

A GBLUP method including dominance genetic effects was also proposed in a previous study [11]. In that model, the additive genomic relationship matrix is same as that in traditional GBLUP. However, the dominance genomic relationship matrix ($\mathbf{D}*$) in the previous study differs from that in the present study and is defined as follows:

$$\mathbf{D}^* = \frac{\mathbf{HH}'}{\sum_{j=1}^{Nm} 2p_j(1-p_j)\{1-2p_j(1-p_j)\}},$$

where the element of $\mathbf{H}$ for the ith individual at the jth marker is calculated as

$$\mathbf{H}_{i,j} = \begin{cases} -2p_j(1-p_j)(A_1A_1) \\ 1-2p_j(1-p_j)(A_1A_2). \\ -2p_j(1-p_j)(A_2A_2) \end{cases}$$

This element corresponds to the heterozygosity coefficient but not the dominance genetic effect. To compare the performance of GBLUP-D in the present study and the model in the previous study [11], predictive ability was calculated in the standard simulation scenario (Table 9). GBLUP-D yielded higher accuracies of additive, dominance, and total genetic effects than the previous model [11]. This might be because the heterozygosity coefficient includes part of the additive genetic effect. In fact, the previous study [12] reported that dominance genetic variance calculated from the previous model [11] was larger than that from GBLUP-D.

Assuming linkage equilibrium and uncorrelated marker effects, the dominance genetic variances in the present study (σ_d^2) and the previous study [11] (σ_{d*}^2) are calculated as follows:

Table 7. Aces of estimates ($r_{\hat{a},a^*}$and $r_{\hat{g},y}$) and regression coefficients ($b_{\hat{a},a^*}$ and $b_{\hat{g},y}$) of $\hat{a}$ on full PIC dataset (a^*) and $\hat{g}$ on phenotypic value (y) in the test population for the PIC pig dataset.

Trait	Method	$r_{\hat{a},a^*}$	$r_{\hat{g},y}$	$b_{\hat{a},a^*}$	$b_{\hat{g},y}$
T4	GBLUP	0.455	0.286	0.637	0.726
	GBLUP-D	0.456	0.286	0.710	0.724
T5	GBLUP	0.379	0.288	0.631	0.786
	GBLUP-D	0.376	0.299	0.654	0.805

Table 8. Accuracies of estimates ($r_{\hat{a},a}$, $r_{\hat{d},d}$, and $r_{\hat{g},g}$) and regression coefficients of estimates on their true values ($b_{\hat{a},a}$, $b_{\hat{d},d}$, and $b_{\hat{g},g}$) in the test population for simulation data with 3 dominance degrees (0.25, 0.5, and 1.0) when genome comprises of 5 chromosomes with 1 Morgan each.

Condition	Method	$r_{\hat{a},a}$	$r_{\hat{d},d}$	$r_{\hat{g},g}$	$b_{\hat{a},a}$	$b_{\hat{d},d}$	$b_{\hat{g},g}$
τ = 0.25	GBLUP	0.672	-	0.636	1.012	-	1.017
N_m = 1,000	GBLUP-D	0.673	0.148	0.641	1.013	0.810	1.018
τ = 0.5	GBLUP	0.646	-	0.502	1.015	-	1.018
N_m = 1,000	GBLUP-D	0.647	0.244	0.528	1.016	1.011	1.015
τ = 1.0	GBLUP	0.612	-	0.490	1.112	-	1.124
N_m = 1,000	GBLUP-D	0.612	0.348	0.525	1.122	1.030	1.120

Table 9. Accuracies of estimates ($r_{\hat{a},a}$, $r_{\hat{d},d}$, and $r_{\hat{g},g}$) and regression coefficients of estimates on their true values ($b_{\hat{a},a}$, $b_{\hat{d},d}$, and $b_{\hat{g},g}$) in the test population for simulation data ($\tau = 0.5, N_m = 1{,}000$).

Method	$r_{\hat{a},a}$	$r_{\hat{d},d}$	$r_{\hat{g},g}$	$b_{\hat{a},a}$	$b_{\hat{d},d}$	$b_{\hat{g},g}$
GBLUP	0.743	-	0.616	0.891	-	0.976
GBLUP-D	0.745	0.339	0.664	0.900	0.893	0.994
Su et al. [17]	0.709	0.239	0.655	0.939	0.569	0.967

$$\sigma_d^2 = \sum_{j=1}^{Nm} \{2p_j(1-p_j)\}^2 \sigma_\delta^2,$$

$$\sigma_{d*}^2 = \sum_{j=1}^{Nm} 2p_j(1-p_j)\{1-2p_j(1-p_j)\}\sigma_\delta^2.$$

In addition, the estimated dominance genetic effects in the present study ($\hat{\mathbf{d}}$) and the previous study [11] ($\hat{\mathbf{d}}^*$) are calculated as follows:

$$\hat{\mathbf{d}} = \sigma_d^2 \mathbf{D}\mathbf{Z}'\mathbf{V}^{-1}(\mathbf{y}-\mathbf{X}\mathbf{b}),$$

$$\hat{\mathbf{d}}^* = \sigma_{d*}^2 \mathbf{D}^*\mathbf{Z}'\mathbf{V}^{-1}(\mathbf{y}-\mathbf{X}\mathbf{b}),$$

where **V** is the variance of **y**. If the distribution of the allelic frequencies is available, $\hat{\mathbf{d}}^*$ can be transformed to $\hat{\mathbf{d}}$.

Epistasis

Increasing knowledge about biological pathways and gene networks highlights the importance of gene–gene interactions, i.e., epistasis; some authors argue that much of the genetic variance in a population is due to such interactions [20–22]. When considering second-order epistasis in GBLUP, the epistatic genomic relationship matrix can be approximately calculated from the Hadamard product of the genomic relationship matrix. For example, additive by additive and additive by dominance interactions are represented as **G#G** and **G#D**, respectively. The present study tried to use linear mixed models including **G#G** and **G#D** for T4 and T5 in pig data. However, the variance components of these epistasis could not be estimated because they were outside parameter space (data not shown). The previous study [11] also included epistatic genomic relationship matrices in GBLUP. In that model, the estimated epistatic variances were almost 0 for daily gain in Danish Duroc pigs. The estimated epistatic variances in the Bayesian model for the percentage of $CD8^+$ cells in publicly available mouse data were almost 0 [23]. Although these studies could not detect epistastic effect, in recent, the marker-generated kinship matrices were suggested in a new mixed model method [24] and nonparametric approaches and machine-learning techniques were recommended to model more complex gene interaction patterns [25–27].

Author Contributions

Conceived and designed the experiments: MN MS. Performed the experiments: MN. Analyzed the data: MN. Contributed reagents/materials/analysis tools: MN. Wrote the paper: MN.

References

1. Muewissen THE, Hayes BJ, Goddard ME (2001) Prediction of total genetic effect using genome-wide dense marker maps. Genetics 157: 1819–1829.
2. Hayes BJ, Bowman PJ, Chamberlain AL, Goddard ME (2009) Invited review: Genomic selection in daily cattle: progress and challenges. J Dairy Sci 92: 433–443.
3. VanRaden PM (2008) Efficient methods to compute genomic predictions. J Dairy Sci 91: 4414–4423.
4. McHugh N, Muewissen TH, Cromie AR, Sonesson AK (2011) Use of female information in dairy cattle genomic breeding programs. J Dairy Sci 94: 4109–4118.
5. Wiggans GR, VanRaden PM, Cooper T. 2011. The genomic evaluation system in the United States: Pasts, present, future. Journal of Dairy Science 94, 3202–3211.
6. VanRaden PM, Van Tassell CP, Wiggans GR, Sonstegard TS, Schnabel RD (2009) Invited review: Reliability of genomic predictions for North American Holstein bulls. J Dairy Sci 92: 16–24.
7. Hayes BJ, Visscher PM, Goddard ME (2009) Increased accuracy of artificial selection by using the realized relationship matrix. Genet Res 91: 47–60.
8. Calus MPL (2010) Genomic breeding value prediction: methods and procedures. Animal 4: 157–164.
9. Garrick DJ, Taylor JF, Fernando RL (2009) Deregressing estimated breeding values and weighting information for genomic regression analyses. Genet Sel Evol 41: 55.
10. Toro MA, Varona L (2010) A note on mate allocation for dominance handling in genomic selection. Genet Sel Evol 42: 33.
11. Su G, Christensen OF, Henryon M, Lund MS (2012) Estimating additive and non-additive genetic variances and prediction genetic merits using genome-wide dense single nucleotide polymorphism markers. PLoS One 7: e45293.
12. Vitezica ZG, Varona L, Legarra A (2013) On the additive and dominance variance and covariance of individuals within the genomic selection scope. Genetics: In press.
13. Da Y, Wang S (2013) Mixed model methods for genomic prediction and estimation of variance components of additive and dominance effects using SNP markers. J Dairy Sci 96 (E-Suppl.1): 620.
14. Wang C, Prakapenka D, Wang S, Runesha HB, Da Y (2013) GVCBLUP 2.1: Acomputing package for genomic prediction and estimation of variance components for additive and dominance effects using SNP markers. J Dairy Sci 96 (E-Suppl.1): 620.
15. Wittenburg D, Melzer N, Reinsch N (2011) Including non-additive genetic effects in Bayesian methods for the prediction of genetic values based on genome-wide markers. BMC Genet 12: 74.
16. Falconer DS, Mackay TFC (1996) Introduction to quantitative genetics. Essex: Longman Group. 126 p.
17. Cleveland MA, Hickey JM, Forni S (2012) A common dataset for genomic analysis of livestock populations. G3 2: 429–435.
18. Johnson DL, Thompson R (1995) Restricted maximum likelihood estimation of variance components for univariate animal models using sparse matrix techniques and average information. J Dairy Sci 78: 449–456.
19. Sun C, VanRaden PM, O'Connell JR, Weigel KA, Gianola D (2013) Mating programs including genomic relationships and dominance effects. J Dairy Sci: In press.
20. Schadt EE, Lamb J, Yang X, Zhu J, Edwards S (2005) An interactive genomics approach to infer causal associations between gene expression and disease. Nat Genet 37: 710–717.
21. Evans DM, Marchini J, Morris AP, Cardon LR (2006) Two-stage two-locus models in genome-wide association. PLoS One 2: e157.
22. Marchini J, Donnelly P, Cardon LR (2005) Genome-wide strategies for detecting multiple loci that influence complex diseases. Nat Genet 37: 413–417.
23. Lee SH, van der Werf JHJ, Hayes BJ, Goddard ME, Visscher PM (2008) Predicting unobserved phenotypes for complex traits from whole-genome SNP data. PLoS Genet 4: e1000231.
24. Xu S (2013) Mapping quantitative trait loci by controlling polygenic background effects. Genetics: In press.
25. Gianola D, Fernando RL, Stella A (2006) Genomic-assisted prediction of genetic value with semiparametric procedure. Genetics 173: 1761–1776.
26. Gonzalez-Recio O, Gianola D, Long N, Weigel KA, Rosa GJM (2008) Nonparametric methods for incorporating genomic information into genetic evaluations: an application to mortality in broilers. Genetics 178: 2305–2313.
27. Yao C, Spurlock DM, Armentano LE, Page Jr CD, VandeHarr JM, et al. (2013) Random forests approach for identifying additive and epistatic single nucleotide polymorphisms associated with residual feed intake in dairy cattle. J Dairy Sci 96: 6716–6729.

8

Forecasting Seizures in Dogs with Naturally Occurring Epilepsy

J. Jeffry Howbert[1], Edward E. Patterson[2], S. Matt Stead[3], Ben Brinkmann[3], Vincent Vasoli[3], Daniel Crepeau[3], Charles H. Vite[4], Beverly Sturges[5], Vanessa Ruedebusch[1], Jaideep Mavoori[1], Kent Leyde[1], W. Douglas Sheffield[1], Brian Litt[6], Gregory A. Worrell[3]*

1 NeuroVista Corp., Seattle, Washington, United States of America, **2** Veterinary Medical Center, University of Minnesota, St. Paul, Minnesota, United States of America, **3** Mayo Systems Electrophysiology Laboratory, Mayo Clinic, Rochester, Minnesota, United States of America, **4** School of Veterinary Medicine, University of Pennsylvania, Philadelphia, Pennsylvania, United States of America, **5** Veterinary School, University of California Davis, Davis, California, United States of America, **6** Department of Bioengineering, University of Pennsylvania, Philadelphia, Pennsylvania, United States of America

Abstract

Seizure forecasting has the potential to create new therapeutic strategies for epilepsy, such as providing patient warnings and delivering preemptive therapy. Progress on seizure forecasting, however, has been hindered by lack of sufficient data to rigorously evaluate the hypothesis that seizures are preceded by physiological changes, and are not simply random events. We investigated seizure forecasting in three dogs with naturally occurring focal epilepsy implanted with a device recording continuous intracranial EEG (iEEG). The iEEG spectral power in six frequency bands: delta (0.1–4 Hz), theta (4–8 Hz), alpha (8–12 Hz), beta (12–30 Hz), low-gamma (30–70 Hz), and high-gamma (70–180 Hz), were used as features. Logistic regression classifiers were trained to discriminate labeled pre-ictal and inter-ictal data segments using combinations of the band spectral power features. Performance was assessed on separate test data sets via 10-fold cross-validation. A total of 125 spontaneous seizures were detected in continuous iEEG recordings spanning 6.5 to 15 months from 3 dogs. When considering all seizures, the seizure forecasting algorithm performed significantly better than a Poisson-model chance predictor constrained to have the same time in warning for all 3 dogs over a range of total warning times. Seizure clusters were observed in all 3 dogs, and when the effect of seizure clusters was decreased by considering the subset of seizures separated by at least 4 hours, the forecasting performance remained better than chance for a subset of algorithm parameters. These results demonstrate that seizures in canine epilepsy are not randomly occurring events, and highlight the feasibility of long-term seizure forecasting using iEEG monitoring.

Editor: Maxim Bazhenov, University of California, Riverside, United States of America

Funding: NeuroVista Inc. had a role in study design, analysis, decision to publish, and preparation of the manuscript. The National Institute of Neurological Disorders and Stroke NIH (U01-NS073557, R01- NS630391, U24-NS063930), Mayo Clinic, and NeuroVista Inc. supported this work. No additional external funding received for this study.

Competing Interests: Drs. Worrell, Litt, Vite, and Patterson have received research funding from NeuroVista Inc. for portions of this project. Drs. Worrell and Litt have served as paid consultants for NeuroVista. NeuroVista Inc. participated in the study design, analysis, decision to publish, and preparation of the manuscript. Drs. Howbert, Sheffield, Leyde, and Mavoori served as employees of NeuroVista during the period of the research activity. Drs. Worrell, Litt, and Mr. Leyde hold patents pertaining to seizure forecasting devices. Full details of the 45 patents are available upon request. The remaining authors have no additional conflicts of interest. There are no further patents, products in development or marketed products to declare.

* E-mail: Worrell.Gregory@mayo.edu

These authors contributed equally to this work.

Introduction

Epilepsy is a common neurological disorder affecting 0.5–1% of the world's population [1], and according to the World Health Organization accounts for nearly 1% of the entire global burden of disease [2]. Pharmacotherapy with anti-epileptic drugs is the mainstay of epilepsy treatment, but 20–40% of patients continue to have seizures despite medications [3]. A significant, if not the most important, cause of epilepsy related disability for patients is the uncertainty of when seizures occur [4–6]. Even patients with infrequent seizures report persistent anxiety about when their next seizure will strike [6]. In addition, patients take medications daily that produce cognitive and physical side effects for events that may occur infrequently. The ability to forecast seizures would make individualized epilepsy treatment possible, and patients could be warned of their impending seizures and take medications only when needed to prevent seizures. Because of the potential clinical impact, seizure forecasting has stimulated intense interest [7].

Multiple lines of investigation support the hypothesis that ictogenesis, the process of seizure generation, is not random. The recurrent seizures that define focal epilepsy originate from a localized brain region, and in most patients are associated with a stereotypic electroencephalography (EEG) discharge with characteristic spectral pattern [8,9]. The spatial reproducibility of seizure onsets is the basis of successful epilepsy surgery, where focal resection of the brain tissue generating seizures can cure epilepsy [10]. The stereotypical pattern of seizure onset recorded with intracranial EEG (iEEG) is critical for 1st generation devices that use algorithms to detect focal seizures and deliver responsive stimulation to abort them [11]. In addition to the spatial and spectral reproducibility of seizures, multiple studies show that seizures tend to cluster in time [12,13] and exhibit underlying periodicities [14,15]. This spatiotemporal clustering of seizures suggests a fixed network generating seizures and the potential for spatial and temporal seizure forecasting. Successful seizure forecasting requires the existence of a pre-ictal state associated with an increased probability of seizure generation and a physiological signal that distinguishes the pre-ictal (ictogenic) state from the inter-ictal state [15–17]. Physiological evidence that spontaneous seizures arise from periods of increased seizure probability, i.e. a pre-ictal state, comes from multiple lines of investigation. Some of the earliest clinical descriptions of epilepsy reported on precursory symptoms extending for hours and even days prior to seizures [18]. Subsequent studies have shown that in some patients, self-reported prodromes forecast seizures better than chance [19]. Physiological changes reported to occur prior to seizures include changes in cerebral blood flow [20,21], blood oxygenation [22], blood oxygen-level dependent signal [23], and cortical excitability [24,25]. The vast majority of seizure forecasting investigations have used EEG for quantifying interictal and pre-ictal brain states(for reviews see Mormann et al. [16]; Andrzejak et al. [26]).

Seizure forecasting, however, has remained controversial. Many early studies investigating seizure forecasting with intracranial EEG (iEEG) were later demonstrated to be statistically flawed [16,26]. However, when using appropriate statistical tests some patients continued to show evidence for a pre-ictal state [27–29]. In many patients seizures are relatively rare events, and large data sets with multiple seizures and associated long inter-ictal periods are needed to rigorously evaluate forecasting algorithms [16,29,30]. The surgical evaluation of drug resistant epilepsy using intracranial electrodes provides a unique opportunity to directly investigate the generation of spontaneous seizures. Unfortunately, iEEG data sets from patients undergoing evaluation for epilepsy surgery are of relatively short duration and compromised by the acute effect of surgery and relatively rapid drug tapers used to capture adequate seizures for localization within a short period of time. Most of these clinical recordings only span a few days to a week. The inability to record long-term continuous electrophysiology and spontaneous seizures has significantly hindered progress in seizure forecasting. A similar data limitation complicates other areas of science directed at forecasting rare events, such as earthquakes [31], that also require long recordings to get adequate number of events.

We previously described an implantable device capable of acquisition of high quality, continuous iEEG over many months in dogs with naturally occurring epilepsy [32,33]. The same device has recently been used in a first-in-human trial in Australia [7,29]. Canine epilepsy [34] appears to be a good model of human epilepsy with similar epidemiology, clinical features, electrophysiology [35,36], and response to anti-epileptic drugs [37]. Approximately 65% of canine seizures are characterized as focal onset with or without secondary generalization [38], and approximately 25% of dogs are not controlled with medications [39–42].

Here we evaluate a computationally simple seizure forecasting algorithm based on iEEG spectral power in multiple bands. Three dogs with naturally occurring epilepsy were instrumented with an implanted device recording 16 channels of continuous iEEG. Spontaneously occurring seizures were automatically detected and visually verified to create accurate long-term seizure catalogs. Long-term, continuous, iEEG records (ranging from 6.5 to 15 months) containing multiple seizures (15–83 events) were evaluated. A seizure forecasting algorithm using multiple spectral iEEG power band features and a variable seizure warning time was demonstrated to forecast seizures significantly better than chance. The results support the feasibility of long-term seizure forecasting in naturally occurring canine epilepsy. The ability to forecast seizures in canine epilepsy will allow exploration of new therapeutic approaches to epilepsy, such as responsive therapy to prevent seizures before they occur.

Subjects and Methods

All studies had prior approval of the University of Minnesota Institutional Animal Care and Use Committee where the animals are maintained.

Subjects

Seven mixed hounds with naturally occurring epilepsy and spontaneous seizures were implanted with the NeuroVista Seizure Advisory System previously described [32,33]. All dogs had normal neurological examinations and MRI. The dogs were housed in the University of Minnesota canine epilepsy monitoring unit and continuously monitored (24 hours/day) with video and iEEG. The dogs were maintained on anti-epileptic medications during this study. Three dogs had an adequate number of seizures and prolonged interictal recordings suitable for analysis (Table 1).

Surgical protocol

As described previously [32], dogs were anesthetized using a standardized protocol for intracranial surgery. Bilateral craniectomies were performed using standard aseptic procedures and two silicone strip electrodes each containing 4 contacts were placed bilaterally in the subdural space (Figure 1A). Lead tails were tunneled subcutaneously to the telemetry unit implanted in a dorsal tissue pocket on the dog.

Device (Figure 1A)

A custom implantable iEEG acquisition system was used to acquire long-term continuous iEEG in three dogs [32,33]. The system has three major components: (1) Implantable Lead Assembly (ILA) consisting of four silicone strips each containing four platinum-iridium contacts (4 mm^2 surface area) separated by 20 mm; (2) Implantable Telemetry Unit (ITU); and (3) external Personal Advisory Device (PAD). The iEEG signals are recorded from the ILA contacts, filtered, amplified, and digitized (sampling rate 400 Hz) within the ITU and then wirelessly transmitted to the external PAD device. The ITU is charged daily for approximately 1 hour via an external battery powered device. The PAD was kept on the dog's back within a harness, and collected continuous iEEG data wirelessly from the ITU. The PAD has embedded seizure detection algorithms, and includes a user interface with functional lights for seizure warning, audible alarms, text messaging, and email algorithm outputs [33].

Data curation: For each dog 16-channel iEEG recordings (Figure 1B) were wirelessly transmitted to the PAD and stored on a flash memory card. Continuous video and iEEG data were archived to a central storage each week. A high sensitivity automated seizure detection algorithm was used to detect candidate seizures [43]. All candidate detections were visually reviewed and correlated with the continuous video to verify clinical seizure activity. The entire iEEG record was annotated by expert readers and all subclinical and clinical seizures verified. Canine data from this study are freely available on the iEEG portal (https://www.ieeg.org/).

Table 1. Subject IDs, imaging results, and specifics of iEEG records for three canines with naturally occurring epilepsy implanted with the NeuroVista Seizure Advisory System.

Subject ID (Breed)	MRI Brain	Recording duration, days	Number all seizures	Number lead seizures
002 Mixed	Normal	197*	27	27
004 Mixed	Normal	330	15	8
007 Mixed	Normal	451	83	18
Group totals (mean ± std)		978 (326±127)	125 (41.7±36.3)	53 (17.7±9.5)

Lead seizures were defined as seizures preceded by at least 4 hours of non-seizure.
*Full record was 475 days in duration; only final 197 days used for forecasting to avoid post-surgical non-stationarities in iEEG.

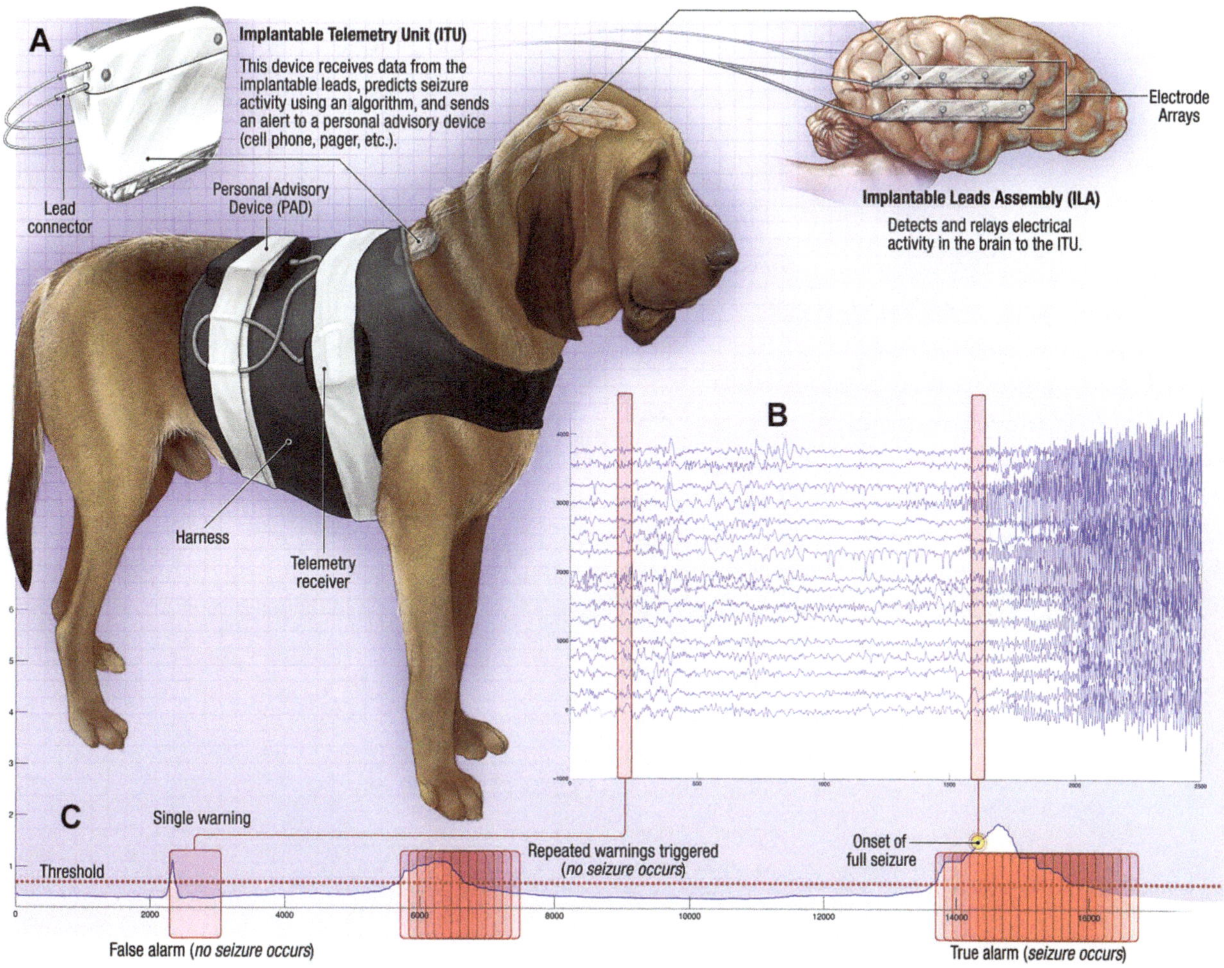

Figure 1. Seizure Advisory System (SAS) in Canines with Epilepsy. (A) The implantable device for recording and storing continuous iEEG includes: Implantable Lead Assembly (ILA) placed in the subdural space (right), Implantable Telemetry Unit (ITU), and Personal Advisory Device (PAD). The system acquires 16 channels of iEEG and wirelessly transmits the data to the PAD. Data is stored on a flash drive and uploaded weekly via internet to a central data storage site. (B) Sixteen channels of intracranial EEG recorded with SAS. A focal onset, secondarily generalized seizure is shown. The top 1–8 channels are from the left hemisphere and 9–16 from the right hemisphere, as shown on the brain schematic above. The onset of the seizure is from left hemisphere (underlined) electrodes 3 & 4. (C) Schematic of temporal profile of forecast seizure probability and threshold defining the pre-ictal state, i.e. period of increased seizure probability. When the forecast probability exceeds the defined threshold a fixed duration warning is triggered. Consecutive warnings that occur within the duration of a prior warning are combined, allowing for variable duration warnings. i) Single warning triggered without seizure (false positive warning). ii) Multiple consecutive warnings combined into prolonged warning without seizure (false positive warning). iii) Compounded warning prior to seizure onset (true positive seizure warning prior to electrographic seizure onset).

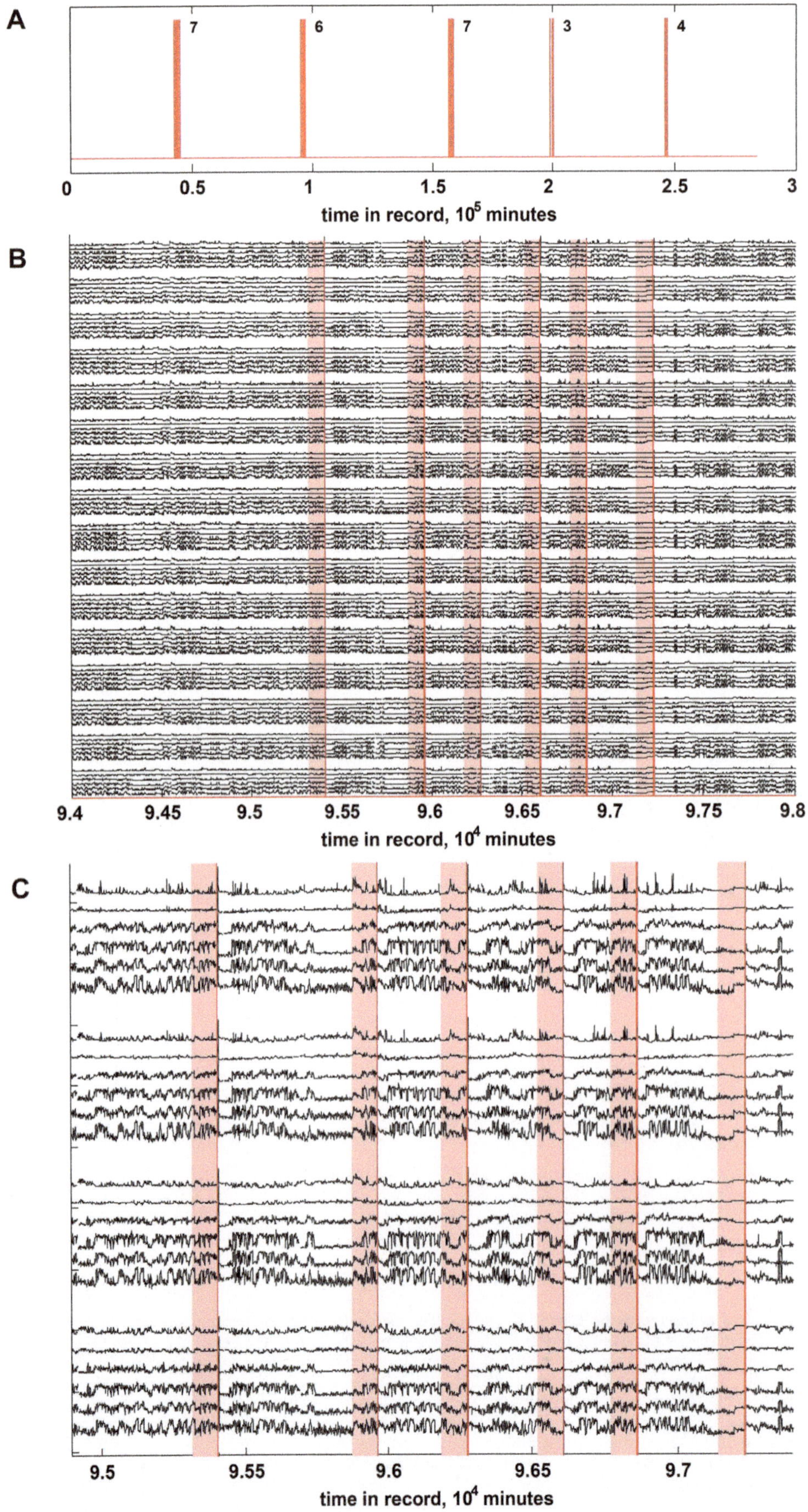
A
7
6
7
3
4
0
0.5
1
1.5
2
2.5
3
time in record, 10^5 minutes
B
9.4
9.45
9.5
9.55
9.6
9.65
9.7
9.75
9.8
time in record, 10^4 minutes
C
9.5
9.55
9.6
9.65
9.7
time in record, 10^4 minutes

Figure 2. Temporal profile of seizures. (A) Full temporal record for canine 002, showing time of occurrence of the 27 clinically verified seizures (vertical red lines). The seizures fall into 5 clusters; each cluster is annotated with the number of seizures in the cluster. (B) Temporal profile of 96 power-in-band (PIB) features for canine 002, spanning approximately 2.8 days in the vicinity of seizure cluster 2 in (A). Each grouping of 6 traces shows the 6 PIB features derived from one iEEG recording channel; from bottom to top these features capture the power-in-band of the delta (δ: 0.1–4 Hz), theta (θ: 4–8 Hz), alpha (α: 8–12 Hz), beta (β: 12–30 Hz), gamma-low (γ-low: 30–70 Hz) and gamma-high (γ-high: 70–180 Hz) frequency bands, respectively. The grouping for channel 1 is at the bottom of the figure, and channel 16 at the top. Vertical red lines locate the occurrence of individual seizures. Light red shading indicates the 90-minute period preceding each seizure, which was labeled as pre-ictal for training purposes; everything else (white background) was treated as the inter-ictal state. (C) Expansion of traces for channels 9–12 in (B.

Feature extraction

Seizure forecasting requires a physiological signal that is characteristic of the hypothesized pre-ictal state. Here we investigated the spectral power of multichannel iEEG as a source of candidate features to distinguish the pre-ictal and inter-ictal states. Average reference iEEG was used for computation. In each of the 16 iEEG channels, the iEEG record (sampling rate 400 Hz) was partitioned into non-overlapping 1-minute blocks, each block Fourier transformed, and the resulting power spectrum (0.1–200 Hz) divided into 6 frequency bands: delta (0.1–4 Hz), theta (4–8 Hz), alpha (8–12 Hz), beta (12–30 Hz), low-gamma (30–70 Hz), and high-gamma (70–180 Hz). Within each frequency band the power was summed over band frequencies to produce a *power-in-band* (PIB) feature. These features were aggregated into a feature vector containing 96 PIB values (16 channels×6 bands) (Figures 2.1, 2.2, 2.3). The resulting feature vectors, one per 1-minute block in the original record, were collected over both pre-ictal and inter-ictal periods to form the training and testing data sets for the algorithm.

Classifier training and testing

Logistic regression classifiers were trained to discriminate labeled pre-ictal and inter-ictal blocks using combinations of PIB features. For training purposes, blocks between 90 minutes preceding a seizure and the seizure itself were given a pre-ictal label, and all other blocks were labeled as inter-ictal. When applied to test data, the output of the trained classifier was a relative seizure risk for each test block on a continuous scale between 0 and 1 (Figures 3.1, 3.2). Classifier performance was assessed via 10-fold cross-validation, where folds were formed by dividing the entire record into 10 contiguous sub-records.

Forecasting algorithm

The seizure forecasting algorithm triggers a seizure warning when the classifier output exceeds a defined seizure risk threshold. The threshold is chosen adaptively so that total time in warning approximately matches a predefined target time in warning. Once triggered, the warning persists for a defined period of 90 minutes. Additional warnings that begin prior to termination of the previous warning are combined with it into a single variable duration warning period (Figure 1C). The variable duration of warning prior to a seizure encompasses the concept that the pre-ictal state is a period of increased seizure probability, and that seizures may occur at any time during the warning period [27,29]. To distinguish seizure forecasting from simple seizure detection, a 5 minute forecast horizon is used. The beginning of the warning period must precede the seizure by the forecast horizon for it to be a valid forecast. That is, seizures not preceded by a warning beginning at least 5 minutes prior to seizure onset are false negative forecasts. If a seizure does not occur at some point during the warning period the warning is counted as a false positive forecast.

Statistical analysis

Performance of the seizure forecasting algorithm was statistically evaluated using a Poisson-process chance prediction algorithm, as described previously by Snyder et al. [27]. The Poisson-process algorithm generates chance predictions for direct comparison to a candidate seizure forecasting algorithm. The Poisson-process algorithm uses the same parameters, including warning duration time, warning persistence rules, and forecasting horizon, as the candidate algorithm, and is constrained to have same total time in warning to control for specificity. A candidate algorithm must perform significantly ($p<0.05$) better than a matched chance predictor in order to claim evidence for seizure forecasting [44,27].

Results

In three dogs, continuous, long-term iEEG recordings were obtained and all clinical and subclinical seizure activity annotated. Table 1 shows the recording duration and number of seizures. The average duration of recordings was 326±127 days with a total of 125 seizures (41.7±36.3). Seizures separated by at least 4 hours from any preceding seizure were labeled lead seizures.

A simple forecasting algorithm using multiple spectral power bands as features to classify inter-ictal and pre-ictal data segments was tested on the three large datasets. The forecasting sensitivity, duration of recording spent in warning, number of false positive per day, and p-values are shown in Table 2. With the forecasting horizon set to 5 minutes, and triggered seizure warning duration of 90 minutes, forecasting runs with a range of different target total time in warning (0.1 to 0.5 of record) were evaluated (Table 2). The algorithm demonstrated seizure forecasting better than chance ($p<0.05$) for a range of total warning times for all 3 dogs when considering all seizures. Seizure clusters were observed in all 3 dogs (e.g. Figure 2.1). If the effect of clusters was decreased by considering only lead seizures, the forecasting performance remained better than chance for a subset of algorithm parameters. Dog 002 in particular showed very good seizure forecasting (74% of seizures forecasted, average false positives (FP) 2.8/day, $p<0.00005$, at total warning time = 0.3). Higher sensitivity was achieved at the expense of additional FP (e.g. 90% of seizures forecasted, average FP 3.0/day, and $p<0.00005$ at total warning time = 0.4). For Dog 004 seizure forecasting was better than chance when considering all seizures (73% of seizures forecasted, average FP 2.0/day, and $p<0.0007$ at total warning time = 0.3), but when considering only lead seizures, the performance did not reach significance. For Dog 007 seizure forecasting was significantly better than chance for both all seizures and lead seizures only, at total warning time = 0.3 (89% and 56% of seizures forecasted, with $p<0.00005$ and $p<0.05$ respectively; average FP 1.4/day).

Discussion

In 3 canines with naturally occurring epilepsy we investigated the feasibility of long-term seizure forecasting using iEEG. The

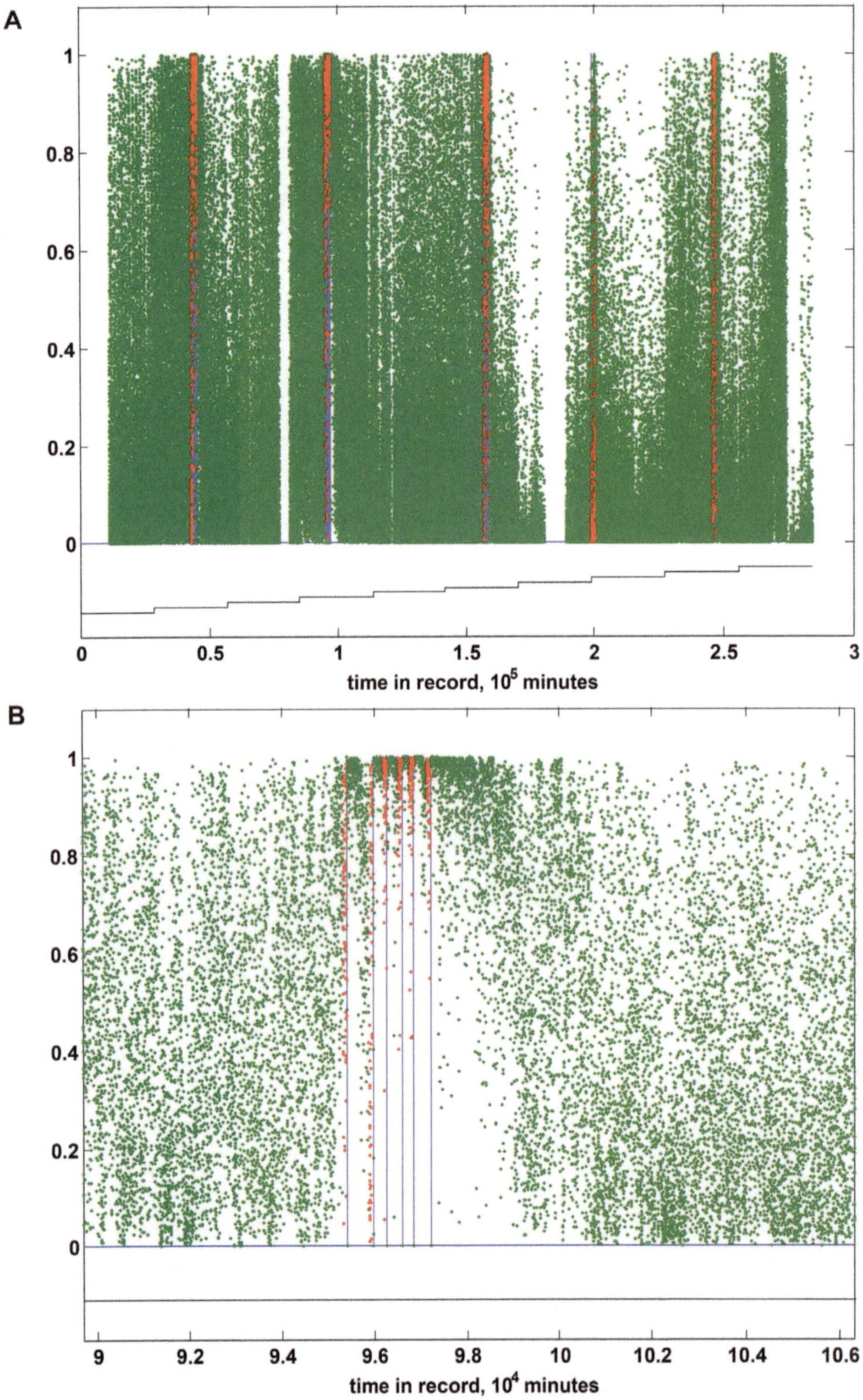

Figure 3. Seizure probability versus time. (A) The seizure probability for 1-minute blocks of full iEEG record of canine 002, as predicted by a logistic regression classifier trained on 96 power-in-band (PIB) features. Green and red dots indicate blocks labeled inter-ictal and pre-ictal, respectively, for classifier training, and vertical blue lines indicate occurrences of clinically verified seizures. Gaps in the plot correspond to periods of invalid data. The staircase underneath the plot delineates the 10 cross-validation partitions used during classifier training and testing. (B) Seizure probabilities for 1-minute blocks of iEEG record of canine 002 in vicinity of seizure cluster 2 (second from left in Figure 3 (A)), as predicted by a logistic regression classifier trained on 96 power-in-band (PIB) features. Green and red dots indicate blocks labeled inter-ictal and pre-ictal, respectively, for classifier training, and vertical blue lines indicate occurrences of clinically verified seizures.

Table 2. Seizure forecasting results for three canines with naturally occurring epilepsy implanted with the NeuroVista Seizure Advisory System.

ID	TIW	S_n	p	$S_{n\text{-lead}}$	$p_{n\text{-lead}}$	False Positive/day
002	0.1	0.482	**0.0000**	0.482	**0.0000**	1.293
002	0.15	0.593	**0.0000**	0.593	**0.0000**	1.818
002	0.2	0.667	**0.0000**	0.667	**0.0000**	2.257
002	0.3	0.741	**0.0000**	0.741	**0.0000**	2.792
002	0.35	0.741	**0.0001**	0.741	**0.0001**	2.910
002	0.4	0.889	**0.0000**	0.889	**0.0000**	3.074
002	0.5	0.889	**0.0001**	0.889	**0.0001**	3.186
004	0.1	0.000	0.2435	0.000	0.6081	0.811
004	0.15	0.133	1.0000	0.125	1.0000	0.794
004	0.2	0.467	**0.0141**	0.250	1.0000	1.079
004	0.3	0.733	**0.0007**	0.500	0.2534	1.954
004	0.35	0.733	**0.0035**	0.500	0.4691	2.335
004	0.4	0.733	**0.0183**	0.500	0.7290	2.670
004	0.5	0.800	0.0700	0.625	0.7407	3.026
007	0.1	0.759	**0.0000**	0.222	0.7548	0.658
007	0.15	0.819	**0.0000**	0.278	0.5582	0.791
007	0.2	0.843	**0.0000**	0.389	0.1658	0.991
007	0.3	0.892	**0.0000**	0.556	**0.0421**	1.427
007	0.35	0.892	**0.0000**	0.556	0.0895	1.617
007	0.4	0.904	**0.0000**	0.556	0.2270	1.695
007	0.5	0.916	**0.0000**	0.611	0.2459	1.927

A logistic regression classifier was trained and used to predict iEEG data (1 minute blocks) as either pre-ictal or inter-ictal, using power-in-band (PIB) features. Each classifier used 10 of 96 available PIB features, chosen during the training process via forward selection. Forecasts and subsequent statistical analyses were based on different target proportions of total time in warning, ranging from 0.1 to 0.5. A 10-fold cross-validation scheme was used for all phases of feature selection, classifier training, seizure forecasting, and statistical analysis. For each dog a range of values for time in warning (**TIW**) are considered. The Sensitivity (**S_n**) and p-value (p) are reported separately for all seizures and for lead seizures only (**$S_{n\text{-lead}}$**, $p_{n\text{-lead}}$) and have been adjusted to account for the performance of the chance prediction algorithm.

ability to record, annotate, curate, and analyze large iEEG data sets spanning multiple months was demonstrated. In each of the three dogs studied, the results demonstrate seizure forecasting significantly better than would be expected by chance (using a generic Poisson-process algorithm) when considering all seizures. The analysis was performed on a massive high fidelity data set of 978 days of iEEG containing 125 recorded seizures. The large volume of interictal data and seizures made it possible to rigorously validate a simple seizure forecasting algorithm utilizing iEEG spectral power to classify brain states, and differentiate inter-ictal from pre-ictal states. A key component of the algorithm is the use of variable duration seizure warnings, rather than rigidly defining the duration of pre-ictal states [27]. The variable duration warning results from a simple technique that consolidates multiple overlapping warnings into a single longer duration warning (Figure 1C). The seizure forecasting algorithm performance was rigorously tested by benchmarking the performance against a chance prediction model, constrained to have an equal proportion of time spent in warning [27,44]. Even when restricting the analysis to include only seizures separated by at least 4 hours (lead seizures), which diminishes statistical power by reducing the number of seizures from 125 to 53, seizure forecasting at levels better than chance was demonstrated in 2 of the 3 dogs.

As recently reviewed by Mormann, et al [16], previous studies have often failed to adequately evaluate the performance of proposed algorithms. Significant limitations of previous studies were the small data sets with few seizures and limited interictal iEEG. In the current study care was taken to avoid common pitfalls [16]: 1) The forecasting algorithm was tested on, continuous long-term recordings spanning multiple months. The long records contain adequate inter-ictal data and seizure numbers to definitively evaluate forecasting performance. 2) The sensitivity and specificity were rigorously tested and directly compared to a chance prediction algorithm constrained to have the same persistence parameters, forecasting horizon, and total time in warning. 3) Algorithm parameters were selected prior to analysis, and utilized across all animals. 4) The forecasting algorithm was optimized on training data (in-sample), but algorithm performance was calculated only on independent data (out-of-sample).

The results from three dogs demonstrate seizure forecasting performance at levels significantly better than chance when considering all seizures recorded (Table 2). The results show that canine epileptic seizures are not random events, and support the feasibility of seizure forecasting. The optimized level of performance for each dog (Table 2) supports the claim that clinically relevant seizure forecasting may be possible in humans, but highlights the need for long continuous iEEG recordings to capture an adequate number of seizures.

Seizure forecasting performance can likely be improved in the future. For example, the optimal recording bandwidth for differentiating interictal and pre-ictal states is not known. The current device has a relatively narrow bandwidth (sampling rate 400 Hz), but recent research has shown that the dynamic range of EEG activity generated from human brain spans a much wider spectrum, DC – 1000 Hz [45]. In addition, in this study we did not know the seizure onset zone location. The electrodes were placed according to a standardized protocol, similar to the approach in the recent human trial [29]. Previous studies have shown, however, that the EEG activity within and around the seizure onset zone is different [46]. Electrodes in the seizure onset zone could potentially improve seizure forecasting performance, and this is an area of ongoing research.

How this technology will translate to human subjects with epilepsy was recently evaluated in a clinical trial with encouraging results [29]. The optimal parameters for a clinical application of seizure forecasting likely depends on the application. For example, if seizure forecasting is used to trigger a therapy that does not disturb normal brain function (say low dose drug or sub-threshold stimulation, as used in the Neuropace investigational device [11]), then the algorithm could be adjusted for high sensitivity and a low false negative rate. This would lead to more false positives and more time in warning, but if the approach is able to prevent seizures with low levels of therapy without side effects it would be useful. This approach could reduce the exposure to the cognitive and physical side effects of anti-epileptic medications and render the drugs more efficacious. More challenging is a device solely for seizure warning, i.e. providing patients with a warning about impending seizures in order to reduce the potential for injury, and limiting social anxieties associated with the persistent fear of not knowing when the next seizure will strike. In this scenario, patients would likely demand a very high sensitivity (low false negative rate), because any missed seizure would be very disruptive. The device could prove dangerous if patients use the device to guide their ability to participate in certain activities, e.g. driving. Conversely, patients may not tolerate frequent false positive warnings as this would adversely affect their ability to participate in activities and could increase anxiety. Thus patients relying on a seizure forecasting device would likely have a low tolerance for missed seizures, i.e. false negatives, as well as frequent false positives. These possibilities are areas of current research.

Author Contributions

Conceived and designed the experiments: WDS JJH EEP KL BL GAW. Performed the experiments: JJH EEP CHV BS VR JM. Analyzed the data: JJH JM GAW SMS VV DC KL BL. Contributed reagents/materials/analysis tools: JJH EEP SMS VV DC CHV BS VR BB KL WDS GAW. Wrote the paper: JJH GAW BB.

References

1. Linehan J, Berg AT (2011) Epidemiologic aspects of epilepsy. In: The Treatment of Epilepsy: Principle and Practice. Williams & Wilkins. pp. 165–172.
2. Murray, Lopez, AD (1994) Global Comparative Assessments in the Health Sector: Disease Burden, Expenditure, Intervention Packages. Geneva, Switzerland: World Health Organization.
3. Kwan P, Brodie MJ (2000) Early identification of refractory epilepsy. N Engl J Med 342: 314–319.
4. Schulze-Bonhage A, Kuhn A (2008) Unpredictability of Seizures and the Burden of Epilepsy. Seizure Prediction in Epilepsy: From Basic Mechanisms to Clinical Applications.
5. Schulze-Bonhage A, Sales F, Wagner K, Teotonio R, Carius A, et al. (2010) Views of patients with epilepsy on seizure prediction devices. Epilepsy Behav 18: 388–396.
6. Fisher RS (2000) Epilepsy from the Patient's Perspective: Review of Results of a Community-Based Survey. Epilepsy Behav 1: S9–S14.
7. Fisher RS (2012) Therapeutic devices for epilepsy. Ann Neurol 71: 157–168.
8. Alarcon G, Binnie CD, Elwes RD, Polkey CE (1995) Power spectrum and intracranial EEG patterns at seizure onset in partial epilepsy. Electroencephalogr Clin Neurophysiol 94: 326–37.
9. Lee SA, Spencer DD, Spencer SS (2000) Intracranial EEG seizure-onset patterns in neocortical epilepsy. Epilepsia 41: 297–307.
10. Engel J, Pedley TA, Aicardi J, Dichter MA, Moshé S (2007) Epilepsy: a comprehensive textbook. Lippincott Williams & Wilkins.
11. Morrell MJ, RNS System in Epilepsy Study Group (2011) Responsive cortical stimulation for the treatment of medically intractable partial epilepsy. Neurology 77: 1295–1304.
12. Haut SR, Lipton RB, LeValley AJ, Hall CB, Shinnar S (2005) Identifying seizure clusters in patients with epilepsy. Neurology 65: 1313–1315.
13. Haut SR (2006) Seizure clustering. Epilepsy Behav 8: 50–55.
14. Durazzo TS, Spencer SS, Duckrow RB, Novotny EJ, Spencer DD, et al. (2008) Temporal distributions of seizure occurrence from various epileptogenic regions. Neurology 70: 1265–1271.
15. Dudek FE, Staley KJ (2011) Seizure probability in animal models of acquired epilepsy: a perspective on the concept of the preictal state. Epilepsy Res 97: 324–331.
16. Mormann F, Andrzejak RG, Elger CE, Lehnertz K (2007) Seizure prediction: the long and winding road. Brain 130: 314–333.
17. Stacey W, Le Van Quyen M, Mormann F, Schulze-Bonhage A (2011) What is the present-day EEG evidence for a preictal state? Epilepsy Res 97: 243–251.
18. Gowers SWR (1885) Epilepsy and other chronic convulsive diseases. London: W. Wood and Company.
19. Haut SR, Hall CB, LeValley AJ, Lipton RB (2007) Can patients with epilepsy predict their seizures? Neurology 68: 262–266.
20. Weinand ME, Carter LP, el-Saadany WF, Sioutos PJ, Labiner DM, et al. (1997) Cerebral blood flow and temporal lobe epileptogenicity. J Neurosurg 86: 226–232.
21. Baumgartner C, Serles W, Leutmezer F, Pataraia E, Aull S, et al. (1998) Preictal SPECT in temporal lobe epilepsy: regional cerebral blood flow is increased prior to electroencephalography-seizure onset. J Nucl Med 39: 978–982.
22. Adelson PD, Nemoto E, Scheuer M, Painter M, Morgan J, et al. (1999) Noninvasive continuous monitoring of cerebral oxygenation periictally using near-infrared spectroscopy: a preliminary report. Epilepsia 40: 1484–1489.
23. Federico P, Abbott DF, Briellmann RS, Harvey AS, Jackson GD (2005) Functional MRI of the pre-ictal state. Brain 128: 1811–1817.
24. Badawy R, Macdonell R, Jackson G, Berkovic S The peri-ictal state: Cortical excitability changes within 24 h of a seizure. Brain 132: 1013–1021.
25. Badawy RA, Vogrin SJ, Lai A, Cook MJ (2013) Are patterns of cortical hyper-excitability altered in catamenial epilepsy? Ann Neurol. (In Press)
26. Andrzejak RG, Chicharro D, Elger CE, Mormann F (2009) Seizure prediction: Any better than chance? Clin Neurophysiol 120(8): 1465–1478.
27. Snyder DE, Echauz J, Grimes DB, Litt B (2008) The statistics of a practical seizure warning system. J Neural Eng 5: 392–401.
28. Park Y, Luo L, Parhi KK, Netoff T (2011) Seizure prediction with spectral power of EEG using cost-sensitive support vector machines. Epilepsia 52: 1761–1770.
29. Cook MJ, O'Brien TJ, Berkovic SF, Murphy M, Morokoff A, et al. (2013) Prediction of seizure likelihood with a long-term, implanted seizure advisory system in patients with drug-resistant epilepsy: a first-in-man study. Lancet Neurol 12: 563–571.
30. Elger CE, Mormann F (2013) Seizure prediction and documentation-two important problems. Lancet Neurol 12: 531–532.
31. Main I (1999) Is the reliable prediction of individual earthquakes a realistic scientific goal. Nature Debates. http://www.nature.com/nature/debates/earthquake/equake_contents.html
32. Davis KA, Sturges BK, Vite CH, Ruedebusch V, Worrell G, et al. (2011) A novel implanted device to wirelessly record and analyze continuous intracranial canine EEG. Epilepsy Res 96: 116–122.
33. Coles L, Patterson E, Sheffield W, Mavoori J, Higgin J, et al. (2013) Feasibility Study of a Caregiver Seizure Alert System in Canine Epilepsy. Epilepsy Behavior. (In Press)
34. Potschka H, Fischer A, von Rüden EL, Hülsmeyer V, Baumgärtner W (2013) Canine epilepsy as a translational model? Epilepsia 54: 571–579.
35. Pellegrino FC, Sica RE (2004) Canine electroencephalographic recording technique: findings in normal and epileptic dogs. Clin Neurophysiol 115: 477–487.
36. Berendt M, Høgenhaven H, Flagstad A, Dam M (1999) Electroencephalography in dogs with epilepsy: similarities between human and canine findings. Acta Neurol Scand 99: 276–283.
37. Chandler K (2006) Canine epilepsy: what can we learn from human seizure disorders? Vet J 172: 207–217.
38. Berendt M, Gram L (1999) Epilepsy and seizure classification in 63 dogs: a reappraisal of veterinary epilepsy terminology. J Vet Intern Med 13: 14–20.
39. Dewey CW, Guiliano R, Boothe DM, Berg JM, Kortz GD, et al. (2004) Zonisamide therapy for refractory idiopathic epilepsy in dogs. J Am Anim Hosp Assoc 40: 285–291.
40. Platt SR, Adams V, Garosi LS, Abramson CJ, Penderis J, et al. (2006) Treatment with gabapentin of 11 dogs with refractory idiopathic epilepsy. Vet Rec 159: 881–884.
41. Volk HA, Matiasek LA, Luján Feliu-Pascual A, Platt SR, Chandler KE (2008) The efficacy and tolerability of levetiracetam in pharmacoresistant epileptic dogs. Vet J 176: 310–319.
42. von Klopmann T, Rambeck B, Tipold A (2007) Prospective study of zonisamide therapy for refractory idiopathic epilepsy in dogs. J Small Anim Pract 48: 134–138.
43. Gardner A, Krieger A, Vachtsevanos G, Litt B (2007) One-Class Novelty Detection for Seizure Analysis from Intracranial EEG. Journal of Machine Learning Research 7: 1025–1044.
44. Andrzejak RG, Mormann F, Kreuz T, Rieke C, Kraskov A, et al. (2003) Testing the null hypothesis of the nonexistence of a preseizure state. Phys Rev E Stat Nonlin Soft Matter Phys 67: 010901.
45. Worrell GA, Jerbi K, Kobayashi K, Lina JM, Zelmann R, et al. (2012) Recording and analysis techniques for high-frequency oscillations. Prog Neurobiol 98: 265–278
46. Warren CP, Hu S, Stead M, Brinkmann BH, Bower MR, et al. (2010) Synchrony in normal and focal epileptic brain: the seizure onset zone is functionally disconnected. J Neurophysiol 104: 3530–3539

9

Evidence for Two Numerical Systems That Are Similar in Humans and Guppies

Christian Agrillo[1]*, Laura Piffer[1], Angelo Bisazza[1], Brian Butterworth[2]

1 Department of General Psychology, University of Padova, Padova, Italy, **2** Institute of Cognitive Neuroscience, University College of London, London, United Kingdom

Abstract

Background: Humans and non-human animals share an approximate non-verbal system for representing and comparing numerosities that has no upper limit and for which accuracy is dependent on the numerical ratio. Current evidence indicates that the mechanism for keeping track of individual objects can also be used for numerical purposes; if so, its accuracy will be independent of numerical ratio, but its capacity is limited to the number of items that can be tracked, about four. There is, however, growing controversy as to whether two separate number systems are present in other vertebrate species.

Methodology/Principal Findings: In this study, we compared the ability of undergraduate students and guppies to discriminate the same numerical ratios, both within and beyond the small number range. In both students and fish the performance was ratio-independent for the numbers 1–4, while it steadily increased with numerical distance when larger numbers were presented.

Conclusions/Significance: Our results suggest that two distinct systems underlie quantity discrimination in both humans and fish, implying that the building blocks of uniquely human mathematical abilities may be evolutionarily ancient, dating back to before the divergence of bony fish and tetrapod lineages.

Editor: Sarah Frances Brosnan, Georgia State University, United States of America

Funding: This research was funded by research grant "Progetto di Eccellenza CARIPARO 2007", and "Progetto Giovani Studiosi 2010" from University of Padova to Christian Agrillo. The funders had no role in study design, data collection and analysis, decision to publish, or preparation of the manuscript.

Competing Interests: The authors have declared that no competing interests exist.

* E-mail: christian.agrillo@unipd.it

Introduction

Numerousness, like shape, size and color, is a basic property of our perceptual world. It has long been recognized that adults, infants and non-human animals can instantly extract the numerical information from a visual scene without counting the elements [1,2,3,4]. In humans, this ability is thought to be based on two distinct non-verbal systems that operate over different parts of the number range [5,6,7]. One is a system for representing approximate numerosities as analog magnitudes, and is usually referred to as the 'analog magnitude system' (ANS) [8,9,10] that has virtually no upper limit but is subject to a ratio limit in accordance with Weber's Law, which states that the capacity to discriminate between two quantities becomes increasingly accurate as the ratio between them increases. The second system is referred to as the 'object-file system' and is held to depend on a mechanism for representing and tracking small numbers of individual objects [3,7,11]. If this object-tracking system operates by keeping track of individual elements, it is precise but allows for the parallel representation of a small number of objects (usually three-four elements in adults). It is often assumed that this is the system that supports 'subitizing' – the accurate reporting of the numerosity of small sets without serial counting [6,7]. The two mechanisms appear to differ in many respects, including speed, accuracy and cognitive load [11,12].

The lack of a ratio effect is the main signature that allows experimental differentiation of the object-file system from the analog magnitude system [3]: the performance of an adult is very similar when discriminating 3 vs. 4 or 1 vs. 4 objects, whereas we are much more accurate in discriminating 5 from 20 objects than 15 from 20 objects. Not all the studies, however, reported a different ratio-effect between small and large numbers [13,14,15]. For instance, in a task requiring to apply an ordinal numerical rule, Cantlon and Brannon [16] also found evidence of a ratio-effect in the small number range. Recent neuropsychological, electrophysiological and brain imaging data suggest that these two non-verbal numerical systems probably have distinct neural substrates [17,18,19,20,21].

The analog magnitude system appears to be shared among many vertebrates. When required to determine the larger of two sets of elements, infants, macaques, dogs, swordtails and mosquitofish give approximate responses and their capacity to discriminate is strongly influenced by the numerical ratio [4,16,22,23,24]. Some have proposed that human infants and non-human vertebrates may also share with adults a distinct mechanism for precisely representing quantities up to four. However, evidence of separate systems in human infants and animals is less clear. In some studies, infants were found to discriminate different ratios in the small and large number range. For instance they were found to discriminate 2 vs. 3 but not 4 vs. 6 items in a habituation task despite the identical ratio difference [25]. Moreover, a study showed that children with Williams syndrome report a specific impairment in large number discrimination while the discrimination capacity in the small number range was unaffected [18]. Yet, vanMarle and Wynn [26] found that in infants the discrimination of auditory events was ratio-dependent also in the small number range, suggesting that infants use a single system of analog magnitude in the auditory domain.

Regarding non-human primates, chimpanzees have been shown to increase their reaction times to estimate the number of dots in a large number range but not in a small number range [27], but another study combining the data from four great apes found that numerical ratio was the best performance predictor in the range 0–6 [28]. In support of two-system hypothesis, rhesus monkeys successfully selected the greater group of apple slices with comparison of 1 vs. 2, 2 vs. 3 and 3 vs. 4 [29] while they discriminated between 4 and 8 lemons (1:2 numerical ratio) but not between 4 and 6 (2:3). However two recent studies reported that in rhesus monkeys accuracy was strongly affected by numerical ratio for both small and large quantities, in agreement with the existence of a single non-verbal mechanism over the whole numerical range [16,30].

Several lines of evidence suggest that separate systems for small and large numbers may exist in other vertebrates too. In a recent work on attack/retreat decisions in free-ranging dogs [31], Bonanni and collaborators reported that dogs spontaneously assessed large quantities as noisy magnitudes. In contrast dogs approached aggressively with the same probability when they outnumbered opponents by a 1:2, 2:3 or 3:4 ratio, suggesting that dogs discriminate two small quantities using an object-file mechanism. Among birds, New Zealand robins can discriminate 1 vs. 2, 2 vs. 3 and 3 vs. 4 but not 4 vs. 5 and 5 vs. 6; however these birds can also distinguish between large numbers provided that the numerical ratio is at least 1:2 [32]. Studies on domestic chicks and bees have demonstrated that both species can successfully discriminate 2 vs. 3 while they fail to distinguish the same numerical ratio when two large numbers are presented, such as 4 vs. 6 [33,34]. However, this kind of evidence for two numerical systems may not be conclusive. In some cases there are alternative explanations. For example, if chicks and bees can only represent small numbers, then they will not be able to discriminate 4 from 6.

In recent years, numerical competence has been investigated in several teleost species using either operant conditioning or spontaneous preferences [22,23,35,36,37,38]. Two of these studies imply the possibility that separate systems for large and small numbers may exist even in fish. Mosquitofish have been found to discriminate between shoals differing in numerosity when the paired numbers were 1 vs. 2, 2 vs. 3 and 3 vs. 4 but they succeeded with only up to a 1:2 numerical ratio (4 vs. 8 or 8 vs. 16) when they had to discriminate between large numbers [22]. In the second study it has been shown that one-day old guppies can discriminate between small quantities of social companions (<4), showing an inborn ability to elaborate small quantities, while the capacity to discriminate large quantities (4 vs. 8) emerges later, as a consequence of both maturation and social experience. This developmental dissociation indirectly suggests the existence of different systems for small and large quantities also in guppies [37].

Overall, the existence of a phylogenetically shared analog magnitude system appears generally accepted, but authors disagree as to whether a single analog magnitude mechanism accounts for discrimination over the whole numerical range, or a distinct system operates over the small number range.

In the present study, we used guppies to test one of the predictions of the hypothesis of two separate systems; according to this hypothesis fishes' ability to discriminate two large numbers should become more accurate as the difference between them increases while, in discriminating quantities <4, the performance should not be affected by the numerical ratio.

Guppies were required to choose the more numerous of two available groups of conspecifics. As reference, a group of undergraduate students were required to estimate the larger of two groups of dots while prevented from verbal counting. Both species were presented with the same five numerical ratios (0.25/0.33/0.50/0.67/0.75) both within the small number range (1–4) and beyond it. Furthermore, we examined the possible role of learning on performance of the fish, comparing subjects with or without previous experience of social groups.

Materials and Methods

Ethics Statement

The Experiment involving fish complies with all laws of the country (Italy) in which it was performed (D.M. 116192) and was approved by 'Ministero della Salute' (permit number: 6726-2011). The experiment with undergraduates was approved by the ethics committee of the Department of General Psychology of University of Padova and was conducted according to the Declaration of Helsinki. Before testing, all participants gave their written consent.

Undergraduate experiment

In this experiment we adopted a procedure commonly used to measure non-verbal numerical abilities in adults, namely a computerized numerical judgement with sequential presentation of the stimuli [17,36,39]. Undergraduates were required to estimate the larger of two groups of dots while being prevented from using verbal counting.

Participants. A total of 18 undergraduate students (three males) between the ages of 18 and 31 (mean age: 21.33) took part in the present study for course credits. The experiments were carried out at the Department of General Psychology, University of Padova. All participants had normal or corrected vision.

Stimuli and procedure. The stimuli consisted of 240 pairs of arrays composed of different numbers of black dots. The dots differed in size and appeared in the center of the screen on a white background. The number of dots presented on the screen ranged from 1 to 24. Half of the pairs were controlled for continuous variables (cumulative surface area, density, luminance and overall space occupied by arrays), while the other half were not. Twenty further pairs, with identical features, were created for the initial training phase. The stimuli were displayed on a 17-inch monitor, using E-Prime software, in a darkened room.

After a period of dark adaptation, a short familiarization training phase with feedback was presented. The participants initially read the experimental instructions on the screen. A fixation cross then appeared in the center of the screen for 1000 ms, then a group of dots was displayed in the centre of the screen for 150 ms (Fig. 1). Following a 500 ms delay, the participants were shown another group of dots for 150 ms. They were required to estimate which one of the two groups was more numerous by pressing one of two keys on the keyboard. In half of the stimuli the larger group was presented first, in the other half the smaller group was presented first. They were instructed to make their responses as quickly and as accurately as possible. Furthermore, to prevent verbal processing of the stimuli, verbal suppression was introduced during the test by asking the participants to continuously repeat 'abc'. No feedback was provided during the test.

Five numerical ratios were presented (0.25, 0.33, 0.50, 0.67, 0.75) for small (1 vs. 4, 1 vs. 3, 1 vs. 2, 2 vs. 3 and 3 vs. 4) and large (6 vs. 24, 6 vs. 18, 6 vs. 12, 6 vs. 9, 6 vs. 8) numerical contrasts. Reaction time is the common measure of numerical acuity in human studies, especially when, as in our study, participants must

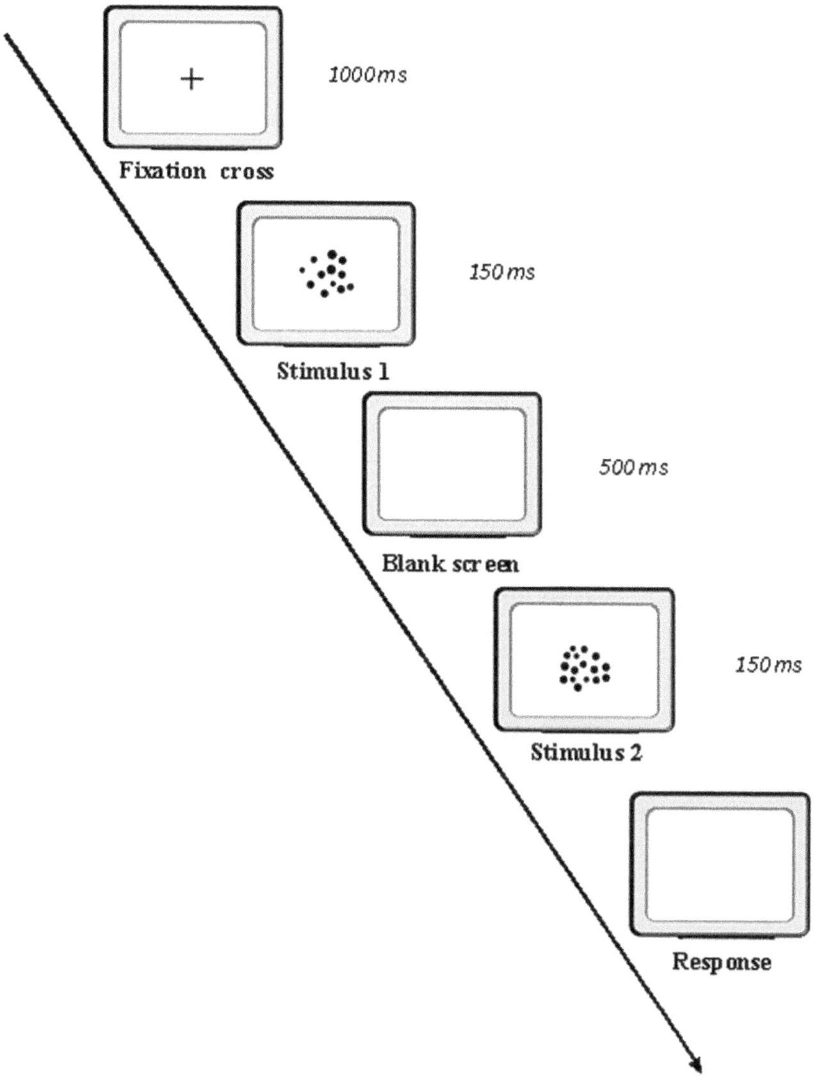

Figure 1. The experimental design used in the undergraduate experiment. The participants were sequentially presented with two groups of dots and had to estimate which group was more numerous.

perform easy discriminations [17,40,41]. In these cases, in fact, participants commit few errors and variation in accuracy is potentially obscured by a ceiling effect. However since accuracy (proportion of time spent near the larger shoal) was the only variable measured in the fish experiment, for comparison, we also analyzed accuracy in the human experiment.

Fish experiment

Like many other social fish, single guppies placed in an unknown environment show a strong tendency to join social companions and, if choosing between two shoals, they exhibit a preference for the larger one [37,42]. This spontaneous tendency was used in this experiment to measure the ability of guppies to discriminate between two numerosities. During their life guppies might have different opportunity of familiarizing with large or small shoals and this could potentially affect the experiment. Because of this, alongside a sample of 140 adults, we tested a sample of 200 immature fish reared in pairs and therefore with no previous experience of social groups.

Subjects. The experienced subjects were adult females because they are more gregarious than males. They were reared in groups of 15 or more individuals. Seventy fish (14 in each numerical contrast) were tested in small quantity discriminations, and 70 were tested in large quantity discriminations. The inexperienced subjects were juvenile fish tested at the onset of their numerical abilities. A recent study showed that one-day-old guppies could discriminate the larger shoal when the choice was between numbers in the small number range, whereas the ability to discriminate large quantities appeared later, at approximately day 40 [37]. Thus, 100 one-day-old fish (20 in each numerical contrast) were tested in small quantity discrimination tasks, and 100 40-day-old fish were tested in large quantity discrimination tasks.

Stimuli and procedure. The experimental apparatus was composed of three adjacent tanks (Fig. 2). The central one, the 'subject tank', housed the test fish (36×60×35 cm). At the two ends two 'stimulus tanks' (36×10×10 cm) faced the subject tank. The apparatus was filled with 10 cm of water. The apparatus for

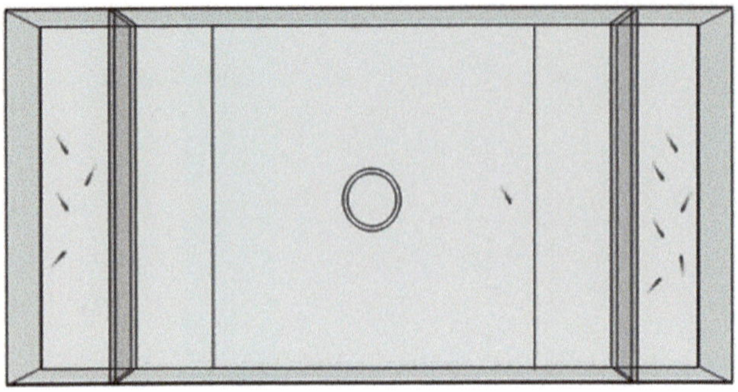

Figure 2. The experimental apparatus used in the fish experiment. Fish were individually placed into the middle of the apparatus where two shoals containing different numbers of fish were visible at the ends.

the juvenile fish was similar to that used for the adult fish but reduced in size. The central 'subject tank' was 20×17.5×25 cm and the two 'stimulus tanks' were semi-octagonal shaped, with a side of 6.3 cm. This apparatus was filled with 4 cm of water and was partially modified when testing the 40-day-old fish by enlarging the stimulus tanks (semi-octagonal sides 8.3 cm long) and increasing the level of water (6 cm).

Two shoals containing different numbers of fish were placed into the stimulus tanks. The subjects were individually introduced into a transparent plastic cylinder (10 cm in diameter for the adults, and 3.5 cm for the juveniles) in the middle of the subject tank and allowed to acclimatize for two minutes. After this period the subject was observed for 15 minutes. Shoal preference was calculated as the time spent by the subject within a distance of 11 cm (4 cm when testing juveniles) from the glass facing either of the stimulus tanks. Subjects that did not visit either stimulus sector at least three times or spent less than 50% of the time in a choice area were considered inactive; they were discarded and replaced by another fish.

The same five numerical ratios of the students' experiment were presented to fish both for small (1 vs. 4, 1 vs. 3, 1 vs. 2, 2 vs. 3 and 3 vs. 4) and large (4 vs. 16, 4 vs. 12, 4 vs. 8, 4 vs. 6 and 6 vs. 8) numerical contrasts.

At the end of these trials, we increased the sample size for ratio 0.67 by testing 64 additional adult females, 32 in a small quantity discrimination (2 vs. 3) and 32 in a large quantity discrimination (4 vs. 6).

Results

Undergraduate experiment

Small numbers. In accordance with the theory, reaction time was not affected by numerical ratio (ANOVA $F_{(4,68)} = 1.474$, $P = 0.220$, Fig. 3). Control of continuous variables did not affect

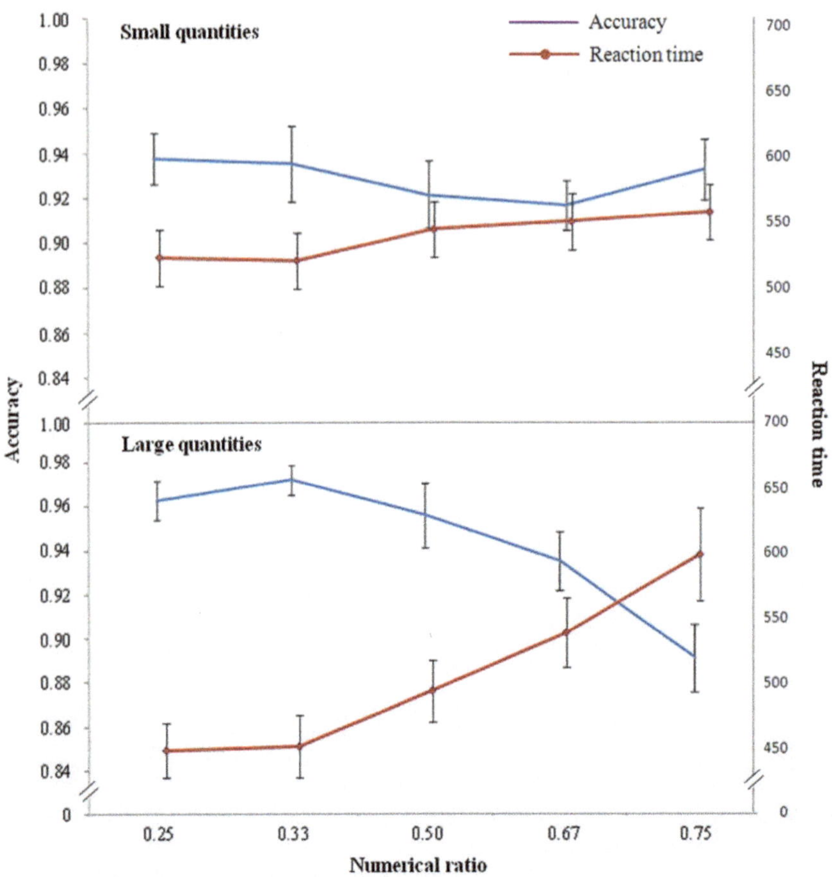

Figure 3. The results of the undergraduate experiment. Accuracy (proportion of correct responses) and reaction time are plotted against the numerical ratio of the contrasts for both large and small number (1–4) ranges. The performance of the participants showed ratio sensitivity for large numbers and ratio insensitivity in the small number range. Bars represent the standard error.

performances ($F_{(1,17)}=0.421$, $P=0.525$; interaction $F_{(4,68)}=0.822$, $P=0.516$). To test if there was a significant slope, we performed a polynomial trend analysis [43]. No significant linear or larger order trend was found ($F_{(1,17)}=2.644$, $P=0.122$). A likelihood ratio analysis (see [44] for details) also reflected the absence of ratio effect ($\lambda=3.67$).

Comparable differences were observed in accuracy. Accuracy was not affected by numerical ratio ($F_{(4,68)}=0.572$, $P=0.684$), continuous variables did not affect the performance ($F_{(1,17)}=1.285$, $P=0.273$) and the interaction was not significant ($F_{(4,68)}=0.383$, $P=0.820$). No significant trend was found ($F_{(1,17)}=0.378$, $P=0.547$). A likelihood ratio analysis also reflected the absence of ratio effect ($\lambda=1.15$).

Large numbers. In the large quantity range the reaction time increased with decreasing numerical ratio ($F_{(4,68)}=31.889$, $P<0.001$), while the continuous variable factor was not significant ($F_{(1,17)}=2.880$, $P=0.108$; interaction $F_{(4,68)}=2.469$, $P=0.053$, Fig. 3). There was a significant trend (linear trend: $F_{(1,17)}=57.302$, $P<0.001$; quadratic trend: $F_{(1,17)}=8.250$, $P=0.011$).

Similarly, accuracy decreased with decreasing numerical ratio ($F_{(4,68)}=8.564$, $P<0.001$), control of the continuous variables did not affect performance ($F_{(1,17)}=0.205$, $P=0.657$) and no interaction was found ($F_{(4,68)}=0.359$, $P=0.837$). There was a significant trend (linear trend: $F_{(1,17)}=24.348$, $P<0.001$; quadratic trend: $F_{(1,17)}=8.303$, $P=0.010$).

On the whole, accuracy did not differ between the small number range and the large number range ($t_{(17)}=1.877$, $P=0.078$). However, for the ratio of 0.75 the participants were significantly more accurate in the small number range than the large number range ($t_{(17)}=2.197$, $P=0.042$).

Fish experiment

Data were analysed separately for small number and large number range with a two (experience: juveniles/adults) by five (numerical ratio: 0.25/0.33/0.50/0.67/0.75) between-subject ANOVA.

Small numbers. The proportion of time spent near the larger shoal was not influenced by either numerical ratio ($F_{(4,169)}=0.047$, $P=0.996$) or experience ($F_{(1,169)}=0.030$, $P=0.864$), and the interaction was not significant ($F_{(4,169)}=0.242$, $P=0.914$, Fig. 4). No significant trend was found ($F_{(4,\ 169)}=0.045$, $P=0.876$). Likelihood ratio analysis also reflected the absence of ratio effect ($\lambda=3.68$).

Large numbers. In the large number range, the proportion of time spent near the larger shoal was influenced by numerical ratio ($F_{(4,169)}=3.190$, $P=0.015$) but not by experience ($F_{(1,169)}=0.300$, $P=0.585$; interaction: $F_{(4,169)}=0.341$, $P=0.850$, Fig. 4). There was a significant linear trend ($F_{(4,\ 169)}=3.312$, $P<0.001$).

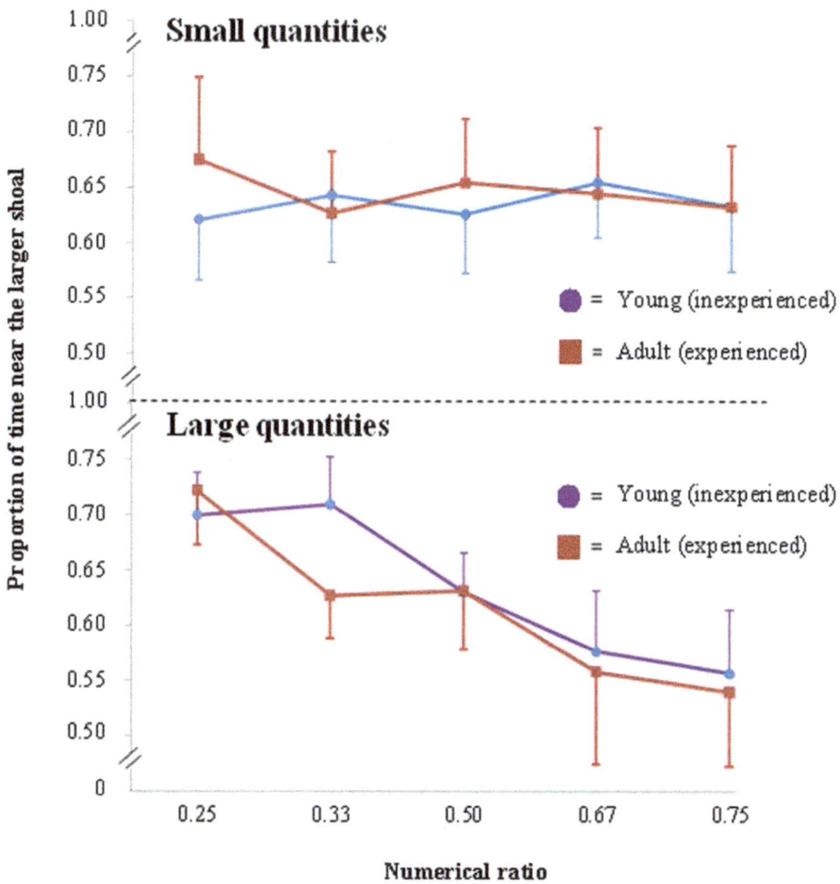

Figure 4. The results of the fish experiment. The proportion of time spent near the larger shoal is plotted against the numerical ratio of the contrasts for both large and small number (1–4) ranges. The performance of the fish (both experienced and inexperienced fish) adhered to the same patterns as for humans in the two numerical ranges, with ratio sensitivity only being shown in the large number range. Bars represent the standard error.

Within the small number range, preference for the larger group was statistically significant for all ratios (0.25: $t_{(33)} = 3.265$, $P = 0.003$; 0.33: $t_{(33)} = 3.260$, $P = 0.003$; 0.50: $t_{(33)} = 3.537$, $P < 0.001$; 0.67: $t_{(33)} = 4.003$, $P < 0.001$; 0.75: $t_{(33)} = 3.230$, $P = 0.003$); in the large number range, the fish spent significantly more time near the larger shoal for ratios 0.25 ($t_{(33)} = 6.922$, $P < 0.001$), 0.33 ($t_{(33)} = 5.788$, $P < 0.001$) and 0.50 ($t_{(33)} = 4.466$, $P < 0.001$), but not for ratios 0.67 ($t_{(33)} = 1.481$, $P = 0.148$) or 0.75 ($t_{(33)} = 1.159$, $P = 0.255$).

No single ratio significantly differed between the two ranges (two sample t-test, df = 66 all $P > 0.169$). Yet all comparisons had a very low statistical power ($0.052 \leq x \leq 0.279$) and therefore lack of significance could be due to type II error. Sample size calculation indicates that, for each comparison of our experiment, a minimum of 90 subjects was needed in order to reach a 80% power to detect a 20% difference between treatment groups with a two-sided test and alpha = 0.05 [45].

To test if lack of significance of between-range comparisons was due to inadequate sample size we increased the sample size for one numerical ratio (0.67), by testing 64 additional adult fish, 32 in a small number discrimination (2 vs. 3) and 32 in a large number discrimination (4 vs. 6). The difference between the 2 vs. 3 and the 4 vs. 6 contrasts became statistically significant (independent t-test $t_{(130)} = 2.688$, $P = 0.008$). As observed with humans, as discrimination became more difficult, the fish tended to be more accurate in the small number range; they could in fact discriminate between two numbers with a 0.67 or 0.75 ratio in the small number range, whereas they required a 2:1 ratio when tested with large numbers.

Discussion

Previous studies have reported remarkable similarities in the performance of non-verbal numerical tasks among humans, apes and monkeys, suggesting the existence of the same basic numerical systems among primates [3,29,46,47,48]. Here we provide evidence of a similar correspondence in numerical abilities between humans and teleost fishes.

When tested in the same numerical tasks, the students and guppies showed almost identical performance patterns. In both species, the ability to discriminate between large numbers (>4) was approximate and strongly dependent on the ratio between the numerosities. In contrast, in both fish and students, discrimination in the small number range was not dependent on ratio and discriminating 3 from 4 was as easy as discriminating 1 from 4. Likelihood ratio analyses indicate that the lack of ratio effect is 3.68 times more likely than the alternative hypothesis in fish and 3.67 times in students. As a consequence, the discrimination of larger numerical ratios, 0.67 and 0.75, is easy in the small number range for both species, but becomes relatively more difficult (among students) or even impossible (among fish) when confronted with large numbers.

It is possible that, in fish, the different discriminative ability in the two numerical ranges is the consequence of a different encounter rate with large or small social groups and thus a different familiarity with large and small numbers. However, the observation that juveniles raised in pairs, and therefore with no previous experience in comparing social groups, showed the same pattern as experienced adults rules this possibility out and indicates that numerical systems are probably innate in fish. The observation that some numerical abilities are exhibited by guppies at birth is certainly remarkable and is line with evidence that both chicks and newborns display some rudimentary numerical skills [49,50]. This reinforces the proposition by some authors of a core knowledge of number, a system of innate numerical representation shared among the different non-human species [3,51].

Since numerosity normally co-varies with other physical attributes such as the total area occupied by objects, one may argue that in the fish experiment subjects were using these cues instead of numerical information to solve the task. This possibility was not checked in the present study. However in two previous studies we showed that guppies and mosquitofish can discriminate between two schools of fish using the numerosity information only, both within the small number range or outside it [37,52] and that mosquitofish can discriminate between sets of geometric figures in both numerical ranges even after all continuous variables were controlled for. In particular a recent study using a training procedure has shown no difference in the learning rate between fish trained to use numerical information only and fish trained to use continuous quantities only, suggesting that the number per se is not more cognitively demanding than continuous quantities [53]. On the other hand students showed a very similar performance whether continuous variables were controlled or not, thus making the comparisons fully legitimate.

One can argue that the guppy and student experiments differ in many respects. In particular students were tested in a sequential presentation, whereas fish saw the simultaneous presentation of the numerosity pairs. However in this respect the two experiments may differ less than may appear at first glance. In our test situation the fish could never see the two stimuli binocularly simultaneously. It was possible for a subject to observe both shoals only when swimming perpendicularly to the main axis, which occurred very rarely during a test. Moreover, in this position, each stimulus was seen with a different eye; in this situation, as a consequence of the lateralization of social recognition [54,55] and reduced inter-hemispheric transfer of information [56,57], fish cannot be expected to guess the larger group.

The difference in ratio-dependence suggests the existence in fish, as in humans, of two distinct non-verbal mechanisms of numerical representation, one for numbers 1–4 and one for large quantities. Yet the hypothesis that a precise object-file mechanism does underlie small number discrimination also predicts higher performance in the small number range, a finding not evident in our study. However, the lack of a statistical difference between ranges observed in fish experiment may be due to the limit of the procedure adopted. Previous studies using this procedure have shown that accuracy in selecting the larger shoal never exceed 70% even with very easy numerical ratios (guppies [37,58], mosquitofish [22], topminnows [59], angelfish [60]). Due to large measurement variance combined with small sample size in each numerical contrast, these statistical tests suffered from low power and hence lack of significance may be attributable to a great probability of making a type II error. Sample size calculation indicates that in our experiment a threefold sample size was needed to obtain an adequate statistical power when two treatments had to be compared. As confirmation, a statistical difference emerged between the large and small number range after we increased the sample size in the 0.67 ratio. Regarding this latter evidence, previous work has reported that macaques, mosquitofish, chicks and bees could distinguish two from three items but failed with the same ratio in the large number range [22,29,33,34,61], even though none of these studies could provide a statistical test to support a difference in performance between the two ranges. Our result highlights the possibility that, as in other research fields [62,63,64], many studies are underpowered to detect statistical differences among subgroups.

The results reported here differ from those found in a very recent study on angelfish which discriminated 2 vs. 3 but not 3 vs.

4 fish [60]. Since both guppies and mosquitofish appear to discriminate 3 from 4 fish [22,37] there might be taxonomic or ecological differences in the numerical abilities among fish species. Other explanations are however possible. Since poeciliids do not form tight shoals, during the experiment, all stimulus fish of one group are usually simultaneously visible to the subject. Angelfish, by contrast, are characterized by a wide lateral body surface and tend to form coherent and synchronized schools; thus, as the number of individuals increases some individuals could often block the sight of other shoalmates, making it difficult for a subject to accurately estimate group numerosity.

Some recent studies found that accuracy was affected by numerical ratio for both small and large numbers leading some researchers to question the existence of two separate numerical systems [16,65]. It is unclear why some studies find that performance in the range 1–4 is independent of ratio and some do not. A possible explanation for this literature inconsistency, recently proposed by some authors (e.g. [26,66,67]), is that small quantities may be represented by both analog magnitude and object-file mechanisms, and that recruitment of one or the other system may depend on context and previous experience. In line with this hypothesis, a recent study found that, unlike controls, a sample of participants with an expertise that requires years of training in estimation of magnitudes showed the typical signature of the analog magnitude system, ratio effect, in the small number range too [Agrillo and Piffer, unpublished].

However that may be, the lack of a ratio effect in the 1–4 numerical range does not necessarily entail the existence of a separate system. As pointed out by Gallistel and Gelman [9], the difference in ratio effect between large and small numbers could occur because there is so little error in the analog magnitude representations of 1, 2, 3, and 4 that they are highly distinguishable from one another, and thus coarse behavioral measures of the underlying processes (as they often are with both human and non-human experiments), fail to evidence a ratio dependence when in fact such a relationship may exist.

Our results do not help to clarify this issue, being compatible with both hypotheses. However, even assuming the existence of a single system of analog magnitude with a different ratio sensitivity in the range 1–4 and beyond it, the similarity in human and guppy experiments in the steepness of the slope in both ranges is again strongly suggestive of similar systems of numerical representation.

Other studies have provided evidence to support strong similarities between teleosts and primates. Swordtails, mosquitofish, angelfish and climbing perch appear to adhere to Weber's Law when discriminating between two large quantities [22,23,38,68] and mosquitofish trained to discriminate between large sets of geometric figures were found to be equally efficient in discriminating 4 vs. 8 items or 100 vs. 200 items, exactly like the college students tested with the same stimuli [36]. Naturally, as with most comparative data, it is always possible that a strong similarity in cognitive abilities is the product of convergent evolution and that similar performance reflect very dissimilar underlying mechanisms.

It might seem surprising to discover similar numerical abilities in humans and in guppies, especially when considering that the brain size of a guppy is less than a thousandth of that of primates. However, it is clear from recent literature that the cognitive abilities of fish have been greatly underestimated and that teleosts are capable of complex behaviors such as individual recognition, transmission of cultural traditions, cooperation, copying behavior and deception, which have traditionally been associated with the evolution of large cortical areas in mammals and birds [69,70]. On the other hand, it is also possible that a cognitive function such as numerical discrimination, which is apparently complex, may actually be based on relatively simple neural circuits, as suggested by a recent neural network study [71].

On a more general note, the evolution of numerical abilities in animal species is still a largely unexplored field and future research is needed to understand the origin of the quantitative systems shown by vertebrates. If numerical abilities have evolved many times independently in different taxa it would be challenging to understand which selective constraints have shaped them in a converging fashion. On the other hand, the results of this comparative study admits the possibility of common mechanisms between primates and basal vertebrates, suggesting that the evolutionary emergence of numerical abilities may be very ancient, possibly dating back to before the teleost-tetrapod divergence.

Author Contributions

Conceived and designed the experiments: CA AB. Performed the experiments: CA LP. Analyzed the data: LP BB. Wrote the paper: CA AB BB.

References

1. Burr D, Ross J (2008) A visual sense of number. Curr Biol 18: 425–428.
2. Nieder A (2005) Counting on neurons: The neurobiology of numerical competence. Nat Rev Neurosci 6: 177–190.
3. Feigenson L, Dehaene S, Spelke ES (2004) Core systems of number. Trends Cogn Sci 8: 307–314.
4. Xu F, Spelke ES (2000) Large number discrimination in 6-month-old infants. Cognition 74: B1–B11.
5. Aoki T (1977) On the counting process of patterned dots. Toh Psychol Folia 36: 15–22.
6. Atkinson J, Campbell FW, Francis MR (1976) The magic number 4±0: a new look at visual numerosity judgements. Perception 5: 327–334.
7. Trick LM, Pylyshyn ZW (1994) Why are small and large numbers enumerated differently: A limited-capacity preattentive stage in vision. Psychol Rev 101: 80–102.
8. Dehaene S (1997) The number sense. New York: Oxford University Press.
9. Gallistel CR, Gelman R (1992) Preverbal and verbal counting and computation. Cognition 44: 43–74.
10. Nieder A, Dehaene S (2009) Representation of number in the brain. Ann Rev Neurosci 32: 185–208.
11. Kahneman D, Treisman A, Gibbs BJ (1992) The reviewing of object files - object-specific integration of information. Cogn Psychol 24: 175–219.
12. Meck WH, Church RM (1983) A mode control model of counting and timing processes. J Exp Psych: Anim Behav Proc 9: 320–334.
13. Whalen J, Gallistel CR, Gelman R (1999) Nonverbal counting in humans: The psychophysics of number representation. Psychol Sci 10: 130–137.
14. Beran MJ, Taglialatela LA, Flemming TM, James FM, Washburn DA (2006) Nonverbal estimation during numerosity judgements by adult humans. Quart J Exp Psych 59: 2065–2082.
15. Cordes S, Gelman R, Gallistel CR, Whalen J (2001) Variability signatures distinguish verbal from nonverbal counting for both large and small numbers. Psychon Bull Rev 8: 698–707.
16. Cantlon JF, Brannon EM (2006) Shared system for ordering small and large numbers in monkeys and humans. Psychol Sci 17: 401–406.
17. Ansari D, Lyons IM, van Eitneren L, Xu F (2007) Linking visual attention and number processing in the brain: The role of the temporo-parietal junction in small and large symbolic and nonsymbolic number comparison. J Cogn Neurosci 19: 1845–1853.
18. Demeyere N, Lestou V, Humphreys GW (2010) Neuropsychological evidence for a dissociation in counting and subitizing. Neurocase 16: 219–237.
19. Hyde DC, Spelke ES (2009) All numbers are not equal: An electrophysiological investigation of small and large number representations. J Cogn Neurosci 21: 1039–1053.
20. Vetter P, Butterworth B, Bahrami B (2011) A candidate for the attentional bottleneck: Set-size specific modulation of right TPJ during attentive enumeration. J Cogn Neurosci 23: 728–736.
21. Piazza M, Mechelli A, Butterworth B, Price CJ (2002) Are subitizing and counting implemented as separate or functionally overlapping processes? Neuroimage 15: 435–446.
22. Agrillo C, Dadda M, Serena G, Bisazza A (2008) Do fish count? Spontaneous discrimination of quantity in female mosquitofish. Anim Cogn 11: 495–503.

23. Buckingham JN, Wong BBM, Rosenthal GG (2007) Shoaling decisions in female swordtails: how do fish gauge group size? Behaviour 144: 1333–1346.
24. Ward C, Smuts BB (2007) Quantity-based judgments in the domestic dog (*Canis lupus familiaris*). Anim Cogn 10: 71–80.
25. Starkey P, Cooper RG (1980) Perception of numbers by human infants. Science 210: 1033–1035.
26. vanMarle K, Wynn K (2009) Infants' auditory enumeration: Evidence for analog magnitudes in the small number range. Cognition 111: 302–316.
27. Tomonaga M, Matsuzawa T (2002) Enumeration of briefly presented items by the chimpanzee (*Pan troglodytes*) and humans (*Homo sapiens*). Anim Learn Behav 30: 143–157.
28. Hanus D, Call J (2007) Discrete quantity judgments in the great apes (*Pan paniscus, Pan troglodytes, Gorilla gorilla, Pongo pygmaeus*): The effect of presenting whole sets versus item-by-item. J Comp Psych 121(3): 241–249.
29. Hauser MD, Carey S, Hauser LB (2000) Spontaneous number representation in semi-free-ranging rhesus monkeys. Proc R Soc Lond B 267: 829–833.
30. Cantlon JF, Brannon EM (2007) Basic math in monkeys and college students. PLoS Biol 5: 2912–2919.
31. Bonanni R, Natoli E, Cafazzo S, Valsecchi P (2011) Free-ranging dogs assess the quantity of opponents in intergroup conflicts. Anim Cogn 14: 103–115.
32. Hunt S, Low J, Burns KC (2008) Adaptive numerical competency in a food-hoarding songbird. Proc R Soc Lond B 267: 2373–2379.
33. Rugani R, Regolin L, Vallortigara G (2008) Discrimination of small numerosities in young chicks. J Exp Psych: Anim Behav Proc 34: 388–399.
34. Gross HJ, Pahl M, Si A, Zhu H, Tautz J, et al. (2009) Number-based visual generalisation in the honeybee. PLoS ONE 4: e4263.
35. Agrillo C, Dadda M, Serena G, Bisazza A (2009) Use of number by fish. PLoS ONE 4(3): e4786.
36. Agrillo C, Piffer L, Bisazza A (2010) Large number discrimination by mosquitofish. PLoS ONE 5(12): e15232.
37. Bisazza A, Piffer L, Serena G, Agrillo C (2010) Ontogeny of numerical abilities in fish. PLoS ONE 5(11): e15516.
38. Gómez-Laplaza LM, Gerlai R (2010) Can angelfish (*Pterophyllum scalare*) count? Discrimination between different shoal sizes follows Weber's law. Anim Cogn. pp 1–9.
39. Shuman M, Kanwisher N (2004) Numerical magnitude in the human parietal lobe: Tests of representational generality and domain specificity. Neuron 44: 557–569.
40. Buckley P, Gillman C (1974) Comparisons of digits and dot patterns. J Exp Psychol 103: 1131–1136.
41. Revkin SK, Piazza M, Izard V, Cohen L, Dehaene S (2008) Does subitizing reflect numerical estimation? Psychol Sci 19: 607–614.
42. Ledesma JM, McRobert SP (2008) Shoaling in juvenile guppies: The effects of body size and shoal size. Behav Proc 77: 384–388.
43. Keppel G (1991) Design and analysis: A researcher's handbook. Englewood Cliffs: Prentice Hall.
44. Glover S, Dixon P (2004) Likelihood ratios: A simple and flexible statistic for empirical psychologists. Psychon Bull Rev 11: 791–806.
45. Sokal RR, Rohlf FJ (1995) Biometry. New York: Freeman and Company.
46. Beran MJ (2008) The evolutionary and developmental foundations of mathematics. PLoS Biology 6: 221–223.
47. Feigenson L, Carey S, Hauser MD (2002) The representations underlying infants' choice of more: Object files versus analog magnitudes. Psychol Sci 13: 150–156.
48. Matsuzawa T (2009) Symbolic representation of number in chimpanzees. Curr Opin Neurobio 19: 92–98.
49. Izard V, Sann C, Spelke ES, Streri A (2009) Newborn infants perceive abstract numbers. Proc Natl Acad Sci USA 106: 10382–10385.
50. Rugani R, Fontanari L, Simoni E, Regolin L, Vallortigara G (2009) Arithmetic in newborn chicks. Proc R Soc Lond B 276: 2451–2460.
51. Spelke ES, Kinzler KD (2007) Core knowledge. Develop Sci 10: 89–96.
52. Dadda M, Piffer L, Agrillo C, Bisazza A (2009) Spontaneous number representation in mosquitofish. Cognition 112: 343–348.
53. Agrillo C, Piffer L, Bisazza A (2011) Number versus continuous quantity in numerosity judgments by fish. Cognition 119: 281–287.
54. Dadda M, Zandona E, Agrillo C, Bisazza A (2009) The costs of hemispheric specialization in a fish. Proc R Soc Lond B 276: 4399–4407.
55. Sovrano VA, Rainoldi C, Bisazza A, Vallortigara G (1999) Roots of brain specializations: preferential left-eye use during mirror-image inspection in six species of teleost fish. Behav Brain Res 106: 175–180.
56. Ingle D (1968) Interocular integration of visual learning by goldfish. Brain Behav Evol 1: 58–85.
57. Mark R (1966) The tectal commissure and interocular transfer of pattern discrimination in cichlid fish. Exp Neurol 16: 215–225.
58. Piffer L, Agrillo C, Hyde DC (2011) Small and large number discrimination in guppies. Anim Cogn;on-line first: 10.1007/s10071-10011-10447-10079.
59. Agrillo C, Dadda M (2007) Discrimination of the larger shoal in the poeciliid fish *Girardinus falcatus*. Ethol Ecol Evol 19: 145–157.
60. Gòmez-Laplaza LM, Gerlai R (2011) Spontaneous discrimination of small quantities: shoaling preferences in angelfish (*Pterophyllum scalare*). Anim Cogn 14: 565–574.
61. Flombaum JI, Junge JA, Hauser MD (2005) Rhesus monkeys (*Macaca mulatta*) spontaneously compute addition operations over large numbers. Cognition 97: 315–325.
62. Maxwell SE (2004) The persistence of underpowered studies in psychological research: Causes, consequences, and remedies. Psychol Meth 9: 147–163.
63. Maggard MA, O'Connell JB, Liu JH, Etzioni DA, Ko CY (2003) Sample size calculations in surgery: Are they done correctly? Surgery 134: 275–279.
64. Jennions MD, Moller AP (2003) A survey of the statistical power of research in behavioral ecology and animal behavior. Behav Ecol 14: 438–445.
65. Beran MJ (2007) Rhesus monkeys (*Macaca mulatta*) enumerate large and small sequentially presented sets of items using analog numerical representations. J Exp Psych: Anim Behav Proc 33: 42–54.
66. Cordes S, Brannon EM (2008) Quantitative competencies in infancy. Develop Sci 11: 803–808.
67. Cordes S, Brannon EM (2009) Crossing the divide: Infants discriminate small from large numerosities. Develop Psych 45: 1583–1594.
68. Binoy VV, Thomas KJ (2004) The climbing perch (*Anabas testudineus Bloch*), a freshwater fish, prefers larger unfamiliar shoals to smaller familiar shoals. Curr Sci 86: 207–211.
69. Brown C, Laland KN (2003) Social learning in fishes: A review. Fish Fish 4: 280–288.
70. Bshary R, Wickler W, Fricke H (2002) Fish cognition: A primate's eye view. Anim Cogn 5: 1–13.
71. Stoianov I, Zorzi M (2012) Emergence of a "Visual Number Sense" in hierarchical generative models. Nat Neurosci;Online first, doi: 10.1038/nn.2996.

10

The Liver X Receptor Agonist GW3965 Improves Recovery from Mild Repetitive Traumatic Brain Injury in Mice Partly through Apolipoprotein E

Dhananjay R. Namjoshi[1], Georgina Martin[1], James Donkin[1], Anna Wilkinson[1], Sophie Stukas[1], Jianjia Fan[1], Michael Carr[1], Sepideh Tabarestani[1], Kelli Wuerth[2], Robert E. W. Hancock[2], Cheryl L. Wellington[1]*

1 Department of Pathology and Laboratory Medicine, The University of British Columbia, Vancouver, British Columbia, Canada, **2** Department of Microbiology and Immunology, The University of British Columbia, Vancouver, British Columbia, Canada

Abstract

Traumatic brain injury (TBI) increases Alzheimer's disease (AD) risk and leads to the deposition of neurofibrillary tangles and amyloid deposits similar to those found in AD. Agonists of Liver X receptors (LXRs), which regulate the expression of many genes involved in lipid homeostasis and inflammation, improve cognition and reduce neuropathology in AD mice. One pathway by which LXR agonists exert their beneficial effects is through ATP-binding cassette transporter A1 (ABCA1)-mediated lipid transport onto apolipoprotein E (apoE). To test the therapeutic utility of this pathway for TBI, we subjected male wild-type (WT) and apoE−/− mice to mild repetitive traumatic brain injury (mrTBI) followed by treatment with vehicle or the LXR agonist GW3965 at 15 mg/kg/day. GW3965 treatment restored impaired novel object recognition memory in WT but not apoE−/− mice. GW3965 did not significantly enhance the spontaneous recovery of motor deficits observed in all groups. Total soluble $A\beta_{40}$ and $A\beta_{42}$ levels were significantly elevated in WT and apoE−/− mice after injury, a response that was suppressed by GW3965 in both genotypes. WT mice showed mild but significant axonal damage at 2 d post-mrTBI, which was suppressed by GW3965. In contrast, apoE−/− mice showed severe axonal damage from 2 to 14 d after mrTBI that was unresponsive to GW3965. Because our mrTBI model does not produce significant inflammation, the beneficial effects of GW3965 we observed are unlikely to be related to reduced inflammation. Rather, our results suggest that both apoE-dependent and apoE-independent pathways contribute to the ability of GW3965 to promote recovery from mrTBI.

Editor: Makoto Makishima, Nihon University School of Medicine, Japan

Funding: This work was supported by an operating grant (MOP 67088) to C.L.W. from the Canadian Institutes of Health Research (CIHR). D.R.N. was supported by Alzheimer Society Research Program (ASRP, Alzheimer Society of Canada) doctoral award. J.D. was supported by an ASRP post-doctoral award. S.S. was supported by a CIHR Vanier studentship. J.F. was supported by a Child & Family Research Institute studentship. The funders had no role in study design, data collection and analysis, decision to publish, or preparation of the manuscript.

Competing Interests: The authors have declared that no competing interests exist.

* E-mail: *cheryl@cmmt.ubc.ca

Introduction

Traumatic brain injury (TBI) is a leading cause of death and disability in North America with an estimated annual incidence of 500 cases per 100,000 persons [1]. Motor vehicle accidents, sports injuries, and falls are the most common causes of TBI in civilians [2]. TBI prevalence in military personnel is estimated to be at least 20% and encompasses both blast-related and impact injuries [3]. The mechanical forces experienced during TBI deform brain tissue, causing a primary injury that directly affects blood vessels, axons, neurons, and glia in a focal, multifocal or diffuse pattern. This primary injury initiates a cascade of secondary processes that result in complex cellular, inflammatory, neurochemical, and metabolic alterations in the hours to weeks after injury [4]. Although few options are available to manage the primary injury, the ensuing secondary injury pathways are potentially treatable [5].

Apolipoprotein E (apoE) is the major lipid carrier in the brain [6]. Brain injury increases apoE levels in order to scavenge lipids released by degenerating neurons and myelin, which are later delivered to surviving neurons during reinnervation and synaptogenesis [7,8]. In humans, TBI results in a 14-fold selective reduction in cerebrospinal fluid (CSF) apoE levels by 2 to 5 days following injury and a 2- and 4-fold increase in total and free cholesterol, respectively [9], reflecting the pronounced lipid mobilization, remodeling, and uptake of CSF apoE lipoproteins by the injured brain. In animal models, apoE is induced within hours to weeks after TBI followed by increased neuronal uptake of apoE-containing lipoproteins through the low-density lipoprotein receptor (LDLR) [8,10]. In mice, apoE deficiency compromises recovery from acute neurological injuries, including TBI [11,12].

ApoE also regulates β-amyloid (Aβ) metabolism, a function that underlies its genetic association with Alzheimer's disease (AD). AD is defined neuropathologically by the presence of extracellular amyloid plaques composed of aggregated Aβ peptides and intracellular neurofibrillary tangles (NFTs) consisting of hyperphosphorylated tau [13]. The human apoE4 allele increases AD risk and hastens its onset [14,15], whereas apoE2 delays onset and reduces AD risk [16]. Recent microdialysis studies demonstrated that apoE genotype modulates the Aβ half-life in brain interstitial fluid (ISF), with the rate of Aβ degradation accelerated by apoE2 ($t_{1/2} = 0.56$ h) and prolonged by apoE4 ($t_{1/2} = 1.1$ h) compared to apoE3 ($t_{1/2} = 0.71$ h) [17]. Although AD risk and age of onset are indisputably modified by apoE genotype, the relationship between apoE genotype and TBI outcome is complex. Some studies report that apoE4 carriers have significantly poorer outcomes compared to non-carriers [18–20], while others find no association [21,22]. Many variables pose challenges to resolving these uncertainties, including relatively short follow-up times after TBI and the diversity of subjects with respect to the age, gender, and injury severity. Despite these challenges, a meta-analysis of 14 studies with 2,527 subjects concluded that apoE4 does not affect initial injury severity but rather may compromise recovery at 6 months post-injury [23].

Several epidemiological studies suggest that TBI may increase the risk of dementia, particularly AD, although this association is not always observed (reviewed in [24]). TBI has also been associated with an earlier onset of AD [24]. Both NFTs and amyloid plaques are found in post-mortem TBI brain tissue. However, the widespread white matter involvement in TBI that is not present in AD results in noteworthy differences in the pattern and distribution of their common neuropathological features. For example, TBI neuropathology is largely that of tau deposition, as only approximately 30% of TBI patients also contain Aβ deposits [25]. Aβ plaques can appear within hours and remain detectable up to 47 years after a single moderate-severe TBI [26]. NFTs are also detectable after a single TBI and are remarkably prominent in mild, repetitive, concussive injuries [26]. Despite the low prevalence of amyloid deposits in the post-mortem TBI brain, amyloid precursor protein (APP) accumulation in damaged axons is a striking histological hallmark of diffuse axonal injury (DAI) [27,28]. Increased APP in the post-TBI brain is hypothesized to trigger a burst of Aβ production that can deposit in amyloid plaques [29]. Intriguingly, microdialysis experiments in brain-injured humans have demonstrated that ISF Aβ levels correlate positively with the patient's Glasgow Coma Score [30,31], suggesting that Aβ release is associated with recovery of synaptic function, as has been demonstrated in animals [32].

Improving apoE function may therefore facilitate both neuronal repair and Aβ clearance after TBI, thus potentially offering both acute and long-term benefits. One method to enhance apoE function is to promote its lipidation by ATP-binding cassette transporter A1 (ABCA1), which is the rate-limiting step in generating apoE-containing lipoprotein particles in the central nervous system (CNS). ABCA1 deficiency leads to poorly-lipidated apoE in the CNS [33,34], and increases amyloid load in AD mice [35–37]. Conversely, overexpression of ABCA1 in the CNS increases apoE lipidation and greatly decreases amyloid deposits in AD mice [38]. Transcription of ABCA1 and apoE is induced by agonists of Liver X receptors (LXR), which regulate many genes involved in lipid metabolism and inflammation [39,40]. Genetic deficiency of LXRs increases amyloid burden in AD mice [41]. Synthetic LXR agonists including TO901317 and GW3965 cross the blood-brain barrier (BBB), induce ABCA1 and apoE expression, improve memory and reduce Aβ levels in AD mice [42–44]. Importantly, ABCA1 is required for several beneficial effects of GW3965 in AD mice [42], including increased CSF apoE levels, reduced amyloid load, and improved memory. These observations provide a compelling rationale for testing the therapeutic potential of LXR agonists for TBI. Indeed, TO901317 reduces Aβ accumulation and promotes cognitive recovery in a controlled cortical impact (CCI) model of moderate-severe brain injury in mice [45].

Approximately 80% of human TBI are mild injuries without skull fracture and loss of consciousness [46]. It is increasingly appreciated that repetitive mild TBI, commonly experienced by athletes in high-contact sports may lead to chronic traumatic encephalopathy, which is characterized by cognitive, executive, and motor function disturbances, tau deposition and, in some cases, amyloid deposits similar to those found in AD [47].

To determine the therapeutic utility of LXR agonists for this type of brain injury, we established a mouse model of mild repetitive TBI (mrTBI), wherein a gravity-driven weight drop device was used to deliver two consecutive injuries 24 h apart. Here we report that GW3965 improves cognitive recovery and suppresses axonal damage in an apoE-dependent manner. Surprisingly, apoE was not required for GW3965 to suppress the transient increase in Aβ levels. Our results provide additional support for the therapeutic potential of LXR agonists in TBI, and demonstrate that both apoE-dependent and apoE-independent pathways contribute to their beneficial effects.

Results

1. ApoE is Required for GW3965 to Improve Cognitive Function after mrTBI

Non-spatial, working memory was assessed using the novel object recognition (NOR) task at 2, 7, and 14 days following mrTBI. Discrimination between the novel and familiar objects was determined using the discrimination index (DI) [48]. During training, all animals displayed an equivalent time exploring each of the two identical objects (Figure 1A,B, DI ~ 0) indicating that genotype, surgical procedures, treatment, and time did not affect baseline exploratory behavior. Sham-operated WT (Figure 1A, gray bar, $p<0.001$ compared to training, two-way ANOVA) and apoE−/− (Figure 1B, gray bar, $p<0.01$ compared to training, two-way ANOVA) mice retained preference to novelty as indicated by positive DI values. Untreated WT (Figure 1A, open bars) and apoE−/− (Figure 1B, open bars) failed to discriminate between the novel and familiar objects ($p>0.05$ compared to training, two-way ANOVA) at all post-mrTBI time points. Although GW3965-treated WT mice showed impaired memory at 2d post-mrTBI (Figure 1A, black bars, $p>0.05$ compared to training, two-way ANOVA) object recognition was restored by day 7 and 14 post-mrTBI ($p<0.001$ compared to training, two-way ANOVA). Finally, GW3965 treatment failed to improve NOR performance in apoE−/− mice at any post-mrTBI time points (Figure 1B, black bars, $p>0.05$ compared to training, two-way ANOVA). NOR performance was not affected by motor impairment as indicated by no significant difference between the total path length covered by sham and injured WT as well as apoE−/− mice during NOR testing (Figure S1, $p>0.05$, two-way ANOVA).

2. Spontaneous Recovery of Motor Dysfunction after mrTBI is Unaffected by GW3965

mrTBI significantly impaired the motor performance of WT (Figure 2A, $p<0.05$, two-way repeated measures ANOVA) and apoE−/− (Figure 2B, $p<0.05$, two-way repeated measures ANOVA) mice at 1, 2, and 7 d post-injury as indicated by reduced failing latencies on an accelerating rotarod, compared to the baseline performance. By 14 d, however, motor performance had fully recovered in WT and apoE−/− animals (Figure 2A,B,

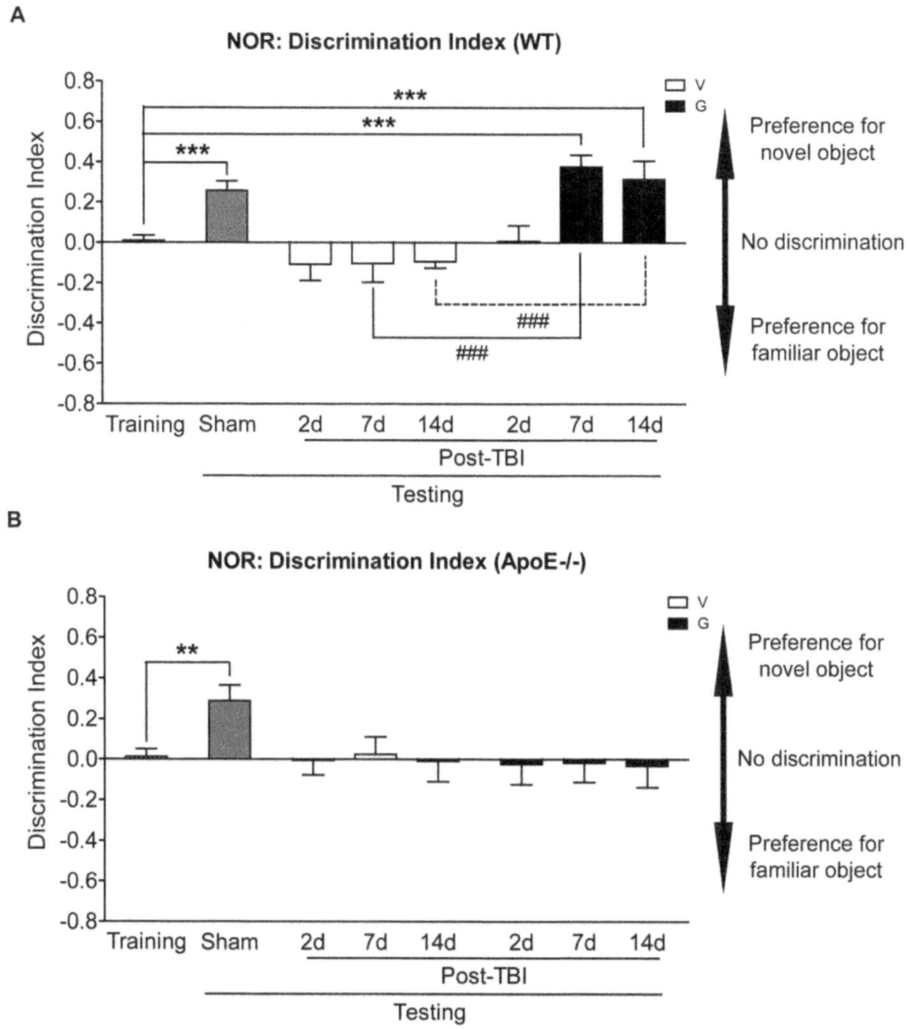

Figure 1. ApoE is required for GW3965 to improve NOR performance after mrTBI. NOR memory was evaluated in untreated (V) and GW3965-treated (G) WT (**A**) and apoE −/− (**B**) mice at 2, 7, and 14 days post-mrTBI. Bars represent discrimination index (DI) scores. Cohorts were: sham (WT, n = 23, pooled; apoE−/−, n = 18, pooled) and 2 day (WT: untreated, n = 18, GW3965-treated, n = 15; apoE−/−: untreated, n = 10, GW3965-treated, n = 10), 7 day (WT: untreated, n = 13, GW3965-treated, n = 10; apoE−/−: untreated, n = 11, GW3965-treated, n = 11), and 14 day (WT: untreated, n = 14, GW3965-treated, n = 14; apoE−/−: untreated, n = 9, GW3965-treated, n = 9) post-mrTBI. DI scores were analyzed using two-way ANOVA and Bonferroni post hoc test. **: $p<0.01$, ***: $p<0.001$, ###: $p<0.001$.

$p>0.05$, two-way repeated measures ANOVA). GW3965 treatment of WT or apoE−/− mice did not significantly reduce the severity of motor impairment at any time point, nor did it accelerate the rate of spontaneous recovery in either genotype (Figure 2A,B, curly brackets). In contrast to GW3965, loss of apoE−/− led to significantly more severe motor impairment compared to WT animals at 1 and 2 d post-mrTBI in both untreated and treated groups (Figure 2C,D, $p<0.05$, two-way repeated measures ANOVA).

3. GW3965 Prevents mrTBI-induced Elevation of Endogenous Aβ Levels

We determined levels of soluble endogenous $A\beta_{40}$ and $A\beta_{42}$ species from ipsilateral half brains of WT and apoE−/− mice at 2, 7, and 14 d after mrTBI. Compared to sham controls, injured WT mice showed a sustained ~1.5-fold increase in $A\beta_{40}$ (Figure 3A, open bars, $p<0.05$, two-way ANOVA), and a transient ~1.7 fold increase in $A\beta_{42}$ that returned to baseline by 14 d post-injury (Figure 3B, open bars, $p<0.05$, two-way ANOVA). In contrast to untreated controls, GW3965 suppressed the increase of both $A\beta_{40}$ and $A\beta_{42}$ in WT mice (Figure 3A,B, black bars, $p>0.05$, two-way ANOVA) as compared to the sham. Similarly, apoE−/− mice exhibited a sustained ~1.5-fold increase in $A\beta_{40}$ (Figure 3C, open bars, $p<0.05$, two-way ANOVA), and a transient ~1.4-fold increase in $A\beta_{42}$ after mrTBI that returned to baseline by 14 d post-injury (Figure 3D, open bars, $p<0.05$, two-way ANOVA). Intriguingly, GW3965 effectively suppressed the increase in both $A\beta_{40}$ and $A\beta_{42}$ in apoE−/− mice (Figure 3C,D, black bars, $p>0.05$, two-way ANOVA) as compared to the sham. The GW3965 treatment effect was significant in both WT and apoE−/− mice for $A\beta_{40}$ (Figure 3A,C, $p<0.05$, two-way ANOVA), but only in WT mice for $A\beta_{42}$ (Figure 3B, $p<0.05$, two-way ANOVA). Furthermore, compared to WT mice, loss of apoE did not lead to significantly greater accumulation of Aβ species, with the single exception of $A\beta_{40}$ levels 7 d post-mrTBI in both vehicle (Figure 3A,C, #, $p<0.05$, two-way ANOVA) and GW3965-treated groups (Figure 3A,C, §, $p<0.05$, two-way

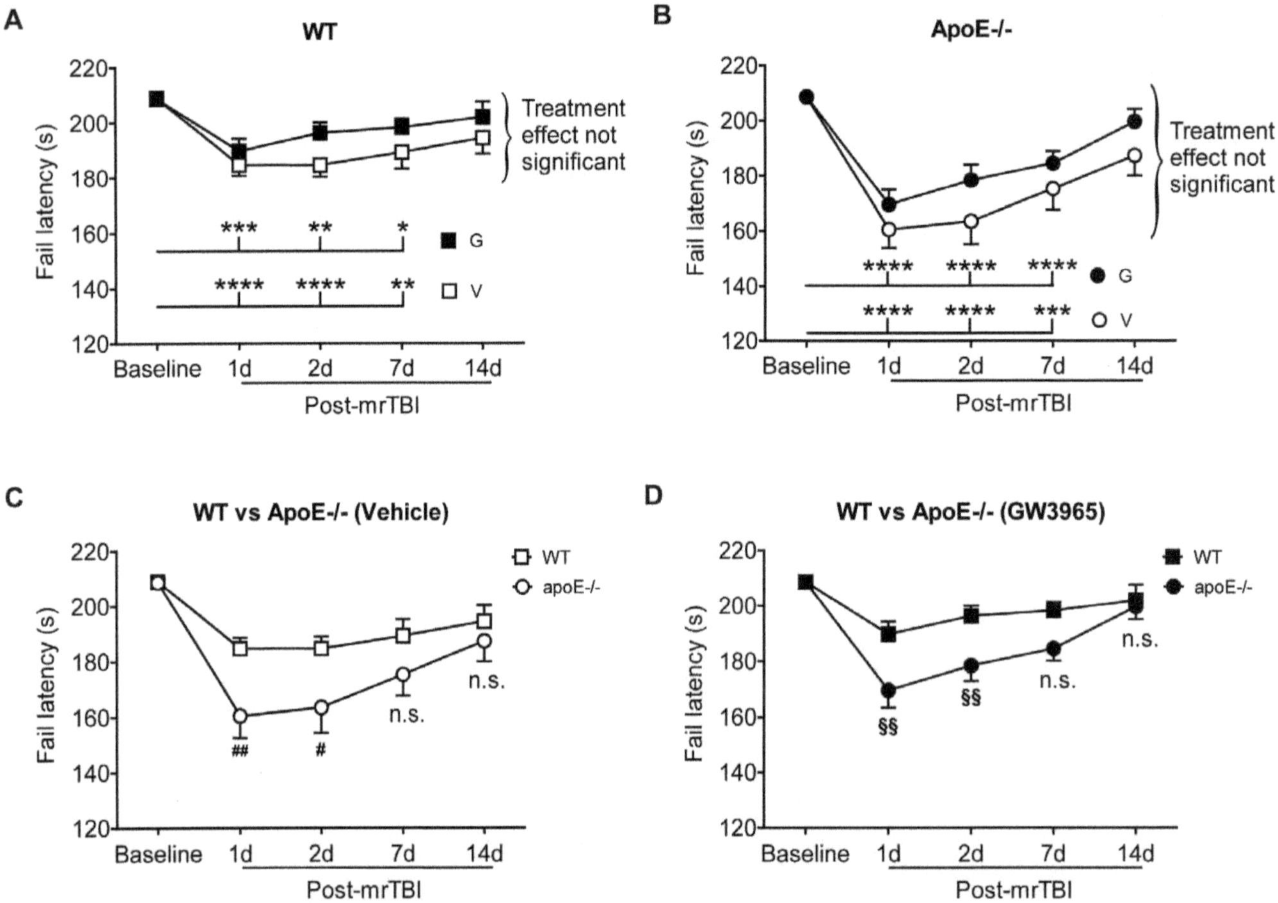

Figure 2. mrTBI-induced motor impairment recovers spontaneously independent of GW3965 and apoE. Motor performance of WT (n = 38) and apoE−/− (n = 30) mice was evaluated by the accelerating rotarod task (0 to 30 rpm in 210 s). (**A**) Rotarod latencies of untreated (V, open squares) and GW3965-treated (G, black filled squares) WT mice before and following mrTBI. (**B**) Rotarod latencies of untreated (V, open circles) and GW3965-treated (G, black filled circles) apoE−/− mice before and following mrTBI. Asterisks in A and B denote significant differences between baseline and post-mrTBI latencies within each group. For both genotypes, the treatment effect was not significant at any post-TBI time point (curly brackets). (**C**) Rotarod latencies of untreated WT (open squares) and apoE−/− (open circles) mice before and following mrTBI. (**D**) Rotarod latencies of GW3965-treated WT (black filled squares) and apoE−/− (black filled circles) mice before and following mrTBI. # and § in C and D denote significant differences between the latencies of WT and apoE−/− mice at the respective time points. Cohorts were: Untreated WT mice: (1 d = 34, 2 d = 34, 7 d = 24; 14 d = 14); GW3965-treated WT mice: (1 d = 35, 2 d = 35, 7 d = 25, 14 d = 14); untreated apoE−/− mice: (1 d = 24, 2 d = 24, 7 d = 20, 14 d = 9), and GW3965-treated apoE−/− mice: (1 d = 27, 2 d = 22, 7 d = 20, 14 d = 9). Data were analyzed using two-way repeated measures ANOVA followed by a Bonferroni post hoc test. *: $p<0.05$, **: $p<0.01$, ***: $p<0.001$, ****: $p<0.0001$, #: $p<0.05$, ##: $p<0.01$, §§: $p<0.01$.

ANOVA). APP holoprotein and APP-CTFα levels remained unchanged in both untreated and treated WT and apoE−/− mice following mrTBI (Figure S2, $p>0.05$, two-way ANOVA), indicating that increased APP processing does not account for the changes in Aβ levels.

4. GW3965 Enhances ABCA1 Induction after mrTBI

To determine whether lipid mobilization pathways are induced in our mrTBI model, we determined the levels of ABCA1, apoE, and LDLR in treated and untreated WT and apoE−/− mice. Compared to sham controls, untreated WT brains showed a transient 1.6-fold increase in ABCA1 levels on post-mrTBI day 7 (Figure 4A, open bars, $p<0.01$, two-way ANOVA) that returned to baseline by 14 d. This finding suggests that endogenous LXR agonists may be transiently induced after TBI. As ABCA1 is a sensitive LXR target, GW3965 augmented this response, leading to a 1.5–2.4-fold increase in ABCA1 protein levels over the time points examined (Figure 4A, black bars, $p<0.05$, two-way ANOVA). These results are consistent with our previous observations of ABCA1 upregulation in GW3965-treated APP/PS1 mice [42]. In contrast to WT animals, untreated apoE−/− mice did not show a transient elevation of ABCA1 after mrTBI (Figure 4B, open bars, $p>0.05$, two-way ANOVA), suggesting that the endogenous signals that lead to ABCA1 induction after TBI may be compromised in the absence of apoE. Nevertheless, similar to the WT mice GW3965 treatment significantly increased ABCA1 levels in injured apoE−/− mice (Figure 4B, black bars, $p<0.05$, two-way ANOVA), suggesting that GW3965 may bypass apoE and directly increase ABCA1 expression after mrTBI. Neither mrTBI nor GW3965 treatment led to significant changes in apoE levels in WT mice (Figure S3A, $p>0.05$, two-way ANOVA), a result consistent with the relative insensitivity of apoE compared to ABCA1 as a LXR target [42]. Similar to apoE, LDLR levels remained unaffected after mrTBI in both treated and untreated WT (Figure S3B) and apoE−/− (Figure S3C) mice ($p>0.05$, two-way ANOVA).

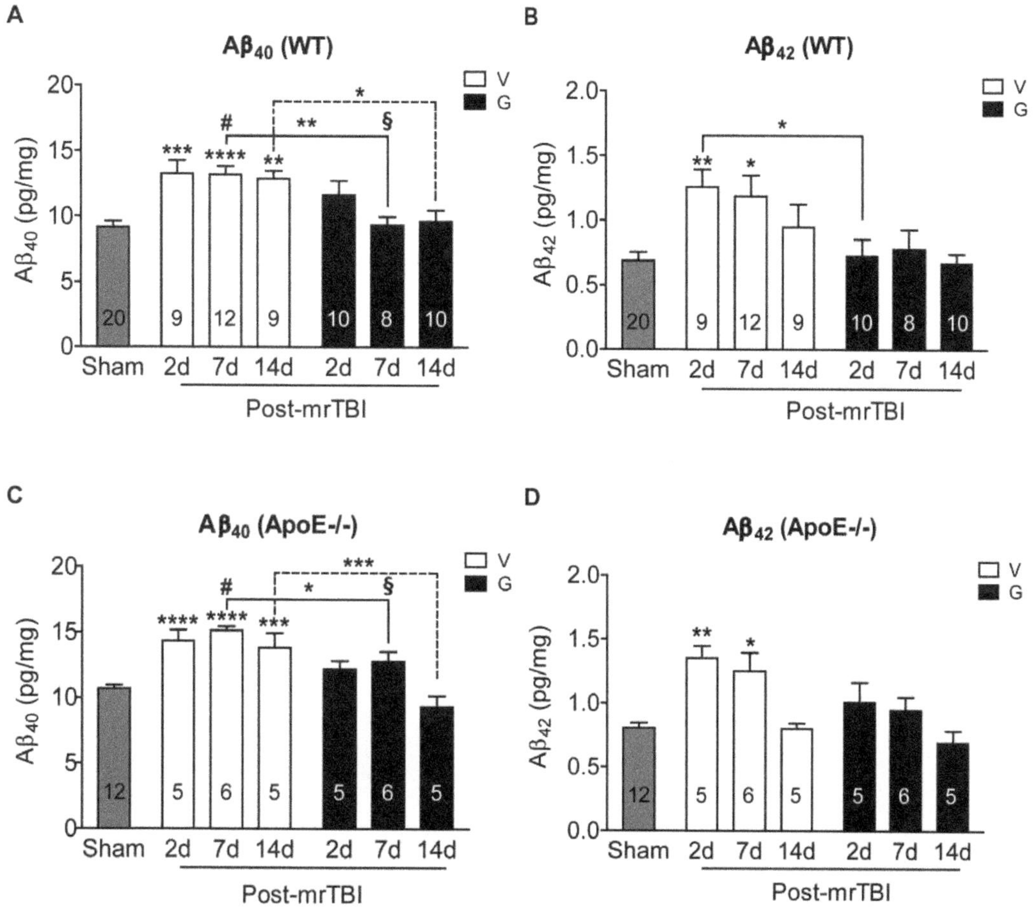

Figure 3. GW3965 prevents mrTBI-induced accumulation of endogenous Aβ in WT and apoE−/− mice. Total, soluble murine $A\beta_{40}$ and $A\beta_{42}$ levels were measured from ipsilateral half brains of WT (**A, B**) and apoE−/− (**C, D**) mice, respectively, at 2, 7, and 14 d post-mrTBI. Data from sham animals within each genotype were pooled (grey bars). Numbers inside the bars indicate sample size. Asterisks above individual bars indicate significant difference compared to the respective sham levels. *: $p<0.05$, **: $p<0.01$, ***: $p<0.001$, ****: $p<0.0001$. # indicates a significant difference ($p<0.05$) between $A\beta_{40}$ levels of untreated WT and apoE−/− brains at 7 d post-mrTBI. § indicates a significant difference ($p<0.05$) between $A\beta_{40}$ levels of GW3965-treated WT and apoE−/− brains at 7 d post-mrTBI. Data were analyzed by two-way ANOVA followed by a Bonferroni post hoc test. Legend: S: sham-operated mice, gray bars, V: untreated mice, open bars, G: GW3965-treated mice, black bars.

5. ApoE is Required for GW3965-mediated Suppression of Axonal Damage after mrTBI

Histological assessment of axonal damage using silver staining revealed argyrophilic structures in cell bodies and axons in several ipsilateral white matter regions, including the corpus callosum, cingulum, external and internal capsules, and optic tracts, with strikingly greater accumulation observed in apoE−/− mice (Figure 5, Figure S4, S5, arrows). Notably, intense argyrophilia was observed in the bilateral optic tracts of WT and apoE−/− mice, suggesting the involvement of strong coup-contrecoup forces in our model. In comparison, very mild silver staining was observed in grey matter regions including the sensory-motor cortex directly under the impact site as well as the piriform cortex, farthest from the impact site (not shown).

In untreated WT mice, axonal damage was significantly elevated in the corpus callosum, cingulum, external capsule, and internal capsule at 2 d post mrTBI (Figure 6A–D, open bars, $p<0.05$, two-way ANOVA). In this group, silver staining intensity returned to baseline by 7 d, suggesting that neuronal repair pathways were efficiently activated in these regions. More robust and sustained axonal damage was observed in the optic tracts of untreated WT mice (Figure 6E, open bars, $p<0.05$, two-way ANOVA). In WT mice, GW3965 effectively suppressed axonal damage in all regions examined (Figure 6A–E, black bars, $p<0.05$, two-way ANOVA).

Untreated apoE−/− mice showed extensive axonal damage that peaked at 7 d post-injury and remained elevated at 14 d in the corpus callosum, external and internal capsules, and optic tract (Figure 6F–J, open bars, $p<0.05$, two-way ANOVA). In contrast to WT mice, GW3965 failed to suppress axonal damage in apoE−/− animals (Figure 6F–J, black bars, $p<0.05$, two-way ANOVA). In all regions examined, axonal damage was significantly more extensive in apoE−/− mice compared to WT controls (Figure 6F–J, $p<0.05$, two-way ANOVA; # and § denote comparison of untreated and treated WT and apoE−/− mice, respectively).

6. Our mrTBI Model Produces Negligible Inflammation

To investigate the extent to which our mrTBI model induces inflammation, we examined post-injury microglial activation using Iba1 immunohistochemistry and measured IL-6, TNFα, and MCP-1 levels in ipsilateral half brains. Examination of coronal sections revealed uniform distribution of Iba1-immunoreactive microglia with negligible changes in the immunoreactivity in the

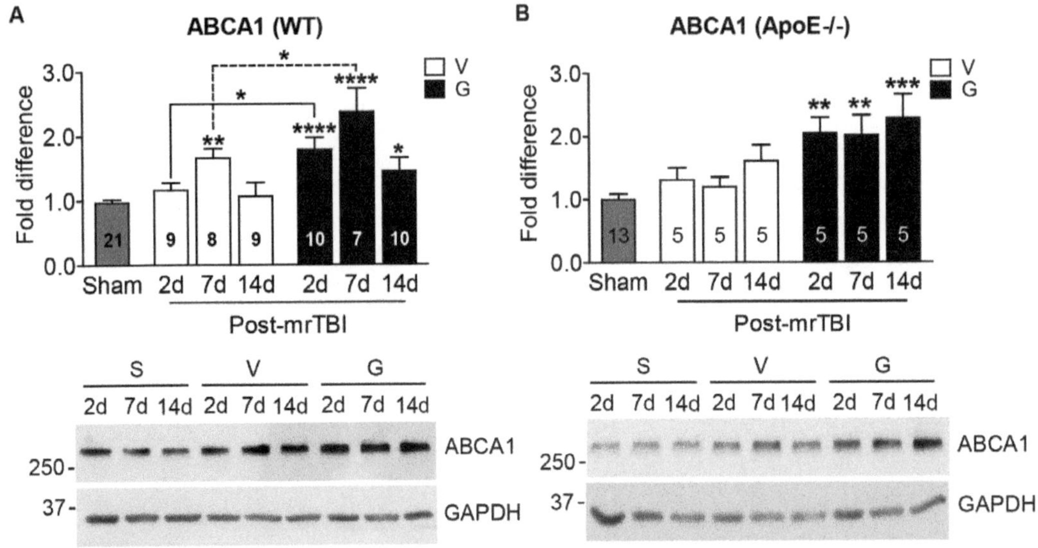

Figure 4. GW3965 augments ABCA1 levels in WT and apoE−/− mice following mrTBI. ABCA1 protein levels were determined in ipsilateral half brains of WT (**A**) and apoE−/− (**B**) mice following mrTBI using Western blots, with representative blots shown below the graphs. Data are expressed as fold difference normalized to sham values. Data from sham animals within each genotype were pooled (grey bars). Numbers inside the bars indicate sample size. Asterisks above individual bars indicate significant difference compared to the respective sham levels. *: $p<0.05$, **: $p<0.01$, ***: $p<0.001$, ****: $p<0.0001$. Data were analyzed by two-way ANOVA followed by a Bonferroni post hoc test. Legend: S: sham V: untreated mice, open bars, G: GW3965-treated mice, black bars.

injured WT and apoE−/− mice in either the ipsilateral cortex (Figure 7) or hippocampus (Figure S6) compared to sham controls. Further, GW3965 treatment did not reduce Iba1 staining below baseline (Figure 7, S6). Although no animal in this study had any obvious skull fracture, approximately 4–8% of mice showed evidence of microcontusions in the cortex directly under the impact site (Figure 8, arrows). Increased Iba1 immunoreactivity was clearly associated with these localized areas of more severe injury (Figure 8). Consistent with this model producing little inflammation, IL-6 levels in the ipsilateral brains were unchanged in both WT and apoE−/− mice, with or without GW3965 (Figure S7, $p>0.05$, two-way ANOVA). TNFα and MCP-1 levels were below the detection limit (not shown).

Discussion

The goal of this study was to evaluate the ability of GW3965 to promote recovery in a model of mrTBI specifically designed to mimic repeated concussion. We found that therapeutic administration of GW3965 improved NOR performance, suppressed Aβ accumulation, and reduced axonal damage after mrTBI. Loss of apoE exacerbated the severity of motor impairment and axonal damage and eliminated the ability of GW3965 to restore NOR performance and to promote axonal recovery. These results are consistent with the role of apoE in neuronal repair and synaptic restoration. ApoE levels did not change after TBI or after GW3965 treatment, which suggests that injury severity was not sufficient to elevate apoE as well as reflects the poor sensitivity of apoE as an LXR target compared to ABCA1 [42]. However, it is possible that apoE may show localized upregulation in regions with more severe damage where microglial activation is pronounced. Future studies will test whether ABCA1-mediated lipidation of apoE contributes to the beneficial effects of GW3965 after mrTBI.

Surprisingly, apoE was not required for GW3965 to suppress the transient increase in Aβ levels induced in our model. Further studies will be required to characterize these apoE-independent pathways that promote Aβ clearance after TBI. This will be an important endeavor, as axonal APP accumulation is a hallmark of TBI [27,28]. Theoretically, the Aβ produced after TBI could trigger Aβ-dependent toxic pathways that exacerbate damage [29]. In support of this γ-secretase inhibitors, which block Aβ production, reduce cognitive and pathological changes following CCI in mice [49]. However, the relationship between Aβ levels and TBI recovery is complex. In WT and AD mice, PBS-soluble Aβ levels in brain ISF microdialysates decrease by up to 50% within 90 min after CCI with impact depths ranging from 1–2 mm [32]. Intriguingly, Schwetye et al. [32] found no evidence of transient increases in Aβ levels after CCI, which differs from our findings and those from other investigators using relatively crude tissue lysates (see references in the Schwetey et al. [32] paper/discussion). Intra-axonal Aβ accumulation is observed in 3XTg-AD and APP/PS1 mice from 1–24 h after moderate-severe CCI [50,51], again suggesting that distinct pools of Aβ may differ in their responses after injury. Future studies are needed to clarify the factors that influence Aβ dynamics after TBI, including apoE.

Although our mrTBI model clearly leads to behavioral deficits and axonal damage, it is not severe enough to trigger significant inflammation except in localized regions near microcontusions. Importantly, inclusion or exclusion of mice with microcontusions did not influence the group outcomes for NOR, rotarod, and silver stain data. These results suggest that Iba1-positive staining of reactive microglia may be a potent secondary effect of structural brain damage, although it is possible that the time points examined in this study may have missed a peak of more global inflammation in our model.

Interest in understanding the pathophysiology of mild closed-head TBI has led to the development of various impact and blast models. However, there is considerable variation in outcomes. For example, many groups have reported mild cognitive deficits without motor impairment in weight drop-based mild TBI mouse models [52–56]. In contrast, the Shohami group observed motor

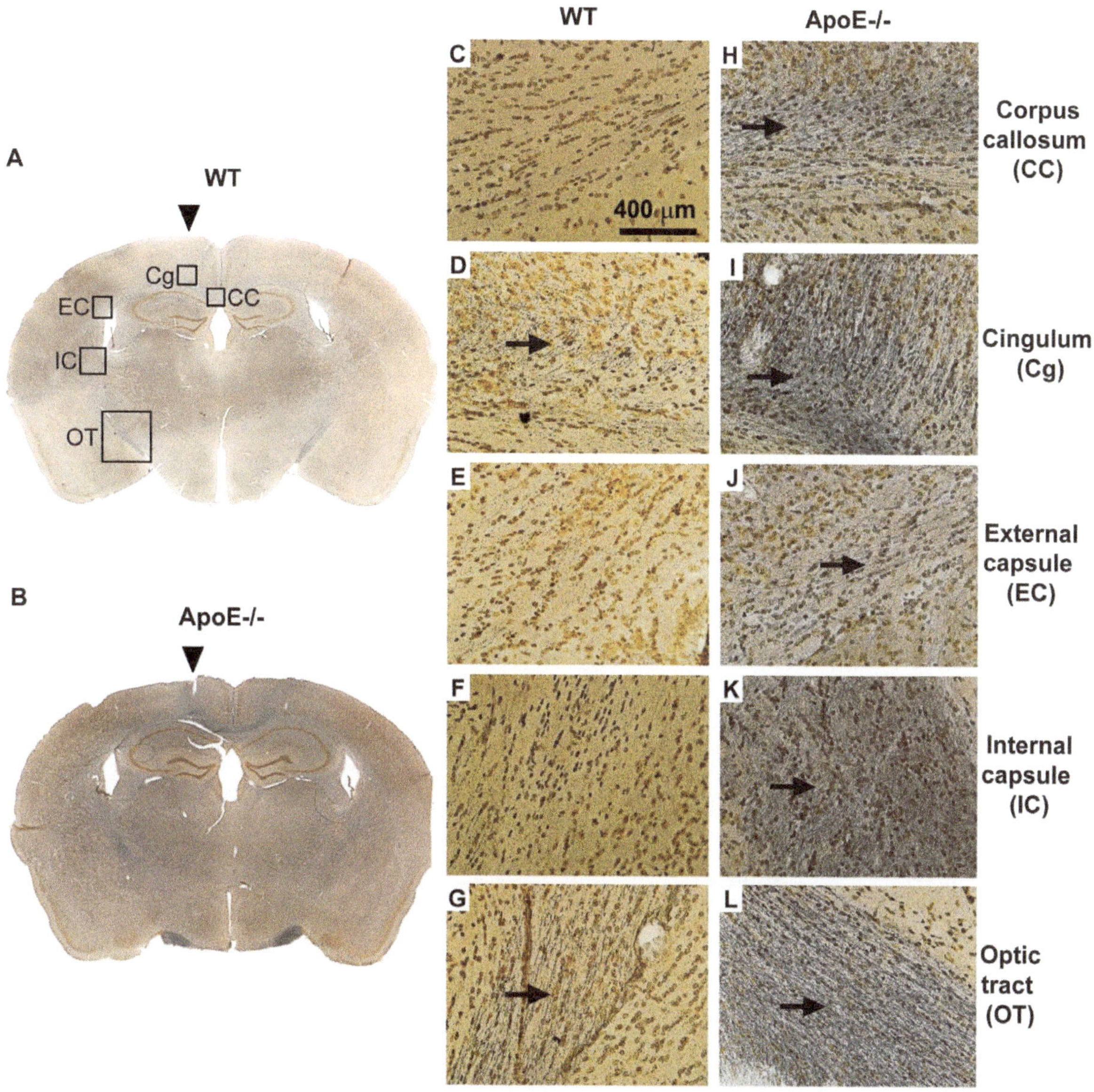

Figure 5. Loss of apoE exacerbates axonal injury after mrTBI. Axonal damage following mrTBI was assessed with silver staining (arrows). The left panel depicts representative images of silver-stained coronal sections at approximately −1.58 mm from bregma [66] of untreated WT (**A**) and apoE−/− (**B**) mouse brains harvested at 7 d post-mrTBI. White matter areas with prominent silver staining are indicated by the black squares. Injury location is indicated by the arrowhead. The right panel (**C–L**) depicts representative 40x-magnified images of silver staining in five white matter areas in the brains of untreated WT (**C–G**) and apoE−/− (**H–L**) mice harvested at 7 d post-mrTBI.

deficits in both severe and mild TBI, but found that severe injury was required for cognitive impairment [57]. Using an electromagnetically-controlled piston to produce mrTBI, Shitaka et al observed cognitive deficits, white matter damage, and robust microglial activation without significant structural damage or APP immunoreactivity [58]. Our mrTBI model also shows sustained cognitive and transient motor deficits, axonal damage, and transient Aβ accumulation. The most noteworthy difference between our model and that of Shitaka et al. [58] is that we observed very little microglial activation, suggesting that marked inflammation is not required to lead to behavioral and axonal changes. Our model also does not lead to significant increases in apoE or LDLR levels, which have been observed in models of moderate-severe TBI [8,10,59]. No changes in LDLR levels is consistent with studies showing no alternation of LDLR levels after CCI or TO901317 treatment [45]. We observed no evidence of white matter abnormalities by 7T magnetic resonance imaging (not shown).

One caveat of our model is the lack of a systematic biomechanical assessment of injury. Although input parameters such as weight mass and drop height are reported in most closed head injury studies including ours, the reproducibility of these mechanical inputs and the response of the animal's head to the forces applied are not well understood. Li et al. [60] recently studied the biomechanical parameters of the weight-drop based Marmarou rat TBI model [61] and found that DAI severity was related to the linear and angular response of the rat head but not with the drop height. Goldstein et al. recently demonstrated that a single blast injury in mice can generate significantly impaired memory performance, long-term potentiation, and axonal con-

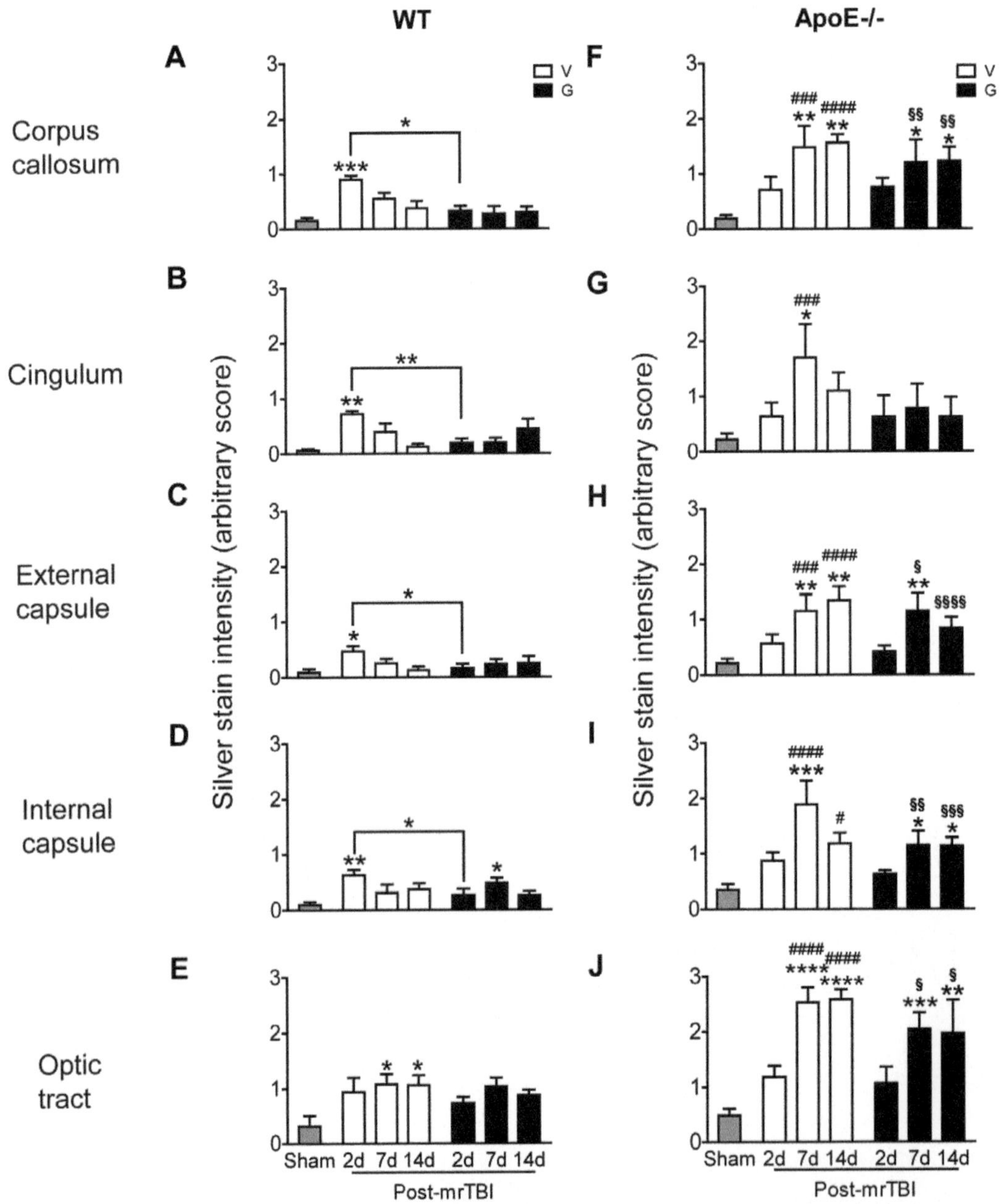

Figure 6. ApoE is required for GW3965 to suppress axonal damage after mrTBI. Silver-stained images of white matter areas in WT and apoE−/− brains were analyzed semiquantitatively using an arbitrary silver staining scale extending from 0 (<10% argyrophilic structures covering the image field) to 3 (>70% argyrophilic structures covering the image field). The bar graphs represent mean ± SEM silver stain intensity score (arbitrary value) of WT (**A–E**) and apoE−/− (**F–J**) brains in the corpus callosum (**A, F**), cingulum (**B, G**), external capsule (**C, H**), internal capsule (**D, I**), and optic tracts (**E, J**). Sample sizes were: sham, n = 6/genotype (pooled); untreated mrTBI, n = 5/time point/genotype; GW3965-treated mrTBI, n = 5/time point/genotype. Asterisks above individual bars indicate significant differences compared to the respective sham values (grey bars). *: $p<0.05$, **: $p<0.01$, ***: $p<0.001$, ****: $p<0.0001$. ### ($p<0.001$) and #### ($p<0.0001$) represent significant differences in silver stain scores between untreated WT and apoE−/− mice. § ($p<0.05$), §§ ($p<0.01$), §§§ ($p<0.001$), and §§§§ ($p<0.0001$) represent significant differences in silver stain scores between GW3965-treated WT and apoE−/− mice. Data were analyzed by two-way ANOVA followed by a Bonferroni post hoc test. Legend: V- untreated mice, open bars, G- GW3965-treated mice, black bars.

ductance accompanied by tau pathology, myelinated axonopathy, microvasculopathy, and neuroinflammation [62]. Intriguingly, immobilizing the animal's head to prevent blast-induced head oscillations prevented memory deficits [62]. It is possible that the very mild pathology in our model is due to relatively little head movement after impact.

We and others have demonstrated that synthetic LXR agonists effectively enhance the ability of lipidated apoE to reduce Aβ levels and restore cognitive function in AD mice [42,63]. Given that TBI also involves altered apoE and Aβ metabolism, it is of considerable interest to determine whether LXR agonists may also have potential therapeutic benefits for TBI. The importance of addressing this question is two-fold. First, LXR treatment may minimize neuronal damage and promote acute recovery by reducing inflammation and promoting neuronal repair. Second, by facilitating Aβ clearance in the first weeks after injury, LXR

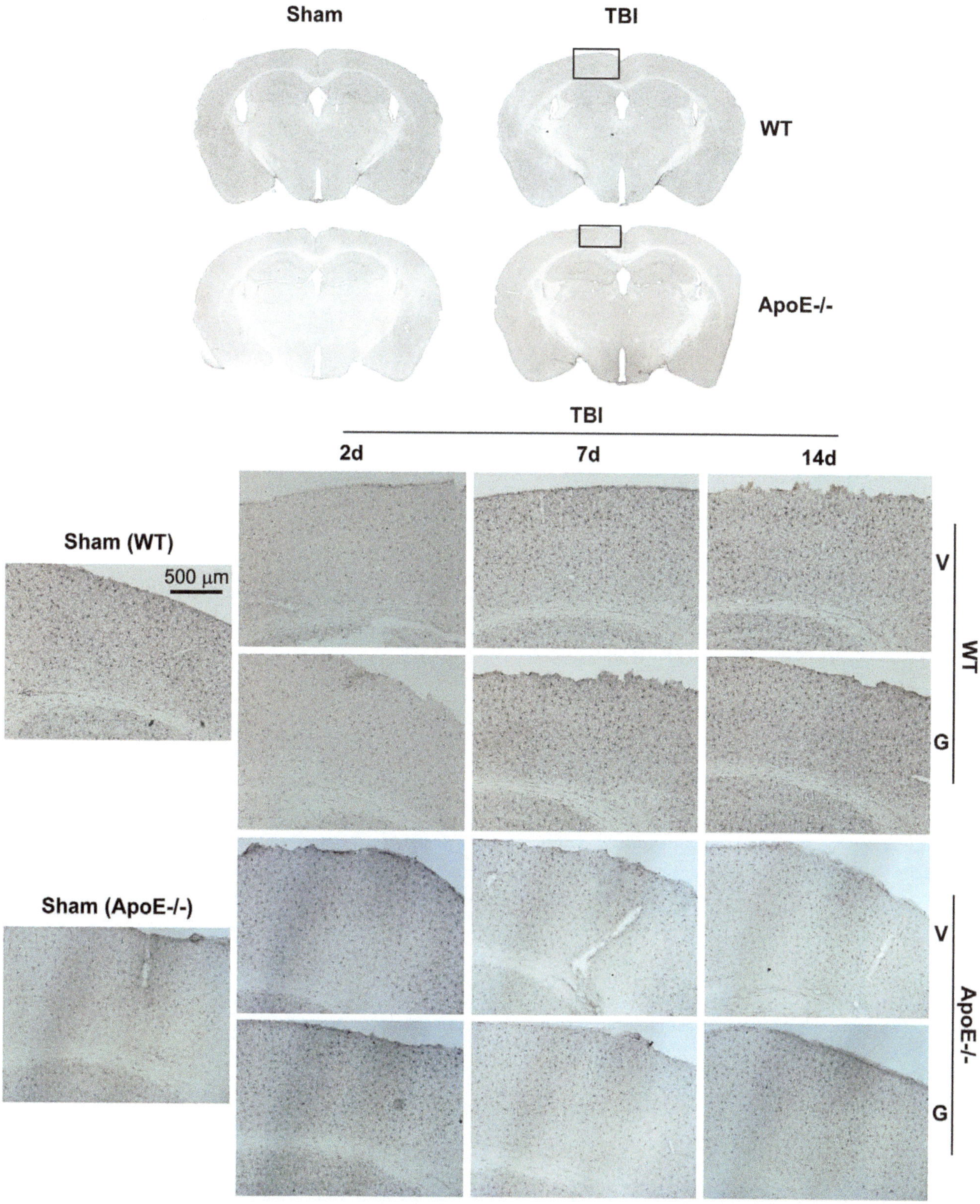

Figure 7. Microglia are not significantly activated by mrTBI in either WT or apoE−/− mice. Microglial activation in sham and injured WT and apoE−/− mice was assessed with Iba1 immunohistochemistry. The top panel depicts representative images of Iba1-stained coronal sections at approximately −1.82 mm from bregma [66]. The bottom panel depicts 10X-magnified images of ipsilateral cortex underlying the injury site (indicated by the black rectangle). Legend: V- untreated mice, G- GW3965-treated mice.

treatment may also reduce the increased long-term risk of AD years or decades later. Although current LXR agonists have safety issues such as hypertriglyceridemia and hepatic steatosis that preclude their present clinical use [64,65], ongoing drug discovery

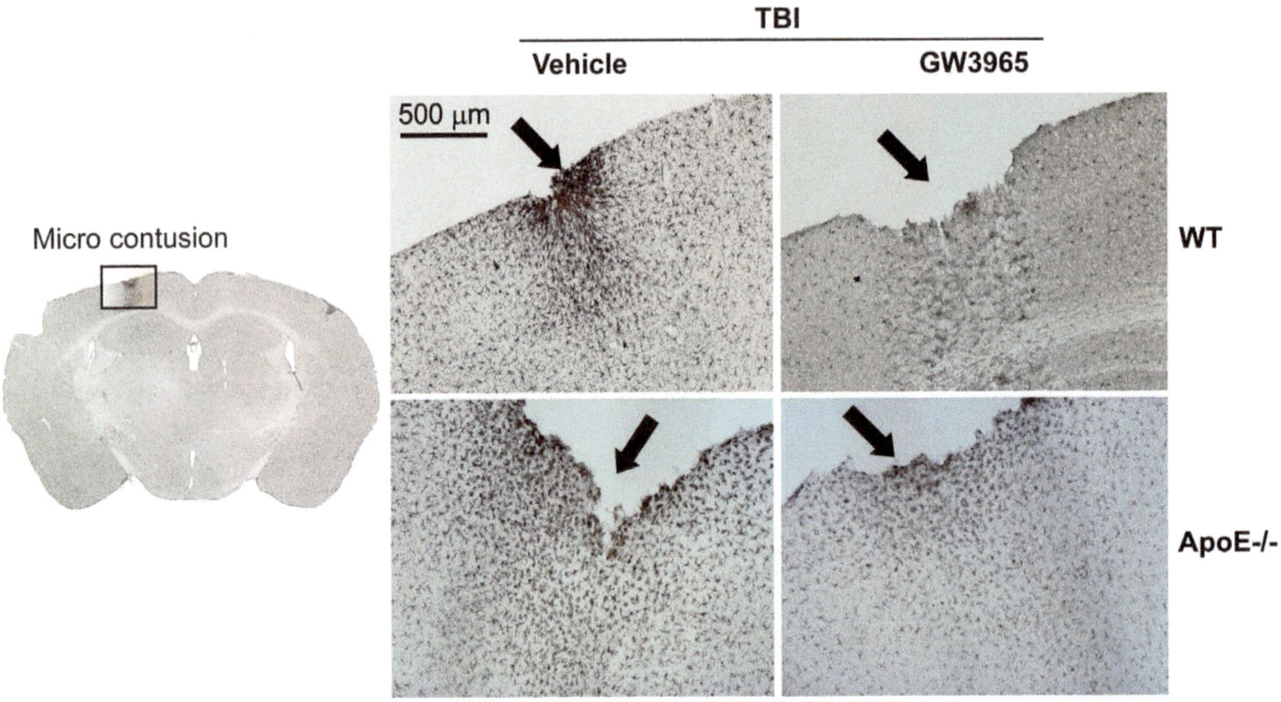

Figure 8. Pronounced microglial activation is localized only around contused areas. In this study, approximately 4–8% of brains subjected to mrTBI showed micro contusions in the absence of gross skull fracture. The left panel shows a Iba1-stained coronal section at approximately −1.82 mm from bregma [66] with a micro contusion in the cortex below impact site (black square). The right panel shows representative 10X-magnified images of Iba1-stained untreated WT and apoE−/− contused cortices. Microcontusions are denoted by black arrows. Note the pronounced localized activation of microglia around the contusion.

efforts may lead to a safe and effective compound. Furthermore, the metabolic risks of LXR agonists may be clinically tolerable if short-term treatment could improve functional recovery as well as decrease long-term AD risk. Building on the findings of Loane et al. [45], our study provides additional support for the potential of LXR agonists to treat mrTBI.

Our study also illustrates that the mechanisms by which LXR agonists promote recovery after TBI may not directly correspond to their effects in AD models. For example, outcomes such as cognitive recovery and axonal damage are apoE-dependent, but surprisingly, Aβ clearance is independent of apoE in our model. Additionally, because our model produces negligible inflammation, the beneficial effects of LXR agonists on TBI recovery may be independent from their established anti-inflammatory effects. Future investigations will be designed to identify the LXR targets operative in our mrTBI model, assess their efficacy to reduce tau pathology, and evaluate their therapeutic utility in models of moderate-severe TBI.

Materials and Methods

Animals

Male C57Bl/6 (WT) and apoE−/− mice were obtained from Jackson Laboratories (Bar Harbor, ME). Animals were housed with a reverse 12 h light −12 h dark cycle for at least 10 days before TBI. All animal procedures were approved by the University of British Columbia Committee on Animal Care and adhered to the Canadian Council for Animal Care guidelines.

Mild Repetitive Traumatic Brain Injury (mrTBI)

WT (body weight: 29.6±0.24 g) and apoE−/− (body weight: 30±0.26 g) mice at 4 months of age were subjected to two mild closed-head injuries spaced 24 h apart using a gravity-driven weight-drop device obtained from Dr. Esther Shohami (The Hebrew University of Jerusalem). All surgical procedures were conducted using aseptic technique and animals were kept warm using a heating pad. Animals were anaesthetized with isoflurane (induction: 3–4%, maintenance: 1.5%–2%) in oxygen (0.9 L/min). Lubricating eye ointment was applied to prevent corneal drying. The scalp was shaved and scrubbed three times with 4% chlorhexidine solution and 70% ethanol. Ketamine (15 mg/kg) and xylazine (1.75 mg/kg) mixture [subcutaneous (s.c.)], meloxicam (1 mg/kg, s.c.), and bupivacaine (0.125 mg/kg, s.c., under the scalp) were administered for analgesia. Sterile, warm saline (1 ml/100 g body weight) was administered s.c. for hydration. After confirming a surgical plane of anesthesia, a midline longitudinal incision was made in the scalp to expose the calvarium. After clearing the parietal bone, a sterile piece of gauze (2 mm×2 mm), pre-moistened with bupivacaine solution (2.5 mg/ml) was overlaid over the left parietal bone to minimize the incidence of skull fracture. Anesthesia was temporarily discontinued and the animal was quickly moved under the weight-drop device. The head was manually held in place such that the tip of the Teflon-tipped cone (tip diameter: 2 mm) of the injury device rested over the piece of gauze in the left mid-parietal bone (approximately between −1 and −3 mm from bregma and 1 mm left of the sagittal suture [66]). A 95 g brass weight was dropped from a pre-determined height to result in injury across the left hemisphere. The precise drop height (cm) of the weight was adjusted to be one-third the weight (g) of the animal (Mean drop height: WT, 10±0.08 cm; apoE−/−, 10.1±0.09 cm). Mean body weight and drop height did not defer significantly across the genotypes, time points, and treatment groups (Figure S8). Anesthesia was reinstituted following injury. After removing the gauze and confirming the absence of skull fracture, the incision was closed and the animals were returned to their cages. Twenty-four hours after the first injury, the incision was reopened under isoflurane anesthesia and a second identical TBI was induced as described above. Mice suffering from skull fracture after injury were immediately euthanized. Recovery was monitored until animals moved freely and fed normally. Sham controls received isoflurane anesthesia, pre-surgical analgesia, and skin incision only. Mice within each group were randomly assigned to one of three assessment time points, viz., 2, 7, and 14 days after the second injury. The time of the 2nd TBI was defined as time zero.

GW3965 Treatment

Injured mice were randomized into treated and untreated groups (WT mice: N = 45/group, apoE−/− mice: N = 30/group). Mice in the treated group received an intraperitoneal (i.p.) bolus of GW3965 (2-[3-[3-[[2-chloro-3-(trifluoromethyl)phenyl]methyl-[2,2-di(phenyl)ethyl]amino]propoxy]phenyl]acetic acid) (20 mg/kg) 30 min after the 2nd injury. GW3965 treatment was continued thereafter by feeding mice with standard rodent chow (Research Diets Inc., New Brunswick, NJ) in which GW3965 was compounded at 120 mg/kg until the end of the study, resulting in an average daily dose of 15 mg/kg upon determination of average daily food intake. Because we had previously demonstrated that GW3965 doses of 2.5 mg/kg/day and 33 mg/kg/day provided cognitive and neuropathological benefits in APP/PS1 mice [42], we aimed to achieve a dose between these two extremes. Financial limitations precluded a complete dose-response study. Because GW3965 has a plasma half-life of 2 h [67], delivering GW3965 in chow provides a more constant plasma levels than by once-daily oral gavage. Injured mice in the untreated group received a single i.p. bolus of equivalent volume of vehicle (DMSO) 30 min after 2nd injury. These mice were thereafter fed standard rodent chow. Sham-operated mice (WT mice: N = 24; apoE−/− mice: N = 18) received neither GW3965 nor vehicle and were fed standard rodent chow. We have previously determined that GW3965 at a dose of 33 mg/k/day does not cause hepatotoxicity in mice [42].

Cognitive Assessment

Medial temporal lobe function was assessed using the novel object recognition (NOR) task performed at 2, 7, or 14 days following mrTBI [68]. This task was selected to facilitate comparisons to our previous NOR study of GW3965 efficacy in APP/PS1 mice [42], and because motor impairments observed after TBI may confound tasks such as Morris Water Maze. Each animal was assessed with the NOR task at only one time point after sham or mrTBI surgery. Twenty-four hours prior to the NOR task, mice were habituated to an empty open field (14″×24″×14″) for 10 min in a brightly-lit room. During training, mice were placed in the center of the open field containing two identical objects located in the north and south quadrant, spaced equidistant from the walls. The time the animal explored each of the two objects during a 5 min period was quantified, with exploring defined as sniffing, climbing on, or touching the object and at a body length or less away from the object while oriented toward it. Trials were tracked using an overhead digital camera and video tracking system (ANY-maze, Stoelting Co, Wood Dale, IL). Between animals, the open field and objects were cleaned with 70% ethanol to eliminate olfactory cues. Four hours after training, mice were tested by replacing one of the identical objects with a novel object of similar size but different shape and color. The side of the open field with the novel and familiar objects was alternated for each mouse to avoid a side preference. Novel object recognition was quantified using the discrimination index (DI) calculated as: $DI = (T_N - T_F)/(T_N+T_F)$, where T_N and T_F represent exploration time for the novel and familiar objects, respectively [48]. DI values range from +1 to −1, where a positive score indicates preference for the novel object, a negative score indicates preference for the familiar object, and zero indicates no discrimination.

Motor Function Assessment

Motor performance was evaluated using an accelerating rotarod (Model 7650, Ugo Basile, Collegeville, PA). Twenty-four hours before the 1st TBI, mice were trained to walk on accelerating rotarod (0 to 30 rpm in 210 s) to establish baseline failing latency. Unless the animal was scheduled for sacrifice, rotarod latencies were repeatedly assessed at 1, 2, 7, and 14 d post-mrTBI. At each test day, every mouse underwent three accelerating speed trials (0 to 30 rpm in 210 s) separated by an inter-trial interval of at least 15 min. The animal was recorded as failing during the 210 s period if: (1) the animal fell completely off the device before 210 s, or (2) the animal gripped the rod and made one complete passive revolution without actively walking on the rod. The average failing latency was calculated from the three trials for each animal.

Biochemical Analyses

Tissue collection. Mice were anesthetized with an i.p. injection of 150 mg/kg ketamine (Bimeda-MTC) and 20 mg/kg xylazine (Bayer) and transcardially perfused with ice-cold phosphate-buffered saline (PBS) containing heparin (0.5 units/ml) for 7 min. For biochemistry, brains were longitudinally bisected, rapidly frozen over dry ice and stored at −80°C until analysis (WT mice N: injured = 10/group/time point, sham = 6/time point; apoE−/− mice N: injured = 5/group/time point, sham = 4/time point). For histology, anesthetized mice were perfused with heparinized PBS and whole brains were immersion-fixed in 10% neutral buffered formalin for 48 h and cryoprotected in 30% sucrose in PBS at 4°C (N: injured = 5/treatment group/time point/genotype, sham = 2/time point/genotype).

Protein extraction. Frozen ipsilateral half brains were thawed over ice and homogenized in 1.5 ml of ice-cold RIPA lysis buffer (5 mM EDTA, 50 mM NaCl, 10 mM sodium pyrophosphate, 50 mM NaF, and 1% NP40 alternative, pH: 7.4) containing complete protease inhibitor (Roche Applied Science) with a Tissuemite homogenizer at full speed for 20 s. The homogenate was sonicated at 20% output for 10 s followed by centrifugation at 4°C for 10 min at 9,000 rpm in a microcentrifuge (Eppendorf). The supernatant was separated and used for analysis of ABCA1, apoE, APP, APP-CTF, LDLR, IL-6, TNF-α, MCP-1, Aβ40, and Aβ42. Aliquots of all lysates were immediately frozen at −80°C until analysis. Protein concentration was determined by DC Protein assay (Bio-Rad, Hercules, CA).

Western blot analysis. RIPA lysate (80 μg protein) was electrophoresed through 10% SDS-polyacrylamide gels, transferred to PVDF membrane (Millipore), and immunodetected using following antibodies: monoclonal anti-ABCA1 (AC10, 1:1000, a gift from Dr. M. R. Hayden), murine-specific apoE (clone M-20, 1:1000, Santa Cruz Biotechnology, Santa Cruz, CA) and anti-LDLR (1:1000, R&D Systems). APP and APP-CTF were analyzed by resolving RIPA lysate through 4–12% NuPAGE Bis-Tris Gradient gels (catalogue # NP0335, Invitrogen) and immunoblotted using anti-APP (clone 22C11, 1:2000, Chemicon) and anti-

APP-CTF (clone C1/6.1, 1:1000, Covance) antibodies. All membranes were probed with anti-GAPDH antibody (clone 6C5, 1:5000, Chemicon) as a loading control. Blots were developed using enhanced chemiluminescence (Amersham ECL, GE Healthcare) according to the manufacturer's recommendations. Films were scanned into TIFF format at 600 dpi resolution and pixel counts were determined using ImageJ (version 1.46, NIH). Levels of ABCA1, apoE, APP, APP-CTF, and LDLR were normalized to GAPDH and expressed as fold difference compared to sham controls.

Aβ eLISA. Endogenous $A\beta_{40}$ and $A\beta_{42}$ levels in ipsilateral half-brain lysates were quantified by ELISA (KMB3481 and KMB3431, Invitrogen) following the manufacturer's instructions. Levels of soluble $A\beta_{40}$ and $A\beta_{42}$ were normalized to total protein.

IL-6, TNF-α, and MCP-1 ELISA. IL-6, TNF-α and MCP-1 levels in ipsilateral half-brain lysates were determined by ELISA. Briefly, 96-well plates were coated and incubated overnight at 4°C with primary antibody against murine IL-6 (clone MP5-20F3, 1:1000, eBioscience), TNF-α (clone TN3-19.12, 1:1000, eBioscience) and MCP-1 (clone 4E2, 1:1000, eBioscience). Brain lysates were incubated with primary antibody at room temperature for 1 h. Following washing, samples were incubated with biotinylated secondary antibodies for IL-6 (clone MP5-32C11, 1:1000, eBioscience), TNF-α (Catalogue # 13–7341, 1:1000, eBioscience), and MCP-1 (clone 2H5, 1:1000, eBioscience) at room temperature for 1 h. Samples were developed with avidin-horseradish peroxidase (eBioscience) and TMB substrate.

Histology

Coronal brain sections (40 μm) were cut on a cryostat starting from the genu of corpus callosum (~1.1 mm anterior to bregma) to the posterior hippocampus (~3.3 mm posterior to Bregma). Sections were collected in 1X PBS with sodium azide.

Silver staining. Three to four coronal brain sections, separated by 400 μm, from the bregma to the posterior hippocampus, were stained using FD NeuroSilver Kit II (FD NeuroTechnologies Inc, Columbia, MD) according to the manufacturer's protocol (N: injured = 5/genotype/time point/group, sham = 2/genotype/time point) with the following modifications: 1) Sections were stored in 4% paraformaldehyde in saline-free phosphate buffer at 4°C for a minimum of 5 d prior to staining. 2) Step 4 was decreased from 2×2 min incubations to 1×2 min incubation to increase staining intensity. 3) Steps 3, 4, and 5 were performed with vigorous shaking. From each section, 40X-magnified images of ipsilateral corpus callosum, cingulum, external capsule, internal capsule, and optic tracts were captured with a Zeiss Axio Imager A1 microscope attached to an Olympus DP72 camera and digitized using cellSens® Standard digital imaging software (version 1.4.1, Build 8624, OLYMPUS). A semiquantitative scale was developed to score staining intensity, ranging from 0 (0 to <10% staining) to 3 (>70% staining). The scale was based on the extent and intensity of argyrophilic structures (cell bodies and axonal fibers) within an image field. Every image was independently scored by two trained raters who were blinded to injury, genotype, and treatment status. Scores from the two raters were averaged. A minimum of 3 sections per brain were scored and combined to define the average silver stain score of each brain region.

Iba1 staining. Microglial activation induced by mrTBI was assessed with Iba1 immunohistochemistry. Briefly, 3–4 coronal brain sections, separated by 400 μm, starting from the bregma to the posterior hippocampus were washed in Tris-buffered saline (TBS), treated with 0.3% hydrogen peroxide for 10 min and blocked with 3% normal goat serum (NGS) in TBS containing 0.25% Triton-X (TBS-X) for 30 min. Sections were incubated overnight in TBS-X containing anti-Iba1 polyclonal antibody (Catalogue # 019–19741, 1:1000, Wako Chemicals, Richmond, VA) and 1% NGS. Following washing, sections were incubated in biotinylated goat-anti-rabbit secondary antibody (1:1000 in TBS-X) for 1 h. Sections were visualized with horseradish peroxidase (Vectastain Elite ABC Kit PK-6120, Vector Laboratories (Canada) Inc, Burlington, ON) and DAB substrate. Sections were mounted on Superfrost® Plus (Fisher Scientific) slides, dehydrated in ascending concentrations (50, 70, 90, and 95%) of alcohol, cleared with xylene, and coverslipped in DPX mounting medium (Electron Microscopy Sciences, Hatfield, PA). Images were captured and digitized as above.

Statistical Analyses

All data are presented as mean ± SEM. For analysis of NOR data, because all mice equally explored two similar objects during the training phase (DI ~ 0), DI scores during training for animals of the same genotype were pooled. Similarly, DI scores during testing were pooled for sham mice within each genotype. DI scores were compared using two-way ANOVA followed by a Bonferroni post-hoc test. For rotarod analysis, baseline latencies of mice from all time points within genotype and treatment groups were pooled. Rotarod data were analyzed for time and treatment as well as time and genotype effects using two-way repeated measures ANOVA followed by a Bonferroni post-hoc test, as animals were tested repeatedly until sacrifice. For biochemical and histological analyses, data from sham mice from all time points within each genotype were pooled. Biochemistry and silver stain data were analyzed for time and treatment as well as for time and genotype effects using two-way ANOVA followed by a Bonferroni post-hoc test. A p value of <0.05 was considered significant. All statistical analyses were performed using GraphPad Prism (version 5.04, GraphPad Software Inc).

Supporting Information

Figure S1 NOR performance was not affected by motor impairment. To assess whether NOR performance was affected by motor impairment; the total path length (m) covered by WT and apoE−/− mice during testing and training was measured. **(A, B)**, total path length covered by WT mice during training and testing, respectively. **(C, D)**, total path length covered by apoE−/− mice during training and testing, respectively. The path lengths covered by WT and apoE−/− mice were not significantly different from sham animals during testing, indicating that NOR was not affected by motor impairment. *: $p<0.05$ and **: $p<0.01$. Data were analyzed by two-way ANOVA followed by a Bonferroni post hoc test. Legend: V- untreated mice, open bars, G- GW3965-treated mice, black bars.

Figure S2 APP and APP-CTF-α levels remain unchanged following mrTBI. APP holoprotein **(A, C)** and APP-CTF-α **(B, D)** protein levels in ipsilateral half brains were analyzed by Western blotting, with representative blots shown for WT (A, B) and apoE−/− (C, D) mice. Data are expressed as fold difference relative to sham values. Data from sham animals within each genotype were pooled (grey bars). Numbers inside bars indicate sample size. Data were analyzed by two-way ANOVA followed by a Bonferroni post hoc test. Legend: S: sham-operated mice, gray bars, V: untreated mice, open bars, G: GW3965-treated mice, black bars.

Figure S3 ApoE and LDLR levels are unaffected by mrTBI or GW3965. Levels of apoE and LDLR protein in WT mice **(A, B)** and LDLR protein in apoE−/− mice **(C)** were determined in ipsilateral half brains following mrTBI using Western blotting, with representative blots shown on the right. Data are expressed as fold difference normalized to sham values. Data from sham animals within each genotype were pooled. Numbers inside bars indicate sample size. Data were analyzed by two-way ANOVA followed by Bonferroni post hoc test. Legend: S: sham-operated mice, gray bars, V: untreated mice, G: GW3965-treated mice.

Figure S4 mrTBI leads to mild axonal damage in WT mice. Axonal damage following mrTBI was assessed with silver staining. The far left column depicts representative images of silver-stained coronal sections at approximately −1.82 mm from bregma [66] of sham **(A)**, untreated **(B)**, and GW3965-treated **(C)** WT brains harvested at 2 d post-mrTBI. Locations of the white matter areas where silver staining was most prominent are indicated by black squares. Injury location is indicated by the arrowhead. Columns on the right depict representative 40X-magnified images of silver staining in five white matter areas in sham-operated **(D–H)**, untreated TBI **(I–M)**, and GW3965-treated **(N–R)** brains of WT mice harvested at 2 d post-mrTBI.

Figure S5 Loss of apoE exacerbates axonal damage after mrTBI. Axonal damage following mrTBI was assessed with silver staining (arrows). The far left column depicts representative images of silver-stained coronal sections at approximately −1.82 mm from bregma [66] of sham **(A)**, untreated **(B)**, and GW3965-treated **(C)** apoE−/− brains harvested at 7 d post-mrTBI. Locations of the white matter areas where silver staining was most prominent are indicated by black squares. Injury location is indicated by the arrowhead. The columns on the right depict representative 40X-magnified images of silver staining in five white matter areas in sham-operated **(D-H)**, untreated **(I-M)**, and GW3965-treated **(N-R)** brains of apoE−/− mice harvested at 7 d post-mrTBI.

Figure S6 mrTBI induces negligible hippocampal microglial activation. Microglial activation in sham and injured WT and apoE−/− hippocampi was assessed with Iba1 immunohistochemistry. The top panel depicts representative images of Iba1-stained coronal sections at approximately -1.82 mm from bregma [66]. The bottom panel depicts 10X-magnified images ipsilateral hippocampus (indicated by black rectangle). Legend: V- untreated mice, G- GW3965-treated mice.

Figure S7 IL-6 levels remain unchanged following mrTBI. Interleukin- 6 (IL-6) levels were measured from ipsilateral half brains of WT and apoE−/− mice harvested at 2, 7, and 14 d post-mrTBI. Data from sham animals within each genotype were pooled. Cohorts were: sham (WT, n = 9, pooled; apoE−/−, n = 10, pooled), 2 day (WT: untreated, n = 5, GW3965-treated, n = 5; apoE−/−: untreated, n = 5, GW3965-treated, n = 5), 7 day (WT: untreated, n = 4, GW3965-treated, n = 4; apoE−/−: untreated, n = 6, GW3965-treated, n = 6), and 14 day (WT: untreated, n = 5, GW3965-treated, n = 5; apoE−/−: untreated, n = 5, GW3965-treated, n = 5) post-mrTBI. Data were analyzed by two-way ANOVA followed by a Bonferroni post hoc test. Legend: Sham WT: open bars, sham apoE−/−: grey bars, vehicle-treated WT: open striped bars, vehicle-treated apoE−/−: grey striped bars, GW3965-treated WT: open stippled bars, GW3965-treated apoE−/−: grey stippled bars.

Figure S8 Animal body weight and drop height did not differ across all groups. To induce mrTBI, a 95 g brass weight was dropped on the animal's skull using a height (cm) adjusted to one third of the animal's body weight (g) (see Methods and Materials). The figure depicts body weight **(A, B)** and drop height **(C, D)** of WT and apoE−/− mice, respectively, across post-mrTBI time points and treatment groups. Neither drop height nor body weight differed significantly across genotypes, time points or treatment groups. Data were analyzed by two-way ANOVA followed by a Bonferroni post hoc test. Legend: V: untreated mice, open bars, G: GW3965-treated mice, black bars.

Acknowledgments

We are grateful to Dr. Jon Collins (GlaxoSmithKline R&D) for the gift of GW3965 and to Dr. David Brody for guidance and critical feedback throughout this work.

Author Contributions

Conceived and designed the experiments: DRN JD CLW. Performed the experiments: DRN GM AW SS JF KW. Analyzed the data: DRN MC ST KW. Contributed reagents/materials/analysis tools: CLW REWH. Wrote the paper: DRN CLW.

References

1. Corrigan JD, Selassie AW, Orman JA (2010) The epidemiology of traumatic brain injury. J Head Trauma Rehabil 25: 72–80.
2. Faul MM, Xu L, Wald MM, Coronado VG (2010) Traumatic brain injury in the united states: emergency department visits, hospitalizations and death 2002–2006. Atlanta: Centres for Disease Conrtol and Prevention.
3. Elder GA, Cristian A (2009) Blast-related mild traumatic brain injury: mechanisms of injury and impact on clinical care. Mount Sinai J Med 76: 111–118.
4. Werner C, Engelhard K (2007) Pathophysiology of traumatic brain injury. Br J Anaesth 99: 4–9.
5. Graham DI, McIntosh TK, Maxwell WL, Nicoll JA (2000) Recent Advances in Neurotrauma. J Neuropathol Exp Neurol 59: 641–651.
6. Ladu MJ, Reardon C, Van Eldik L, Fagan AM, Bu G, et al. (2000) Lipoproteins in the central nervous system. Ann New York Acad Sci 903: 167–175.
7. Laskowitz DT, Horsburgh K, Roses AD (1998) Apolipoprotein E and the CNS response to injury. J Cereb Blood Flow Metab 18: 465–471.
8. Poirier J (1994) Apolipoprotein E in animal models of CNS injury and Alzheimer's disease. Trends Neurosci 17: 525–530.
9. Kay AD, Day SP, Kerr M, Nicoll JA, Packard CJ, et al. (2003) Remodeling of cerebrospinal fluid lipoprotein particles after human traumatic brain injury. JNeurotrauma 20: 717–723.
10. Iwata A, Browne KD, Chen XH, Yugushi T, Smith DH (2005) Traumatic brain injury induces biphasic upregulation of ApoE and ApoJ protein in rats. J Neurosci Res 82: 103–114.
11. Chen Y, Lomnitski L, Michaelson DM, Shohami E (1997) Motor and cognitive deficits in apolipoprotein E-deficient mice after closed head injury. Neuroscience 80: 1255–1262.
12. Han SH, Chung SY (2000) Marked hippocampal neuronal damage without motor deficits after mild concussive-like brain injury in apolipoprotein E-deficient mice. Ann N Y Acad Sci 903: 357–365.
13. Hyman BT, Trojanowski JQ (1997) Consensus recommendations for the postmortem diagnosis of Alzheimer disease from the National Institute on Aging and the Reagan Institute Working Group on diagnostic criteria for the neuropathological assessment of Alzheimer disease. J Neuropathol Exp Neurol 56: 1095–1097.
14. Corder EH, Saunders AM, Strittmatter WJ, Schmechel DE, Gaskell PC, et al. (1993) Gene dose of apolipoprotein E type 4 allele and the risk of Alzheimer's disease in late onset families. Science 261: 921–923.

15. Poirier J, Davignon J, Bouthillier D, Kogan S, Bertrand P, et al. (1993) Apolipoprotein E polymorphism and Alzheimer's disease. The Lancet 342: 697–699.
16. Corder EH, Saunders AM, Risch NJ, Strittmatter WJ, Schmechel DE, et al. (1994) Protective effect of apolipoprotein E type 2 for late onset Alzheimer disease. Nat Genet 7: 180–184.
17. Castellano JM, Kim J, Stewart FR, Jiang H, DeMattos RB, et al. (2011) Human apoE isoforms differentially regulate brain amyloid-beta peptide clearance. Sci Transl Med 3: 89ra57.
18. Lichtman SW, Seliger G, Tycko B, Marder K (2000) Apolipoprotein E and functional recovery from brain injury following postacute rehabilitation. Neurology 55: 1536–1539.
19. Smith C, Graham DI, Murray LS, Stewart J, Nicoll JA (2006) Association of APOE e4 and cerebrovascular pathology in traumatic brain injury. JNeurolNeurosurgPsychiatry 77: 363–366.
20. Teasdale GM, Nicoll JA, Murray G, Fiddes M (1997) Association of apolipoprotein E polymorphism with outcome after head injury. The Lancet 350: 1069–1071.
21. Jiang Y, Sun XC, Gui L, Tang WY, Zhen LP, et al. (2008) Lack of association between apolipoprotein E promoters in epsilon-4 carriers and worsening on computed tomography in early stage of traumatic brain injury. Acta Neurochir Suppl 105: 233–236.
22. Pruthi N, Chandramouli BA, Kuttappa TB, Rao SL, Subbakrishna DK, et al. (2010) Apolipoprotein E polymorphism and outcome after mild to moderate traumatic brain injury: a study of patient population in India. Neurol India 58: 264–269.
23. Zhou W, Xu D, Peng X, Zhang Q, Jia J, et al. (2008) Meta-analysis of APOE4 allele and outcomes after traumatic brain injury. J Neurotrauma 25: 279–290.
24. May S, Martin G, Wellington C (2011) Brain injury and dementia: Is there a connection? J Neurology Neurosci 2: 1–7.
25. Roberts GW, Gentleman SM, Lynch A, Murray L, Landon M, et al. (1994) Beta amyloid protein deposition in the brain after severe head injury: implications for the pathogenesis of Alzheimer's disease. J Neurol Neurosurg Psychiatry 57: 419–425.
26. Johnson VE, Stewart W, Smith DH (2012) Widespread tau and amyloid-beta pathology many years after a single traumatic brain injury in humans. Brain Pathol 22: 142–149.
27. Gentleman SM, Nash MJ, Sweeting CJ, Graham DI, Roberts GW (1993) Beta-amyloid precursor protein (beta APP) as a marker for axonal injury after head injury. Neurosci Lett 160: 139–144.
28. Sherriff FE, Bridges LR, Gentleman SM, Sivaloganathan S, Wilson S (1994) Markers of axonal injury in post mortem human brain. Acta Neuropathol (Berl) 88: 433–439.
29. Johnson VE, Stewart W, Smith DH (2010) Traumatic brain injury and amyloid-beta pathology: a link to Alzheimer's disease? Nat Rev Neurosci 11: 361–370.
30. Magnoni S, Esparza TJ, Conte V, Carbonara M, Carrabba G, et al. (2012) Tau elevations in the brain extracellular space correlate with reduced amyloid-beta levels and predict adverse clinical outcomes after severe traumatic brain injury. Brain 135: 1268–1280.
31. Magnoni S, Brody DL (2010) New perspectives on amyloid-beta dynamics after acute brain injury: moving between experimental approaches and studies in the human brain. Arch Neurol 67: 1068–1073.
32. Schwetye KE, Cirrito JR, Esparza TJ, Mac Donald CL, Holtzman DM, et al. (2010) Traumatic brain injury reduces soluble extracellular amyloid-β in mice: A methodologically novel combined microdialysis-controlled cortical impact study. Neurobiology of Disease 40: 555–564.
33. Wahrle SE, Jiang H, Parsadanian M, Legleiter J, Han X, et al. (2004) ABCA1 is required for normal CNS apoE levels and for lipidation of astrocyte-secreted apoE. J Biol Chem 279 40987–40993.
34. Hirsch-Reinshagen V, Zhou S, Burgess BL, Bernier L, McIsaac SA, et al. (2004) Deficiency of ABCA1 impairs apolipoprotein E metabolism in brain. J Biol Chem 279: 41197–41207.
35. Hirsch-Reinshagen V, Maia LF, Burgess BL, Blain JF, Naus KE, et al. (2005) The absence of ABCA1 decreases soluble apoE levels but does not diminish amyloid deposition in two murine models of Alzheimer's Disease. J Biol Chem 280: 43243–43256.
36. Koldamova R, Staufenbiel M, Lefterov I (2005) Lack of ABCA1 considerably decreased brain apoE level and increases amyloid deposition in APP23 mice. J Biol Chem 280: 43224–43235.
37. Wahrle S, Jiang H, Parsadanian M, Hartman RE, Bales KR, et al. (2005) Deletion of Abca1 increases Abeta deposition in the PDAPP transgenic mouse model of Alzheimer disease. J Biol Chem 280 43236–43242.
38. Wahrle SE, Jiang H, Parsadanian M, Kim J, Li A, et al. (2008) Overexpression of ABCA1 reduces amyloid deposition in the PDAPP mouse model of Alzheimer disease. J Clin Invest 118: 671–682.
39. Wojcicka G, Jamroz-Wisniewska A, Horoszewicz K, Beltowski J (2007) Liver X receptors (LXRs). Part I: structure, function, regulation of activity, and role in lipid metabolism. Postepy Hig Med Dosw (Online) 61: 736–759.
40. Jamroz-Wisniewska A, Wǿjcicka G, Horoszewicz K, Beltowski J (2007) Liver X receptors (LXRs). Part II: Non-lipid effects, role in pathology, and therapeutic implications. Postepy Hig Med Dosw (Online) 61: 760–785.
41. Zelcer N, Khanlou N, Clare R, Jiang Q, Reed-Geaghan EG, et al. (2007) Attenuation of neuroinflammation and Alzheimer's disease pathology by liver x receptors. Proc Natl Acad Sci U S A 104: 10601–10606.
42. Donkin JJ, Stukas S, Hirsch-Reinshagen V, Namjoshi D, Wilkinson A, et al. (2010) ATP-binding cassette transporter A1 mediates the beneficial effects of the liver-X-receptor agonist GW3965 on object recognition memory and amyloid burden in APP/PS1 mice. J Biol Chem 285: 34144–34154.
43. Fitz NF, Cronican A, Pham T, Fogg A, Fauq AH, et al. (2010) Liver X receptor agonist treatment ameliorates amyloid pathology and memory deficits caused by high-fat diet in APP23 mice. J Neurosci 30: 6862–6872.
44. Jiang Q, Lee CY, Mandrekar S, Wilkinson B, Cramer P, et al. (2008) ApoE promotes the proteolytic degradation of Abeta. Neuron 58: 681–693.
45. Loane DL, Washington PM, Vardanian L, Pocivavsek A, Hoe HS, et al. (2011) Modulation of ABCA1 by an LXR agonist reduces beta-amyloid levels and improves outcome after traumatic brain injury. J Neurotrauma 28: 225–236.
46. Decuypere M, Klimo P, Jr. (2012) Spectrum of traumatic brain injury from mild to severe. Surg Clin North Am 92: 939–957, ix.
47. Gavett BE, Stern RA, McKee AC (2011) Chronic traumatic encephalopathy: a potential late effect of sport-related concussive and subconcussive head trauma. Clin Sports Med 30: 179–188.
48. Antunes M, Biala G (2012) The novel object recognition memory: neurobiology, test procedure, and its modifications. Cogn Process 13: 93–110.
49. Loane DJ, Pocivavasek A, Moussa CE, Thompson R, Matsuoka Y, et al. (2009) Amyloid precursor protein secretases as therapeutic targets for traumatic brain injury. Nat Med 15: 377–379.
50. Tran HT, LaFerla FM, Holtzman DM, Brody DL (2011) Controlled cortical impact traumatic brain injury in 3xTg-AD mice causes acute intra-axonal amyloid-β accumulation and independently accelerates the development of tau abnormalities. J Neurosci 31: 9513–9525.
51. Tran HT, Sanchez L, Esperza TJ, Brody DL (2011) Distint temporal and anatomical distributions of amyloid-β and tau abnormalities following controlled cortical impact in transgenic mice. PLoS One 6: e2575.
52. Tang YP, Noda Y, Hasegawa T, Nabeshima T (1997) A concussive-like brain injury model in mice (I): impairment in learning and memory. J Neurotrauma 14: 851–862.
53. DeFord SM, Wilson MS, Rice AC, Clausen T, Rice LK, et al. (2002) Repeated mild brain injuries result in cognitive impairment in B6C3F1 mice. J Neurotrauma 19: 427–438.
54. Creeley CE, Wozniak DF, Bayly PV, Olney JW, Lewis LM (2004) Multiple episodes of mild traumatic brain injury result in impaired cognitive performance in mice. Acad Emerg Med 11: 809–819.
55. Zohar O, Rubovitch V, Milman A, Schreiber S, Pick CG (2011) Behavioral consequences of minimal traumatic brain injury in mice. Acta Neurobiol Exp (Wars) 71: 36–45.
56. Pan W, Kastin AJ, Rigai T, McLay R, Pick CG (2003) Increased hippocampal uptake of tumor necrosis factor alpha and behavioral changes in mice. Exp Brain Res 149: 195–199.
57. Tsenter J, Beni-Adani L, Assaf Y, Alexandrovich AG, Trembovler V, et al. (2008) Dynamic changes in the recovery after traumatic brain injury in mice: effect of injury severity on T2-weighted MRI abnormalities, and motor and cognitive functions. J Neurotrauma 25: 324–333.
58. Shitaka Y, Tran HT, Bennet RE, Sanchez L, Levy MA, et al. (2011) Repetitive closed-skull traumatic brain injury in mice causes persistent multifocal axonal injury and microglial reactivity. J Neuropathol Exp Neurol 70: 551–567.
59. Petit-Turcotte C, Aumont N, Blain JF, Poirier J (2007) Apolipoprotein E receptors and amyloid expression are modulated in an apolipoprotein E-dependent fashion in response to hippocampal deafferentation in rodent. Neuroscience 150: 58–63.
60. Li Y, Zhang L, Kallakuri S, Zhou R, Cavanaugh JM (2011) Quantitative relationship between axonal injury and mechanical response in a rodent head impact acceleration model. J Neurotrauma 28: 1767–1782.
61. Marmarou A, Foda MA, van den Brink W, Campbell J, Kita H, et al. (1994) A new model of diffuse brain injury in rats. Part I: Pathophysiology and biomechanics. J Neurosurg 80: 291–300.
62. Goldstein LE, Fisher AM, Tagge CA, Zhang XL, Velisek L, et al. (2012) Chronic traumatic encephalopathy in blast-exposed military veterans and a blast neurotrauma mouse model. Sci Transl Med 4: 143ra160.
63. Riddell DR, Zhou H, Comery TA, Kouranova E, Lo CF, et al. (2007) The LXR agonist TO901317 selectively lowers hippocampal Abeta42 and improves memory in the Tg2576 mouse model of Alzheimer's disease. Mol Cell Neurosci 34: 621–628.
64. Grefhorst A, Elzinga BM, Voshol PJ, Bl"sch T, Kok T, et al. (2002) Stimulation of lipogenesis by pharmacological activation of the liver X receptor leads to production of large, triglyceride-rich very low density lipoprotein particles. J Biol Chem 277: 34182–34190.
65. Groot PH, Pearce NJ, Yates JW, Stocker C, Sauermelch C, et al. (2005) Synthetic LXR agonists increase LDL in CETP species. J Lipid Res 46: 2182–2191.
66. Paxinos G, Franklin KBJ (2001) The Mouse Brain In Stereotaxic Coordinates. San Diego: Elsevier/Academic Press.
67. Miao B, Zondlo S, Gibbs S, Cromley D, Hosagrahara BP, et al. (2004) Raising HDL cholesterol without inducing hepatic steatosis and hypertriglyceridemia by a selective LXR modulator. J Lipid Res 45: 1410–1417.
68. Ennaceur A, Delacour J (1988) A new one-trial test for neurobiological studies of memory in rats. 1: Behavioral data. Behav Brain Res 31: 47–59.

11

Enhanced Functional Recovery in MRL/MpJ Mice after Spinal Cord Dorsal Hemisection

Sandrine Thuret*¤a, Michaela Thallmair¤b, Laura L. Horky¤c, Fred H. Gage*

Laboratory of Genetics, The Salk Institute for Biological Studies, La Jolla, California, United States of America

Abstract

Adult MRL/MpJ mice have been shown to possess unique regeneration capabilities. They are able to heal an ear-punched hole or an injured heart with normal tissue architecture and without scar formation. Here we present functional and histological evidence for enhanced recovery following spinal cord injury (SCI) in MRL/MpJ mice. A control group (C57BL/6 mice) and MRL/MpJ mice underwent a dorsal hemisection at T9 (thoracic vertebra 9). Our data show that MRL/MpJ mice recovered motor function significantly faster and more completely. We observed enhanced regeneration of the corticospinal tract (CST). Furthermore, we observed a reduced astrocytic response and fewer micro-cavities at the injury site, which appear to create a more growth-permissive environment for the injured axons. Our data suggest that the reduced astrocytic response is in part due to a lower lesion-induced increase of cell proliferation post-SCI, and a reduced astrocytic differentiation of the proliferating cells. Interestingly, we also found an increased number of proliferating microglia, which could be involved in the MRL/MpJ spinal cord repair mechanisms. Finally, to evaluate the molecular basis of faster spinal cord repair, we examined the difference in gene expression changes in MRL/MpJ and C57BL/6 mice after SCI. Our microarray data support our histological findings and reveal a transcriptional profile associated with a more efficient spinal cord repair in MRL/MpJ mice.

Editor: Brian D. McCabe, Columbia University, United States of America

Funding: ST was supported by the Paralyzed Veterans of America Spinal Cord Research Foundation (www.pva.org) and the Paralysis Project of America (www.paralysisproject.org). MT was supported by the Deutsche Forschungsgemeinschaft (www.dfg.de/). Additional support was provided by the Christopher and Dana Reeve Foundation (www.christopherreeve.org). The funders had no role in study design, data collection and analysis, decision to publish, or preparation of the manuscript.

Competing Interests: The authors have declared that no competing interests exist.

* E-mail: sandrine.1.thuret@kcl.ac.uk (ST); gage@salk.edu (FHG)

¤a Current address: Institute of Psychiatry, King's College London, London, United Kingdom
¤b Current address: Brain Research Institute, ETH and University Zurich, Zurich, Switzerland
¤c Current address: Department of Radiology, Brigham and Women's Hospital, Boston, Massachusetts, United States of America

Introduction

Injury to the mammalian spinal cord typically results in very limited regeneration of damaged axons, causing permanent functional deficits [1]. Central nervous system (CNS) axonal regeneration appears to be impeded partly by myelin inhibitors [2] and by formation of a post-lesion scar barrier [3]. However, if axons can traverse the injury site, there is evidence that they may re-grow into unscarred regions [4,5].

The main components of the scar are gliotic astrocytes and microglia/macrophages [6]. Upon injury, astrocytes proliferate, up-regulate the expression of glial fibrillary acidic protein (GFAP) and form a dense network of glial processes at the injury site. In addition to creating this physical barrier, astrocytic gliosis will result in the production of a variety of cytokines, cell adhesion and extra-cellular matrix molecules. Some of these molecules, such as chondroitin sulfate proteoglycans (CSPG), have been shown to inhibit axonal growth [7] and, if degraded in the glial scar at the injury site, axonal regeneration as well as functional recovery is promoted [8]. Similarly, suppression of glial scar formation with a 2-Gy dose of radiation after SCI improves functional recovery [9].

Wound healing of mammalian tissue is a long process and is always accompanied by scarring. In general, adult mammals are poor healers compared to lower vertebrates such as some tailed amphibians, which can completely regenerate their spinal cord without scar formation [10]. Among mammals, the MRL/MpJ mouse strain has a unique capacity for wound healing and regeneration, as shown by the closure of ear punches with normal tissue architecture and cartilage replacement. This characteristic is in contrast to the C57BL/6 mouse, which heals only part of the ear punch hole and forms a scar [11]. Furthermore, the MRL/MpJ mouse has been shown to regenerate injured heart tissue in contrast to the C57BL/6 mouse [12]. Neither of these regenerative repair processes is normally seen in adult mice, and the healing response of MRL/MpJ mice is reminiscent of amphibian regeneration with scarless healing [11,12]. More recently, Chadwick and colleagues [13] showed that, after amputation, MRL mouse digit tips were found to distally re-grow more quickly and reform nails either partially or completely to a greater degree than C57BL/6 mice. Gourevitch and colleagues [14] further observed a blastema-like structure along apparent chondrogenesis, consistent with a histological profile of a regenerative response to injury.

Because the MRL/MpJ mouse has shown enhanced healing abilities without scarring, we investigated whether injured MRL/MpJ spinal cord would show enhanced healing and possibly axon regeneration without glial scar formation. To assess putative regenerative repair in the MRL/MpJ spinal cord, we performed a dorsal spinal cord hemisection and used C57BL/6 mice as control animals.

Materials and Methods

Animals

The MRL/MpJ is an inbred mouse derived from the interbreeding of 4 progenitor inbred lines: LG/J (75%), AKR/J (12.6%), C3H/Hej (12.1%) and C57BL/B6 (0.3%) [15]. Over 20 genetic loci have been identified and their healing phenotype is a complex multigenic and sexually dimorphic trait [16,17,18,19].

We chose C57BL/6 animals as controls because they have been used as a control strain in all published work related to the enhanced healing abilities of MRL/MpJ mice, including all gene and loci identification studies [16,17,18,19,20,21]. Using the C57BL/6 mice as a control strain thereby allows us to compare our gene expression data to already published literature. The intact C57BL/6 mice performed better than the MRL/MpJ animals in the rotarod and the grid test. Therefore, the motor recovery results were analyzed after normalizing the performance of the injured MRL/MpJ and C57BL/6 mice to their respective baseline levels (intact mice). It is likely that the MRL/MpJ mice are not as agile as the C57BL/6 mice because of their higher body weight (10-week-old female C57BL/6 weighed 20.1 g±0.5 g, MRL/MpJ weighed an average of 31.7 g±0.4 g). The only available inbred strains of mice matching the weight of MRL/MpJ mice, the C3HeB/FeJ strain, is blind (Mouse Phenome Database, http://www.jax.org) and therefore would not have been suitable for behavioral testing.

Ten-week-old female MRL/MpJ and C57BL/6 mice were purchased from the Jackson Laboratory, ME. They were housed in groups of 5 or singly (after SCI) and had access to water and food ad libitum at the Salk Institute animal facilities. All animal procedures were in accordance with NIH guidelines and approved by the Salk Institutional Animal Care and Use Committee. Intraperitoneal injections of ketamine (100 mg/kg) and xylazine (10 mg/kg) were used as anesthesia for all surgical procedures and before perfusion.

SCI and post-operative care

Under appropriate aseptic techniques, 10-week-old female mice (22 MRL/MpJ and 22 C57BL/6) received a partial dorsal laminectomy at a single thoracic vertebral segment (T9). The spinal cord was exposed and Lidocaine (2%, 10 µl, Bimeda, Riverside, MO) was applied to the dura for 1 minute to anesthetize the region to be transected. The dorsal half of the spinal cord was cut 0.5 mm deep with a pair of microscissors to sever the main parts of the CST, including the dorsal lateral component of the CST (the diameter of the spinal cord at T9 was not significantly different between the two strains, allowing injuries of equal sizes for both strains). The overlaying muscle layers were sutured and the skin was stapled closed. Postoperatively, all animals were placed singly in a cage on warming pads and were provided with moist food pellets on the bottom of the cage. Temperature and respiration were monitored until animals were fully awake. Then cages were placed back on the housing system rack. Bladders were manually expressed twice a day until normal voiding reflexes returned. An antibiotic (enrofloxacin, 2.27 mg/kg, Baytril, Bayer, KS) was given once daily for 6 days. Animals were transcardially perfused with saline, followed by 4% paraformaldehyde at different time points post-SCI.

Behavioral analyses

Intact MRL/MpJ and C57BL/6 mice (n = 10 in each group) were used to establish the baseline performance of each strain in the rotarod, grid and footprint test (baseline group). Also, injured groups of 10 MRL/MpJ and 10 C57BL/6 mice were tested 4 days before the injury and up to 88 days after the injury.

Rotarod test. The Rotamex 4/9 (Columbus Instrument, OH) was used to evaluate fore-and hindlimb motor coordination and balance. Each mouse was placed on the slowly moving rotarod (5 rpm) and the trial was started when the animal found its balance. The rotarod accelerated from 5 rpm to a maximum of 70 rpm within 3 minutes. The latency to fall off the rotarod was recorded. Mice underwent 3 consecutive trails. The mean latency for 3 trails was used for statistical analysis.

Grid test. This test evaluated the ability of the mice to navigate over a wire mesh grid (2.5×2.5 cm grid spaces, 28×35 cm total area). Each mouse was videotaped while on the grid until it achieved a total of 20 steps. A foot-fault was scored each time the animal misplaced its hindlimb to protrude entirely or partially through the grid.

Bladder function. The bladder of each injured animal was voided twice daily until it recovered autonomic bladder function. The amount of urine voided (or the level of "bladder fullness") was estimated and ranked from 3 to 0: 3 = dysfunctional; full bladder, medium to high pressure needed to manually express it and a large amount of urine expressed, 2 = medium amount of urine expressed after medium pressure, 1 = small amount of urine expressed after slight pressure, and 0 = empty bladder and no urine expressed.

CST tracing

Six MRL/MpJ and 6 C57BL/6 received bilateral CST tracing at 40 days post-injury (dpi), 10 MRL/MpJ and 10 C57BL/6 at 95 dpi, and 2 intact MRL/MpJ and 2 intact C57BL/6 also received bilateral CST tracing. A hole was drilled on each side of the skull overlying the sensorimotor cortex. The anterograde neuronal tracer biotin dextran amine (BDA) (10% BDA in 0.01 M phosphate buffer, pH 7.4; Molecular Probes, Eugene, OR) was injected (2 µl) at 4 injection sites into the sensorimotor area (anterior-posterior coordinates from bregma in mm: 1.0/1.5, 0.5/1.5, −0.5/1.5, −1.0/1.5, at a depth of 0.5 mm) using a Hamilton syringe. Two weeks after BDA injection, the animals were perfused and tissue was collected for histology.

BrdU injections

Bromodeoxyuridine (BrdU; Sigma) was dissolved in 0.9% NaCl and filtered at 0.2 µm. To label dividing cells in intact mice, 12 animals received a daily intraperitoneal injection for 6 consecutive days at a dose of 50 mg/kg body weight (10 mg/ml BrdU). Six mice were sacrificed 1 day and 6 mice were sacrificed 4 weeks after the last injection to assess rate of proliferation, differentiation and survival of mitotic cells. SCI mice received the first injection 2 hours after the injury. Six lesioned mice received a single injection 2 hours after injury and were sacrificed 1 day later. Twenty injured mice received one injection per day for 6 days. Of these, 10 animals were sacrificed at 54 dpi and 10 were sacrificed at 109 dpi.

Tissue processing

Spinal cords were cut into 7-mm segments and embedded in cryomolds (Fisher Scientific, Pittsburgh, PA) with O.C.T.

mounting medium (Tissue Tek, Torrance, CA). Mounted spinal cord segments were stored at −80°C until processed further. Spinal cords were sectioned sagittally on a cryostat at 20 μm thickness and stored on slides at −20°C.

CST detection and analysis

The spinal cord sections were incubated with an avidin–biotin–peroxidase complex (ABC Elite, Vector Laboratories, Burlingame, CA), and the BDA tracer was visualized by a nickel-enhanced diaminobenzidine (DAB) reaction [22]. BDA-labeled fibers were quantified at 0.5 mm and 2 mm caudal and at 5 mm rostral to the lesion site. To account for inter-individual tracing variability, numbers were normalized to the amount of fibers 5 mm rostral to the lesion site, where the CST was intact, and are presented as percentages [8].

BrdU immunohistochemistry and quantification

For the stereological quantification of BrdU-labeled cells, cryostat sections were stained with rat anti-BrdU primary antibody (1:100; Accurate Chemicals) for DAB immunohistochemistry. BrdU-labeled cells in intact mice were quantified as described before [23]. To obtain the number of BrdU-labeled cells per mm^3 in injured spinal cords, labeled cells were counted at the injury epicenter, every 60 μm through the entire width of the spinal cord. For each sagittal section, the area surrounding the injury epicenter (5 mm surrounding the injury epicenter) was traced and recorded using a mechanical stage attached to an Olympus BH2 microscope, a Dage MTI CCD-300TIFG Video camera (Michigane City, IN) and the StereoInvestigator software (MicroBrightfield, Colchester, VT). The number of labeled cells within each traced area was recorded.

Multiple marker immunofluorescence and quantification

Sections were processed for multiple markers to determine the phenotype of BrdU-labeled cells. Primary antibodies detecting immature and mature astrocytes (S-100β polypeptide), mature astrocytes (GFAP), oligodendrocyte progenitors (NG2), microglia (OX-42), and immature (TUJ1) and mature neurons [neuronal nuclear antigen clone A60 (NeuN)] were used. Cryostat sections were pretreated for BrdU detection as described previously [23]. Three compatible primary antibodies were incubated on the sections for 2 days at 4°C in TBS +0.1% Triton X-100 +5% donkey serum. Primary antibodies were used at the following concentrations: rat anti-BrdU (1:100; Accurate Chemicals), rabbit anti-S-100β (1:10,000; SWant, Bellinzona, Switzerland), mouse anti-NeuN (1:10; clone A60; Dr. R. Mullin, Salt Lake City, UT), rabbit anti-GFAP (1:1000; Dako, Carpinteria, CA), rabbit anti-NG2 (1:500; Chemicon, Temecula, CA) and mouse anti-OX-42 (1:200; Serotec, UK). Sections were rinsed twice in 0.1 M TBS, pH 7.5, and blocked in 0.1 M TBS +0.1% Triton X-100 for 15 min before application of secondary antibodies. The following secondary antibodies were applied in 0.1 M TBS and 0.1% Triton X-100, each at a concentration of 1:250 for 2 hr in the dark: donkey anti-rat IgG conjugated to FITC (1:250; Jackson ImmunoResearch, West Grove, PA), donkey anti-mouse conjugated to Cy2 or Cy5 (1:250; Jackson ImmunoResearch), and donkey anti-rabbit conjugated to Cy2 or Cy5 (1:250; Jackson ImmunoResearch). Incubation with secondary antibodies was followed by three rinses (15 min) in TBS. Slides were then immediately coverslipped using polyvinyl alcohol-1,4 diazabicyclo[2.2.2]octane and kept in the dark at 4°C until analysis.

Multiple label immunofluorescent images were collected and quantified using confocal microscopy (Bio-Rad MRC 1000, Hercules, CA). Single confocal plane images of BrdU or phenotype markers (see above) were collected and combined to assess co-localization. A cell was scored for double-staining when a well-defined BrdU-labeled nucleus was associated with an immunopositive (e.g., GFAP, S100b etc.) cell body. The entire cell was followed through the *z*-axis, and only cells with a well-circumscribed, immunopositive soma were considered positive for a particular phenotype. A total of 100 BrdU cells were randomly counted at the injury epicenter.

TdT-mediated dUTP nick end labelling (TUNEL) assay

Apoptosis was examined by TUNEL assay, using a commercially available kit (Oncor, Apoptag). The modified protocol developed by Whiteside et al. was followed to simultaneously counterstain with Hoechst 33342 [24]. TUNEL positive cells were only counted as apoptotic cells if the nucleus had fragmented into several membrane-bounded, highly condensed bodies and if it exhibited shrinkage of both cytoplasm and nucleus, creating pericellular space. Any artifacts of TUNEL labeling, i.e., non-nuclear material, was disregarded.

Detection and quantification of micro-cavities

The sections were counterstained with GFAP antibody (rabbit anti-GFAP, 1:1000; Dako). To obtain the number of micro-cavities per mm^3, micro-cavities were counted at the injury epicenter, every 60 μm through the entire width of the spinal cord. For each section the area surrounding the injury epicenter was traced and recorded as described above. The number of micro-cavities within the traced area was recorded.

Statistical evaluation

All the behavioral data and quantification of traced CST fibers were analyzed using a two-factor ANOVA. Post hoc analysis was carried out using Bonferroni-corrected individual comparisons (Microsoft Excel and XLSTAT; Addinsoft, NY). Statistical evaluation of bladder recovery was assessed by a non-parametric Mann Whitney test. Statistical evaluation of BrdU cell counts and micro-cavity quantification was performed using a Student's t-test. In all analyses, a P-value of less than 0.05 was chosen as significance threshold. All data are presented as the mean ± standard error of the mean (SEM).

Experimental design for the microarray gene expression analysis

For MRL/MpJ and C57BL/6 mice, three groups were examined at 4 dpi: intact control, laminectomy control and hemisection at T9. For each condition, 3 mice were sacrificed and a 1-cm spinal cord segment containing the injury epicenter was collected.

RNA extraction and microarray procedures

The procedure was performed as described in detail before [25]. Total RNA was extracted from each sample individually using TRIzol reagent (Invitrogen, Carlsbad, CA). First-strand cDNA was generated by using a T7-linked-$(dT)_{24}$ primer. After second-strand synthesis, the double-stranded cDNA was used for in vitro transcription using the ENZO BioArray High-Yield RNA transcript labeling kit (ENZO, Farmingdale, NY). The labeled cRNA was purified and fragmented. Each fragmented, biotin-labeled cRNA sample (20 μg) was hybridized to an Affymetrix mouse 430A chip. After scanning, array images were assessed by eye to confirm scanner alignment and the absence of significant bubbles or scratches. Chips had to meet the following criteria to be included in the data analysis: the number of probe pairs called present had to be greater than or equal to 30%, all spiked-in bacterial controls had to be called present, background had to be <200, the number of outliers had to be <500, and the 3V/5V ratio of GAPDH and actin had to be <2.

Microarray data analysis and quantitative PCR validation

The chips were analyzed using two different complementary analytical tools. Data were preprocessed (*.dat files to *.cel files) using Affymetrix Microarray Analysis Suite (MAS) 4.0. Data were subsequently analyzed using the dChip 1.3 PM-MM [26] and the Drop Method [27]. dChip approximates expression values by modeling the whole body of chip data to an ideal system, and Drop determines the significance of change between the triplicate groups statistically with minimal assumptions.

To detect up- or down-regulated genes following SCI, Hemisection chips were compared to Laminectomy control chips for both strains of mice. Each group had 3 chips, so each comparison had a 3×3 design, with a possibility of 9 pairwise comparisons for each group. To obtain a list of genes that were differentially expressed after SCI in MRL/MpJ animals, but not significantly changed in C57BL/6 mice, we compared the gene lists of MRL/MpJ and C57BL/6 mice after imposing very strict criteria (dChip fold change>1.2 AND Drop method confidence = 95%) for the MRL/MpJ chips and very loose criteria (dChip fold change>1.2 OR Drop method confidence = 50%) for the C57BL/6 chips. To detect differentially expressed genes between the 2 uninjured strains, Intact control chips from MRL/MpJ were compared to Intact control chips from C57BL/6. For statistical significance we imposed dChip fold change>1.2 and Drop method confidence = 95%.

Expression changes of genes of interest were verified by quantitative RT-PCR (Q-PCR). cDNA was prepared from RNA samples used for GeneChip analysis, and reverse transcription was performed using Superscript First-strand cDNA Synthesis System (Invitrogen). Primer Express v1.5 software was used to design oligonucleotide primers (Applied Biosystems, Foster City, CA) (see methods Table S1 for sequences of primers), which were constructed by Sigma-Aldrich. Q-PCR was performed with 3.33 ng of cDNA, SYBR Green Master Mix (Applied Biosystems), and 0.2 μM of each primer in a 25-μl reaction. All reactions were performed in triplicate using an ABI Prism 7700 Sequence Detection System (Applied Biosystems). Data analysis was performed according to the protocol provided by Applied Biosystems using Sequence Detector Systems v1.7 software. Intact spinal cord cDNA was used to create standard curves for both strains. Expression of each gene was calculated based on the standard curve for a given primer set. The relative amount of calculated message was normalized to the level of a control gene (18s rRNA). The Q-PCR results are summarized in table 1

Results

Enhanced functional recovery of MRL/MpJ mice after SCI

The locomotor performance of injured mice was assessed at several time points up to 88 dpi. Due to the disparity in performance between intact MRL/MpJ and C57BL/6 mice, a

Table 1. Fold changes for the 10 most up- and down-regulated genes in MRL/MpJ compare to C57BL/6 mice after spinal cord hemisection.

Up-regulated Genes	MRL/MpJ Fold Change	C57Bl/6 Fold Change	Down-regulated Genes	MRL/MpJ Fold Change	C57Bl/6 Fold Change
interferon activated gene 202B [GeneID: 26388]	35.66 [48.80]	1.00 [0.98]	Mus musculus adult male olfactory brain cDNA, RIKEN full-length enriched library, clone:6430537M22 product:unknown EST, full insert sequence.	−4.43	−1.82
allograft inflammatory factor 1 [GeneID: 11629]	23.83 [32.15]	3.14 [2.32]	potassium voltage-gated channel, shaker-related subfamily, beta member 1 [GeneID: 16497]	−3.75 [−4.52]	−1.61 [−2.96]
histidine ammonia lyase [GeneID: 15109]	18.4 [15.34]	2.09 [3.43]	calcitonin-related polypeptide, beta [GeneID: 116903]	−3.55 [−4.82]	−1.50 [−1.03]
Mus musculus, clone MGC:28142 IMAGE:3982042, mRNA, complete cds	7.88	2.70	vesicle-associated membrane protein 1 [GeneID: 22317]	−3.41 [−2.98]	−1.34 [−0.63]
cDNA sequence BC032204 [GeneID: 108101]	5.75	1.85	sulfotransferase family 4A, member 1 [GeneID: 29859]	−3.36 [−4.23]	−1.63 [−1.06]
apolipoprotein B editing complex 1[GeneID: 11810]	4.68 [6.98]	1.68 [1.85]	regulator of G-protein signaling 4 [GeneID: 19736]	−3.16 [−2.98]	−1.35 [−0.85]
cDNA sequence BC032204 [GeneID: 108101]	4.45	1.81	Parvalbumin [GeneID: 19293]	−3.10 [−4.50]	−1.32 [−1.28]
Rho, GDP dissociation inhibitor (GDI) beta [GeneID: 11857]	4.44 [3.96]	1.85 [1.78]	creatine kinase, mitochondrial 1, ubiquitous [GeneID: 12716]	−3.02 [−2.20]	−1.87 [−1.00]
beta-glucuronidase [GeneID: 110006]	4.27 [3.98]	1.63 [1.86]	cartilage acidic protein 1 [GeneID: 72832]	−3.00 [−5.02]	−1.30 [−1.32]
cysteine rich protein 61 [GeneID: 16007]	4.18 [6.40]	1.86 [2.59]	solute carrier family 17 (sodium-dependent inorganic phosphate cotransporter), member 6 [GeneID: 140919]	−2.94 [−1.98]	−1.23 [−0.75]

The complete data set is available at http://genechip.salk.edu/rawdata/Thuret_MRL.zip.
Expression changes of genes of interest were also verified by quantitative RT-PCR (Q-PCR). When applicable, the Q-PCR fold changes are presented in brackets [ ].

control group of intact, age-matched MRL/MpJ and C57BL/6 animals underwent the same behavioral tests as the injured mice to establish the baseline performance for both mice strains.

The rotarod test evaluates fore- and hindlimb motor coordination and balance. Intact mice of both strains improved and reached a plateau at 18 days, showing the same performance up to 88 days (Fig. 1a). However, C57BL/6 mice performed better from the start (72.02 s±4.73) and were able to stay longer on the rotarod (at 88 days, 91.1 s±4.5) than the MRL/MpJ (at start: 45.56 s±5.03; at 88 days, 65.1 s±5.4; Fig.1a). In both mouse strains, the rotarod perfomance decreased after SCI (MRL/MpJ, −13.0 s±4.0 and C57BL/6, −19.0 s±3.5). Interestingly, the injured MRL/MpJ mice reached their baseline level by the end of 88 days of testing, whereas the performance of the injured C57BL/6 animals was still reduced (23.5 s±3.9 below their baseline performance at 88 days; Fig. 1a). To eliminate the strain differences in rotarod performance from the recovery curve after SCI, the performance of the injured MRL/MpJ and C57BL/6 mice was normalized to their respective baseline levels (Fig. 1b). There was no difference in the normalized rotarod performance of MRL/MpJ and C57BL/6 mice at 4 days after injury, indicating that both strains were affected equally by the SCI. Interestingly, MRL/MpJ mice recovered rotarod performance over time whereas C57BL/6 did not recover over the 88 days of testing [repeated measures ANOVA: effect of group ($F_{1,\ 18} = 149.2$, $P<0.0001$), time ($F_{13,\ 234} = 5.5$, $P<0.0001$), interaction group× time ($F_{13,\ 234} = 3.1$, $P<0.0005$)]. Individual comparisons (Bonferroni post-hoc) between groups indicated that injured MRL/MpJ performed significantly better than injured C57BL/6 at 12, 18, 53, 60, 67, 74 and 88 dpi ($P<0.002$).

The grid test was used to assess deficits in descending fine motor control after SCI. Intact animals of both mouse strains improved their ability to cross the grid with fewer foot-faults over time (Fig. 1c). However, intact MRL/MpJ animals made significantly more foot-faults through the entire training period (18 days, 2.4±0.5 foot-faults; $p = 0.0025$) than C57BL/6 (18 days, 0.8±0.15 foot-faults). Following SCI, animals from both strains made more foot-faults when crossing the grid: MRL/MpJ 9.5±2.7 and C57BL/6 mice 3.5±1.6 additional foot-faults. The injured MRL/MpJ animals reached their baseline level by 12 dpi, whereas the injured C57BL/6 never recovered back to their baseline level within the testing period of 18 days (Fig. 1c). To account for differences in baseline performance, the number of foot-faults of the injured mice was expressed as a ratio of the baseline (Fig. 1d). Again, we did not find significant differences in the normalized performance on the grid walk of MRL/MpJ and C57BL/6 4 days after the injury, indicating that both strains were impaired similarly by the SCI. MRL/MpJ mice showed a better performance than C57BL/6 on the grid at 12 and 18 days [repeated measures ANOVA: effect of group ($F_{1,\ 18} = 10.4$, $P<0.002$), time ($F_{3,\ 54} = 3.7$, $P<0.01$), interaction group×time ($F_{3,\ 54} = 2.0$, $P>0.1$); in individual comparisons (Bonferroni post-

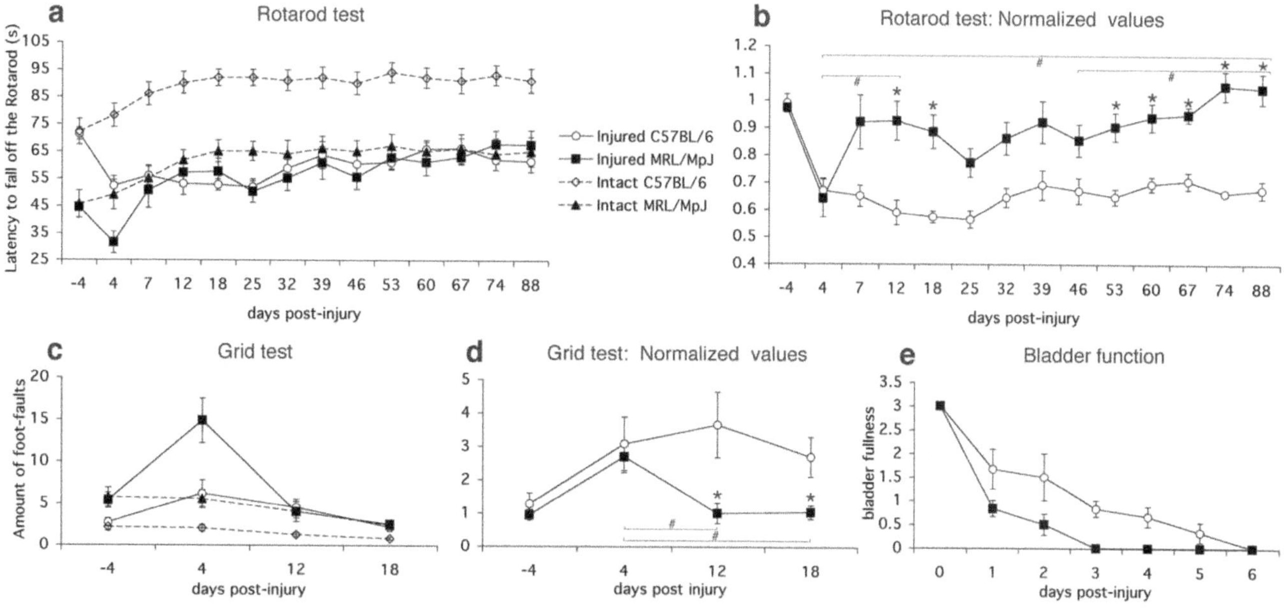

Figure 1. Enhanced functional recovery after SCI in MRL/MpJ mice. a, The latency to fall off the rotarod is depicted as a function of time for intact (baseline) and injured MRL/MpJ and C57BL/6 mice. In both strains intact mice improved their performance over the course of the 88 days of training, but the C57BL/6 performed better at all time points tested. The injured MRL/MpJ reached their baseline level by the end of the 88 days of testing, whereas the injured C57BL/6 did not. **b,** Rotarod performance of injured mice is expressed as a ratio of baseline performance. Injured MRL/MpJ performed significantly better than injured C57BL/6 at 12, 18, 53, 60, 67, 74 and 88 dpi (*$P<0.002$). Injured MRL/MpJ significantly improved between 4 dpi and 88 dpi (#$P<0.002$), whereas injured C57BL/6 did not improve during that period. **c,** The number of foot-faults made during grid walking is reported as a function of time for intact (baseline) and injured MRL/MpJ and C57BL/6 mice. Both intact strains improved their ability to cross the grid with fewer foot-faults, but intact MRL/MpJ made more foot-faults throughout the entire training period. The injured MRL/MpJ reached their baseline level already by 12 dpi, whereas the injured C57BL/6 still had not reached their baseline level by the end of the testing period at 18 dpi. **d,** Grid walking performance of injured mice is expressed as a ratio to baseline performance. Injured MRL/MpJ performed significantly better than injured C57BL/6 at 12 and 18 dpi ($p<0.05$). Injured MRL/MpJ had significantly improved between 4 dpi and 12 dpi, 18 dpi ($p<0.05$), whereas injured C57BL/6 did not improve during that period. **e,** Estimation of bladder fullness is reported as a function of time after SCI in MRL/MpJ and C57BL/6 mice. All MRL/MpJ mice recovered autonomic bladder function by 3 dpi, whereas in C57BL/6 mice, autonomic bladder function did not returned until 6 dpi (Mann Whitney test, p-value = 0.028). Data present mean ± SEM. * denotes significant difference between MRL/MpJ and C57BL/6 and # denotes significant differences in MRL/MpJ performance between time points (P value: two-way ANOVA, Bonferroni post-hoc analysis).

hoc), injured MRL/MpJ performed significantly better than injured C57BL/6 at 12 and 18 dpi, $P<0.05$].

The gait of the lesioned mice was evaluated by analyzing footprint patterns. When assessing base of support and stride length, we found a trend for the injured MRL/MpJ mice to recover faster than the injured C57BL/6 animals, but the differences were not statistically significant (data not shown).

The bladders of the injured animals were manually expressed twice daily until the mice had recovered autonomic bladder function. At 6 hours post-injury (0 dpi), none of the injured animals showed autonomic bladder function. All MRL/MpJ mice recovered autonomic bladder function by 3 dpi, whereas in C57BL/6 mice, autonomic bladder function did not return until 6 dpi [Mann Whitney test of group difference in Area Under Curve summary statistic of mouse bladder function recovery; p-value = 0.028)].

Enhanced regeneration of CST axons in MRL/MpJ mice after SCI

To show that the CST was completely transected by our lesion approach (a 0.5 mm dorsal hemisection), intact MRL/MpJ and intact C57BL6 mice were injected with BDA. Two weeks later these mice underwent a dorsal hemisection and the CST was assessed at 4 dpi (i.e., 18 days after tracing). In both mouse strains the dorsal CST was BDA-labeled and fully transected. Axons were observed rostral as well as caudal to the injury site, but there were no BDA-labeled fibers in the injury epicenter (Fig. 2 a, b).

To analyze the regeneration of the CST after SCI, injured animals were injected with BDA at 40 or 95 dpi. Irrespective of the time point of BDA tracing, we found that C57BL/6 CST axons had retracted from the lesion site and we observed very few or no fibers caudal to the lesion (Fig. 2c, e, g). In contrast, MRL/MpJ CST fibers were present at the lesion site as well as caudal to the injury. Moreover, in some cases, CST axons crossed through the lesion site (Fig. 2d, f, h). At both time points, MRL/MpJ mice displayed significantly more fibers at and caudal to the lesion site compared to C57BL/6 mice [Fig. 2i, j; two-way ANOVA revealed significant effect of group (54 dpi: $F_{1,\ 10}=6.8$, $P<0.05$; 109 dpi: $F_{1,\ 18}=11.2$, $P<0.005$)]. Individual comparisons (Bonferroni post-hoc) between groups indicated that injured MRL/MpJ had significantly more regenerating fibers at 0.5 mm rostral to the lesion site ($P<0.02$): 54 dpi: 19.3±9.9% for MRL/MpJ versus 1.3±1.3% for C57BL/6; 109 dpi: 19.5±9.2% for MRL/MpJ versus 3.2±1.5% for C57BL/6.

In both mouse strains, no significant difference in fiber quantification was found between 54 dpi and 109 dpi, demonstrating that the regenerated fibers were stable and did not retract.

Reduced astrocytic response and micro-cavity formation in MRL/MpJ mice after SCI

At 54 dpi the dorsal surface of the injury site in the C57BL/6 mice was convex, bulging and filled with densely packed GFAP-positive astrocytes. Numerous GFAP-positive astrocytes were delineating the injury site forming a glial scar (Fig. 3a, c). In MRL/MpJ mice, the dorsal surface of the injury site was concave and contained GFAP-negative cells surrounded by a few GFAP-positive astrocytes. The incision site itself was no longer visible in MRL/MpJ spinal cord (Fig. 3b, d). Furthermore, the lesion epicenter of C57BL/6 mice contained 10 times more micro-cysts than the lesion site of MRL/MpJ mice (211.9±23.3 micro-cavities/mm^3 vs. 26.1±2.7 micro-cavities/mm^3; Student's t-test, $p<0.002$; Fig. 3.e, f, g).

At 109 dpi the tissue looked similar to 54 dpi in C57BL/6 mice: the injury site in the C57BL/6 mice was still visible, filled with GFAP-positive astrocytes and micro-cavities (Fig. 3h). In MRL/MpJ mice, the dorsal surface of the injury site was flat and fewer GFAP-positive astrocytes were present (Fig. 3i) when compared to the C57BL/6 injury site. The lesion site could be located because of the higher cell division at the injury site compared to the surrounding spinal cord tissue.

Cell proliferation is different in the intact and injured spinal cord in MRL/MpJ mice

Previous studies have identified progenitor cells throughout the intact and injured rodent spinal cord [23,28,29,30]. Therefore, we examined whether differences between C57BL/6 and MRL/MpJ endogenous spinal cord progenitor cell proliferation could account for the enhanced healing properties of MRL/MpJ mice.

Quantification of BrdU-positive cells in intact spinal cord 1 day after six days of BrdU application revealed a higher level of proliferation in MRL/MpJ mice ($p<0.05$; Fig. 4b,c). Survival of the progeny of dividing progenitor cells was assessed by evaluating BrdU-positive cells 4 weeks after the last BrdU injection and comparing the number of labeled cells with the number obtained at 1 day after the last injection. The total number of surviving BrdU-positive cells after 4 weeks in MRL/MpJ spinal cord was not significantly different from C57BL/6 mice ($p>0.05$; Fig. 4c). The cell fate of the BrdU-positive cells 4 weeks post-injection was also assessed in the intact spinal cord. As reported previously for the rat intact spinal cord [23], there was no sign of ongoing neurogenesis detected in MRL/MpJ or C57BL/6 intact spinal cords. In both strains, the majority of the BrdU-positive cells was immature glia (co-labeled with NG2) or mature astrocytes (co-labeled with S100b/GFAP). No significant differences were detected between MRL/MpJ and C57BL/6 (data not shown).

As described previously [28,29,30], a substantial increase in cellular proliferation occurs after SCI in and around the injury site (Fig. 4d–f). We injected BrdU 2 h after SCI and once daily for 6 dpi. When compared to intact mice, the number of proliferating cells at the site of injury at 1 dpi was 23.5-fold higher for C57BL/6 mice and only 9.6 times higher for MRL/MpJ mice ($p<0.05$), showing that MRL/MpJ injured cord contained 1.5 times fewer BrdU-labeled nuclei than C57BL/6 lesioned cord. The number of BrdU-labeled nuclei in MRL/MpJ mice was still 1.5 times lower at 54 dpi (Fig. 4 e, f), whereas no significant difference was found at 109 dpi.

Apoptotic cells were detected at 1 dpi using TUNEL labeling [31]. Numerous TUNEL-positive cells exhibiting typical fragmented and condensed nucleus were detected at the injury epicenter in both mouse strains (MRL/MpJ: 3292±569 apoptotic cells/mm^3; C57BL/6 3645±631 apoptotic cells/mm^3). Similarly, Fluorojade staining specifically detecting neuronal cell death [32] revealed no differences between MRL/MpJ (2499±101 Fluorojade-positive cells/mm^3) and C57BL/6 mice (2590±348 degenerating neurons/mm^3) at 1 dpi at the injury site.

Reduced astrocytic and increased microglial differentiation of proliferating cells in MRL/MpJ injured spinal cord

Since it has been suggested that the astrogliotic scar is in part produced by proliferating precursor cells differentiating into astrocytes post-injury [29,33,34,35], the cell fate of the dividing cells after SCI was examined in both strains.

In intact mice, differentiation of the surviving BrdU-positive cells was examined in the spinal cord 4 weeks after the last BrdU

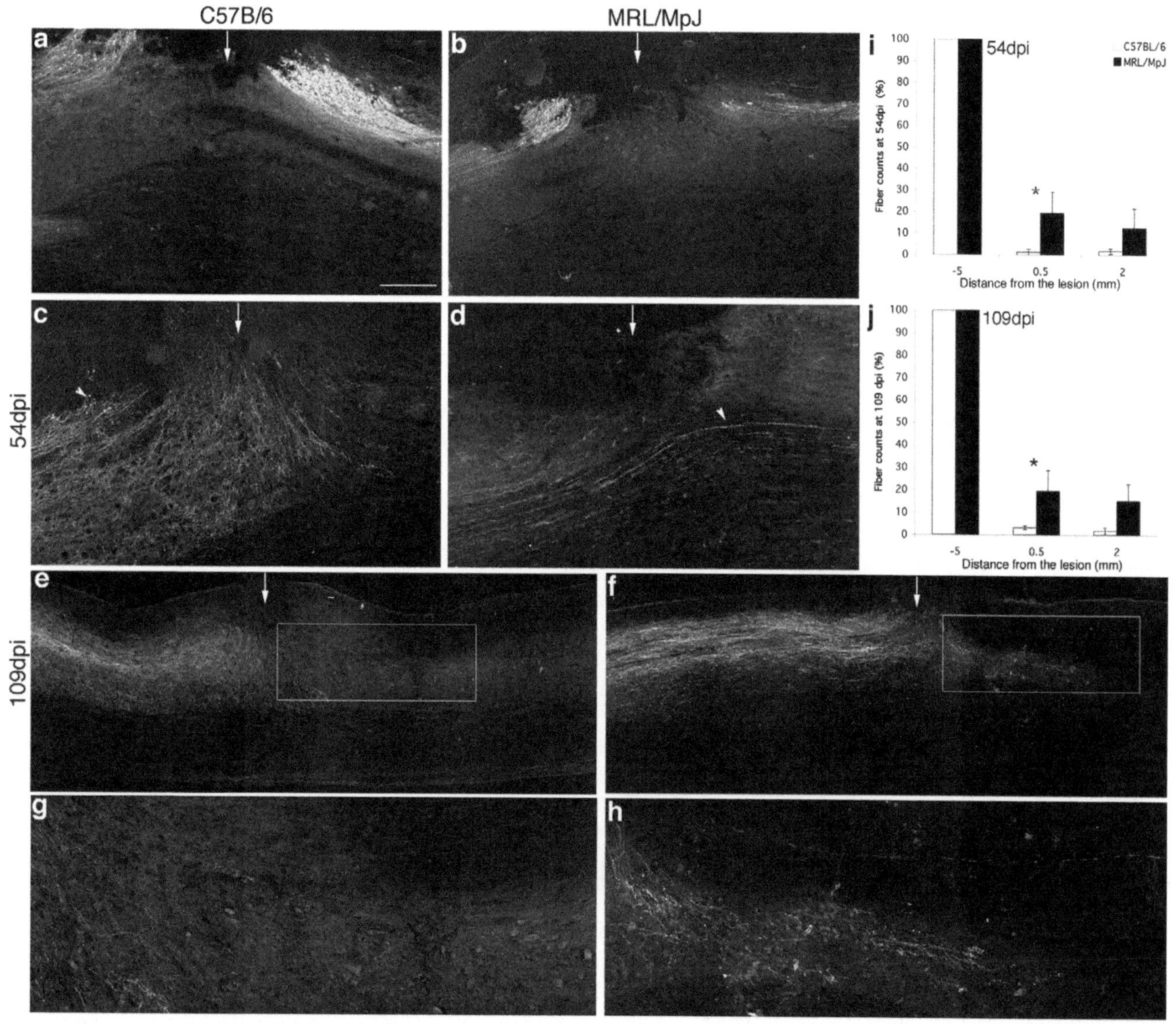

Figure 2. Enhanced regeneration of corticospinal tract (CST) axons in MRL/MpJ mice after SCI. a, b, Sagittal sections of BDA-labeled CST axons in C57BL/6 (**a**) and MRL/MpJ (**b**) mice. Two weeks after tracing, dorsal hemisections were performed. The CST was analyzed at 4 dpi (=18 days after tracing). In both mouse strains, axons are visible rostral and caudal from the injury site. No BDA-labeled fibers were found at the injury site, confirming that the CST was completely transected. **c, d**, Sagittal sections of BDA-labeled CST axons in injured C57BL/6 (**c**) and MRL/MpJ (**d**) mice at 54 dpi. Here, BDA tracing was performed right after SCI. C57BL/6 CST axons retracted from the lesion (arrowhead in **c**). In contrast, some MRL/MpJ CST fibers crossed through the lesion site (arrowhead in (**d**)). **e, f**, Montage of sagittal sections of BDA-labeled CST axons in C57BL/6 (**e**) and MRL/MpJ (**f**) mice at 109 dpi. Higher magnifications are shown in **g** (from boxed area in **e**) and **h** (from boxed area in **f**). We found no BDA-positive fibers below the injury site in C57BL/6 spinal cord (**e, g**), in contrast to MRL/MpJ mice (**f, h**). **i, j**, Data represent percentage of fibers and are reported as a function of the distance from the lesion at 54 dpi (i) and 109 dpi (j). MRL/MpJ mice exhibited significantly more regenerating axons at 0.5 mm caudal from the lesion site. Rostral end of the spinal cord is to the left. The arrows point at the injury site. Asterisks denote significant differences between MRL/MpJ and C57BL/6, $P<0.02$ (two-way ANOVA, Bonferroni Post-hoc). Scale bar, 250 μm (**a–d**); 300 μm (**e, f**); 100 μm (**g, h**).

injection by means of confocal microscopy and immunofluorescent double labeling for BrdU with lineage-specific markers (neuronal markers NeuN and TUJ1, glial progenitor marker NG2, astrocyte markers GFAP and S100b, microglia marker OX-42). No significant differences were found between MRL/MpJ and C57BL/6 mice (data not shown). In accordance with previous publications [23,28], no evidence of neurogenesis was found. Many proliferating cells were oligodendroglial progenitor cells (>40%), approximately 5% were astrocytes and fewer than 1% were microglia.

We then assessed the phenotype of BrdU-positive cells in injured mice. At 54 dpi, no neurogenesis occurred in either of the strains and most of the surviving proliferating cells exhibited a glial phenotype (Fig. 5). The proportions of BrdU-positive cells co-labeled with NG2 (9.7±2.1% vs 4.7±2.0%), S100b (4.0±2.1% vs 2.9±1.4%) and OX-42 (2.0±1.4% vs 1.6±1.4%) in C57BL/6 versus MRL/MpJ mice were not significantly different (Fig. 5s). Nevertheless, BrdU+/GFAP+ cells were significantly higher in C57BL/6 mice (22.5±0.5%) than in MRL/MpJ animals (16.1±0.8%; $p<0.05$; Fig. 5s). When we assessed double-labeled cells (absolute numbers) per volume, we found that C57BL/6 mice had almost 3 times more NG2+/BrdU+ cells (Fig. 5g–l) and 2 times more GFAP+/BrdU+ cells (Fig. 5a–f) than MRL/MpJ mice

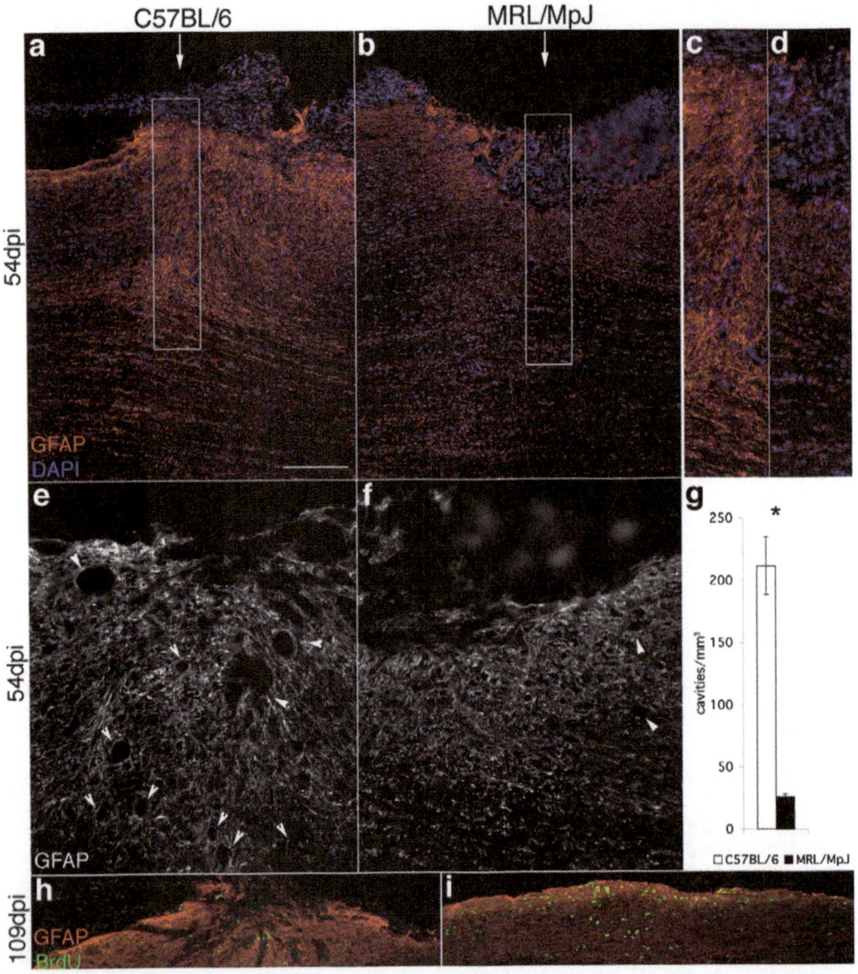

Figure 3. Reduced astrocytic response and micro-cavity formation in MRL/MpJ mice after SCI. a, b, Sagittal sections of the lesion epicenter at 54 dpi labeled with GFAP (red) and counterstained with DAPI (blue) in C57BL/6 (**a**) and MRL/MpJ (**b**) mice. Higher magnifications are shown in **c** (from boxed area in **a**) and **d** (from boxed area in **b**). The dorsal surface of the injury site in the C57BL/6 mice bulged out and was filled with GFAP-positive astrocytes. In white and gray matter, numerous GFAP-positive astrocytes delineated the incision site (**a, c**). In MRL/MpJ mice, the dorsal surface of the injury site was concave and contained GFAP-negative cells surrounded by a few GFAP-positive astrocytes. The incision site was not visible any more (**b, d**). **e, f**, Sagittal sections of the lesion epicenter at 54 dpi stained for GFAP in C57BL/6 (**e**) and MRL/MpJ (**f**) mice. Numerous micro-cavities (arrowheads) are present at the injury site (arrows) in C57BL/6 mice (**e**) compared to none or few in MRL/MpJ mice (**f**). **g**, Bars represent the number of micro-cavities per cm^3 at the injury epicenter at 54 dpi. C57BL/6 injured spinal cord developed significantly more micro-cavities than MRL/MpJ cord. **h, i**, Sagittal sections of the lesion epicenter at 109 dpi labeled with GFAP (red) and BrdU (green) in C57BL/6 (**h**) and MRL/MpJ (**i**) mice. The injury site of the C57BL/6 mice was filled with GFAP-positive astrocytes and micro-cavities were present (**h**). The injury site in the MRL/MpJ mice was not visible any more and could only be located by the presence of BrdU+ cells. The dorsal surface of the injury site was flat and few GFAP-positive astrocytes were present. (**i**). Asterisks denote significant difference between MRL/MpJ and C57BL/6, $P<0.002$ (Student's t-test). Scale bar, 250 μm (**a, b, h, i**); 140 μm (**c, d**); 125 μm (**e, f**).

(Fig. 5u; $p<0.05$), but there was no significant difference between the number of OX-42+/BrdU+ cells. At 109 dpi, the phenotype of BrdU-labeled cells was similar to 54 dpi with C57BL/6 mice, showing still more GFAP+/BrdU+ cells than MRL/MpJ mice (1167.8 ± 172.3 vs 660.9 ± 151.5; $p<0.05$; Fig. 5t, v). But interestingly, at 109 dpi, OX-42+/BrdU+ cells were absent in C57/BL6 mice, whereas they represented $2.7 \pm 0.9\%$ of the MRL/MpJ BrdU-labeled cells ($p<0.05$; Fig. 5n–r, t, v).

Gene expression profiling of the MRL/MpJ and C57BL/6 injured spinal cord

To assess the molecular events occurring in the MRL/MpJ spinal cord after injury and to reveal putative candidate genes responsible for the enhanced spinal cord repair, we used microarray technology. To exclude as many genes as possible that are changed by SCI in general, and to find genes that are strictly differentially regulated in MRL/MpJ mice, we used different criteria when analyzing the gene chips. We used "loose" criteria (dChip1.3 fold change>1.2 OR Drop method confidence = 50%) to reveal differentially expressed genes in C57BL/6 spinal cord 4 days after dorsal hemisection and found 1538 genes. We applied "strict" criteria (dChip1.3 fold change>1.2 AND Drop method confidence = 95%) to find genes differentially expressed in MRL/MpJ spinal cord 4 days after dorsal hemisection and found 745 genes. Of these genes, 272 are exclusively regulated in MRL/MpJ injured spinal cord (Fig. 6a; complete data set available at http://genechip.salk.edu/rawdata/Thuret_MRL.zip). We also compared gene expression in the intact spinal cord of MRL/MpJ and C57BL/6 mice using the "strict" criteria and found 54 genes to be differentially expressed.

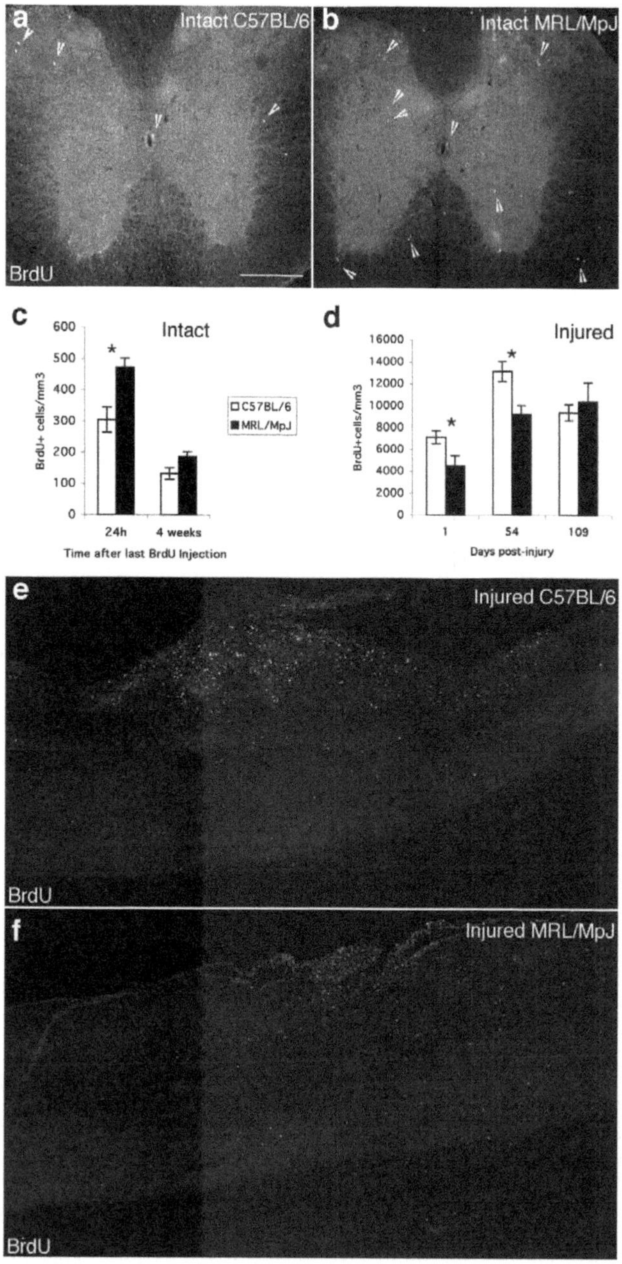

Figure 4. Higher cell proliferation in intact spinal cord but lower increase in cell proliferation after SCI in MRL/MpJ mice. **a**, **b**, Immunohistochemical staining for BrdU (arrowheads) on cross sections at the mid-thoracic level of intact spinal cord in C57BL/6 (**a**) and MRL/MpJ (**b**) mice 24 h after the last of 6 daily BrdU injections. **c**, Graphs show the amount of BrdU-labeled nuclei per mm^3 in the intact spinal cord at the thoracic level 24 h and 4 weeks after the last BrdU injection. MRL/MpJ mice showed a 1.5× higher level of cell proliferation than C57BL/6. Cell survival was assessed at 4 weeks and there was no significant difference between the two strains. **d**, Number of BrdU-labeled nuclei per mm^3 in the injured spinal cord 1, 54 and 109 dpi. BrdU was injected after injury and for the 6 following days. MRL/MpJ spinal cord showed less cell proliferation at 1 and 54 dpi. **e**, **f**, Sagittal sections of the lesion epicenter at 54 dpi stained for BrdU in C57BL/6 (**e**) and MRL/MpJ (**f**) mice. Asterisks denote significant difference between MRL/MpJ and C57BL/6, P<0.05 (Student's t-test). Scale bar, 260 μm.

None of these genes was among the 272 genes that were differentially expressed in MRL/MpJ injured spinal cord exclusively.

We focused our interest on the genes that were exclusively differentially expressed in MRL/MpJ injured spinal cord and classified the 103 upregulated genes (Fig. 6b) and the 169 downregulated genes by biological function (Fig. 6c). More than a quarter of these genes are to date Expressed Sequence Tags (ESTs) or have an unknown function. Following SCI, MRL/MpJ mice appear to upregulate mainly genes implicated in repair mechanisms: 10% of the upregulated genes are involved in cytoskeleton reorganization, 16% take part in protein and ribozome biosynthesis and 24% participate in cell cycle, DNA replication, lipid metabolism, angiogenesis, glycolysis or cell adhesion. Genes involved in inflammatory processes (6%) are also upregulated upon SCI in MRL/MpJ mice. Among the 169 down-regulated genes, 24% are involved in cell proliferation and differentiation, 12% in cation/ion transport and 12% take part in the signal transduction cascade. Genes participating in the protein ubiquitination process are also downregulated (10%), suggesting a diminished cellular degradation upon SCI in MRL/MpJ mice. Table 1 shows the 10 most up- and down-regulated genes in MRL/MpJ injured spinal cord of which expression levels have also been verified by Q-PCR. In particular, the interferon activated gene 202B (*Ifi202b*) -encoding p202b [36]- is increased 35.66 fold in the injured MRL/MpJ spinal cord and is unchanged in C57BL/6 (Table 1). The second highest fold change difference (x7.59) is the allograft inflammatory factor 1 gene (*Aif1*). This gene is an interferon (IFN)-gamma-inducible Ca2+-binding EF-hand protein.

The identification of candidate genes that complement our histological findings reveals potential molecular mechanisms associated with the enhanced spinal cord healing ability of MRL/MpJ mice.

Discussion

The aim of this work was to assess whether MRL/MpJ mice show enhanced spinal cord repair upon injury. We performed a dorsal spinal cord hemisection and used the C57BL/6 mouse as a control strain. We show in this study that MRL/MpJ mice recover faster than C57BL/6 mice in several behavioral tests. This recovery parallels the enhanced regeneration of CST axons in MRL/MpJ mice. Our data suggest that these axons have a facilitated passage across the injury site because of the reduced astrocytic response and diminished micro-cavity formation in MRL/MpJ mice. The reduced astrocytic response is probably due to reduced cell proliferation post-injury and reduced astrocytic differentiation of the proliferating cells. In addition, we observed an increased microglial differentiation in MRL/MpJ mice, suggesting a possible role of the inflammatory response in the MRL/MpJ spinal cord repair mechanisms. Finally, to evaluate the molecular basis of enhanced spinal cord repair, we examined the gene expression changes after SCI in MRL/MpJ and C57BL/6 mice. The microarray data complement our histological findings and unveil a transcriptional profile that reflects the more efficient spinal cord repair in MRL/MpJ mice

Locomotor recovery and axonal regeneration

The dorsal hemisection destroys a portion of supraspinal input to the lumbar spinal cord and produces major, but incomplete, loss of hindlimb function. Moreover, there is a slight possibility for a minority of CST fibers to not have been fully transected in our model. The limited recovery of C57BL/6 animals in motor tasks

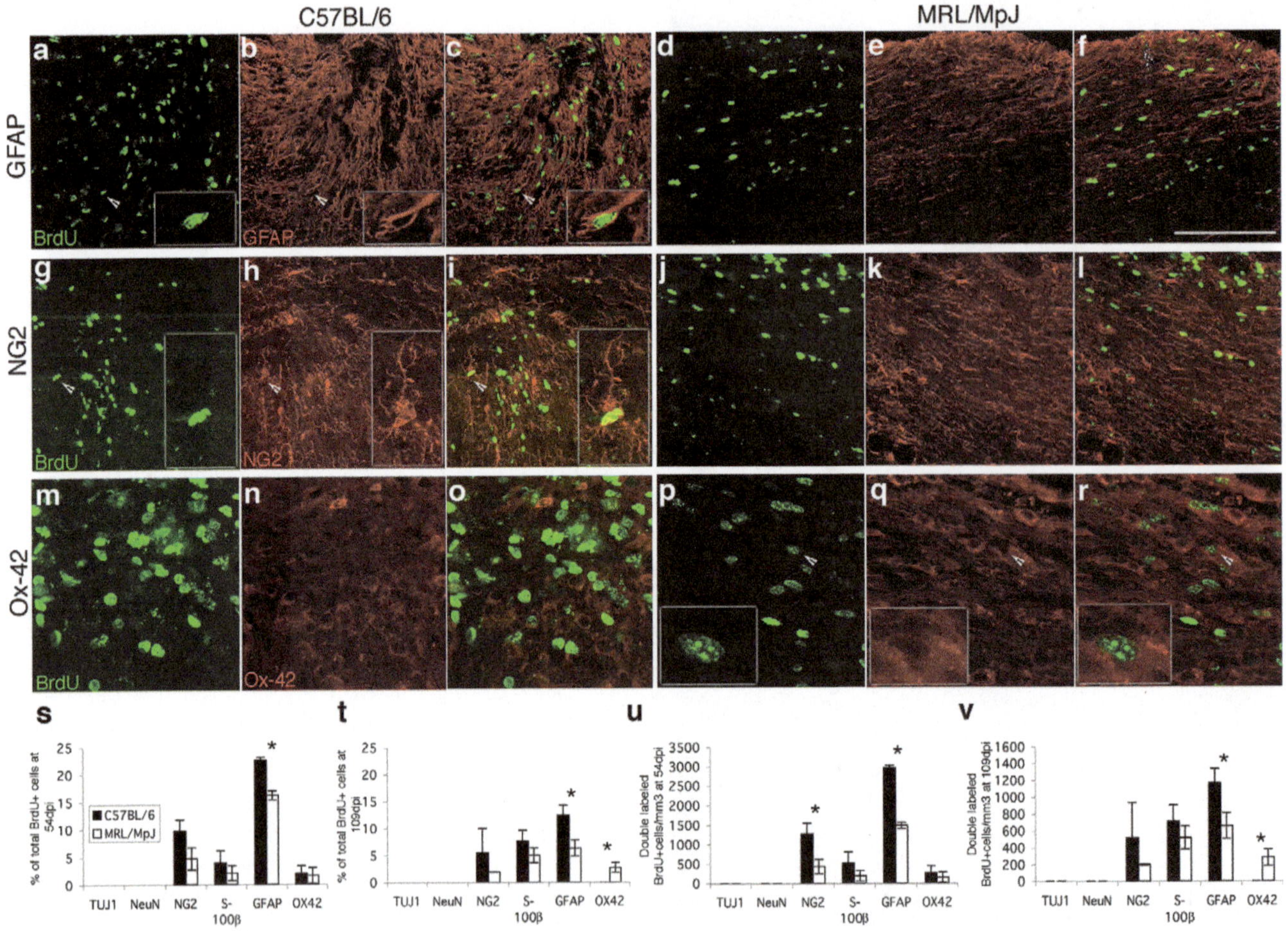

Figure 5. Reduced astrocytic differentiation in MRL/MpJ injured spinal cord. Sagittal sections of the lesion epicenter at 54 dpi (**a–l**) and 109 dpi (**m–r**) labeled for BrdU (green) and lineage markers (red): GFAP (**a–f**), NG2 (**g–l**), OX-42 (**m–r**) in C57BL/6 (**a–c**, **g–i**, **m–o**) and MRL/MpJ (**d–f**, **j–l**, **p–r**) mice. Higher magnifications of cells marked by the arrowheads are shown in boxed inserts. **s–v**, Graphs represent the quantification of BrdU-labeled cells stained for a specific marker at 54 dpi (**s**, **u**) and 109 dpi (**t**, **v**) shown as percentage of all BrdU-labeled cells (**s**, **t**) and as absolute number of co-stained BrdU-labeled nuclei per mm^3 (**u**, **v**). At 54 dpi, C57BL/6 mice showed more NG2/BrdU- and GFAP/BrdU-positive cells than MRL/MpJ mice. At 109 dpi, C57BL/6 mice had still more GFAP/BrdU-expressing cells than MRL/MpJ mice; however, OX-42/BrdU-positive cells were only found in MRL/MpJ spinal cord (2.7±0.9%). Asterisks denote significant difference between MRL/MpJ and C57BL/6, $P<0.05$ (Student's t-test). Scale bar, 100 μm (**a–i**); 50 μm (**p–r**); 50 μm for boxed area (**a–i**); 25 μm for boxed area (**p–r**).

may be attributed to distal rearrangements of spared descending fibers. Recently, it has been demonstrated that local plasticity after SCI creates new circuits between descending CST fibers and cervico-lumbar propriospinal neurons [37]. In MRL/MpJ mice, the recovery of locomotor function was significantly improved. Only in the MRL/MpJ mice did we find CST fiber growth caudal to the injury site, suggesting an important role for regrowing fibers in behavioral recovery. CST axonal regeneration, similar to what we observed in MRL/MpJ mice, has been associated with functional recovery after treatment and in transgenic mice [8,38,39,40,41,42]. It is also possible that the inherent plasticity of the injured spinal cord for creating new circuitries [37] is enhanced in MRL/MpJ mice. We assume that MRL/MpJ CST axons grow because of a more permissive environment at the injury site: (i) we observed fewer micro-cavities providing a better environment for tissue repair and axonal regeneration [43,44] and (ii) we found fewer astrocytes at the injury site, which have been shown to form a physical barrier preventing axonal growth [7]. Thus, we suggest that MRL/MpJ mice have a more permissive environment for axonal regeneration, with a reduced physical barrier and probably a diminished production of growth-inhibitory molecules.

Astrocytic response

For both MRL/MpJ and C57BL/6 mice, the majority of proliferating cells at the injury site differentiated into astrocytes. However, C57BL/6 mice had twice as many newborn astrocytes than MRL/MpJ. Astrocytes are the major contributor to the growth-inhibitory scar after an injury, which is one of the obstacles for successful regeneration. It has been suggested that part of the astrogliotic scar formed after injury is attributable to newly generated astrocytes and not to activation or migration of resident astrocytes [33]. MRL/MpJ injured spinal cord contains fewer newborn astrocytes, probably due to (i) a diminished increase in cell proliferation at the injury site and (ii) a lower astrocytic differentiation of the proliferating cells. X-irradiation treatment eliminates rapidly dividing cells after SCI, suppresses glial scar formation [9,45] and improves functional recovery [9,46]. Taken together, these data suggest that limiting cell proliferation upon

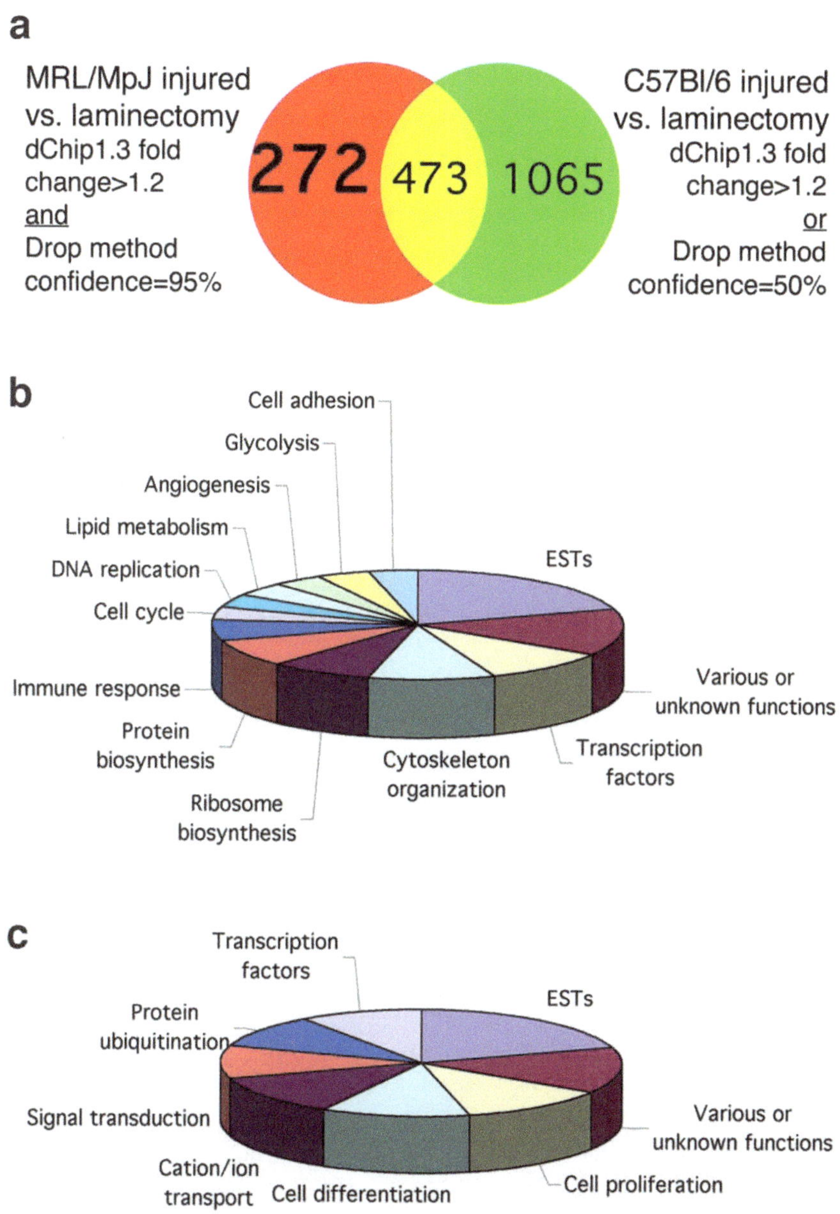

Figure 6. Gene expression profiling of the MRL/MpJ and C57BL/6 injured spinal cord. a, Using loose criteria (dChip1.3 fold change>1.2 OR Drop method confidence = 50%), 1538 genes were found to be differentially expressed in C57BL/6 spinal cord 4 days after hemisection. Using strict criteria (dChip1.3 fold change>1.2 AND Drop method confidence = 95%), 745 differentially expressed genes were found in MRL/MpJ spinal cord 4 days after hemisection, and 272 of these genes were exclusive for the MRL/MpJ injured spinal cord. **b**, Classification by biological function of genes up-regulated exclusively in MRL/MpJ injured spinal cord. **c**, Classification by biological function of genes down-regulated exclusively in MRL/MpJ injured spinal cord.

SCI might be beneficial to reducing astrogliosis and the formation of scar tissue.

Inflammatory response

Our data from the gene profiling analysis at 4 dpi suggest an amplified inflammatory response in injured MRL/MpJ spinal cord. Among the genes exclusively differentially expressed in MRL/MpJ injured spinal cord, the second highest fold change difference (x7.59) is the allograft inflammatory factor 1 gene (*Aif1*; Table 1). This gene encodes an interferon (IFN)-gamma-inducible Ca2+-binding EF-hand protein. In the brain, a subset of microglial cells constitutively expresses AIF-1 [47]. Increased numbers of AIF-1-immunoreactive macrophages/microglial cells were observed in brain tumors and ischemia [48] as well as after SCI [49].

The spinal cord tissue analysis at 109 dpi showed a higher rate of microglia differentiation and/or macrophage recruitment in MRL/MpJ mice, indicating that the immune response might be sustained for the entire repair process. Immune system activity has traditionally been considered harmful for recovery after SCI [48,50,51,52]. However, some studies have shown the potential beneficial effect of a strong inflammatory response. For example, Franzen and colleagues [53] have transplanted macrophages into the injured adult rat spinal cord and suggest that macrophages may exert beneficial effects by degrading myelin products, which inhibit axonal re-growth, and by promoting a permissive extracellular matrix. Moreover, Rapalino and colleagues [54] observed partial recovery of paraplegic rats after implantation of stimulated homologous macrophages. Lastly, Foote and Blakemore [55] have shown that inflammation stimulates remyelination in areas of chronic demyelination. The increased immune response seen in the gene profiling analysis might be due to a more permeable blood-brain-barrier (BBB) because of reduced gliosis. Furthermore, MRL/MpJ mice contain elevated numbers of macrophage progenitor cells circulating in their blood [56], which could possibly pass the more permeable BBB, invade the injury site and proliferate.

Interestingly the Transforming Growth Factor beta is a key regulator in immune responses [57] and its mRNA is upregulated 1.9-fold more in MRL/MpJ injured spinal cord. Injecting TGFb1 was shown to stimulate astrocytes [58]; however, a recent study found an exacerbated astroglial reaction in TGFb1−/− mice [59]. The spinally injured MRL/MpJ mice in our studies showed a reduced astroglial reaction and decreased micro-cavity formation. Importantly, TGFb1 was shown to promote the survival of various CNS neurons *in vitro*, including motoneurons [60]. Additionally, TGFb1 prevents neuronal degeneration caused by hypoxic or excitotoxic injury *in vitro* and rescues hippocampal CA1 neurons from death after transient ischemia *in vivo* [61,62]. Furthermore, anti-TGFb1 treatment following SCI exacerbates secondary damage by increasing the size of cavities in the injured tissue [63], and treatment of SCI rats with TGFb1 reduces lesion volume [64].

The inbred MRL/MpJ strain has a mild autoimmune disorder and is used as a background strain for mutating the Fas gene leading to lupus erythematosus (SLE). The severity of autoimmunity in these mutant mice is strongly affected by the genetic background in which they are placed [65]. The inbred MRL/MpJ strain background creates a severe lupus disease, and even *Fas*-sufficient MRL/MpJ mice suffer from a mild-late-onset SLE [66]. Interestingly, recent evidence suggests that autoimmunity is an endogenous response to CNS injury and that it can be beneficial for repair [67,68]. Further studies showed that passive transfer of T-cells specific to myelin basic protein reduces neuronal loss after SCI and promotes recovery in rats [69,70,71]. Together these data suggest that part of MRL/MpJ functional recovery might be linked to their mild autoimmune disorder.

Brain repair versus spinal cord repair

Hampton and colleagues [72] studied the brain response to injury in MRL/MpJ mice. They did not observe enhanced axonal regeneration after cutting the dopaminergic projection from the substantia nigra to the striatum. Neither did they see a reduced scar after a stab lesion to the cortex. The only common observation between this and the present study was the enhanced and prolonged microglial activity.

In contrast to what we found in the spinal cord, they observed a higher increase in cell proliferation in the cortex after injury. Interestingly, in MRL/MpJ spinal cord we found a dramatic +35.88-fold change for a gene whose over-expression in mammalian cells inhibits cell proliferation: the interferon activated gene 202B (*Ifi202b*) encoding the p202b protein. This inhibition of proliferation might be correlated with the binding and inhibition of the activity of several transcription factors by p202, such as c-Fos, c-Jun, AP2, E2F-1, E2F-4, MyoD, myogenin, and NF-kB p50 and p65 [73,74,75].

In unlesioned MRL/MPJ mice, we observed a lower cell proliferation in the hippocampus than in C57BL/6 animals [76], whereas in the uninjured MRL/MpJ spinal cord we observed a higher cell proliferation. Proliferation in the brain and the spinal cord are differently regulated in intact tissue, therefore it might be also the case following injury. In general, spinal cord and brain seem to react conversely regarding cell proliferation: running increases cell proliferation in the hippocampus but decreases proliferation in the spinal cord of C57BL/6 (our unpublished data).

Another disparity between brain and spinal cord involves the inflammatory response. The acute inflammatory response to traumatic injury is significantly greater in the spinal cord than in the cerebral cortex [77]. Overall, we suspect that the disparity between MRL/MpJ brain and spinal cord repair might be due to a combination of intrinsic differences in cell proliferation and inflammatory response.

Hemisection versus Contusion

More recently, Kostyk and colleagues [78] have assessed the spinal cord regenerative abilities at T9 of MRL/MpJ versus C57/Bl6 mice. They also reported a robust axon growth within the lesion at 28 dpi in MRL/MpJ mice. However, the extent of locomotor recovery was reported to be impaired in the MRL/MpJ mice by 42 dpi. They further described evidence of ongoing degeneration both within and around the lesion site at 42 dpi. Interestingly, Kostyk and colleagues have used a contusion model using rapid displacement of an impounder on the exposed spinal cord following a dorsal laminectomy producing a contusion of moderate severity. These data are consistent with our initial pilot contusion data using the IH-impactor device at 50Kdynes (data not shown), which lead to impaired locomotor recovery in MRL/MpJ mice by 43 dpi. This is in contrast to the present study using a dorsal hemisection model where axonal growth was consistent with functional recovery. It is important to highlight that to inflict a contusion injury using an impounder, it is necessary to make a very large laminectomy. This is in contrast with making a very small laminectomy for a dorsal hemisection. Interestingly, postmortem dissection of our contused mouse samples (data not shown) uncovered a cartilaginous overgrowth within the large laminectomy region inducing a fusion between the remaining neighboring vertebrae leading to extreme spinal column distortion. This distortion possibly explains the poor motor performance of this contusion group. In contrast, such cartilaginous overgrowth was never observed with small laminectomies associated with the dorsal hemisection procedures.

Moreover, a contusion injury is a crude injury as opposed to a micro-scissors hemisection, which is a sharp injury. Interestingly, it has been shown that regeneration of the ear in MRL/MpJ mice depends on the type of wound trauma, where a crude injury (punch) does not heal as well as a sharp injury (clinical biopsy) [79]. Indeed, Rajnoch and colleagues [79] have previously shown that in crude thumb-punched MRL/MpJ ears, an increase in wound area diameter was still observed at the day-28 time point. By contrast, biopsy-punched MRL/MpJ ears showed an 85% wound closure over the same period.

The observations of this study, together with Hampton and colleagues [72] as well as Kostyk and colleagues [78] findings suggest that the occurrence of an enhanced regenerative ability is not exclusively dictated by genetic influences, but is also highly dependent on local cellular interactions and injury type.

Validation of the gene expression profiling after SCI

To help us understand the events occurring in the MRL/MpJ spinal cord after injury, we analyzed gene expression changes after SCI in both strains and compared the results to compile a list of genes differentially expressed exclusively in MRL/MpJ injured spinal cord. Since it is known that genes are differentially regulated in different strains of mice or rats [80,81], we only focused on genes that are specifically expressed in injured MRL/MpJ spinal cord. We have evidences that the MRL/MpJ differentially expressed genes identified in this screen are potentially implicated in spinal cord repair. First, we observed, in both strains some genes that have previously been shown to be differentially regulated after SCI. For example, both mouse strains show expression changes in genes coding for proteins that influence apoptosis: Caspase 3 is upregulated and Caspase 7 is downregulated, as observed previously [25,82]. These studies also found Bax and Bak1 to be upregulated upon SCI, as we have in the present study. For both injured strains, genes coding for cathepsin proteases are upregulated (cathepsins B, C, D, L and S), as previously reported [25,83,84,85]. These data validate that our method is useful for identifying candidate genes that contribute to differences in spinal cord repair between MRL/MpJ and C57BL/6 mice. Second, we found that some of the genes differentially expressed in MRL/MpJ injured spinal cord are located on the quantitative trait loci QTLs that control ear healing in MRL/MpJ mice [16,19,21]. These studies found linkage on chromosome 7. Interestingly, some of the genes differentially expressed in injured MRL/MpJ spinal cord (e.g., genes coding for the mitochondrial uncoupling protein 2 and TGFb1) are located in or near the loci on chromosome 7. Additionally, Masinde et al. [21] found linkage on chromosome 1, where Ifi202b, another of our candidate genes, is located. We found that some genes differentially expressed in the MRL/MpJ healing ear are also present on our gene list. For example we observed that *Decorin* is less up-regulated whereas *DEAD/H* and *Proteasome (prosome, macropain)* are more up-regulated in MRL/MpJ, as previously reported in a different lesion model [21]. We also found that *integrin beta2* and *secretory granule proteoglycan* are more up-regulated in MRL/MpJ upon injury, as shown previously [20]. Since rapid wound healing in MRL/MpJ mice is a genetically controlled quantitative trait [15,19], we expected that the soft tissue regeneration QTLs and genes found in studies on ear regeneration in MRL/MpJ mice would overlap with the genes discovered in this study. The fact that some of the differentially expressed genes are located within the QTL regions or have already been identified provides another validation of our screen. Third, our microarray data complement our histological findings. Indeed, the general tendency is the up-regulation of genes involved in repair mechanisms. Moreover, genes involved in lowering cell proliferation (e.g., *Ifi202b*) and increasing the immune response (e.g., *Aif-1*) are up-regulated in the injured MRL/MpJ spinal cord.

We recognize that this study may underestimate the number of genes involved in MRL/MpJ spinal cord healing because (i) we examined only one time point and some changes in the expression may occur at an earlier or later stage of the repair process, (ii) we used only one strain of mice for control, and (iii) we collected only the lesion epicenter and additional genes might be differently regulated rostral and/or caudal to the lesion. Moreover, it would be interesting in future studies to extract RNA from specific cell types to identify the particular cells responsible for these significant gene expression changes. However, the genes identified here may play a role in the repair processes following SCI and further characterization of these candidates might elucidate some molecular mechanisms of spinal cord repair.

Conclusion and perspectives

This study shows that the MRL/MpJ mice are able to recover faster and better after spinal cord dorsal hemisection than C57BL/6 mice. Our data suggest that a sustained immune response and decreased astrocytic activity are likely to be part of a complex recovery course. Although our data are not able to explain the detailed mechanisms of such a recovery, they do demonstrate that the repair process after SCI is multifaceted. Many pathways are involved, all of which need to be considered and each of which needs to work at an optimal "level" to result in behavioral recovery.

Supporting Information

Table S1 Primer Sequences for Q-PCR.

Acknowledgments

The authors wish to thank J. Brad Aimone for the analysis of the GeneChip Arrays, Lisa Schnell and Martin Schwab for discussion, and Mary Lynn Gage for editing the manuscript.

Author Contributions

Conceived and designed the experiments: ST FHG. Performed the experiments: ST MT LLH. Analyzed the data: ST MT FHG. Wrote the paper: ST MT FHG.

References

1. Thuret S, Moon LD, Gage FH (2006) Therapeutic interventions after spinal cord injury. Nat Rev Neurosci 7: 628–643.
2. Chen MS, Huber AB, van der Haar ME, Frank M, Schnell L, et al. (2000) Nogo-A is a myelin-associated neurite outgrowth inhibitor and an antigen for monoclonal antibody IN-1. Nature 403: 434–439.
3. Fitch MT, Doller C, Combs CK, Landreth GE, Silver J (1999) Cellular and molecular mechanisms of glial scarring and progressive cavitation: in vivo and in vitro analysis of inflammation-induced secondary injury after CNS trauma. J Neurosci 19: 8182–8198.
4. Davies SJ, Fitch MT, Memberg SP, Hall AK, Raisman G, et al. (1997) Regeneration of adult axons in white matter tracts of the central nervous system. Nature 390: 680–683.
5. Neumann S, Woolf CJ (1999) Regeneration of dorsal column fibers into and beyond the lesion site following adult spinal cord injury. Neuron 23: 83–91.
6. Stichel CC, Muller HW (1998) Experimental strategies to promote axonal regeneration after traumatic central nervous system injury. Prog Neurobiol 56: 119–148.
7. Fawcett JW, Asher RA (1999) The glial scar and central nervous system repair. Brain Res Bull 49: 377–391.
8. Bradbury EJ, Moon LD, Popat RJ, King VR, Bennett GS, et al. (2002) Chondroitinase ABC promotes functional recovery after spinal cord injury. Nature 416: 636–640.
9. Ridet JL, Pencalet P, Belcram M, Giraudeau B, Chastang C, et al. (2000) Effects of spinal cord X-irradiation on the recovery of paraplegic rats. Exp Neurol 161: 1–14.
10. Ferretti P, Zhang F, O'Neill P (2003) Changes in spinal cord regenerative ability through phylogenesis and development: lessons to be learnt. Dev Dyn 226: 245–256.
11. Clark LD, Clark RK, Heber-Katz E (1998) A new murine model for mammalian wound repair and regeneration. Clin Immunol Immunopathol 88: 35–45.
12. Leferovich JM, Bedelbaeva K, Samulewicz S, Zhang XM, Zwas D, et al. (2001) Heart regeneration in adult MRL mice. Proc Natl Acad Sci U S A 98: 9830–9835.
13. Chadwick RB, Bu L, Yu H, Hu Y, Wergedal JE, et al. (2007) Digit tip regrowth and differential gene expression in MRL/Mpj, DBA/2, and C57BL/6 mice. Wound Repair Regen 15: 275–284.
14. Gourevitch DL, Clark L, Bedelbaeva K, Leferovich J, Heber-Katz E (2009) Dynamic changes after murine digit amputation: The MRL mouse digit shows waves of tissue remodeling, growth, and apoptosis. Wound Repair and Regeneration 17: 447–455.
15. Li X, Gu W, Masinde G, Hamilton-Ulland M, Xu S, et al. (2001) Genetic control of the rate of wound healing in mice. Heredity 86: 668–674.
16. Blankenhorn EP, Troutman S, Clark LD, Zhang XM, Chen P, et al. (2003) Sexually dimorphic genes regulate healing and regeneration in MRL mice. Mamm Genome 14: 250–260.
17. Heber-Katz E, Chen P, Clark L, Zhang XM, Troutman S, et al. (2004) Regeneration in MRL mice: further genetic loci controlling the ear hole closure trait using MRL and M.m. Castaneus mice. Wound Repair Regen 12: 384–392.

18. Masinde GL, Li X, Gu W, Davidson H, Mohan S, et al. (2001) Identification of wound healing/regeneration quantitative trait loci (QTL) at multiple time points that explain seventy percent of variance in (MRL/MpJ and SJL/J) mice F2 population. Genome Res 11: 2027–2033.
19. McBrearty BA, Clark LD, Zhang XM, Blankenhorn EP, Heber-Katz E (1998) Genetic analysis of a mammalian wound-healing trait. Proc Natl Acad Sci U S A 95: 11792–11797.
20. Li X, Mohan S, Gu W, Baylink DJ (2001) Analysis of gene expression in the wound repair/regeneration process. Mamm Genome 12: 52–59.
21. Masinde G, Li X, Baylink DJ, Nguyen B, Mohan S (2005) Isolation of wound healing/regeneration genes using restrictive fragment differential display-PCR in MRL/MPJ and C57BL/6 mice. Biochem Biophys Res Commun 330: 117–122.
22. Thallmair M, Metz GAS, Z'Graggen WJ, Raineteau O, Kartje GL, et al. (1998) Neurite growth inhibitors restrict plasticity and functional recovery following corticospinal tract lesions. Nature Neuroscience 1: 124–131.
23. Horner PJ, Power AE, Kempermann G, Kuhn HG, Palmer TD, et al. (2000) Proliferation and differentiation of progenitor cells throughout the intact adult rat spinal cord. Journal of Neuroscience 20: 2218–2228.
24. Whiteside G, Cougnon N, Hunt SP, Munglani R (1998) An improved method for detection of apoptosis in tissue sections and cell culture, using the TUNEL technique combined with Hoechst stain. Brain Research Protocols 2: 160–164.
25. Aimone JB, Leasure JL, Perreau VM, Thallmair M (2004) Spatial and temporal gene expression profiling of the contused rat spinal cord. Exp Neurol 189: 204–221.
26. Li C, Wong WH (2001) Model-based analysis of oligonucleotide arrays: expression index computation and outlier detection. Proc Natl Acad Sci U S A 98: 31–36.
27. Aimone JB, Gage FH (2004) Unbiased characterization of high-density oligonucleotide microarrays using probe-level statistics. J Neurosci Methods 135: 27–33.
28. Horky LL, Galimi F, Gage FH, Horner PJ (2006) Fate of endogenous stem/progenitor cells following spinal cord injury. J Comp Neurol 498: 525–538.
29. Johansson CB, Momma S, Clarke DL, Risling M, Lendahl U, et al. (1999) Identification of a neural stem cell in the adult mammalian central nervous system. Cell 96: 25–34.
30. Yamamoto S, Yamamoto N, Kitamura T, Nakamura K, Nakafuku M (2001) Proliferation of parenchymal neural progenitors in response to injury in the adult rat spinal cord. Experimental Neurology 172: 115–127.
31. Whiteside G, Cougnon N, Hunt SP, Munglani R (1998) An improved method for detection of apoptosis in tissue sections and cell culture, using the TUNEL technique combined with Hoechst stain. Brain Res Brain Res Protoc 2: 160–164.
32. Schmued LC, Stowers CC, Scallet AC, Xu L (2005) Fluoro-Jade C results in ultra high resolution and contrast labeling of degenerating neurons. Brain Res 1035: 24–31.
33. Kernie SG, Erwin TM, Parada LF (2001) Brain remodeling due to neuronal and astrocytic proliferation after controlled cortical injury in mice. J Neurosci Res 66: 317–326.
34. Okada S, Nakamura M, Mikami Y, Shimazaki T, Mihara M, et al. (2004) Blockade of interleukin-6 receptor suppresses reactive astrogliosis and ameliorates functional recovery in experimental spinal cord injury. J Neurosci Res 76: 265–276.
35. Takahashi M, Arai Y, Kurosawa H, Sueyoshi N, Shirai S (2003) Ependymal cell reactions in spinal cord segments after compression injury in adult rat. J Neuropathol Exp Neurol 62: 185–194.
36. Wang H, Chatterjee G, Meyer JJ, Liu CJ, Manjunath NA, et al. (1999) Characteristics of three homologous 202 genes (Ifi202a, Ifi202b, and Ifi202c) from the murine interferon-activatable gene 200 cluster. Genomics 60: 281–294.
37. Bareyre FM, Kerschensteiner M, Raineteau O, Mettenleiter TC, Weinmann O, et al. (2004) The injured spinal cord spontaneously forms a new intraspinal circuit in adult rats. Nat Neurosci 7: 269–277.
38. Demjen D, Klussmann S, Kleber S, Zuliani C, Stieltjes B, et al. (2004) Neutralization of CD95 ligand promotes regeneration and functional recovery after spinal cord injury. Nat Med 10: 389–395.
39. Goldshmit Y, Galea MP, Wise G, Bartlett PF, Turnley AM (2004) Axonal regeneration and lack of astrocytic gliosis in EphA4-deficient mice. J Neurosci 24: 10064–10073.
40. Menet V, Prieto M, Privat A, Gimenez y Ribotta M (2003) Axonal plasticity and functional recovery after spinal cord injury in mice deficient in both glial fibrillary acidic protein and vimentin genes. Proc Natl Acad Sci U S A 100: 8999–9004.
41. Li S, Liu BP, Budel S, Li M, Ji B, et al. (2004) Blockade of Nogo-66, myelin-associated glycoprotein, and oligodendrocyte myelin glycoprotein by soluble Nogo-66 receptor promotes axonal sprouting and recovery after spinal injury. J Neurosci 24: 10511–10520.
42. Schwab ME (2004) Nogo and axon regeneration. Curr Opin Neurobiol 14: 118–124.
43. Ramer LM, Au E, Richter MW, Liu J, Tetzlaff W, et al. (2004) Peripheral olfactory ensheathing cells reduce scar and cavity formation and promote regeneration after spinal cord injury. J Comp Neurol 473: 1–15.
44. Ohta M, Suzuki Y, Noda T, Ejiri Y, Dezawa M, et al. (2004) Bone marrow stromal cells infused into the cerebrospinal fluid promote functional recovery of the injured rat spinal cord with reduced cavity formation. Exp Neurol 187: 266–278.
45. Zhang SX, Geddes JW, Owens JL, Holmberg EG (2005) X-irradiation reduces lesion scarring at the contusion site of adult rat spinal cord. Histol Histopathol 20: 519–530.
46. Zeman RJ, Feng Y, Peng H, Visintainer PF, Moorthy CR, et al. (2001) X-irradiation of the contusion site improves locomotor and histological outcomes in spinal cord-injured rats. Exp Neurol 172: 228–234.
47. Mittelbronn M, Dietz K, Schluesener HJ, Meyermann R (2001) Local distribution of microglia in the normal adult human central nervous system differs by up to one order of magnitude. Acta Neuropathol 101: 249–255.
48. Popovich PG, Wei P, Stokes BT (1997) Cellular inflammatory response after spinal cord injury in Sprague-Dawley and Lewis rats. J Comp Neurol 377: 443–464.
49. Schwab JM, Frei E, Klusman I, Schnell L, Schwab ME, et al. (2001) AIF-1 expression defines a proliferating and alert microglial/macrophage phenotype following spinal cord injury in rats. J Neuroimmunol 119: 214–222.
50. Popovich PG, Guan Z, Wei P, Huitinga I, van Rooijen N, et al. (1999) Depletion of hematogenous macrophages promotes partial hindlimb recovery and neuroanatomical repair after experimental spinal cord injury. Exp Neurol 158: 351–365.
51. Blight AR (1992) Macrophages and inflammatory damage in spinal cord injury. J Neurotrauma 9 Suppl 1: S83–91.
52. Jones TB, Ankeny DP, Guan Z, McGaughy V, Fisher LC, et al. (2004) Passive or active immunization with myelin basic protein impairs neurological function and exacerbates neuropathology after spinal cord injury in rats. J Neurosci 24: 3752–3761.
53. Franzen R, Schoenen J, Leprince P, Joosten E, Moonen G, et al. (1998) Effects of macrophage transplantation in the injured adult rat spinal cord: a combined immunocytochemical and biochemical study. J Neurosci Res 51: 316–327.
54. Rapalino O, Lazarov-Spiegler O, Agranov E, Velan GJ, Yoles E, et al. (1998) Implantation of stimulated homologous macrophages results in partial recovery of paraplegic rats. Nat Med 4: 814–821.
55. Foote AK, Blakemore WF (2005) Inflammation stimulates remyelination in areas of chronic demyelination. Brain 128: 528–539.
56. Davis TA, Lennon G (2005) Mice with a regenerative wound healing capacity and an SLE autoimmune phenotype contain elevated numbers of circulating and marrow-derived macrophage progenitor cells. Blood Cells Mol Dis 34: 17–25.
57. Letterio JJ, Roberts AB (1997) TGF-[beta]: A Critical Modulator of Immune Cell Function. Clinical Immunology and Immunopathology 84: 244–250.
58. Logan A, Berry M, Gonzalez AM, Frautschy SA, Sporn MB, et al. (1994) Effects of transforming growth factor beta 1 on scar production in the injured central nervous system of the rat. Eur J Neurosci 6: 355–363.
59. Makwana M, Jones LL, Cuthill D, Heuer H, Bohatschek M, et al. (2007) Endogenous transforming growth factor beta 1 suppresses inflammation and promotes survival in adult CNS. J Neurosci 27: 11201–11213.
60. Martinou JC, Le Van Thai A, Valette A, Weber MJ (1990) Transforming growth factor beta 1 is a potent survival factor for rat embryo motoneurons in culture. Brain Res Dev Brain Res 52: 175–181.
61. Henrich-Noack P, Prehn JH, Krieglstein J (1996) TGF-beta 1 protects hippocampal neurons against degeneration caused by transient global ischemia. Dose-response relationship and potential neuroprotective mechanisms. Stroke 27: 1609–1614; discussion 1615.
62. Prehn JH, Backhauss C, Krieglstein J (1993) Transforming growth factor-beta 1 prevents glutamate neurotoxicity in rat neocortical cultures and protects mouse neocortex from ischemic injury in vivo. J Cereb Blood Flow Metab 13: 521–525.
63. King VR, Phillips JB, Brown RA, Priestley JV (2004) The effects of treatment with antibodies to transforming growth factor beta1 and beta2 following spinal cord damage in the adult rat. Neuroscience 126: 173–183.
64. Tyor WR, Avgeropoulos N, Ohlandt G, Hogan EL (2002) Treatment of spinal cord impact injury in the rat with transforming growth factor-beta. J Neurol Sci 200: 33–41.
65. Vidal S, Kono DH, Theofilopoulos AN (1998) Loci predisposing to autoimmunity in MRL-Fas lpr and C57BL/6-Faslpr mice. J Clin Invest 101: 696–702.
66. Theofilopoulos AN, Dixon FJ (1985) Murine models of systemic lupus erythematosus. Adv Immunol 37: 269–390.
67. Schwartz M, Kipnis J (2001) Protective autoimmunity: regulation and prospects for vaccination after brain and spinal cord injuries. Trends Mol Med 7: 252–258.
68. Yoles E, Hauben E, Palgi O, Agranov E, Gothilf A, et al. (2001) Protective autoimmunity is a physiological response to CNS trauma. J Neurosci 21: 3740–3748.
69. Hauben E, Butovsky O, Nevo U, Yoles E, Moalem G, et al. (2000) Passive or active immunization with myelin basic protein promotes recovery from spinal cord contusion. J Neurosci 20: 6421–6430.
70. Hauben E, Agranov E, Gothilf A, Nevo U, Cohen A, et al. (2001) Posttraumatic therapeutic vaccination with modified myelin self-antigen prevents complete paralysis while avoiding autoimmune disease. J Clin Invest 108: 591–599.
71. Ibarra A, Hauben E, Butovsky O, Schwartz M (2004) The therapeutic window after spinal cord injury can accommodate T cell-based vaccination and methylprednisolone in rats. Eur J Neurosci 19: 2984–2990.
72. Hampton DW, Seitz A, Chen P, Heber-Katz E, Fawcett JW (2004) Altered CNS response to injury in the MRL/MpJ mouse. Neuroscience 127: 821–832.

73. Min W, Ghosh S, Lengyel P (1996) The interferon-inducible p202 protein as a modulator of transcription: inhibition of NF-kappa B, c-Fos, and c-Jun activities. Mol Cell Biol 16: 359–368.
74. Choubey D, Gutterman JU (1997) Inhibition of E2F-4/DP-1-stimulated transcription by p202. Oncogene 15: 291–301.
75. Datta B, Min W, Burma S, Lengyel P (1998) Increase in p202 expression during skeletal muscle differentiation: inhibition of MyoD protein expression and activity by p202. Mol Cell Biol 18: 1074–1083.
76. Thuret S, Toni N, Aigner S, Yeo GW, Gage FH (2009) Hippocampus-dependent learning is associated with adult neurogenesis in MRL/MpJ mice. Hippocampus 19: 658–669.
77. Schnell L, Fearn S, Schwab ME, Perry VH, Anthony DC (1999) Cytokine-induced acute inflammation in the brain and spinal cord. J Neuropathol Exp Neurol 58: 245–254.
78. Kostyk SK, Popovich PG, Stokes BT, Wei P, Jakeman LB (2008) Robust axonal growth and a blunted macrophage response are associated with impaired functional recovery after spinal cord injury in the MRL/MpJ mouse. Neuroscience 156: 498–514.
79. Rajnoch C, Ferguson S, Metcalfe AD, Herrick SE, Willis HS, et al. (2003) Regeneration of the ear after wounding in different mouse strains is dependent on the severity of wound trauma. Dev Dyn 226: 388–397.
80. Dimou L, Schnell L, Montani L, Duncan C, Simonen M, et al. (2006) Nogo-A-deficient mice reveal strain-dependent differences in axonal regeneration. J Neurosci 26: 5591–5603.
81. Schmitt C, Miranpuri GS, Dhodda VK, Isaacson J, Vemuganti R, et al. (2006) Changes in spinal cord injury-induced gene expression in rat are strain-dependent. Spine J 6: 113–119.
82. Citron BA, Arnold PM, Sebastian C, Qin F, Malladi S, et al. (2000) Rapid upregulation of caspase-3 in rat spinal cord after injury: mRNA, protein, and cellular localization correlates with apoptotic cell death. Exp Neurol 166: 213–226.
83. Ellis RC, Earnhardt JN, Hayes RL, Wang KK, Anderson DK (2004) Cathepsin B mRNA and protein expression following contusion spinal cord injury in rats. J Neurochem 88: 689–697.
84. Ellis RC, O'Steen WA, Hayes RL, Nick HS, Wang KK, et al. (2005) Cellular localization and enzymatic activity of cathepsin B after spinal cord injury in the rat. Exp Neurol 193: 19–28.
85. Hashimoto M, Koda M, Ino H, Yoshinaga K, Murata A, et al. (2005) Gene expression profiling of cathepsin D, metallothioneins-1 and -2, osteopontin, and tenascin-C in a mouse spinal cord injury model by cDNA microarray analysis. Acta Neuropathol 109: 165–180.

12

The Effects of Low Levels of Dystrophin on Mouse Muscle Function and Pathology

Maaike van Putten, Margriet Hulsker, Vishna Devi Nadarajah, Sandra H. van Heiningen, Ella van Huizen, Maarten van Iterson, Peter Admiraal, Tobias Messemaker, Johan T. den Dunnen, Peter A. C. 't Hoen, Annemieke Aartsma-Rus*

Department of Human Genetics, Leiden University Medical Center, Leiden, the Netherlands

Abstract

Duchenne muscular dystrophy (DMD) is a severe progressive muscular disorder caused by reading frame disrupting mutations in the *DMD* gene, preventing the synthesis of functional dystrophin. As dystrophin provides muscle fiber stability during contractions, dystrophin negative fibers are prone to exercise-induced damage. Upon exhaustion of the regenerative capacity, fibers will be replaced by fibrotic and fat tissue resulting in a progressive loss of function eventually leading to death in the early thirties. With several promising approaches for the treatment of DMD aiming at dystrophin restoration in clinical trials, there is an increasing need to determine more precisely which dystrophin levels are sufficient to restore muscle fiber integrity, protect against muscle damage and improve muscle function. To address this we generated a new mouse model (*mdx-Xist*$^{\Delta hs}$) with varying, low dystrophin levels (3–47%, mean 22.7%, stdev 12.1, $n=24$) due to skewed X-inactivation. Longitudinal sections revealed that within individual fibers, some nuclei did and some did not express dystrophin, resulting in a random, mosaic pattern of dystrophin expression within fibers. *Mdx-Xist*$^{\Delta hs}$, *mdx* and wild type females underwent a 12 week functional test regime consisting of different tests to assess muscle function at base line, or after chronic treadmill running exercise. Overall, *mdx-Xist*$^{\Delta hs}$ mice with 3–14% dystrophin outperformed *mdx* mice in the functional tests. Improved histopathology was observed in mice with 15–29% dystrophin and these levels also resulted in normalized expression of pro-inflammatory biomarker genes, while for other parameters >30% of dystrophin was needed. Chronic exercise clearly worsened pathology, which needed dystrophin levels >20% for protection. Based on these findings, we conclude that while even dystrophin levels below 15% can improve pathology and performance, levels of >20% are needed to fully protect muscle fibers from exercise-induced damage.

Editor: Ronald Cohn, Johns Hopkins Univ. School of Medicine, United States of America

Funding: This work was supported by the FP6 funded TREAT-NMD network of excellence (LSHM-CT-2006-036825), the FP7 BIO-NMD project (FP7-HEALTH-241665), the Dutch organization for scientific research (NWO/ZonMW, 95100102), and the Dutch Duchenne Parent Project (www.duchenne.nl) and utilizes the infrastructure of the Center for Biomedical Genetics (the Netherlands) and the Center for Medical Systems Biology (the Netherlands). The funders had no role in study design, data collection and analysis, decision to publish, or preparation of the manuscript.

Competing Interests: The authors have declared that no competing interests exist.

* E-mail: a.m.rus@lumc.nl

Introduction

Duchenne Muscular Dystrophy (DMD) is an X-linked recessive disorder affecting 1:3500 new born boys. It is characterized by muscle fiber wasting, functional impairment and eventual death due to respiratory and heart failure. The underlying causes are frame-shifting and nonsense mutations in the *DMD* gene, resulting in the absence of functional dystrophin protein. Intact dystrophin anchors the intracellular cytoskeleton to the extracellular matrix and thereby prevents membrane damage during muscle contraction [1,2]. An allelic, less severe form of the disease, Becker muscular dystrophy (BMD) is caused by mutations that maintain the open reading frame and allow synthesis of internally deleted, partially functional dystrophin proteins [3,4].

There is no cure for DMD, but many potential therapeutic compounds currently tested in clinical trials aim at restoration of (a BMD like) dystrophin [5–12]. These trials at best resulted in the synthesis of low levels of dystrophin protein. However, it is not yet known how these levels will affect disease pathology, and which levels are needed to maintain muscle fiber integrity, to prevent against exercise-induced damage, or to improve muscle function. In addition, it is as yet unknown whether low dystrophin levels will stabilize and/or delay disease progression [8]. In perspective of further optimization of currently tested potential therapeutic compounds, detailed studies in this direction are necessary.

Early studies primarily involved female *DMD* carriers, heterozygous *mdx* mice and isolated BMD patients expressing less than 50% dystrophin. Female *DMD* mutation carriers express dystrophin in approximately 50% of the fibers (when seen in a transverse cross section) due to random X-inactivation early in life [13,14]. During life, the proportion of dystrophin positive fibers increases due to positive selection. It appears that this is sufficient to maintain skeletal muscle function and fiber integrity in both human and mice. Although skeletal muscles appear to escape damage, human carriers are at risk for *DMD*-associated dilated cardiomyopathy [15], while no heart failure is observed in heterozygous *mdx* mice [16,17]. Nevertheless, symptomatic carriers have been described, often expressing less than 50% of normal dystrophin levels due to either gross chromosomal rearrangements [18], or unfortunate skewed X-inactivation in which the degree of skewedness correlated with disease severity [13,19–21]. In addition, BMD patients generally have lower levels of dystrophin and these levels seem to correlate with disease severity, where levels <10% are observed in very severe patients and levels >20% in moderate/mild patients [4,22]. Based on a case study involving one patient, it appears that dystrophin levels as low as 30% can be sufficient to largely prevent a muscle phenotype [23]. The amount of revertant fibers has also been reported to correlate positively with disease severity [24]. However, reports on BMD and skewed X-inactivation cases involve low numbers of patients and detailed analysis (e.g. assessment of dystrophin levels in various muscles) is not possible for obvious reasons. Thus, there is an increasing need for a mouse model expressing low levels of dystrophin to allow the detailed study of the effects of low dystrophin levels on disease pathology.

So far, several attempts have been made to achieve this. Unfortunately, each had limitations. First, transgenic mice expressing low levels of full-length or truncated dystrophin of murine or human origin have been generated to test whether this results in a less severe phenotype [25,26]. However, from the 13 mouse lines generated, only two expressed dystrophin levels lower than 50% of wild type in both the quadriceps and diaphragm. The other lines had higher dystrophin levels or a combination of both higher and lower levels than wild type. From the two potential lines, one expressed low dystrophin levels (15% quadriceps and <5% diaphragm), while the other line expressed higher dystrophin levels (40% quadriceps and 20% diaphragm), thus allowing detailed analysis of only two different dystrophin levels. Dunant et al. generated *mdx* mice in which dystrophin expression was driven by a 1.35-kb MCK promoter resulting in abundant expression in fast-twitch muscles (~50%) and very low expression in slow-twitch fibers. The heart did not express dystrophin at all [27]. The fiber type specific dystrophin expression makes this model less suitable for detailed analysis. In another study, Stillwell et al. generated chimeric mice expressing various dystrophin levels depending on the amount of wild type cells incorporated in *mdx* blastocytes. Mice expressed dystrophin in skeletal muscles in a very high (60–100%) or low (<5%) manner [28]. This, combined with the fact that this method is extremely labour intensive, makes this approach less favourable for extended studies. The final model known to express low dystrophin levels is the *mdx*^{3cv} mouse. Studies in the *mdx*^{3cv} mouse revealed that even ~5% of dystrophin increased specific tetanic force and partly prevented against eccentric contractions induced damage in young but not in old mice. However, it was insufficient to prevent histopathology, since severity did not differ from dystrophin negative *mdx*^{4cv} mice [29]. Although *mdx*^{3cv} mice are easy to generate, dystrophin levels do not exceed 5%, making studies on the effect of higher dystrophin levels e.g. 20–30% impossible.

Taken together, these models give an indication that <50% dystrophin is sufficient for improvement of many pathological aspects, however, they are not ideal to assess the effects of low dystrophin levels on disease pathology in detail. Our study describes a mouse model expressing low dystrophin levels, based on non-random X-inactivation. In mammals, X-inactivation is regulated by the X-inactivation center in which *Xist* is the key player [30,31]. Mutations in the *Xist* promoter can skew the randomness of X-inactivation resulting in preferential inactivation (60–90%) of the X-chromosome containing the mutated *Xist* gene [32]. Here, non-dystrophic female *Xist*$^{\Delta hs}$ mice with a mutation in the *Xist* promoter were crossed with dystrophic *mdx* males. This resulted in female *mdx-Xist*$^{\Delta hs}$ mice in which the X-chromosome containing the wild type dystrophin is preferentially inactivated, leading to a wide range of low dystrophin levels in individual mice (Figure 1). Within individual muscle fibers of these pups, a mosaic pattern of dystrophin expression was observed, in which nuclei either did or did not express dystrophin. Experiments revealed that low dystrophin levels have a beneficial effect on muscle integrity, histology, function and serum biomarkers in a dose dependent manner.

Results

Mdx-Xist$^{\Delta hs}$ mice express varying levels of dystrophin

To study whether low dystrophin levels reduce the dystrophic pathology of *mdx* mice, *mdx-Xist*$^{\Delta hs}$ mice carrying a wide range of low dystrophin levels based on non-random X-inactivation were bred (Figure 1A). For each *mdx-Xist*$^{\Delta hs}$ mouse, dystrophin levels in the quadriceps muscle were quantified by Western blot (Figure 1B). These varied between 3–47% of wild type levels (mean 22.7, stdev 12.1, $n = 24$) (Figure 1C), which correlated well ($R = 0.97$) with the percentages of dystrophin positive fibers in cross sections of the same muscle, as assessed by blinded manual counting (Figure 1D). Dystrophin levels of individual muscles differed (assessed in $n = 6$ *mdx-Xist*$^{\Delta hs}$ mice), with the tibialis anterior and biceps expressing slightly higher levels than the quadriceps, gastrocnemius and triceps. Lowest dystrophin levels were found in the diaphragm and heart (Figure S1A–B). The dystrophin level of the quadriceps was used to group the mice. Longitudinal sections of the quadriceps revealed that dystrophin expression along a single whole fiber was patchy, with nuclei either expressing dystrophin or not (Figure 1E). This indicates that dystrophin proteins cannot freely diffuse along the muscle fiber and that nuclei where the normal *DMD* gene is not inactivated serve only certain regional domains in the muscle fiber. Transverse cross sections of several skeletal muscles and heart revealed that dystrophin positive fibers were most often grouped together and that these small groups were spread over the muscle (Figure S1C).

Dystrophin is part of a larger complex of dystrophin-glycoprotein associated proteins, such as β-dystroglycan and nNOS. As expected, dystrophin positive fibers of the *mdx-Xist*$^{\Delta hs}$ mice were also positive for β-dystroglycan and nNOS (Figure S2A). In absence of dystrophin, the homolog utrophin can partly take over its function. Muscles of adult *mdx* mice have significantly higher utrophin levels than those of adult C57BL/10ScSnJ, as utrophin is overexpressed in the absence of dystrophin [33]. We assessed how dystrophin expression influences utrophin expression for six *mdx-Xist*$^{\Delta hs}$ mice expressing various levels of dystrophin. In *mdx-Xist*$^{\Delta hs}$ mice with very low dystrophin levels (<15%), utrophin levels were increased as confirmed by Western blot (Figure S2B, lane 2 and 6). In contrast, *mdx-Xist*$^{\Delta hs}$ mice with intermediate dystrophin levels (>15%) had low utrophin levels (lanes 1, 3–5). We did not observe a decrease in utrophin intensity in dystrophin positive fibers analyzed by immunohistology (data not shown).

Dystrophin level dependent improvement in muscle function

To determine the effect of low dystrophin levels on functional performance, *Xist*$^{\Delta hs}$, *mdx* and *mdx-Xist*$^{\Delta hs}$ mice were subjected to a functional test regime consisting of four functional tests per week for 12 weeks, during which also body weight and creatine kinase (CK) levels were monitored. After sacrificing, quadriceps muscles were isolated and dystrophin levels were determined by Western blot. K-means clustering was used to divide *mdx-Xist*$^{\Delta hs}$ mice into three groups in an unbiased way. This resulted in the following division: <15%, 15–30% and >30% dystrophin. *Mdx* mice were

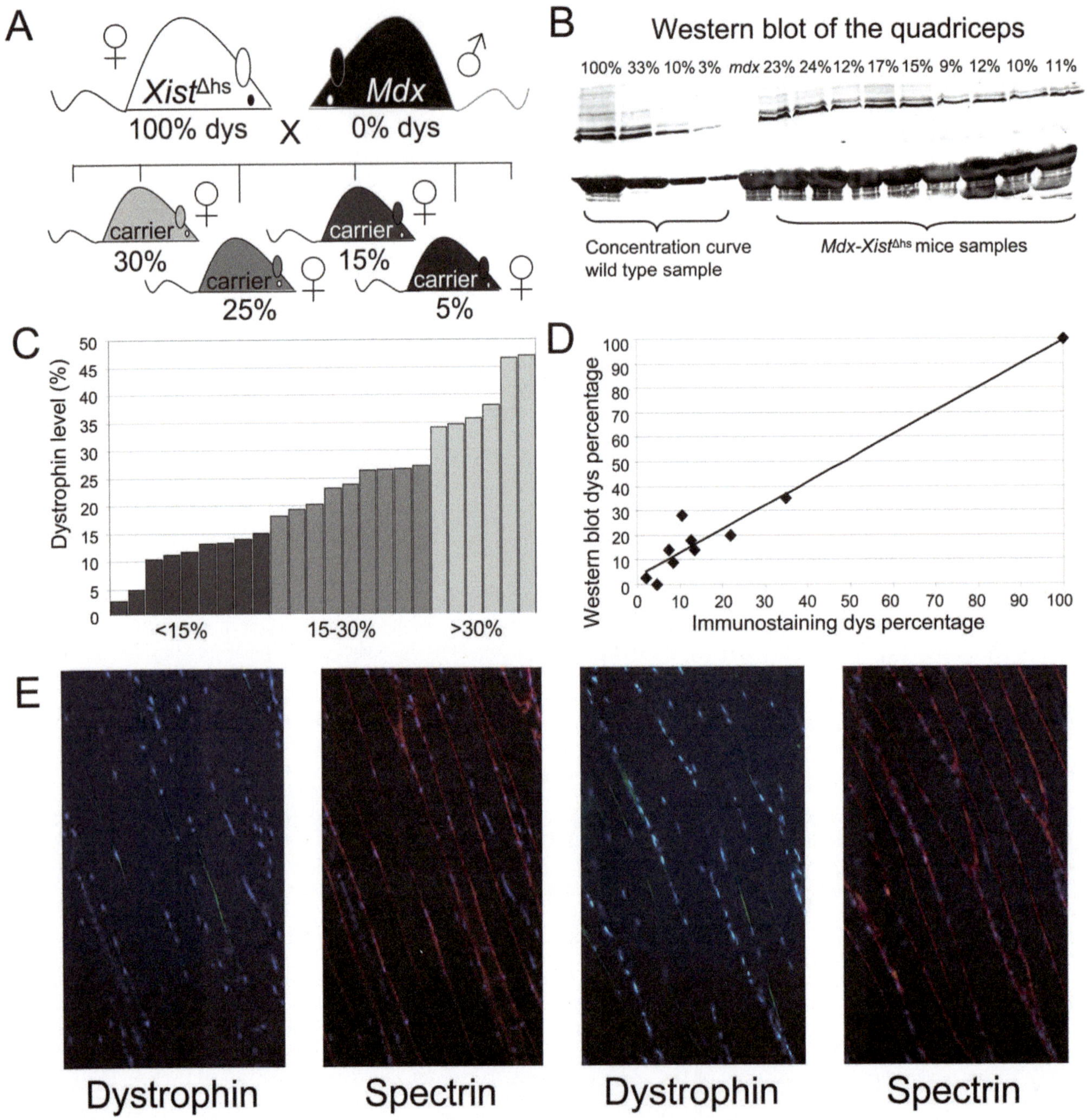

Figure 1. ***Mdx-Xist*$^{\Delta hs}$ mice.** A. To breed mice with low dystrophin levels, female *Xist*$^{\Delta hs}$ mice, carrying a mutation in the *Xist* promoter which coordinates X-inactivation, were crossed with dystrophin negative *mdx* males. During embryogenesis, the maternal X-chromosome encoding a functional dystrophin gene will be preferentially (60–90%) inactivated as a result of the mutated *Xist* promoter. The *Xist*$^{\Delta hs}$ mice were a kind gift from N. Brockdorff (MRC Clinical Sciences Center, London, UK, current affiliation Department of Biochemistry, University of Oxford, UK). B. Picture of a representative Western blot. The percentage of dystrophin was determined for the quadriceps of all *mdx-Xist*$^{\Delta hs}$ mice by Western blot (2–9 technical replicates per mouse). The percentage of individual mice was determined using a concentration curve made from wild type samples. Myosin was used as a loading control. C. Skewed X-inactivation resulted in dystrophin levels of 3–47% (mean 22.7, stdev 12.1, $n=24$) (as determined by Western blot) in the female *mdx-Xist*$^{\Delta hs}$ offspring. Each bar represents the dystrophin level of an individual mouse. The dystrophin levels of the individual mice belonging to the three dystrophin groups can be appreciated from this graph. D. Dystrophin levels determined by Western blot and manual counting of dystrophin positive fibers demonstrate a strong correlation (R = 0.97). E. Longitudinal sections of a quadriceps stained with dystrophin (green) and spectrin (red). From the pictures it can be appreciated that dystrophin expression is not homogeneously expressed across the fiber but rather confined to certain nuclear domains.

significantly ($P<0.001$) heavier at the end of the functional test regime than the other mouse strains (mean weight 26, 24.5 and 18.2 gram for *mdx*, *mdx-Xist*$^{\Delta hs}$ and *Xist*$^{\Delta hs}$, respectively). An increased weight compared to wild type has been described before for *mdx* mice [34]. *Xist*$^{\Delta hs}$ mice had a stable body weight from week 11 onwards, while the other models showed a mild increase over

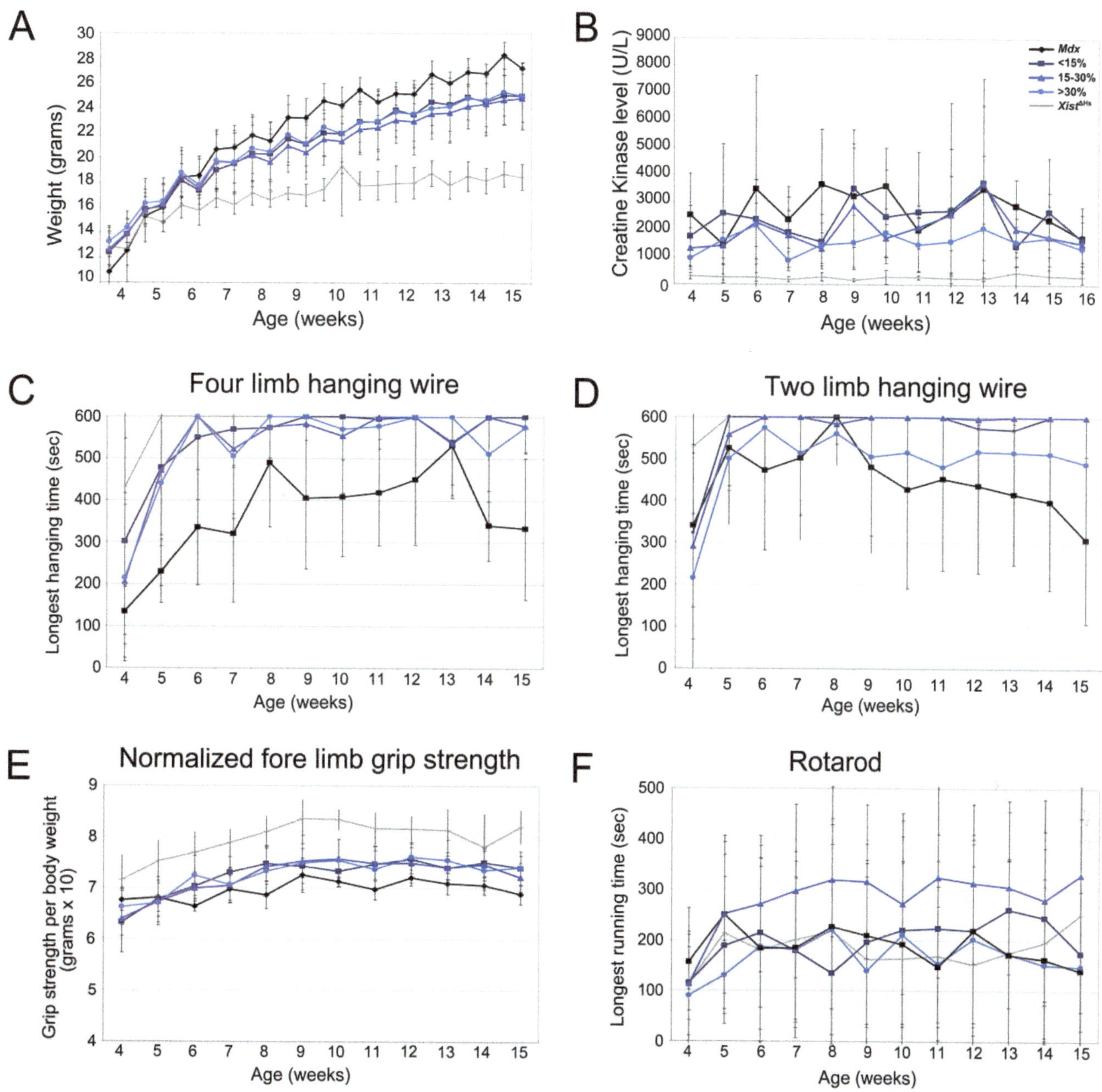

Figure 2. Body weight, CK levels and functional performance of the mouse models. A. Normalization towards wild type levels was observed for bodyweight of the *mdx-Xist*$^{\Delta hs}$ mice. At the end of the functional test regime *mdx* mice were significantly heavier ($P<0.001$) than the other mice. B. CK levels of *mdx* and *mdx-Xist*$^{\Delta hs}$ mice with low dystrophin levels were both significantly ($P<0.001$) higher compared to the levels of *mdx-Xist*$^{\Delta hs}$ mice with >30% dystrophin and *Xist*$^{\Delta hs}$ mice. C. Four limb hanging wire performance was comparable for the *Xist*$^{\Delta hs}$ and *mdx-Xist*$^{\Delta hs}$ mice, while *mdx* mice performed significantly ($P<0.001$) worse. D. The best performance on the two limb hanging wire test was obtained for the *Xist*$^{\Delta hs}$, *mdx-Xist*$^{\Delta hs}$ with <15% and 15–30% while the *mdx* and *mdx-Xist*$^{\Delta hs}$ with >30% dystrophin performed significantly ($P<0.001$) worse. Bad performance of *mdx-Xist*$^{\Delta hs}$ with >30% was due to the bad performance of one individual mouse. E. Normalized fore limb grip strength was significantly ($P<0.001$) higher in the *Xist*$^{\Delta hs}$ mice compared to the other strains and all the *mdx-Xist*$^{\Delta hs}$ mice were significantly ($P<0.001$) stronger than the *mdx* mice. F. No difference in rotarod running was observed, except for the *mdx-Xist*$^{\Delta hs}$ mice with 15–30% dystrophin, which performed significantly ($P<0.001$) better than all other groups.

time. The body weights of the *mdx-Xist*$^{\Delta hs}$ mice from the three dystrophin level groups were very similar and fell in between *mdx* and *Xist*$^{\Delta hs}$ mice, but significantly ($P<0.001$) differed from both (Figure 2A).

CK levels are a measure for muscle condition and are <500 U/L in healthy mice. CK levels of the *Xist*$^{\Delta hs}$ mice never exceeded this threshold and were, averaging 270 U/L, significantly ($P<0.001$) lower than levels of the other mouse strains. *Mdx* mice had the highest CK levels (average 2700 U/L). When analyzed over the complete time course, these were significantly ($P<0.001$, linear regression) higher than those of *mdx-Xist*$^{\Delta hs}$ mice with >30% dystrophin (average 1570 U/L). CK levels of mice with 15–30% dystrophin generally fell between those with <15% and >30%, showing a clear correlation between CK and dystrophin levels. CK levels of mice with <15% of dystrophin were similar to *mdx* suggesting that these dystrophin levels were not enough to decrease CK levels (Figure 2B).

In the four limb hanging wire test, *mdx* mice performed significantly ($P<0.001$) worse compared to all the other groups (Figure 2C). *Xist*$^{\Delta hs}$ mice were able to hang for the maximum period of time for the entire testing period of 12 weeks, whereas all *mdx-Xist*$^{\Delta hs}$ mice performed nearly as well as *Xist*$^{\Delta hs}$ mice. Also in the two limb hanging wire test, *Xist*$^{\Delta hs}$ and *mdx-Xist*$^{\Delta hs}$ mice significantly ($P<0.001$) outperformed *mdx* mice and were able to hang for the maximum allowed time. Performance of *mdx-Xist*$^{\Delta hs}$ mice with >30% dystrophin dropped from an age of 9 weeks compared to the other *mdx-Xist*$^{\Delta hs}$ mice, but this was due to the poor performance of a single mouse. Generally, the decreased performance was more pronounced in *mdx* mice (Figure 2D). The normalized fore limb grip strength measured for *Xist*$^{\Delta hs}$ mice (7.9) was significantly ($P<0.001$) higher than that of *mdx* and *mdx-Xist*$^{\Delta hs}$ mice (Figure 2E). Strength was improved towards wild type levels in all *mdx-Xist*$^{\Delta hs}$ mice (6.5) and significantly ($P<0.001$) lower in *mdx* mice (5.9). Rotarod performance was significantly ($P<0.001$) better in the *mdx-Xist*$^{\Delta hs}$ mice with 15–30% dystrophin compared to all the other groups (Figure 2F). Unfortunately, we observed a high inter group variation, which makes drawing conclusions difficult for this test. Based on these analyses, even the lowest levels of dystrophin analyzed (between 3 and 14%), appear sufficient to improve hanging times and grip strength.

Dystrophin dependent improvement of histopathology and restoration of gene expression

After sacrificing the animals, the quadriceps, heart and diaphragm were dissected and cryosections were made. To determine fiber size and the percentage of centralized nuclei, the quadriceps of all mice was stained with laminin and DAPI. Five pictures were randomly captured of the middle section of the muscle and analysed with Mayachitra Imago (Figure 3A and B). Regenerating fibers (<1000 μm^2) and hypertrophic fibers (>6000 μm^2) were predominantly found in the *mdx* mice (50% and 4.8%, respectively), whereas smaller proportions were observed for the *Xist*$^{\Delta hs}$ mice (20% and 0.9%, respectively) and a dystrophin level dependent normalization towards the wild type fiber size distribution was observed for *mdx-Xist*$^{\Delta hs}$ mice. The percentage of centralized nucleated fibers was significantly higher ($P<0.01$) in *mdx* (51%) compared to *Xist*$^{\Delta hs}$ mice (2%). More than 15% of dystrophin was sufficient to significantly ($P<0.01$) reduce the amount of centralized nuclei by 40%, whereas dystrophin levels above 30% caused a reduction of around 60%.

To determine the percentage of fibrotic/necrotic fibers, sections of the quadriceps, diaphragm and heart were stained with haematoxylin & eosin. For all muscles quantified, except heart, the percentage of fibrosis was significantly higher ($P<0.001$) in *mdx* mice compared to *Xist*$^{\Delta hs}$ mice. In the quadriceps of *mdx-Xist*$^{\Delta hs}$ mice of the different groups no significant differences was observed, although there appears to be a dystrophin-dependent trend towards lower levels of fibrosis in the mice with moderate levels of dystrophin (Figure 3C). The diaphragm was the most severely affected muscle examined, with fibrotic/necrotic tissue percentages of 20% in both *mdx* and *mdx-Xist*$^{\Delta hs}$ mice with <15% dystrophin compared to 8% in *Xist*$^{\Delta hs}$ mice. Again, there was a dystrophin-level dependent pattern of reduced fibrotic tissue which reached significance ($P<0.05$) in mice with >30% dystrophin (Figure 3D). Fibrosis/necrosis levels in heart were below 5% in all mice, which is expected, as fibrosis in heart generally is observed in older mice (data not shown) [35].

To determine whether low levels of dystrophin were able to normalize the expression of genes known to be involved in inflammation, fibrosis, regeneration and heart function RT-qPCR was performed on mRNA of the quadriceps, diaphragm and heart of all mice [36]. The expression of most genes was significantly ($P<0.01$) elevated in *mdx* mice compared to *Xist*$^{\Delta hs}$ mice in the three muscles, and most often normalized in a dystrophin level dependent manner in the *mdx-Xist*$^{\Delta hs}$ mice (Figure 4). Notably, in the diaphragm, mRNA levels of the biomarkers for muscle regeneration (*Tnnt2*, *Bgn*) decreased in mice with <15% dystrophin, while biomarkers for inflammation (*Lgals3*, *Cd68*) only decreased in mice with >30% dystrophin. Biomarkers for fibrosis (*Timp-1*, *Nox2*) decreased in mice with <15% and >15% dystrophin, respectively. No difference in expression of heart function genes was observed, except for *Nppa*, which is due to the young age of the mice.

Dystrophin level dependent protection against exercise induced muscle damage

To determine the protective effect of low dystrophin levels on muscle integrity during chronic exercise, mice were forced to run three times a week on a horizontal treadmill. This was directly followed by a functional test to assess the influence of muscle fatigue on performance. We tested seven *mdx-Xist*$^{\Delta hs}$ mice which had an average dystrophin level of 21% (2%–45% median 25.8), six *mdx* and five C57BL/10ScSnJ females (the wild type sub-strain from which the *mdx* mouse was derived).

CK levels were assessed on Mondays before exercise (Figure 5B) and on Fridays directly after exercise (Figure 5C). For *mdx* and *mdx-Xist*$^{\Delta hs}$ mice CK levels increased significantly ($P<0.01$) upon exercise (4150 U/L before versus 11170 U/L after for *mdx* and 2070 U/L before versus 8850 U/L after for *mdx-Xist*$^{\Delta hs}$), whereas this was not observed for the wild type mice (250 U/L before versus 291 U/L after) (Figure 5A). CK levels of the *mdx-Xist*$^{\Delta hs}$ mice were significantly ($P<0.05$) lower compared to *mdx*, both before and after exercise. Wild type mice had significantly ($P<0.01$) lower CK levels than those of *mdx* and *mdx-Xist*$^{\Delta hs}$ mice. *Mdx-Xist*$^{\Delta hs}$ mice gained significantly ($P<0.01$) less weight over time than *mdx* mice, as they had a higher body weight at the start, but a similar body weight at the end of the test regime (data not shown). Wild type mice were significantly ($P<0.01$) less heavy than *mdx* mice, but did not significantly differ in weight from the *mdx-Xist*$^{\Delta hs}$ mice.

At an age of 8 and 14 weeks, serum levels of matrix metalloproteinase-9 (MMP-9), tissue inhibitors of matrix metalloproteinase-1 (TIMP-1) and osteopontin (OPN) were assessed. These markers have been identified as potential biomarkers to monitor disease progression in DMD patients [37]. For MMP-9, 8 week old *mdx* mice had significantly ($P<0.05$) higher levels when compared to wild type and *mdx-Xist*$^{\Delta hs}$ mice (Figure 5D). No significant difference was observed between *mdx-Xist*$^{\Delta hs}$ and wild type mice. TIMP-1 levels were significantly ($P<0.05$) elevated in both *mdx* and *mdx-Xist*$^{\Delta hs}$ mice compared to wild type mice at the age of 8 and 14 weeks (Figure 5E). Interestingly, a significant ($P<0.05$) difference was also observed between *mdx* and *mdx-Xist*$^{\Delta hs}$ mice aged 8 weeks. OPN levels were significantly ($P<0.05$) elevated in *mdx* compared to both wild type and *mdx-Xist*$^{\Delta hs}$ at 8 weeks. This difference was not observed at 14 weeks of age (Figure 5F).

The chronic treadmill exercise protocol was completed by all mice without any problems. However, we later found that fibrosis was much worse in these exercised mice and that the low dystrophin levels were not enough to prevent worsening of the muscle's condition (see below). Directly after treadmill exercise,

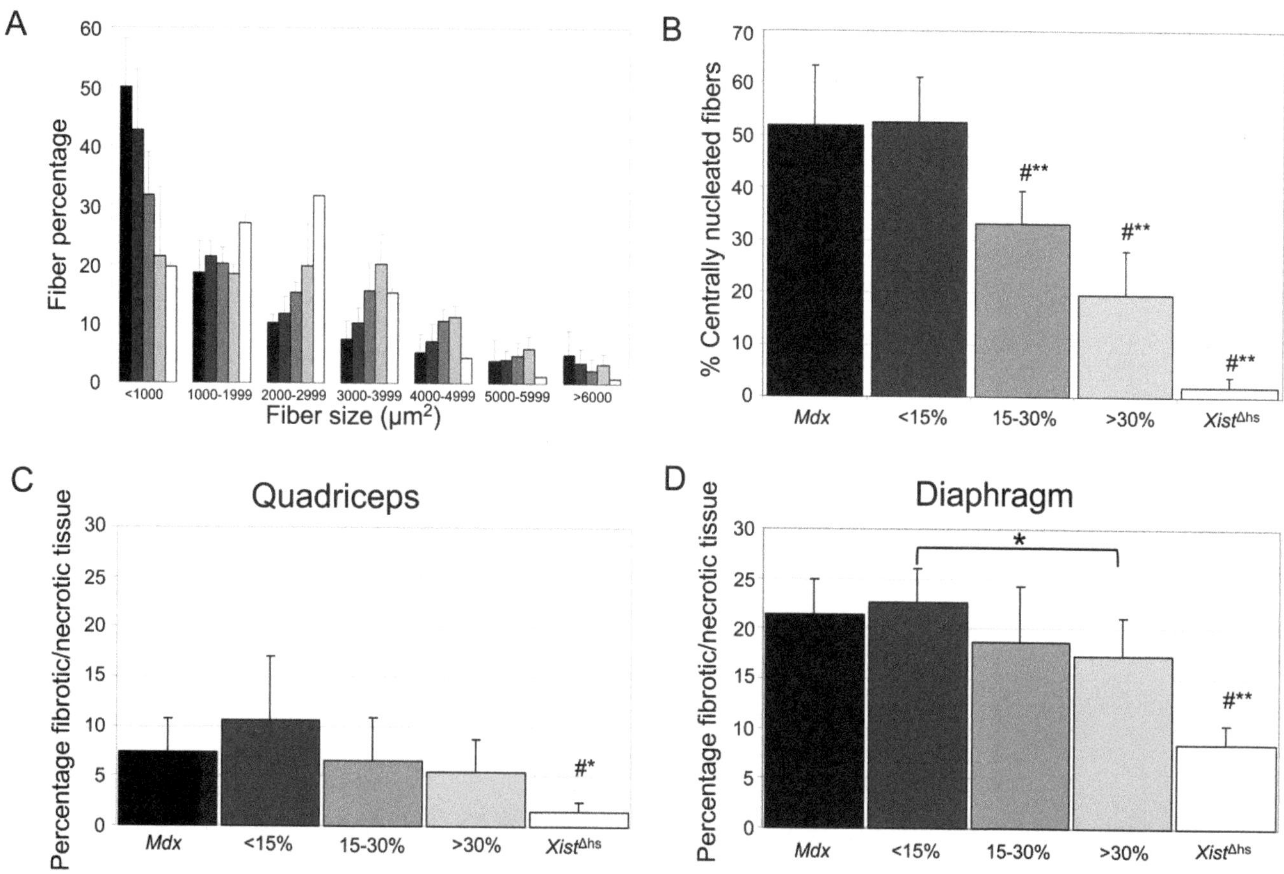

Figure 3. Muscle fiber size, degree of central nucleation and percentage of fibrotic/necrotic tissue. A. Regenerating and hypertrophic fibers were mainly observed in the *mdx* mice. A dystrophin level depend trend towards wild type distribution was observed for *mdx-Xist*$^{\Delta hs}$ mice where <15% dystrophin already resulted in improvement. B. Dystrophin levels between 15–30% and >30% resulted in a reduction of the percentage of centralized nuclei of 40% and 60% respectively. C. The quadriceps of all mice was significantly more severely affected compared to *Xist*$^{\Delta hs}$ mice. D. The diaphragm was the most severely affected muscle with on average 20% fibrotic/necrotic tissue in *mdx* mice. All mice were significantly more severely affected compared to *Xist*$^{\Delta hs}$ mice. *Mdx-Xist*$^{\Delta hs}$ mice with >30% dystrophin had less fibrotic/necrotic tissue than *mdx* and *mdx-Xist*$^{\Delta hs}$ mice with <15% dystrophin, but this difference was only significant between both *mdx-Xist*$^{\Delta hs}$ groups. # indicates a significant difference of that bar with all the other groups. Single asterisks indicate a $P<0.05$ and double asterisks indicate a $P<0.01$.

mice were subjected to two limb hanging wire test, fore limb grip strength or rotarod (Figure 6). For the two limb hanging wire test, the hanging time of *mdx* mice decreased over time, and was significantly ($P<0.001$) lower than those of the *mdx-Xist*$^{\Delta hs}$ and wild type mice. The hanging time of the *mdx-Xist*$^{\Delta hs}$ and wild type mice did not differ significantly. Performance of the grip strength was consistently lower for the *mdx* mice (5.8) than for the other groups (*mdx-Xist*$^{\Delta hs}$ 6.3 and wild type 6.4). Wild type mice performed well at the start of the functional test regime, but their strength slightly decreased over time. Grip strength of *mdx* and *mdx-Xist*$^{\Delta hs}$ mice followed the same curve, and *mdx-Xist*$^{\Delta hs}$ mice performed significantly ($P<0.001$) better than *mdx* mice. Rotarod running time was significantly ($P<0.001$) higher in the wild type mice with an average running time of 430 seconds, which was generally 130 seconds longer than that of both dystrophic mouse strains. The performance of *mdx* and *mdx-Xist*$^{\Delta hs}$ mice did not significantly differ for this test.

Low dystrophin levels did not fully prevent exercise induced pathology

The effect of the chronic exercise regime on histopathology was assessed by measuring the percentage of fibrotic/necrotic tissue for the quadriceps. On average, 18% of the quadriceps of the *mdx* mice consisted of fibrotic/necrotic tissue, compared to less than 5% for wild type mice. *Mdx-Xist*$^{\Delta hs}$ mice had slightly more healthy tissue than *mdx* mice (Figure 7A). When compared to fibrosis observed in mice which underwent the functional test regime without treadmill running, histopathology was much worse. At the end of the chronic exercise regime, Evans blue dye was injected intravenously and the percentage of Evans blue dye positive fibers was manually counted for the quadriceps of all mice. Hardly any Evans blue dye positive fibers were detected in wild type muscle (0.41%), while both in *mdx* and *mdx-Xist*$^{\Delta hs}$ mice 11.11% and 9.23% of positive fibers were found. This suggests that while low dystrophin levels are sufficient to improve muscle function and muscle fiber integrity in the absence of chronic exercise, they apparently are not sufficient to protect against fiber damage induced by the chronic treadmill exercise regime.

The expression of some fibrosis, inflammation and regeneration biomarkers was assessed for the quadriceps. For most genes, *mdx* and *mdx-Xist*$^{\Delta hs}$ mice had significantly ($P<0.05$) increased expression when compared to wild type levels. However, between *mdx* and *mdx-Xist*$^{\Delta hs}$ mice no difference in expression was observed indicating that the dystrophin levels in these mice were not

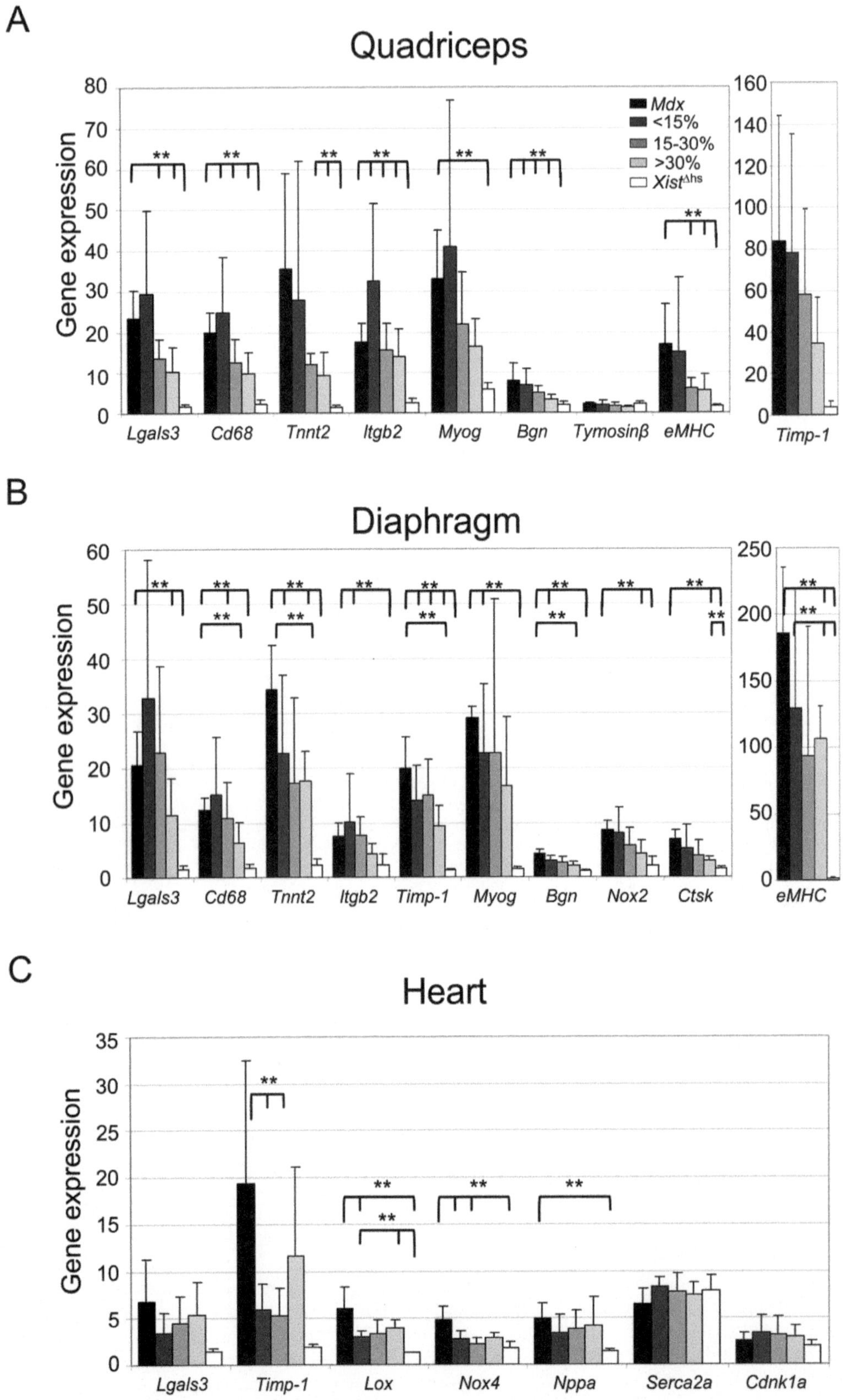

Figure 4. Expression of several genes involved in disease pathology. A and B. For most biomarkers a clear dystrophin level dependent restoration of gene expression levels was observed in the *mdx-Xist*$^{\Delta hs}$ mice, where intermediate dystrophin levels resulted in low expression of genes involved in disease pathology. For some genes, dystrophin levels <15% were enough to reduce gene expression while for other genes >30% was necessary. C. In the heart, even dystrophin levels of <15% decreased expression of fibrotic biomarkers like *Timp-1*. Since the mice were young, no difference in expression levels in heart function was observed, except for *Nppa*. Double asterisks indicate a $P<0.01$.

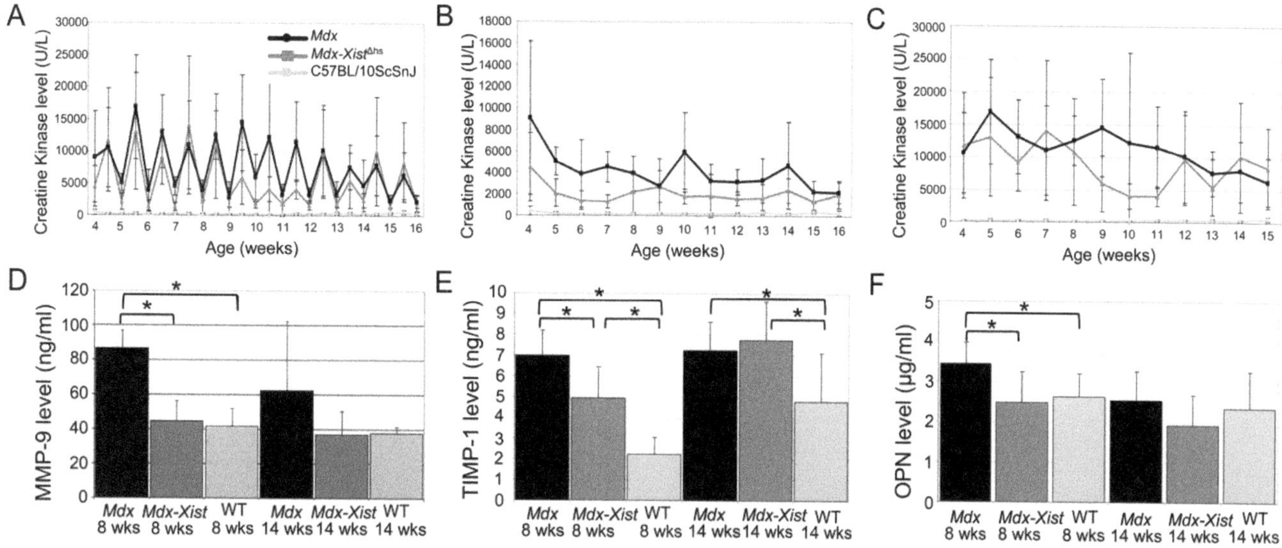

Figure 5. Serum and plasma biomarkers assessed before and directly after treadmill running. A. CK levels assessed before and directly after treadmill running. CK levels collected before exercise were significantly ($P<0.01$) lower than those collected directly after exercise in *mdx-Xist*$^{\Delta hs}$ and *mdx* mice. This exercise induced increase was absent in wild type mice. B. CK levels determined before exercise were significantly ($P<0.01$) elevated in *mdx* and *mdx-Xist*$^{\Delta hs}$ mice compared to wild type, but this was less pronounced for the *mdx-Xist*$^{\Delta hs}$ mice as *mdx* mice had significantly ($P<0.05$) higher CK levels. C. Plasma collected directly after exercise contained extremely high CK levels, in both *mdx* and *mdx-Xist*$^{\Delta hs}$ mice but not in wild type mice. D. Serum levels of MMP-9 were elevated in *mdx* mice compared to *Xist*$^{\Delta hs}$ mice at both 8 and 14 weeks of age, while levels were normalized in *mdx-Xist*$^{\Delta hs}$ mice. E. TIMP-1 levels were elevated in serum of both *mdx* and *mdx-Xist*$^{\Delta hs}$ mice. In wild type mice we found an age related increase of the serum TIMP-1 level. Interestingly, at the age of 8 weeks, levels of *mdx-Xist*$^{\Delta hs}$ mice were significantly lower than those of *mdx* mice. F. *Mdx* mice had significantly elevated OPN levels compared to both wild type and *mdx-Xist*$^{\Delta hs}$ at 8 weeks of age. Single asterisks indicate a $P<0.05$, average dystrophin level of *mdx-Xist*$^{\Delta hs}$ mice was 21% (2%–45% median 25.8).

sufficient to prevent pathology induced by chronic exercise (Figure 7B). In addition, we were able to confirm the differences observed in serum for MMP-9, TIMP-1 and OPN (*Spp1*) on gene expression level.

Discussion

With several potential therapeutic compounds being tested in clinical settings for DMD, there is an increasing need to gain detailed knowledge about which levels of dystrophin are of therapeutic value and which and/or how low dystrophin levels affect the different pathological pathways. This would help the optimization phase of pre-clinical research on therapeutic approaches aiming at dystrophin restoration. It will also facilitate providing patients and parents who are candidates to participate in clinical trials for these approaches with realistic expectations.

Several attempts to generate mice expressing low dystrophin levels have been undertaken previously, unfortunately without success since none of these strains expressed a wide variety of low dystrophin levels in the most interesting range. With no accessibility to a good model, current knowledge and rough estimations are based on human female DMD carriers, heterozygous *mdx* mice, generated transgenic mice expressing lower levels of full-length or truncated dystrophin of murine or human origin and antisense oligonucleotide (AON) treated *mdx* mice.

In this paper, we describe the easy to generate *mdx-Xist*$^{\Delta hs}$ mouse model, which facilitates assessment of the effects of different dystrophin levels on muscle performance and integrity. Mice with very low (<15%), low (15–30%) and moderate dystrophin (>30%) levels showed a dystrophin level dependent reduction of the homolog utrophin. All *mdx-Xist*$^{\Delta hs}$ mice had a normalized body weight and outperformed *mdx* mice in both hanging wire tests and in the grip strength test, indicating that even low dystrophin levels (<15%) can bring about improved performance in this mouse model. The improved functional performance observed is in concordance with that previously described in *mdx*^{3cv} mice and AON treated mice expressing low levels of dystrophin [29,38]. Although, ~5% of dystrophin in the *mdx*^{3cv} mice resulted in increased specific tetanic force and partly resistance to eccentric contractions in young, but not in old mice, histopathology was not different from dystrophin negative *mdx*^{4cv} mice. This again is in line with our observation in *mdx-Xist*$^{\Delta hs}$ mice expressing <15% dystrophin, were we also found no histopathological improvements, while we did find improvements on motor function. Dystrophin levels >30% reduced CK levels significantly while for lower percentages only a non-significant trend was found. Also for the expression of pro-inflammatory biomarkers and histopathology, a dystrophin level dependent improvement was observed where for some genes >15% of dystrophin led to improvement while for other genes >30% dystrophin was needed. In chronically exercised mice with ~20% dystrophin, normalization towards wild type levels was observed for three different disease serum markers (CK, MMP-9 and TIMP-1). This confirms findings that MMP-9 and TIMP-1 are useful serum biomarkers to monitor disease pathology in DMD patients' serum [37]. Even after treadmill exercise the *mdx-Xist*$^{\Delta hs}$ mice performed significantly better than the *mdx* mice for the two limb hanging wire test and grip strength. However, the levels of fibrosis and Evans blue dye uptake were similar for *mdx-Xist*$^{\Delta hs}$ mice and *mdx* mice, indicating that ~20% of dystrophin is not sufficient to protect against forced exercise induced damage. Thus, while even dystrophin levels below 15% can improve pathology and performance, levels of >20% are needed to fully protect muscle fibers from exercise-induced damage.

The fact that most of our observations concur with those of other studies conducted in transgenic mice and AON treated mice

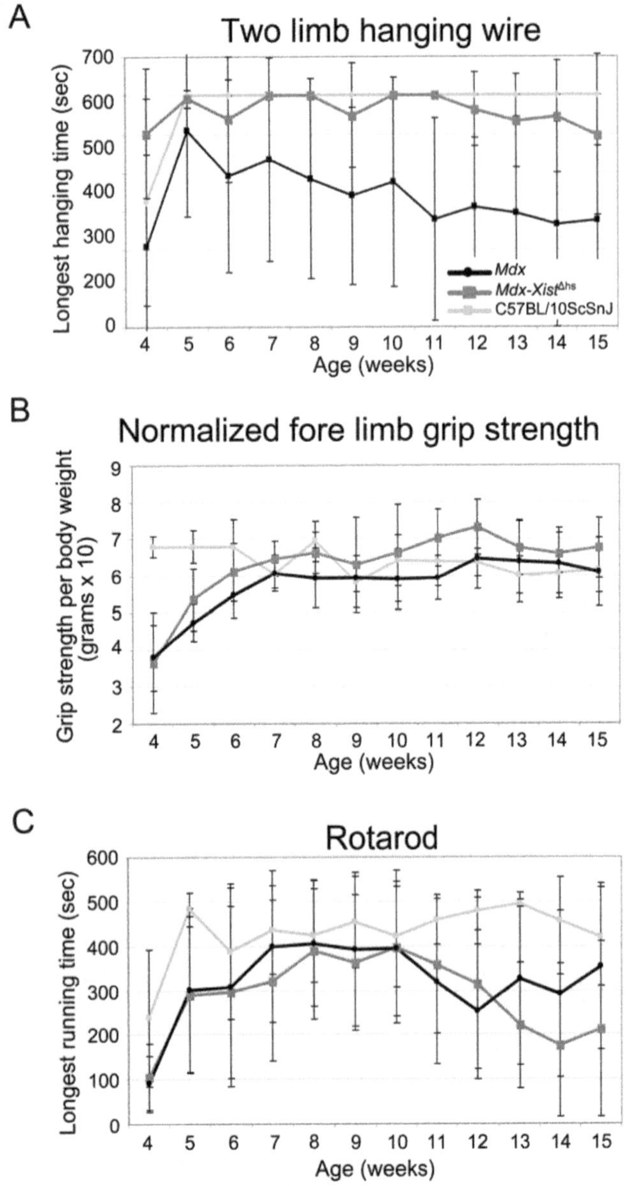

Figure 6. Functional performance measured directly after treadmill exercise. A. Longest hanging time with the two limb hanging test was achieved by wild type and *mdx-Xist*$^{\Delta hs}$ mice which both performed significantly ($P<0.001$) better than *mdx* mice. B. Fore limb grip strength of *mdx-Xist*$^{\Delta hs}$ mice was significantly ($P<0.001$) better than that of *mdx* mice. C. Wild type mice outperformed *mdx* and *mdx-Xist*$^{\Delta hs}$ mice on the rotarod, while no significant difference was observed between *mdx* and *mdx-Xist*$^{\Delta hs}$ mice. Average dystrophin level of *mdx-Xist*$^{\Delta hs}$ mice was 21% (2%–45% median 25.8).

confirms the usefulness of the *mdx-Xist*$^{\Delta hs}$ mouse as a dystrophic mouse strain. It should be noted that the *mdx-Xist*$^{\Delta hs}$ mouse expresses dystrophin in a random manner where nuclei either express wild type levels of dystrophin or no dystrophin at all. This is similar to the pattern what is expected for cell therapy, gene therapy and to what is observed in transverse cross sections after phosphorodiamidate morpholino oligomer (PMO) administration [12], but different from that of 2′-O-methyl phosphorothioate (2OMePS) [11] or gentamicin administration [39], which result in a more evenly low dystrophin expression in nearly all fibers. With detailed knowledge lacking regarding the migration properties of AONs and dystrophin over the membrane, it is unknown whether the mosaic pattern seen in longitudinal cross sections of *mdx-Xist*$^{\Delta hs}$ mice could be expected upon PMO administration. The effect of differences in distribution of dystrophin has never been studied in great detail; i.e. whether 20% of dystrophin in all muscles is better than 100% dystrophin in 20% of muscle fibers. Previously, Phelps et al. showed that mice expressing dystrophin 10-fold greater than wild type levels in a mosaic manner had 34% of centralized nuclei, while mice expressing similar amounts uniformly only had 7% of central nuclei. More detailed studies especially in mice expressing low dystrophin levels are needed and could easily be conducted using our mouse model and comparing them to 2OMePS or gentamicin treated *mdx* mice [25]. Using the *mdx-Xist*$^{\Delta hs}$ mouse as study object is favored above treated *mdx* mice since no expensive treatment is needed to achieve a range of low dystrophin levels. Notably, we observed variation in pathology for mice with similar levels of dystrophin. This phenomenon is also anticipated for patients currently in clinical trials aiming at dystrophin restoration. Our mouse model will allow extended studies to elucidate this discrepancy.

In summary, the Gaussian distribution of low dystrophin levels obtained in the *mdx-Xist*$^{\Delta hs}$ mouse and the ease of generating large numbers of mice, make it a good mouse model for more detailed research on the effect of low levels of dystrophin on several aspects of DMD pathology and effects of future treatment strategies for dystrophinopathies.

Materials and Methods

Ethics statement

All experiments were approved by and performed following the guidelines of the Dier Experimenten Commissie (Animal Experimental Commission) of the Leiden University Medical Center (Permit Numbers: 08096 and 09136). Effort was put in minimizing the amount of distress caused to the animals as much as possible.

Animal care

All studied mouse strains were bred at the animal facility of the LUMC where they were housed in individually ventilated cages with 12-h light-dark cycles. The *Xist*$^{\Delta hs}$ model [32] was a kind gift from Prof Brockdorff (MRC Clinical Sciences Centre London, UK, current affiliation Department of Biochemistry, University of Oxford, UK). Mice were given standard chow and water *ad libitum* and were weighed two times a week. Breeding pairs of *mdx* (C57BL/10ScSn-mdx/J) males and *Xist*$^{\Delta hs}$ females gave birth to *mdx-Xist*$^{\Delta hs}$ females (Figure 1A). Genotyping was performed on DNA obtained from tail tips by PCR analysis to confirm the mutation in the *Dmd* and *Xist* gene (primers and PCR conditions on request).

Functional test regimes

To test the functional abilities of the different mouse strains over time, a group of 5 *mdx*, 24 *mdx-Xist*$^{\Delta hs}$ (0–15% dys $n=9$, 15–30% dys $n=9$ and >30% $n=6$, as assessed by Western blot of quadriceps muscles at the end of the study) and 5 *Xist*$^{\Delta hs}$ females underwent a functional test regime consisting of four different functional tests starting at four weeks of age, as described previously [40]. Plasma creatine kinase (CK) levels were determined once a week at the beginning of the functional test regime. Mice were sacrificed by cervical dislocation after 12 weeks of testing and muscles were isolated.

To determine whether functional test performances could be influenced by chronic treadmill exercise, 6 *mdx*, 7 *mdx-Xist*$^{\Delta hs}$ and 5

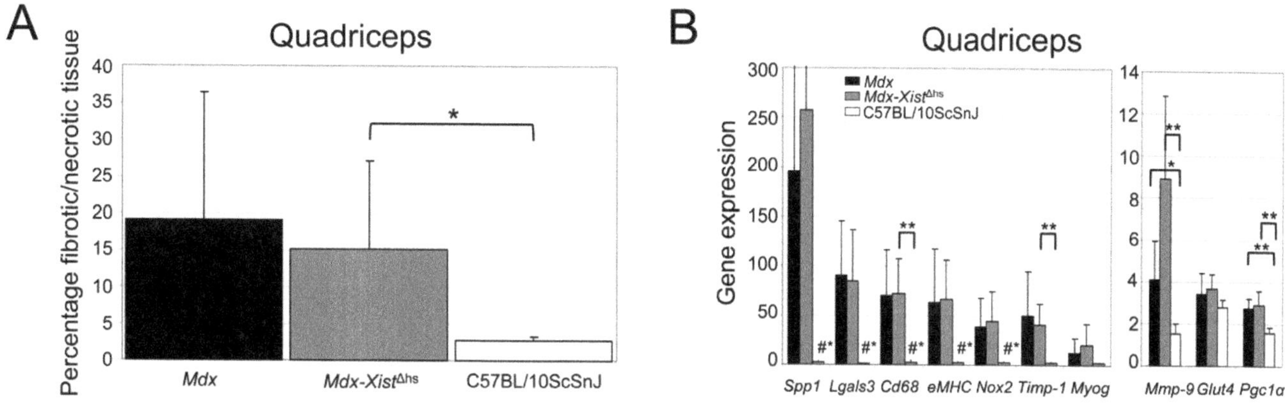

Figure 7. Exercise induced histopathology and biomarker gene expression. A. Wild type mice were less severely affected than *mdx* and *mdx-Xist*$^{\Delta hs}$ mice. Due to high variation between individual *mdx* mice, the difference between *mdx* and wild type mice was not significant. *Mdx-Xist*$^{\Delta hs}$ mice had slightly less fibrotic tissue than *mdx* mice. B. No difference in expression of genes involved in disease pathology was found between *mdx* and *mdx-Xist*$^{\Delta hs}$ mice. Both mouse strains did differ significantly from wild type mice, of which the expression levels were low. Single and double asterisks indicate a $P<0.05$ and $P<0.01$, respectively. # Indicates a significant difference from all other groups, average dystrophin levels of *mdx-Xist*$^{\Delta hs}$ mice was 21% (2%–45% median 25.8).

C57BL/10ScSnJ four week old females had to run 30 minutes on a horizontal treadmill at 12 m/min, directly followed by either fore limb grip strength, rotarod running or the two limb hanging wire test three times a week for 12 weeks. CK levels were determined two times a week; on Mondays before exercise and on Fridays directly after the two limb hanging wire test. At the age of 8 and 14 weeks additional blood was collected via the tail to assess several serum biomarkers (MMP-9, TIMP-1 and OPN) of which we have determined in parallel research that these correlate with disease severity [37]. After the 12 week chronic exercise protocol, mice were forced to run a final time on the horizontal treadmill for 30 minutes at 12 m/min. Evans blue dye (5 μg/μl) was applied intravenously (5 μl/gram of body weight) 20–30 minutes after this exercise. Twenty four hours after the injection, mice were sacrificed by cervical dislocation and muscles were isolated. Standardized operating procedures from the TREAT-NMD network were implemented for grip strength, both hanging wire tests and treadmill running (http://www.treat-nmd.eu/resources/research-resources/dmd-sops/).

Fore limb grip strength test

Fore limb grip strength was assessed by means of a grip strength meter (Columbus Instruments, USA). Mice were tested five times, with three consecutive measurements per trial (15 in total), and a two minute interval between trials. The three highest measured values were averaged to calculate the absolute strength, which was divided by the body weight in grams.

Rotarod

Mice were placed on the Rotarod (Ugo Basile, Italy) that accelerated from 5 to 45 rotations per minute within 15 seconds. The test session ended when a mouse ran for 500 seconds without falling. Mice that fell off before 500 seconds were given a maximum of two more tries. The longest running time was used for analysis.

Two limb hanging wire test

Mice were suspended above a metal cloth hanger secured above a cage and released a few seconds after instinctively grasping the wire with the fore limbs. Depending on the functional ability of the mouse, 2–4 limbs and the tail were used during a 10 minute hanging session. Mice that fell down before the 10 minute time limit were given two more tries. The longest hanging time was used for further analysis.

Four limb hanging wire test

The start position of this test was with all limbs. To this end, mice were placed on a grid, which was then turned upside down above a cage filled with bedding. The session ended after a hanging time of 10 minutes was achieved or otherwise after three sessions. The longest hanging time was used for further analysis.

Blood CK, MMP-9, TIMP-1 and OPN level analysis

For CK determination, blood was collected via a small cut at the end of the tail in a Minicollect tube (0.8 ml Lithium Heparin Sep, Greiner bio-one, Austria). Plasma CK levels were determined with Reflotron CK test strips in the Reflotron plus machine (Roche diagnostics Ltd, UK) at the day of collection.

To detect MMP-9, TIMP-1 and OPN levels, blood was collected via the tail and allowed to clot at room temperature for 10 minutes before spinning (1700 *g* for 10 minutes at 4°C). The serum supernatant was carefully aspirated and stored at −80°C, before use. To determine serum biomarker levels in mouse, the mouse MMP-9 (total) immunoassay kit (catalog number MMPT90) and mouse TIMP-1 immunoassay kit (catalog number MTM100) were purchased from R&D systems (Abingdon, United Kingdom), while the mouse OPN kit (code number 27351) was purchased from IBL (Hamburg, Germany). Experiments were performed as per manufacturer's protocol.

Histological examinations

For all 16 week-old mice the quadriceps, gastrocnemius, tibialis anterior, triceps, biceps, diaphragm and heart were dissected and snap frozen in 2-methylbutane (Sigma Aldrich, the Netherlands) cooled in liquid nitrogen. Cross-sections of 8 μm were cut on Superfrost Plus slides (Thermo Fishes Scientific, Menzel-Gläser, Germany) with a Shandon cryotome (Thermo Fisher Scientific Co., USA) along the entire length of the muscle with an interval of 240 μm between the sections. The excess tissue between the sections was collected in MagNa Lyser Green Beads tubes (Roche diagnostics Ltd, UK) for total RNA and protein isolation.

To determine the percentage of fibrotic/necrotic fibers, sections of the quadriceps were stained with Harris Haematoxylin and Eosin (H&E) (Sigma Aldrich, the Netherlands) and examined with a light microscope (Leica DM LB, Leica Microsystems, the Netherlands) at a 5 times magnification and images were captured with a Leica DC500 camera and Leica IM50 software (Leica Microsystems, the Netherlands) from the entire middle section of the muscle. Analysis was performed in a double-blinded manner, by two independent researchers on stitched pictures with the color deconvolution plugin of the ImageJ software (Rasband, W.S., ImageJ, U. S. National Institutes of Health, Bethesda, Maryland, USA, http://rsb.info.nih.gov/ij/, 1997–2008) as described previously [40]. Quadriceps, tibialis anterior, diaphragm and heart sections were stained with dystrophin (C-20, dilution 1:50, Santa Cruz, Germany) and donkey-anti-goat Alexa 488 (dilution 1:1000, Invitrogen, the Netherlands) and mounted with Vectashield mounting medium including DAPI (Vector Labs, Germany). The percentage of dystrophin positive fibers was manually and independently counted by two persons in a blinded manner for the quadriceps, stained with dystrophin and laminin (ab11575, dilution 1:50 Abcam, UK) where donkey-anti-goat Alexa 488 and donkey-anti-rabbit Alexa 594 (dilution 1:1000, Invitrogen, the Netherlands) were respectively used as a secondary antibody. To determine the dystrophin expression per nucleus, longitudinal sections of the quadriceps were cut and stained with dystrophin, spectrin (dilution 1:200, Thermo Scientific, USA) and DAPI. Co-localization of dystrophin with nNOS (H-299, dilution 1:50, Santa Cruz, Germany) and β-dystroglycan (NCL β-DG, dilution 1:50, Novocastra, UK) was determined in the quadriceps. The donkey-anti-rabbit Alexa 488 (dilution 1:1000, Invitrogen, the Netherlands) and MOM kit (Vector Laboratories, UK) were used as secondary antibody respectively.

Fiber size and the percentage of centrally nucleated fibers were determined with Mayachitra Imago 1 (http://www.mayachitra.com/imago/) on five randomly captured images of the quadriceps stained with laminin and DAPI (van Putten et al. manuscript in preparation). Fibers were segmented based on intensity differences between the membrane and the cytoplasm. The cross sectional area of each segmented fiber was computed by the software. Centrally nucleated fibers were identified based on the maximum intensity of the cytoplasm which is higher in fibers with a centralized nucleus than without. The percentage of Evans blue dye positive fibers was manually counted in a blinded manner for the quadriceps of all mice which were subjected to the chronic treadmill exercise. All fluorescent stained sections were examined with a fluorescent microscope (Leica DM RA2) at a 16 times magnification and six images were randomly captured with a Leica DC350FX snapshot camera.

Western blot analysis

Muscles were homogenized in treatment buffer (100 mM Tris-HCl pH 6.8 and 25% SDS) using MagNa Lyser green beads tubes in the MagNa Lyser (Roche Diagnostics, the Netherlands). Protein concentration was determined using the BCA protein assay kit (Thermoscientific, USA), according to the manufacturer's instructions. Subsequently, the homogenate was complemented to contain 75 mM Tris-HCl pH 6.8, 15% SDS, 5% β-mercaptoethanol, 20% glycerol and 0.001% bromophenol blue. Samples were boiled for 5 minutes and 50 μg was loaded on a hand-made 4–7% gradient polyacrylamide gel and run overnight at 4°C. The gel was blotted to nitrocellulose BA83 (Whatman, Schleicher & Schuell, Germany) for 6 hours at 4°C. The blot was blocked with 5% non-fat dried milk (Campina Melkunie, the Netherlands) in Tris Buffered Saline (TBS) plus 0.05% Tween20 followed by an overnight incubation with NCL-DYS1 (dilution 1:125, Novacastra, UK) or MANCHO3 (dilution 1:50,Tebu-bio, the Netherlands) in TBS plus 0.05% Tween20 to detect dystrophin and utrophin respectively. Myosin (MF20 dilution 1:20000, Developmental Studies Hybridoma Bank, University of Iowa, Iowa City) was used as a loading control. The fluorescent IRDye 800CW goat anti-mouse IgG (dilution 1:5000 for dystrophin and utrophin, dilution 1:10000 for myosin, Li-Cor, USA) was used as a secondary antibody. Blots were visualized and quantified with the Odyssey system and software (Li-Cor, USA) as described previously [41]. For each sample, 2–9 technical replicates were performed. A standard deviation of 15% was allowed between the replicates.

RT-qPCR analysis

Total RNA was isolated with RNA-Bee (Tel-Test, Bio-Connect, the Netherlands) and purified with the NucleoSpin RNA II kit including a DNAse digestion (Bioke, the Netherlands) according to the manufacturer's instructions. The RNA concentration was measured on a Nanodrop (Nanodrop Technologies, USA) and integrity was checked with a total RNA nano chip assay on a labchip assay (Agilent, the Netherlands). cDNA was synthesised with random hexamer primers and gene expression levels were determined by Sybr Green based Real Time qPCR (95°C 10 sec, 60°C 30 sec, 72°C 20 sec. 45 cycles followed by melting curve determination) on the Roche Lightcycler 480 (Roche diagnostics Ltd, UK). Expression of genes involved in inflammation (*Lgals3, Cd68*), fibrosis (*Tnnt2, Itgb2, Timp-1, Nox2, Lox, Nox4, Mmp9, Spp1*) regeneration (*Myog, Bgn, eMHC, Ctsk, Thymosinβ*), heart function (*Nppa, Serca2a, Cdnk1a*) and exercise (*Glut4, Pgc1α*) were analysed. *Gapdh* was used as a reference gene, since the expression of this gene did not differ between different muscles or over time. Primer efficiencies were determined with LinREgPCR version 11.1 [42]. The C_p values were obtained with the second derivative maximum method and analysed.

Statistics

Statistical analyses were performed with statistical software in R (version R2.11.1). *Mdx-Xist*$^{\Delta hs}$ mice were divided in three groups by k-means clustering, based on the median level of dystrophin quantified by Western blot data ($n = 2$–9). Figures 2, 5 and 6 summarize the data, i.e. showing the mean and standard deviation per genotype and age. To overcome applying separate tests for each age between genotypes, which suffers from multiple testing and ignores the age trend, we applied an analysis of covariance (ANCOVA) to the data. ANCOVA was applied to temporal functional performance, body weight and CK level data with age as continuous and genotype as a categorical variable. In choosing the appropriate model we applied the principle of parsimony. Given the application of tests for several different variables, we considered $P<0.01$ as significant.

The two-tailed homoscedastic Student's t-test was conducted for comparison of the single time point histological, serum biomarker and gene expression data. $P<0.05$ was considered significant for all tests, while $P<0.01$ was used for the gene expression data to correct for multiple testing.

Supporting Information

Figure S1 Dystrophin expression in *mdx-Xist*$^{\Delta hs}$ mice. A. Example of a Western blot of some skeletal muscles and heart. In order to determine the expression levels of the different muscles, a concentration curve was made of wild type samples from the corresponding muscle. Myosin was used as a loading control. B. Relative dystrophin levels of skeletal muscles compared to these of the quadriceps. Dystrophin levels of six *mdx-Xist*$^{\Delta hs}$ mice, with dystrophin levels of <15%, 15–30% and >30%, were assessed for skeletal muscles and heart by Western blot. Levels are expressed relative to the quadriceps. Those of gastrocnemius and triceps were similar to quadriceps levels, while those of the tibialis anterior and the biceps were slightly higher. Low levels were observed for the diaphragm and heart. qua: quadriceps, gas: gastrocnemius, ta: tibialis anterior, tri: triceps, bi: biceps, dia: diaphragm, ha: heart. C. Small groups of dystrophin positive fibers were randomly distributed over the muscles. Representative pictures of *mdx-Xist*$^{\Delta hs}$ mice with 8 and 23% dystrophin.

Figure S2 Dystrophin co-localisation with β-dystroglycan and nNOS in *mdx-Xist*$^{\Delta hs}$ mice. A. Dystrophin co-localises with β-dystroglycan (top) and nNOS (bottom). B. For *mdx-Xist*$^{\Delta hs}$ mice dystrophin levels previously determined on several Western blots (representative blot shown in Figure 1B) were compared to utrophin levels. The relative level of utrophin nicely correlated to that of dystrophin. Mice with intermediate dystrophin levels had decreased utrophin levels (lane 2 and 6) while mice with low dystrophin levels had higher utrophin levels (lane 1, 3–5). *Mdx/utrn* mice expressing utrophin on zero, one or two alleles were taken along as controls. Myosin was used as a loading control.

Acknowledgments

We thank prof. Dr. Brockdorff and Tatyana Nesterova (MRC Clinical Sciences Center, London, UK, current affiliation Department of Biochemistry, University of Oxford, UK) for the *Xist*$^{\Delta hs}$ embryos and help with genotyping the mice. We thank Gert-Jan van Ommen for useful discussions and critically reading the manuscript. We are grateful to Jiyun Byun for her expert assistance with Mayachitra Imago.

Author Contributions

Conceived and designed the experiments: JTDD AAR PACH. Performed the experiments: MVP MH VDN SHVH EVH TM. Analyzed the data: MVP MVI PA. Contributed reagents/materials/analysis tools: MVI PA. Wrote the paper: MVP PACH AAR.

References

1. Hoffman EP, Brown RH, Jr., Kunkel LM (1987) Dystrophin: the protein product of the Duchenne muscular dystrophy locus. Cell 51: 919–928.
2. Aartsma-Rus A, van Deutekom JC, Fokkema IF, van Ommen GJ, den Dunnen JT (2006) Entries in the Leiden Duchenne muscular dystrophy mutation database: an overview of mutation types and paradoxical cases that confirm the reading-frame rule. Muscle Nerve 34: 135–144.
3. Monaco AP, Bertelson CJ, Liechti-Gallati S, Moser H, Kunkel LM (1988) An explanation for the phenotypic differences between patients bearing partial deletions of the DMD locus. Genomics 2: 90–95.
4. Hoffman EP, Fischbeck KH, Brown RH, Johnson M, Medori R, et al. (1988) Characterization of dystrophin in muscle-biopsy specimens from patients with Duchenne's or Becker's muscular dystrophy. N Engl J Med 318: 1363–1368.
5. Hirawat S, Welch EM, Elfring GL, Northcutt VJ, Paushkin S, et al. (2007) Safety, tolerability, and pharmacokinetics of PTC124, a nonaminoglycoside nonsense mutation suppressor, following single- and multiple-dose administration to healthy male and female adult volunteers. J Clin Pharmacol 47: 430–444.
6. Kinali M, Arechavala-Gomeza V, Feng L, Cirak S, Hunt D, et al. (2009) Local restoration of dystrophin expression with the morpholino oligomer AVI-4658 in Duchenne muscular dystrophy: a single-blind, placebo-controlled, dose-escalation, proof-of-concept study. Lancet Neurol 8: 918–928.
7. van Deutekom JC, Janson AA, Ginjaar IB, Frankhuizen WS, Aartsma-Rus A, et al. (2007) Local dystrophin restoration with antisense oligonucleotide PRO051. N Engl J Med 357: 2677–2686.
8. Aartsma-Rus A, den Dunnen JT, van Ommen GJ (2010) New insights in gene-derived therapy: the example of Duchenne muscular dystrophy. Ann N Y Acad Sci 1214: 199–212.
9. Mendell JR, Campbell K, Rodino-Klapac L, Sahenk Z, Shilling C, et al. (2010) Dystrophin immunity in Duchenne's muscular dystrophy. N Engl J Med 363: 1429–1437.
10. Malik V, Rodino-Klapac LR, Viollet L, Mendell JR (2010) Aminoglycoside-induced mutation suppression (stop codon readthrough) as a therapeutic strategy for Duchenne muscular dystrophy. Ther Adv Neurol Disord 3: 379–389.
11. Goemans NM, Tulinius M, van den Akker JT, Burm BE, Ekhart PF, et al. (2011) Systemic administration of PRO051 in Duchenne's muscular dystrophy. N Engl J Med 364: 1513–1522.
12. Cirak S, Arechavala-Gomeza V, Guglieri M, Feng L, Torelli S, et al. (2011) Exon skipping and dystrophin restoration in patients with Duchenne muscular dystrophy after systemic phosphorodiamidate morpholino oligomer treatment: an open-label, phase 2, dose-escalation study. Lancet 378: 595–605.
13. Pegoraro E, Schimke RN, Garcia C, Stern H, Cadaldini M, et al. (1995) Genetic and biochemical normalization in female carriers of Duchenne muscular dystrophy: evidence for failure of dystrophin production in dystrophin-competent myonuclei. Neurology 45: 677–690.
14. Schmidt-Achert M, Fischer P, Muller-Felber W, Mudra H, Pongratz D (1993) Heterozygotic gene expression in endomyocardial biopsies: a new diagnostic tool confirms the Duchenne carrier status. Clin Investig 71: 247–253.
15. Ogata H, Nakagawa H, Hamabe K, Hattori A, Ishikawa Y, et al. (2000) A female carrier of Duchenne muscular dystrophy complicated with cardiomyopathy. Intern Med 39: 34–38.
16. Bostick B, Yue Y, Long C, Duan D (2008) Prevention of dystrophin-deficient cardiomyopathy in twenty-one-month-old carrier mice by mosaic dystrophin expression or complementary dystrophin/utrophin expression. Circ Res 102: 121–130.
17. Tanaka H, Ikeya K, Ozawa E (1990) Difference in the expression pattern of dystrophin on the surface membrane between the skeletal and cardiac muscles of mdx carrier mice. Histochemistry 93: 447–452.
18. Boyd Y, Buckle V, Holt S, Munro E, Hunter D, et al. (1986) Muscular dystrophy in girls with X;autosome translocations. J Med Genet 23: 484–490.
19. Azofeifa J, Voit T, Hubner C, Cremer M (1995) X-chromosome methylation in manifesting and healthy carriers of dystrophinopathies: concordance of activation ratios among first degree female relatives and skewed inactivation as cause of the affected phenotypes. Hum Genet 96: 167–176.
20. Lupski JR, Garcia CA, Zoghbi HY, Hoffman EP, Fenwick RG (1991) Discordance of muscular dystrophy in monozygotic female twins: evidence supporting asymmetric splitting of the inner cell mass in a manifesting carrier of Duchenne dystrophy. Am J Med Genet 40: 354–364.
21. Soltanzadeh P, Friez MJ, Dunn D, von NA, Gurvich OL, et al. (2010) Clinical and genetic characterization of manifesting carriers of DMD mutations. Neuromuscul Disord 20: 499–504.
22. Hoffman EP, Kunkel LM, Angelini C, Clarke A, Johnson M, et al. (1989) Improved diagnosis of Becker muscular dystrophy by dystrophin testing. Neurology 39: 1011–1017.
23. Neri M, Torelli S, Brown S, Ugo I, Sabatelli P, et al. (2007) Dystrophin levels as low as 30% are sufficient to avoid muscular dystrophy in the human. Neuromuscul Disord 17: 913–918.
24. Nicholson LV, Johnson MA, Bushby KM, Gardner-Medwin D (1993) Functional significance of dystrophin positive fibres in Duchenne muscular dystrophy. Arch Dis Child 68: 632–636.
25. Phelps SF, Hauser MA, Cole NM, Rafael JA, Hinkle RT, et al. (1995) Expression of full-length and truncated dystrophin mini-genes in transgenic mdx mice. Hum Mol Genet 4: 1251–1258.
26. Rafael JA, Sunada Y, Cole NM, Campbell KP, Faulkner JA, et al. (1994) Prevention of dystrophic pathology in mdx mice by a truncated dystrophin isoform. Hum Mol Genet 3: 1725–1733.
27. Dunant P, Larochelle N, Thirion C, Stucka R, Ursu D, et al. (2003) Expression of dystrophin driven by the 1.35-kb MCK promoter ameliorates muscular dystrophy in fast, but not in slow muscles of transgenic mdx mice. Mol Ther 8: 80–89.
28. Stillwell E, Vitale J, Zhao Q, Beck A, Schneider J, et al. (2009) Blastocyst injection of wild type embryonic stem cells induces global corrections in mdx mice. PLoS One 4: e4759.
29. Li D, Yue Y, Duan D (2008) Preservation of muscle force in Mdx3cv mice correlates with low-level expression of a near full-length dystrophin protein. Am J Pathol 172: 1332–1341.
30. Brockdorff N, Ashworth A, Kay GF, Cooper P, Smith S, et al. (1991) Conservation of position and exclusive expression of mouse Xist from the inactive X chromosome. Nature 351: 329–331.
31. Senner CE, Brockdorff N (2009) Xist gene regulation at the onset of X inactivation. Curr Opin Genet Dev 19: 122–126.
32. Newall AE, Duthie S, Formstone E, Nesterova T, Alexiou M, et al. (2001) Primary non-random X inactivation associated with disruption of Xist promoter regulation. Hum Mol Genet 10: 581–589.
33. van Putten M, Kumar D, Hulsker M, Hoogaars WMH, Plomp JJ, et al. (2012) Comparison of skeletal muscle pathology and motor function of dystrophin and utrophin deficient mouse strains. Neuromuscul Disord;in press.

34. Spurney CF, Gordish-Dressman H, Guerron AD, Sali A, Pandey GS, et al. (2009) Preclinical drug trials in the mdx mouse: assessment of reliable and sensitive outcome measures. Muscle Nerve 39: 591–602.
35. van Erp C, Loch D, Laws N, Trebbin A, Hoey AJ (2010) Timeline of cardiac dystrophy in 3–18-month-old MDX mice. Muscle Nerve 42: 504–513.
36. Turk R, Sterrenburg E, de Meijer EJ, van Ommen GJ, den Dunnen JT, et al. (2005) Muscle regeneration in dystrophin-deficient mdx mice studied by gene expression profiling. BMC Genomics 6: 98.
37. Nadarajah VD, van PM, Chaouch A, Garrood P, Straub V, et al. (2011) Serum matrix metalloproteinase-9 (MMP-9) as a biomarker for monitoring disease progression in Duchenne muscular dystrophy (DMD). Neuromuscul Disord 21: 569–578.
38. Sharp PS, Jee H, Wells DJ (2011) Physiological characterization of muscle strength with variable levels of dystrophin restoration in mdx mice following local antisense therapy. Mol Ther 19: 165–171.
39. Malik V, Rodino-Klapac LR, Viollet L, Wall C, King W, et al. (2010) Gentamicin-induced readthrough of stop codons in Duchenne muscular dystrophy. Ann Neurol 67: 771–780.
40. van Putten M, de Winter CL, van Roon-Mom W, van Ommen GJ, 't Hoen PA, et al. (2010) A 3 months mild functional test regime does not affect disease parameters in young mdx mice. Neuromuscul Disord 20: 273–280.
41. Heemskerk HA, de Winter CL, de Kimpe SJ, van Kuik-Romeijn P, Heuvelmans N, et al. (2009) In vivo comparison of 2′-O-methyl phosphorothioate and morpholino antisense oligonucleotides for Duchenne muscular dystrophy exon skipping. J Gene Med 11: 257–266.
42. Ramakers C, Ruijter JM, Deprez RH, Moorman AF (2003) Assumption-free analysis of quantitative real-time polymerase chain reaction (PCR) data. Neurosci Lett 339: 62–66.

13

Attenuation of Rhes Activity Significantly Delays the Appearance of Behavioral Symptoms in a Mouse Model of Huntington's Disease

Brandon A. Baiamonte[1], Franklin A. Lee[1], Steve T. Brewer[1], Daniela Spano[2,3], Gerald J. LaHoste[1]*

1 Applied Biopsychology Program, Department of Psychology, University of New Orleans, New Orleans, Louisiana, United States of America, 2 Dipartimento di Biochemica e Biotecnologie Mediche, Università di Napoli, Naples, Italy, 3 CEINGE Biotecnologie Avanzate, Naples, Italy

Abstract

Huntington's disease (HD) is a neuropsychiatric disorder characterized by choreiform movement of the limbs, cognitive disability, psychosis and dementia. It is invariably associated with an abnormally long CAG expansion within the IT15 gene on human chromosome 4. Although the mutant huntingtin protein is ubiquitously expressed in HD patients, cellular degeneration occurs predominantly in neurons within the corpus striatum and cerebral cortex. The Ras homolog Rhes is expressed very selectively in the precise brain areas affected by HD. Recent *in vitro* work suggests that Rhes may be a co-factor with mutant huntingtin in cell death. The objective of the present study was to examine whether the inhibition of Rhes would attenuate or delay the symptoms of HD *in vivo*. We used a transgenic mouse model of HD crossed with Rhes knockout mice to show that the behavioral symptoms of HD are regulated by Rhes. $HD^+/Rhes^{-/-}$ mice showed significantly delayed expression of HD-like symptoms in this *in vivo* model. Drugs that block or inhibit the actions of Rhes may be useful as the first treatments for HD.

Editor: Stefano L. Sensi, University G. D'Annunzio, Italy

Funding: The authors have no support or funding to report.

Competing Interests: The authors have declared that no competing interests exist.

* E-mail: glahoste@uno.edu

Introduction

Huntington's disease (HD) is a complex, progressive neurodegenerative disorder that results in deficits of motor function, cognition, and mood [1]. It is incurable, untreatable, and fatal. HD is caused by a genetic defect that is characterized by an unstable expansion of a trinucleotide repeat (CAG, which codes for glutamine) within the IT15 gene on human chromosome 4 [2]. Individuals with ≤35 repeats do not develop the HD phenotype, whereas those with >35 repeats invariably develop HD, usually beginning in middle-age [3]. The mutant gene (mHtt) is transmitted in an autosomal dominant matter, such that, on average, 50% of the offspring of an affected heterozygous individual will have HD. The long stretch of glutamine amino acids present in the mHtt protein (huntingtin) is believed to cause neurodegeneration and the symptoms that constitute HD [4–8].

The exact mechanism underlying neurodegeneration in HD is a matter of some debate [9]. A multitude of mechanisms responsible for this neurodegeneration, such as activation of proteases, protein misfolding, inhibition of protein degradation, transcriptional dysregulation, synaptic dysfunction, mitochondrial dysfunction, and oxidative stress have been proposed [10]. One serious problem for current theories of HD is that mHtt is present in all cells, yet only neurons within the corpus striatum and cerebral cortex of the brain are targeted for degeneration [11,12]. This could be due to the remarkable similarity between the pattern of degeneration and the distribution of the mRNA for Rhes, the Ras homolog enriched in the striatum [13,14]. Based on recent *in vitro* data in cultured cells, Subramaniam et al. [15] suggest that Rhes protein is a co-factor in mHtt-induced neurodegeneration. The rationale underlying this argument is that Rhes inhibits the formation of huntingtin aggregates, which, paradoxically, are believed to have a neuroprotective effect by sequestering soluble, cellular mHtt associated with cytotoxicity [15–18].

The present study was undertaken to determine whether the inhibition of Rhes would attenuate or delay the symptoms of HD *in vivo*. We report that doubly transgenic mice with a human HD gene insertion and Rhes deletion showed significantly delayed expression of HD-like symptoms.

Methods

Prompted by these cellular findings we hypothesized that living, behaving HD mice would not manifest the usual abnormal symptoms in the absence of Rhes. We tested this hypothesis by cross-breeding mice that are transgenic for the human mHtt allele with Rhes knockout (KO) mice generated by homologous recombination [19]. Breeding pairs of mice transgenic for the human mHtt allele were obtained from Jackson Laboratory, Bar Harbor, ME (B6.Cg-Tg(Hdexon1)61 GpB/J; stock # 006741). The most common line of transgenic mice, R6/2 are reproductively incapable of breeding the crucial genotypes necessary to test our hypothesis. We therefore used the R6/1 line which contains ~115 CAG repeats and shows symptoms that develop in mid-life that are homologous to the motor symptoms of HD. In addition, R6/1 mice develop behavioral symptoms later than the R6/2 line.

The slower progression and later onset allows for a detailed analysis of the progression of the disease. Hemizygous males were bred with non-carrier females. For simplicity, we will refer to the Htt haplotype as NC (non-carrier) and to the mHtt-positive haplotype as HD; the Rhes genotypes will be referred to as WT (wild-type), Het (heterozygous), and KO (knockout). The crucial genotype (HD/KO) cannot be generated by simple cross-breeding. Thus, HD/KO mice were generated by secondary breeding of already cross-bred HD/Hets×HD/Hets. Six genotypes were tested: 1) NC/WT; 2) NC/Het; 3) NC/KO; 4) HD/WT; 5) HD/Het; and 6) HD/KO. We used both male and female mice with sample sizes of 7–8.

Ethics Statement

All animals in the experiment were cared for in accordance with the guidelines of the U.S. Public Health Service Policy on Humane Care and Use of Laboratory Animals, and experimental protocols were approved and supervised by the University of New Orleans Institutional Animal Care and Use Committee (Protocol Numbers: UNO-09-011 and 09-012; Assurance Number: A3299-01).

Statistics

The data were examined for statistical outliers and none were found. Therefore, no data were excluded from the analyses. All analyses were performed using SPSS for Windows (version 16.0) with the probability of a Type I error set at 0.05. Data for rotarod performance, limb adduction, brain weight, and brain area were analyzed in separate 2×3×6 mixed design factorial ANOVAs with HD gene (HD+, NC) and Rhes gene (WT, Het, KO) serving as between subjects variables and Testing Age (testing once every month) serving as a within-subjects variable. A 3-way interaction was followed by separate analyses of the HD gene×Rhes gene (2×2) interaction for each day of testing. Significant 2-way interactions on any day were followed by analysis of 1) the simple effects of HD on rotarod performance, limb adduction, brain weight, and brain area, 2) the simple effects of Rhes on rotarod performance, limb movements, brain weight, and brain area, and 3) interaction contrast to determine if the reduced amount of Rhes expression reversed or partially reversed the effects of HD on rotarod performance, limb movements, brain weight, and brain area. Tukey's HSD test was used for all post-hoc tests to control the family-wise error rate.

Results and Discussion

Motor coordination

The mice were tested monthly from 1 to 6 months of age on a battery of previously validated motor behaviors [20]. Motor coordination was assessed using a circular, vertically rotating treadmill (Rotarod) at a fixed rotation speed of 16 rpm. Mice were placed on the treadmill with their noses opposite to the direction of rotation. The ability to stay on the treadmill without falling was taken as an index of motor coordination and balance. Three 60-sec. trials were given with 60-sec. intertrial intervals. All three groups of mice not carrying the HD allele (NC) performed perfectly well on the Rotarod test for the duration of the experiment (Fig. 1). In contrast, HD/WT mice showed a precipitous decline in motor coordination between months 3 and 4, worsening during the time-course of the study so that by 6 months of age the mean duration of time before falling was 20 sec. (max. duration, 180 sec.), representing an 88% decline in performance. However, HD/KO mice showed normal motor coordination until month 5 at which time their performance had declined significantly from NC mice, but only by 15%. Performance of HD/Het mice began to decline at 4 months of age, ultimately being significantly better than HD/WT and significantly worse than HD/KO mice, indicating a gene-dose effect. The HD mice differed as a function of Rhes genotype with respect to the age at which they showed ~50% decline in motor performance. For the HD mice, we performed linear regression on data from months 3–6 to determine the age at which there was a 50% decline in motor performance. Analysis revealed these ages to be 4.33, 5.17, and 6.42 months. Thus, deletion of Rhes forestalled the symptoms of HD by 2.09 months. This amount of disease delay represents 6.5% of the lifespan of a control C57 mouse, or 25% of HD mice of this line [21,22]. Although it is difficult to compare lifespans between species, it is likely that the delay in HD symptoms observed in the present experiment is clinically significant.

Dystonia

When suspended from the tail normal mice splay their limbs outward; by contrast, mice with HD draw their limbs inward toward the body and clasp paired limbs together, an analogue of dystonia [22,23]. Assessment of limb adduction entailed suspending each mouse by the tail and counting the number of limbs drawn inward during 3 monthly 10-sec trials [20]; clasping per se was not distinguished from adduction. The results were similar to those for the Rotarod test (Fig. 2). NC mice did not show limb adduction for the duration of the experiment. HD/WT mice showed significantly greater adduction than any other genotype from 3 months onward. HD/KO mice showed significantly less limb withdrawal than HD/WT or HD/Het mice during this period. HD/KO mice did not show significant limb adduction (relative to NC mice) until months 5 and 6. The amount of adduction for HD/Het mice was intermediate between HD/WT and HD/KO mice, further supporting a gene-dose effect.

Brain measurements

At the time of sacrifice, brains were flash-frozen in liquid isopentane (−20°C) on an analytical balance. As in clinical HD, and as previously show in mouse models, the brains of HD mice weighed significantly less (16%) than the brains of NC mice (Fig. 3).

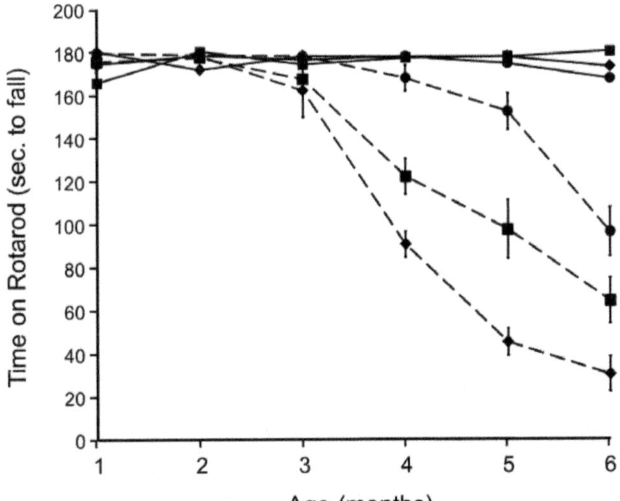

Figure 1. Rhes gene deletion significantly delays the onset of HD-induced motor dysfunction. NC: solid lines; HD: dashed lines; Diamonds: Rhes WT; Squares: Rhes Het; Circles: Rhes KO.

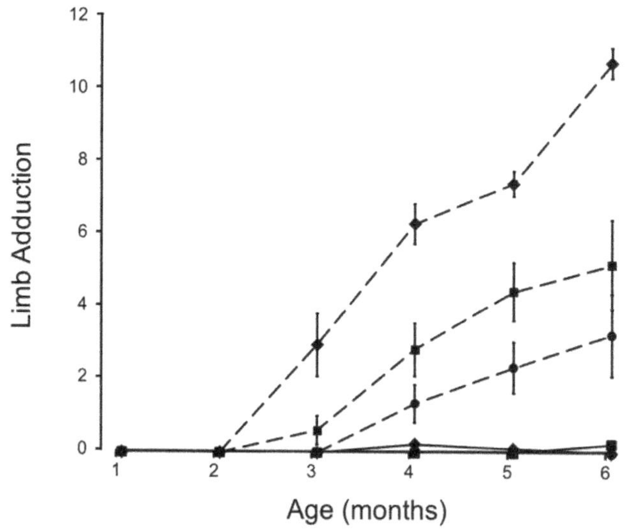

Figure 2. Dystonia as indicated by Limb Adduction. Legend is the same as for Fig. 1. Ordinate gives the average number of limbs drawn inward toward the body (labeled "Limb Adduction") during three trials repeated monthly for the 6 months of observation.

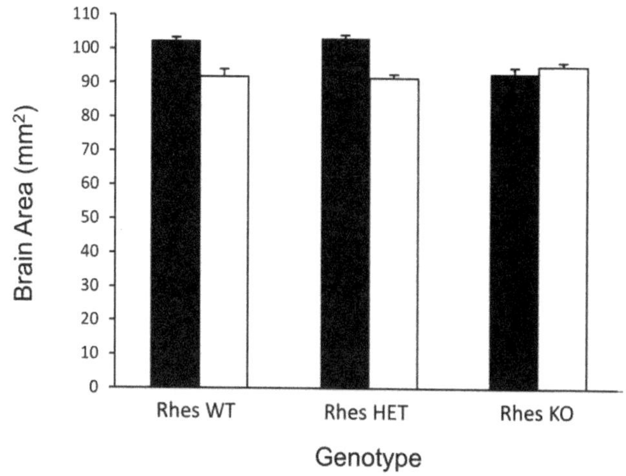

Figure 4. Brain Area (mm^2) as a function of genotype. Legend is the same as in Fig. 3, and the conclusions are the same. Both Rhes and HD were significant factors in diminishing brain weight. However, a combined effect of HD and Rhes KO was not seen in HD/KO mice, indicating a protective effect of Rhes deletion on brain area loss in these mice.

The brains of NC/KO mice weighed significantly less (11%) than those of other NC genotypes. In HD mice, there were no significant differences across the Rhes genotypes. However, there was a significant interaction between HD status and Rhes genotype such that the brain weights of NC/KO and HD/KO were not significantly different.

With respect to a rudimentary measure of whole brain area, the same results were obtained as were found with brain weight. The dimensions of whole, frozen HD brains (measured as the product of antero-posterior distance of the cerebral cortex and the width of the brain at its widest point) were 7% less than the brains of NC mice. The dimensions of NC/KO mice were significantly less (10%) than those of other NC genotypes (Fig. 4). In HD mice, there were no significant differences across the Rhes genotypes.

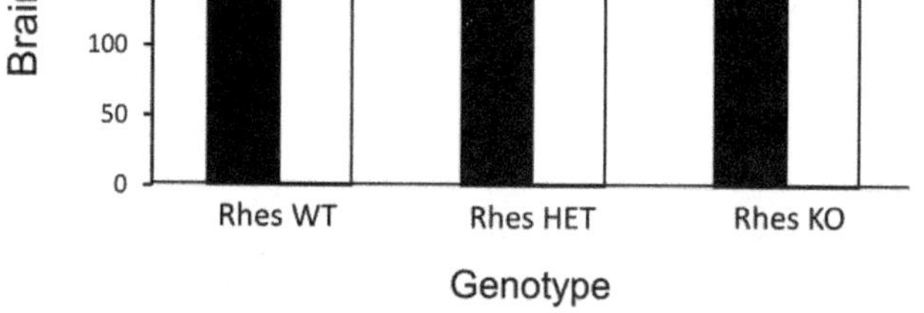

Figure 3. Post-mortem Brain Weight in mg. Black bars = NC; white bars = HD. Both Rhes and HD were significant factors in diminishing brain weight. However, a combined effect of HD and Rhes KO was not seen in HD/KO mice, indicating a protective effect of Rhes deletion on brain weight in these mice.

However, there was a significant interaction between HD status and Rhes genotype such that the brain weights of NC/KO and HD/KO were not significantly different. This pattern of results is identical to what was observed for brain weight.

Body Weights

A 2×3 factorial Analysis of Variance showed that mice carrying the mHtt allele weighed significantly less (20%) than non-carrier mice. However, Rhes genotype was not a significant factor, nor was the interaction between Rhes genotype and HD genotype.

The present study shows that Rhes inhibition attenuates and delays the behavioral symptoms of HD *in vivo*. Our conclusion is based on the use of a Rhes knockout line that shows gene-dosage reduction (heterozygotes) and absence (knockouts) of Rhes protein (unpublished data). Our findings provide evidence of the ability of Rhes to augment HD-induced motor deficits. It is believed that Rhes does this by acting as an E3 ligase in the protein modification process called sumoylation. Subramaniam et al. [15,24] have shown *in vitro* that Rhes interacts with the small ubiquitin-like modifier SUMO, which targets mHtt for sumoylation. In fact, Rhes increases sumoylation by 400% and is the principal determinant of sumoylation in the striatum [18,24]. Increased binding of SUMO to mHtt to the exclusion of ubiquitin could lead to build-up of toxic protein.

Although previous findings suggested that SUMO indeed competed for ubiquitin at the same lysine [18,25–27], it is now known that sumoylation can activate ubiquitination and increase protein clearance [28]. Thus, the interactions among Rhes, SUMOylation and mHtt may not be so simple. Steffan (2010) has proposed that SUMOylation, which can be induced by Rhes, may have an early protective effect in pre-symptomatic HD until age-related declines in protein clearing processes give rise to neurodegeneration and the appearance of symptoms. According to this thesis, Rhes might be expected to have a protective effect in the early, pre-symptomatic stages of HD. While this may be true in human HD, one would predict from this that in our mouse paradigm, HD/KO mice would show an earlier onset of *behavioral* symptoms compared to their HD/Rhes WT littermates, which they did not. We cannot rule out, however, a dual action of Rhes in neurodegeneration *per se*. An additional issue to be raised is that EGFP, contained in the Rhes knockout allele, and thereby confounded with Rhes genotype [19], may influence protein clearance processes [29].

In conclusion, the present findings show that inhibition of Rhes function by means of gene deletion delays the onset of behavioral symptoms in an animal model of HD, thereby identifying Rhes as a clinically significant target for the development of novel therapeutic strategies to treat the symptoms of HD.

Author Contributions

Conceived and designed the experiments: BAB GJL. Performed the experiments: BAB FAL STB. Analyzed the data: BAB. Contributed reagents/materials/analysis tools: DS. Wrote the paper: BAB.

References

1. Huntington G (1872) On chorea. The Medical and Surgical Reporter 26: 320–321.
2. Huntington's Disease Collaborative Research Group (1993) A novel gene containing a trinucleotide repeat that is expanded and unstable on Huntington's disease chromosomes. Cell 72: 971–983.
3. Li SH, Li XJ (2004) Huntingtin-protein interactions and the pathogenesis of Huntington's disease. Trends Genet 20: 146–154.
4. Ross C A. (2002) Polyglutamine pathogenesis. Neuron 35: 819–822.
5. Zoghbi HY, Orr HT (2000) Glutamine repeats and neurodegeneration. Annu Rev Neurosci 23: 217–247.
6. Sathasivam K, Amaechi I, Mangiarini L, Bates GP (1997) Identification of an HD patient with a (CAG)180 repeat expansion and the propagation of highly expanded CAG repeats in lamda phage. Hum Genet 99: 692–695.
7. Rubinsztein DC, Leggo J, Coles R, Almqvist E, Biancalana V, et al. (1996) Phenotypic characterization of individuals with 30–40 CAG repeats in the Huntington's disease (HD) gene reveals HD cases with 36 repeats and apparently normal elderly individuals with 36–39 repeats. Am J Hum Genet 59: 16–22.
8. Andrew SE, Goldberg YP, Kremer B, Telenius H, Theilmann J, et al. (1993) The relationship between trinucleotide (CAG) repeat length and clinical features of Huntington's. Nat Genet 4: 398–403.
9. Landles C, Bates GP (2004) Huntingtin and the molecular pathogenesis of Huntington's disease. Fourth in molecular medicine review series. EMBO Reports 5: 958–963.
10. Gil JM, Rego AC (2008) Mechanisms of neurodegeneration in Huntington's disease. Eur J Neurosci 27: 2803–2820.
11. Vonsattel JP, Difiglia M (1998) Huntington disease. J Neuropathol Exp Neuro 57: 369–384.
12. Pineda JR, Canals JM, Bosch M, Adell A, Mengod G, et al. (2005) Brain-derived neurotrophic factor modulates dopaminergic deficits in a transgenic mouse model of Huntington's disease. J Neurochem 93: 1057–1068.
13. Falk JD, Vargiu P, Foye PE, Usui H, Perez J, et al. (1999) Rhes: A striatal-specific Ras homolog related to Dexras1. J Neurosci Res 57: 782–788.
14. Harrison LM, LaHoste GJ (2006) Rhes, the Ras homolog enriched in striatum, is reduced under conditions of dopamine supersensitivity. Neuroscience 137: 483–492.
15. Subramaniam S, Sixt KM, Barrow R, Snyder SH (2009) Rhes, a striatal specific protein, mediates mutant-Huntingtin cytotoxicity. Science 324: 1327–1330.
16. Kummerle S, Gutekunst CA, Klein AM, Li XJ, Li SH, et al. (1999) Huntingtin aggregates may not predict neuronal death in Huntington's disease. Ann Neurol 46: 842–849.
17. Arrasate M, Mitra S, Schweitzer ES, Segal MR, Finkbeiner S (2004) Inclusion body formation reduces levels of mutant huntingtin and the risk of neuronal death. Nature 431: 805–810.
18. Steffan JS, Agrawal N, Pallos J, Rockabrand E, Trotman LC, et al. (2004) SUMO modification of Huntingtin and Huntington's disease pathology. Science 304: 100–104.
19. Spano D, Branchi I, Rosica A, Pirro MT, Riccio A, et al. (2004) Rhes is involved in striatal function. Mol Cell Biol 24: 5788–5796.
20. Stack EC, Kubilus JK, Smith K, Cormier K, Del Signore SJ, et al. (2005) Chronology of behavioral symptoms and neuropathological sequel in R6/2 Huntington's disease transgenic mice. J Comp Neurol 490: 354–370.
21. Spalding JF, Thomas RG, Tietjen GL (1982) Life span of C57 mice as influenced by radiation dose, dose rate, and age at exposure. Energy Citations Database. Available: http://www.osti.gov/energycitations/product.biblio.jsp?osti_id=6866926. Accessed 28 August 2012.
22. Mangiarini L, Sathasivam K, Seller M, Cozens BA, Harper A, et al. (1996) Exon 1 of the HD gene with an expanded CAG repeat is sufficient to cause a progressive neurological phenotype in transgenic mice. Cell 87: 493–506.
23. Rubinsztein DC (2002) Lessons from animal models of Huntington's disease. Trends Genet 18: 202–209.
24. Subramaniam S, Mealer RG, Sixt KM, Barrow RK, Usiello A, et al. (2010) Rhes, a physiological regulator of sumoylation, enhances cross-sumoylation between the basic sumoylation enzymes E1 and Ubc9. J Biol Chem 285: 20428–20432.
25. Desterro JM, Rodriguez MS, Hay RT (1998) SUMO-1 modification of IκBα inhibits NF-κB activation. Mol Cell 2: 233–239.
26. Hoege C, Pfander B, Moldovan GL, Pyrowolakis G, Jentsch S (2002) RAD6-dependent DNA repair is linked to modification of PCNA by ubiquitin and SUMO. Nature 419: 135–141.
27. Lin X, Liang M, Liang YY, Brunicardi FC, Feng XH (2003) SUMO-1/Ubc9 promotes nuclear accumulation and metabolic stability of tumor suppressor SMAD4. J Biol Chem 278: 31043–31048.
28. Steffan JS (2010) Does Huntingtin play a role in selective macroautophagy? Cell Cycle 9: 3401–3413.
29. Klionsky DJ, Abdalla FC, Abeliovich H, Abraham RT, Acevedo-Arozena A, et al. (2012) Guidelines for the use and interpretation of assays for monitoring autophagy. Autophagy 8: 445–544.

14

PKC Theta Ablation Improves Healing in a Mouse Model of Muscular Dystrophy

Luca Madaro[1], Andrea Pelle[1], Carmine Nicoletti[1], Annunziata Crupi[1], Valeria Marrocco[1], Gianluca Bossi[2], Silvia Soddu[2], Marina Bouché[1]*

1 Unit of Histology, and IIM, Sapienza University, DAHFMO, Rome, Italy, 2 Department of Experimental Oncology, Regina Elena Cancer Institute, Rome, Italy

Abstract

Inflammation is a key pathological characteristic of dystrophic muscle lesion formation, limiting muscle regeneration and resulting in fibrotic and fatty tissue replacement of muscle, which exacerbates the wasting process in dystrophic muscles. Limiting immune response is thus one of the therapeutic options to improve healing, as well as to improve the efficacy of gene- or cell-mediated strategies to restore dystrophin expression. Protein kinase C θ (PKCθ) is a member of the PKCs family highly expressed in both immune cells and skeletal muscle; given its crucial role in adaptive, but also innate, immunity, it is being proposed as a valuable pharmacological target for immune disorders. In our study we asked whether targeting PKCθ could represent a valuable approach to efficiently prevent inflammatory response and disease progression in a mouse model of muscular dystrophy. We generated the bi-genetic mouse model $mdx/\theta^{-/-}$, where PKCθ expression is lacking in *mdx* mice, the mouse model of Duchenne muscular dystrophy. We found that muscle wasting in $mdx/\theta^{-/-}$ mice was greatly prevented, while muscle regeneration, maintenance and performance was significantly improved, as compared to *mdx* mice. This phenotype was associated to reduction in inflammatory infiltrate, pro-inflammatory gene expression and pro-fibrotic markers activity, as compared to *mdx* mice. Moreover, BM transplantation experiments demonstrated that the phenotype observed was primarily dependent on lack of PKCθ expression in hematopoietic cells. These results demonstrate a hitherto unrecognized role of immune-cell intrinsic PKCθ activity in the development of DMD. Although the immune cell population(s) involved remain unidentified, our findings reveal that PKCθ can be proposed as a new pharmacological target to counteract the disease, as well as to improve the efficacy of gene- or cell- therapy approaches.

Editor: Maurilio Sampaolesi, Stem Cell Research Institute, Belgium

Funding: This work was supported by the Italian Ministry for University and Research, by the Sapienza University of Rome, by the French Association against Myopathies and by the Italian Space Agency. Dr. Madaro is recipient of a Duchenne Parent Project, Netherland, Fellowship. The funders had no role in study design, data collection and analysis, decision to publish, or preparation of the manuscript.

Competing Interests: The authors have declared that no competing interests exist.

* E-mail: marina.bouche@uniroma1.it

Introduction

Duchenne muscular dystrophy (DMD) is one of the most common X-linked lethal diseases, and results from the mutation within the gene encoding dystrophin, a large cytoskeletal protein, whose ablation leads to membrane instability [1]. Therapies based on the restoration of dystrophin expression or the administration of dystrophin^{+ve} stem cells are promising, but still in the preclinical phase [2–4]. In this context, one of the barrier to successful gene therapy has been recently identified in cellular immunity [5]. Thus, the monitoring of cellular immune responses should be a priority for any experimental therapy designed to increase the number of dystrophin-positive myofibers in patients with Duchenne's muscular dystrophy. In addition, although mechanical injury and membrane defects are important factors promoting dystrophic pathology, increasing evidences highlight aberrant intracellular signalling cascades that regulate inflammatory and immune processes, as key contributors to the degenerative process [1,6]. Up-regulated inflammatory gene expression and activated immune cell infiltrates are evident during early disease stages in dystrophic muscle, and the identification of specific targets for anti-inflammatory therapies is one of the ongoing therapeutic options. Indeed, glucocorticoids, which have anti-inflammatory properties, are being used to treat DMD with some success; however, the side effects of these drugs often outweigh their benefit [7,8]. Numerous other anti-inflammatory therapies have been proposed to improve healing [9–13]. In this context, it has been recently shown that rapamycin treatment reduced dystrophic phenotype in *mdx* mice and that this effect was associated to a significant reduction in infiltration of T_{eff} cells in skeletal muscle tissue, while T_{reg} cells were preserved [14]. Indeed, a role of lymphocytes activity in the progression of muscular dystrophy is long known, as antibody- or genetic-mediated lymphocyte depletion improved the disease in *mdx*, however, not all studies had produced definitive results, and the implication of lymphocytes and their subtypes in the disease is still to be clearly defined. Among the possible targets, genetic and pharmacological evidences suggest that protein kinase C (PKC) isotypes have essential functions in promoting both early T-cell activation and sustained T-cell adhesion and are proposed as drug targets in adaptive immunity [15]. Among the isotypes, PKCθ is unique in its ability to translocate to the immunological synapses upon T-cell receptor (TCR) activation, regulating NFκB, AP-1, and NFAT transcriptional activity [16]. PKCθ can also directly associate to chromatin, regulating T-cell-specific inducible gene expression program and microRNAs [17]. It is worth noting that PKCθ has been recently

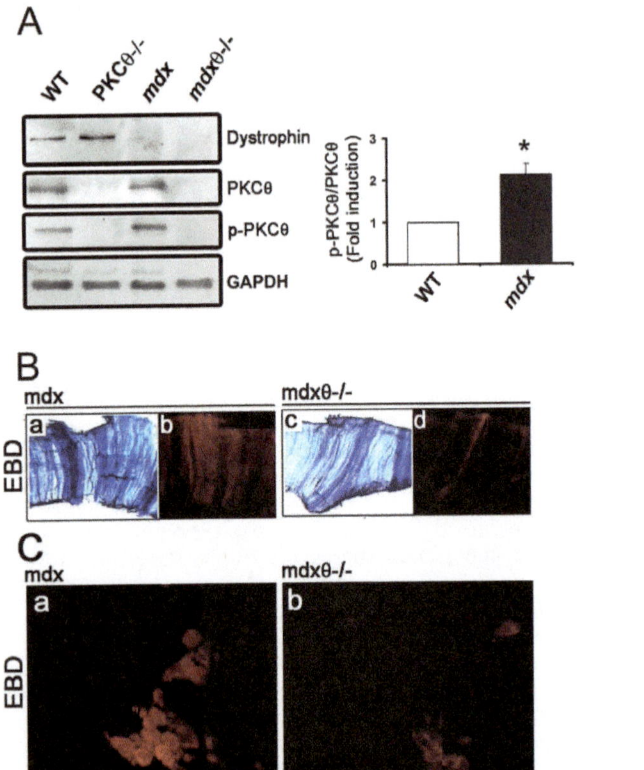

Figure 1. Lack of PKCθ in *mdx* mice reduces muscle degeneration. (**A**) Representative western blot analysis of total protein fraction of TA muscles derived from 2 mo old WT, PKC$\theta^{-/-}$, *mdx* and *mdx/$\theta^{-/-}$* mice, as indicated. The blot was incubated with the indicated primary antibodies. GAPDH expression level is shown in the bottom for equal loading. PKCθ activation in muscle derived from *mdx* (black bar) mice, expressed as fold induction in respect to WT (white bar, assumed as 1), is shown as the ratio of pPKCθ/PKCθ (right panel), as determined by densitometric analysis from three independent experiments (*$p<0.05$). (**B**) EBD uptake in diaphragm derived from 2 mo old *mdx* or *mdx/$\theta^{-/-}$* mice, as indicated, shown under light (**a, c**) and epifluorescence (**b, d**) microscopy. (**C**) EBD uptake in TA muscle derived from 2 mo old *mdx* (**a**) or *mdx/$\theta^{-/-}$* (**b**) mice, as indicated; immunofluorescence analysis of IgG accumulation in *mdx* (**c**) or *mdx/$\theta^{-/-}$* (**d**) mice; bar = 200 μm. (**D**) Mononuclear cells accumulation, revealed as Hoechst staining, around single degenerating fiber, detected as EBD uptake (**a–b**) and IgG immunofluorescence (**c–d**), in TA muscles from *mdx* (**a** and **c**) or *mdx/$\theta^{-/-}$* (**b** and **d**). (**E**) Representative western Blot analysis of IgG accumulation in TA muscles from *mdx* or *mdx/$\theta^{-/-}$* (two mice/genotype), as indicated. Densitometric analysis is shown in the bottom (*mdx*, black bar; *mdx/$\theta^{-/-}$*, grey bar, *$p<0.05$*).

shown to be required for full T_{eff} activation while inhibiting T_{reg}-mediated suppression [18]. Indeed, PKCθ is being proposed as a particularly attractive target for developing ways to selectively manipulate T_{eff} cell functions that are relevant to pathogenesis of different diseases, including asthma, rheumatoid arthritis, multiple sclerosis and colitis [15,19–21], without imparting a severe immunosuppression. Interestingly, PKCθ is the PKC isoform predominantly expressed also in skeletal muscle, where it mediates various cellular responses [22–27]. Although most of these studies demonstrated that PKCθ is actually required for complete histogenesis, differentiation and homeostasis of skeletal muscle, we asked which would be its prevalent role in a context of chronic inflammation, where immune cells activity is a key determinant as in muscular dystrophy. We here demonstrate that immune-cell intrinsic PKCθ activity plays a crucial role in the progression of muscular dystrophy, and targeting PKCθ can be proposed as a valuable therapeutic strategy for the disease.

Results

Lack of PKCθ in mdx mice reduces muscle degeneration and inflammation

To verify whether inhibition of PKCθ may improve healing in muscular dystrophy, we crossed the PKCθ knock out model (PKC$\theta^{-/-}$) with *mdx*, the mouse model of DMD. We first observed, by Western blot analysis, that PKCθ was highly expressed in both WT and *mdx* hindlimb muscle, but a significantly higher portion of it was phosphorylated, as a feature of its activation [16], in the *mdx*, as compared to WT (**Fig. 1A**). As expected, no dystrophin expression was detectable in the *mdx* and *mdx*/$\theta^{-/-}$ muscle, and no PKCθ immunoreactivity in the *mdx*/$\theta^{-/-}$ muscle (**Fig. 1A**). Both mutants were healthy at birth. We thus evaluated myofiber degeneration in diaphragm (as being diaphragm one of the most affected muscle group in *mdx*) and in tibialis anterior (TA) of 2 mo old mice, as Evan's blue dye (EBD) uptake. As shown in **Figure 1B and C** (**a–b**), myofiber degeneration was significantly reduced in both muscles derived from *mdx*/$\theta^{-/-}$ in respect to *mdx*. As expected, those degenerating areas were sites of a robust inflammatory response, as shown by anti-mouse IgG immunofluorescence (**Fig. 1C, c–d**). Indeed, while degenerating fibers in *mdx* muscle were surrounded by many mononucleated cells, degenerating fibers in *mdx*/$\theta^{-/-}$were almost avoided of surrounding mononucleated cells, and mouse IgGs were strictly localized on the fiber itself (**Fig. 1D**). Accordingly, Western Blot analysis revealed that the IgG content in the protein extract from *mdx*/$\theta^{-/-}$ muscle was greatly lower than that from *mdx* (**Fig. 1E**). Hematoxilin/Eosin (H/E) staining of TA muscle sections, showed that lack of PKCθ resulted in significant reduction in cell infiltrate as compared to *mdx* (**Fig. 2A**), leading to an overall maintenance of muscle structure; however, the myofibers variability and the percentage of centro-nucleated myofibers over the total number of fibers (as features of dystrophic muscle) were similar between the two genotypes (**Fig. 2B–C**). As being macrophage infiltration the most prominent immune feature observed in *mdx* [6], TA cryosections were analysed for esterase activity. As shown in **Figure 2D**, the areas of macrophages

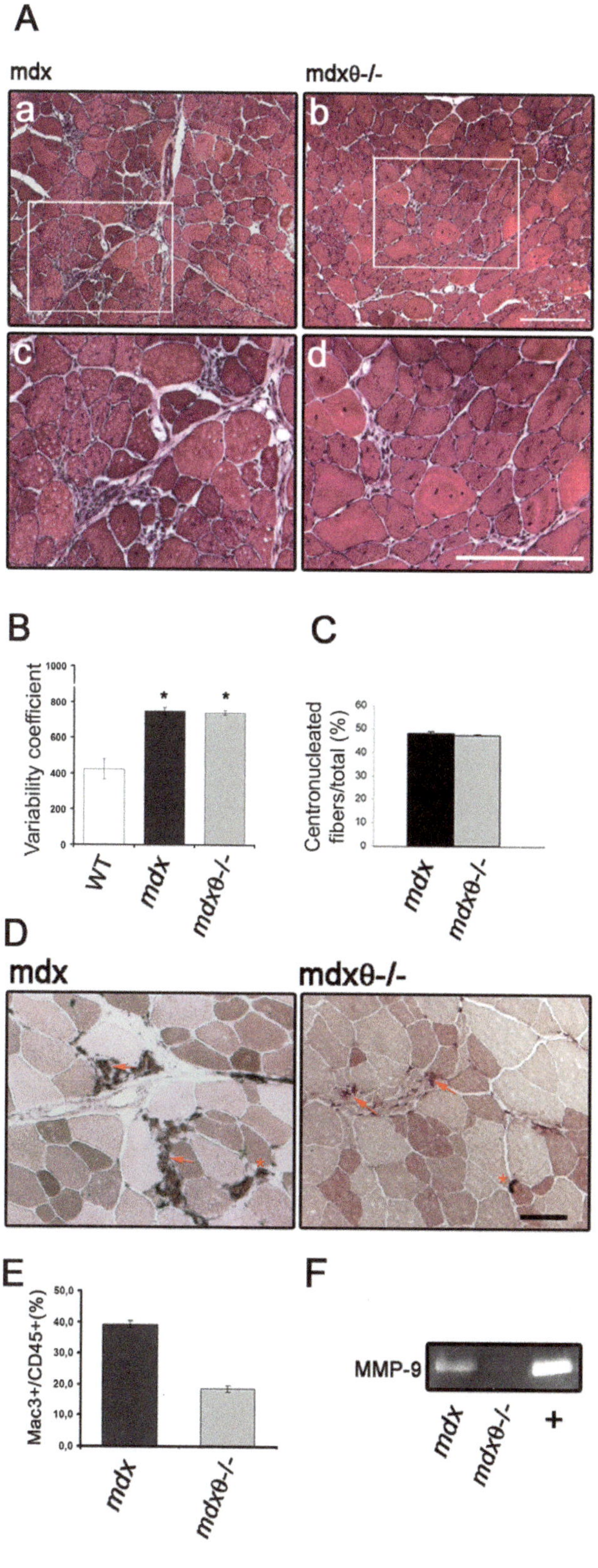

Figure 2. Lack of PKCθ in *mdx* mice reduces cell infiltrate in muscle. (**A**) Hematoxylin/Eosin staining of TA crysosections derived from 2 mo old *mdx* (**a, c**) and *mdx/θ*$^{-/-}$ (**b, d**). The insets in **a** and **b** indicate the areas shown in **c** and **d**, respectively, at higher magnification; bar = 100 μm. (**B**) Myofiber variability coefficient in TA muscles derived from 2 mo old WT, *mdx* and *mdx/θ*$^{-/-}$, determined as described in the material and methods sections. (n = 3/genotype). (**C**) Percentage of centrally nucleated myofibers in TA muscles derived from 2 mo old *mdx* and *mdx/θ*$^{-/-}$, expressed as percentage over the total number of myofibers (n = 3/genotype). (**D**) Esterase histochemical staining of TA cryosections derived from 2 mo old *mdx* and *mdx/θ*$^{-/-}$ mice, as indicated. Arrows indicate cell infiltrates, arrows indicate neuromuscular junctions. Bar = 200 μm (**E**) FACS analysis of CD45^{+ve}/Mac3^{+ve} mononucleated cells isolated from TA muscle derived from 2 mo old *mdx* and *mdx/θ*$^{-/-}$ mice, as indicated, expressed as percentage of the total number of cells examined. The percentage of reduction in *mdx/θ*$^{-/-}$muscle, in respect to *mdx*, is also shown (n = 3/genotype). (**F**) Gel zymography of MMP9 activity in TA muscle derived from 2 mo old *mdx* and *mdx/θ*$^{-/-}$ mice, as indicated; media collected from differentiating muscle cell cultures was used as positive control (+).

infiltrate in 2 mo old *mdx*/θ$^{-/-}$ muscle were strongly reduced; accordingly, FACS analysis of CD45/Mac-3 co-expressing cells, revealed that the level of macrophages was reduced by ≈50% in 2 mo old *mdx*/θ$^{-/-}$ muscle, as compared to *mdx* (**Fig. 2E**). Macrophages are known to store and produce matrix metalloproteinase 9 (MMP-9) in response to different stimuli, such as oxidative stress from necrotic tissue, and represent the major source of MMP-9 [28]. Indeed, the high level of MMP9 activity observed in *mdx*, was strongly prevented in *mdx*/θ$^{-/-}$ muscle, as shown by zymography (**Fig. 2F**). Moreover, as shown in **Figure 3,** lack of PKCθ strongly prevented the hyper-expression/activation of pro-inflammatory signalling pathways; in fact, both the level of expression and of phosphorylation of the p65 subunit of NFkB and of JNK in *mdx*/θ$^{-/-}$ TA muscle were much lower than in *mdx*, very similar to the level observed in WT. The p65 subunit of NFkB is the NFkB subunit mostly hyperactive in *mdx* muscle [11] and JNK is the upstream regulator of AP1 signalling pathway, which is also hyperactive in *mdx* [1].

Lack of PKCθ in mdx mice improves muscle regeneration

To verify whether the observed reduction in muscle wasting and cell infiltrate came along with improvement in muscle repair, muscle regeneration was analysed in the mutant mice by both morphological and biochemical analyses. Immunofluorescence analysis of embryonic myosin (eMyHC) expression, a marker of regenerating myofibers, revealed that lack of PKCθ resulted in an increase on eMyHC expressing myofibers along with a reduction in EBD positive myofibers (**Fig. 4A**). Indeed, quantitative analysis revealed that the extension of "regenerating area", including eMyHC expressing fibers, was significantly higher in *mdx*/θ$^{-/-}$, as compared to *mdx* muscle, parallel to a significant decrease in "degenerating area", including EBD stained fibers (**Fig. 4A**). Accordingly, the level of myogenin expression, a marker of differentiating myoblasts, was strongly increased in *mdx*/θ$^{-/-}$, as compared to *mdx* muscle (**Fig. 4B**). Intriguingly, we previously showed that PKCθ is actually required for myofiber growth both *in vivo* and *in vitro*, as being an upstream regulator of the expression of pro-fusion genes [27]. To unravel the apparent contradiction with our current results, *in vitro* differentiation of primary myoblasts derived from *mdx* and from *mdx*/θ$^{-/-}$ hindlimb muscle was compared. As shown in **Figure 4C**, by 48 hours in DM, *mdx* myoblasts had formed elongated myotubes containing a large number of nuclei; by contrast, *mdx*/θ$^{-/-}$ myoblasts formed thinner myotubes with reduced number of myonuclei, according to our previous observation in PKCθ$^{-/-}$ myoblasts. Taken together, these results suggest that lack of PKCθ in *mdx* makes a more favourable environment for muscle precursor cells to differentiate, rather than enhancing their activation/differentiation ability. Indeed, though the regenerating area *in vivo* was more extensive in *mdx* mice lacking PKCθ, as compared to *mdx*, cross-

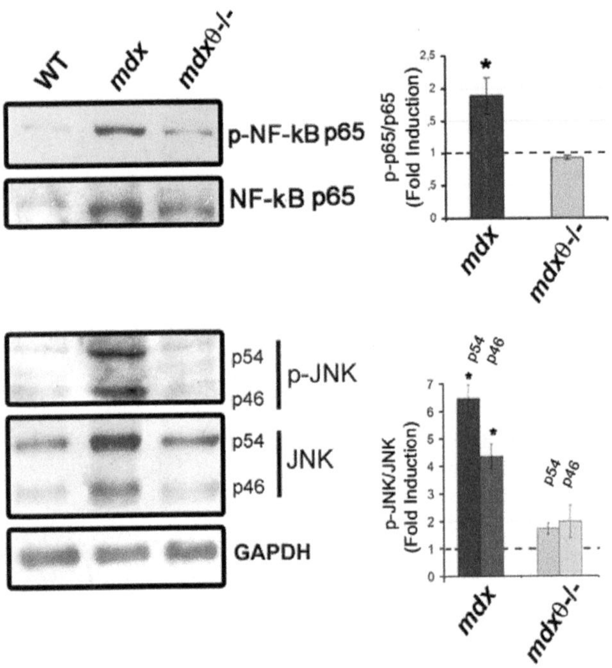

Figure 3. Lack of PKCθ in *mdx* mice prevents up-regulation of pro-inflammatory pathways. Left panel: representative Western Blot analysis of total protein fraction of TA muscles derived from 2 mo old WT, *mdx* and *mdx*/$\theta^{-/-}$ mice, as indicated. The blot was incubated with the anti- p-NF-kB p65, NF-kB p65, - p-JNK, - JNK antibodies, as indicated. GAPDH level of expression was used for normalization. The p-p65NFkB/p65NFkB (top) and of p-JNK/JNK (both p46 and p54) (bottom) ratio, as determined by densitometric analysis from three independent experiments, is shown in the right, expressed as fold induction in respect to WT (assumed as 1, dotted line). **$p<0.01$* in respect to WT.

sectional area of individual eMyHC, regenerating myofibers was medially reduced by ≈15%, as a result of PKCθ.ablation.

Lack of PKCθ in mdx mice preserves exercise performance

To verify whether the improvement in muscle maintenance and regeneration resulted in better performance, a treadmill endurance test was performed and the number of times the mice stopped during the 30 min running, each day of the test, was recorded. As expected, *mdx* mice stopped increasing times during individual running (not shown) and, medially, many more times than WT mice, in each day examined (**Fig. 5**). Importantly, *mdx*/$\theta^{-/-}$ mice behaved very similar to WT, both during the running than in all days examined (**Fig. 5**).

Rescue of PKCθ expression in hematopoietic cells in mdx/$\theta^{-/-}$ mice, by bone marrow transplantation, partly restores mdx mice phenotype

To verify whether the obtained phenotype was actually due to alterations in inflammatory cells activity, 1 mo old *mdx* bone marrow (BM) was transplanted into irradiated, age and sex matching *mdx*/$\theta^{-/-}$ mice. In these mice, hereafter called *mdx*/$\theta^{-/-\mathrm{BM}mdx}$, PKCθ was thus expressed only in the BM-derived cells. As control, another group of age and sex matching *mdx*/$\theta^{-/-}$ mice was transplanted with BM derived from *mdx*/$\theta^{-/-}$, hereafter called *mdx*/$\theta^{-/-\mathrm{BM}mdx/\theta-/-}$. The mice were sacrificed 6 weeks after transplantation and BM repopulation was ensured by genomic PCR analysis, as the appearance of the PKCθ WT PCR product (**Fig. 6A**). Western blot analysis revealed that PKCθ expression was largely rescued in thymus derived from *mdx*/$\theta^{-/-\mathrm{BM}mdx}$ transplanted mice, while the expression of other PKC isoforms was unaltered (**Fig. 6B**), and immunofluorescence analysis revealed that PKCθ expressing cells were detectable in the spleen (**Fig. 6C**), demonstrating that engrafted hematopoietic cells repopulated also mature compartments. H/E staining of TA muscle crysections derived from *mdx*/$\theta^{-/-\mathrm{BM}mdx}$ revealed a significant increase of infiltrating cells (**Fig. 7A, e**), as compared to those derived from *mdx*/$\theta^{-/-\mathrm{BM}mdx\theta-/-}$ (**Fig. 7A, a**), which were mostly macrophages, as shown by esterase histochemistry (**Fig. 7A, f**). Moreover, EBD uptake revealed an increase in degenerating myofibers in the *mdx*/$\theta^{-/-\mathrm{BM}mdx}$ (**Fig. 7A, g**), as compared to *mdx*/$\theta^{-/-\mathrm{BM}mdx\theta-/-}$ (**Fig. 7A, c**), parallel to a reduction in eMyHC expressing fibers (**Fig. 7A, h**) to levels comparable to *mdx*, as shown by quantitative analyses (**Fig. 7B**). Accordingly, iNOS expression, as a marker of macrophage infiltration, increased, as well as both NFkB and JNK expression and activity (**Fig. 7C**). However, the observed worsen phenotype was not translated in worsened exercise performance, as both *mdx*/$\theta^{-/-\mathrm{BM}mdx}$ and *mdx*/$\theta^{-/-\mathrm{BM}mdx\theta-/-}$ behaved similar to WT mice in a treadmill endurance test (**Fig. 7D**).

Discussion

We show in this article that lack of PKCθ in *mdx* prevents muscle wasting while improving muscle structure, regeneration and performance. This phenotype is associated to, and, probably, dependent on, reduced inflammation, which could make a more favourable environment for muscle precursor cells to differentiate.

The *mdx* mouse strain is the most widely used animal model for DMD; although it presents a milder phenotype compared to DMD, the immune cell populations in their muscles resemble those seen in DMD patients [6]. As first, we found that PKCθ is highly phosphorylated/active in *mdx* muscle as compared to WT muscle, and its lack significantly reduced muscle necrosis evaluated as EBD positive fibers in both DIA and TA. Interestingly, not only the percentage of degenerating myofibers was reduced, but also were sites of IgG accumulation. As PKCθ deficient mice have been shown to mount a significantly reduced lung inflammation response to antigen challenge and exhibit reduced inflammation in rodent models of arthritis as well as in autoimmune disease, the possibility that the improved phenotype was dependent on its activity on promoting inflammation is thus reasonable. Indeed, we show that lack of PKCθ in *mdx* resulted in significantly reduced macrophages infiltration, as revealed by both IHC and FACS analysis. Macrophages are the primary immune cells evident in skeletal muscle of *mdx* mice together to T-cells and neutrophils, as being macrophage infiltration the most prominent immune feature observed [6]. The reduction in macrophages is probably responsible for the reduction in the MMP-9 accumulation observed. Matrix metalloproteinases (MMPs), a family of zinc-dependent endopeptidases, have been shown to play an important role in ECM degradation, inflammation, fibrosis, and activation of latent cytokines and cell adhesion molecules in different patho-physiological conditions, including muscular dystrophies [28]. In particular, the expression of MMP-9 is increased in dystrophic muscle and its inhibition, either genetic or pharmacological, considerably reduces inflammatory response, fibrosis, and enhances the myofiber regeneration in *mdx* mice [28]. Moreover, we show that lack of PKCθ prevented the hyper-activation of the pro-inflammatory pathways NFkB and AP1, known to be up-regulated in *mdx*. It is worth noting that chronic activation of NF-kB signalling is required for DMD pathology by acting both on

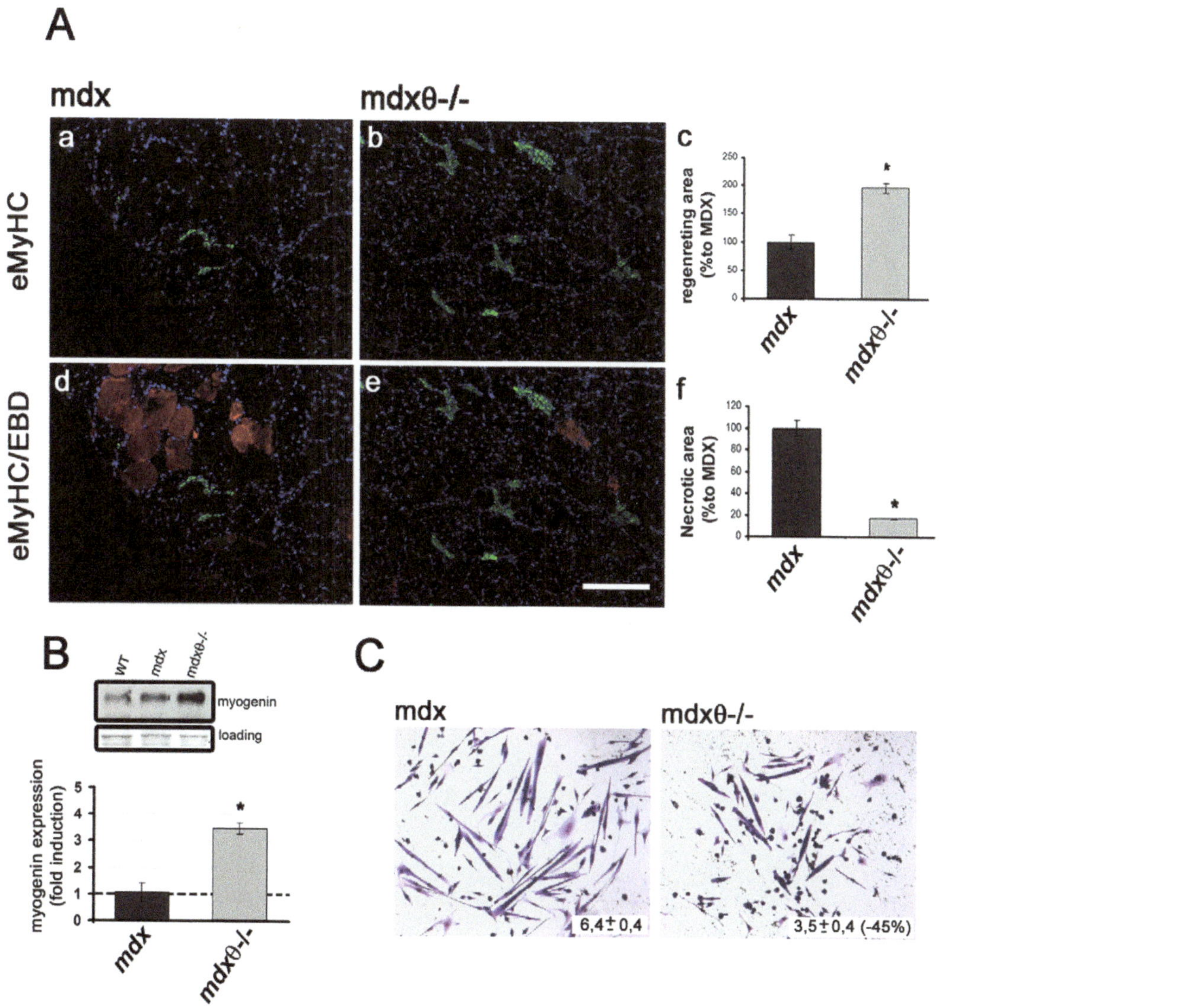

Figure 4. Lack of PKCθ in *mdx* mice improves muscle regeneration. (**A**) eMyHC immunofluorescence (green) in TA cryosections derived from *mdx* (**a** and **d**) and $mdx/\theta^{-/-}$ (**b** and **e**) mice, as indicated. Merge with EBD uptake (red) is shown in **d** (*mdx*) and **e** ($.mdx/\theta^{-/-}$). Bar = 200 μm. Extension of regenerating, eMyHC^{+ve}, area (**c**) and of necrotic, EBD^{+ve}, area (**f**) in $mdx/\theta^{-/-}$, expressed as the percentage in respect to the respective areas in *mdx* (assumed as 1); $*p<0.01$, n = 3/genotype. Bar = 200 μm. (**B**) Representative Western Blot analysis of total protein fraction of TA muscles derived from 2 mo old WT, *mdx* and $mdx/\theta^{-/-}$ mice, as indicated, incubated with the α-myogenin antibody; Red Ponceau staining of the membrane is shown for equal loading. Up-regulation of myogenin expression in $mdx/\theta^{-/-}$, in respect to *mdx* (assumed as 1), muscles, as determined by densitometric analysis of three independent experiments is shown in the bottom (n = 3/genotype) (**C**) Representative Wright staining of *mdx*- and $mdx/\theta^{-/-}$- muscle derived cells, as indicated, cultured in DM for 48 hrs. The mean number of nuclei contained within each myotube is shown, as well as the percentage of reduction in $mdx/\theta^{-/-}$ in respect to *mdx*, as determined from three independent experiments.

immune cells and damaged skeletal muscles to promote inflammation and to inhibit myogenic differentiation of muscle precursors [29]. Genetic and pharmacological studies demonstrated that inhibition of NFkB signalling reduced inflammation and necrosis enhancing regeneration and implicated the NFkB signalling pathway as a potential therapeutic target for this disease [11,30]. Also AP-1 activity is increased early in *mdx* muscle, and is supposed to be implicated in the activation of many inflammatory cytokines and chemokines [31]. Thus, the observed reduction in these pathways in *mdx* muscle lacking PKCθ may contribute to favour muscle healing. It is well known that the inflammatory response negatively contributes to the limited ability of dystrophic muscle to regenerate. Indeed, muscle regeneration is the initial response to muscle damage, but, in Duchenne muscular dystrophy, muscle progenitor cells activation/differentiation is limited [32,33]. This fact depends mainly to the exhaustion of muscle progenitor cells pool, due to the continuous cycles of degeneration/regeneration [32,33], but also chronic inflammation contributes to make an unfavourable environment for their activation/differentiation. As we show here, lack of PKCθ in *mdx* actually favoured muscle regeneration, as demonstrated by the increase in eMyHC expressing fibers and by myogenin up-regulation. In this context, sustained myogenin expression in $mdx/\theta^{-/-}$ muscle may theoretically contribute to myofiber metabolic changes, which, in turn, may be involved in muscle wasting prevention and physiology improvement. In fact, several observations suggested that myogenin may participate in at least a part of a fast-to-slow fiber-type transition [34,35]. Although further

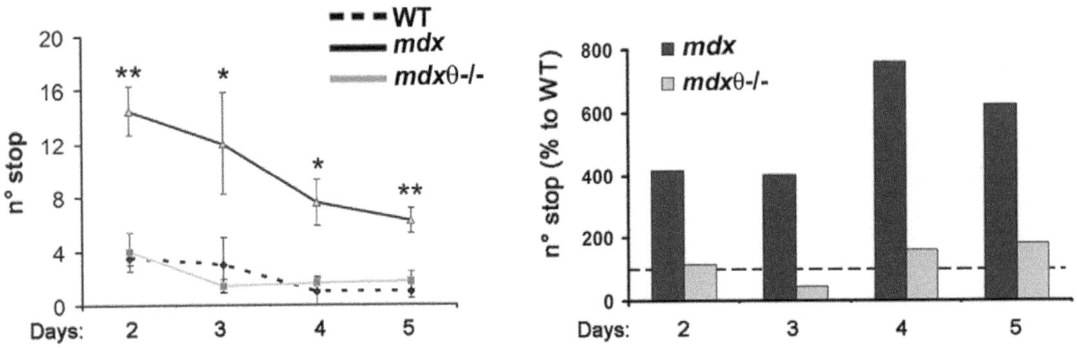

Figure 5. Lack of PKCθ in *mdx* mice preserves exercise performance. Treadmill exercise test performed on 2 mo old WT, *mdx* and *mdx*/$\theta^{-/-}$ mice (n = 6 each genotype) for a 30 min running, twice a week, for 3 weeks. The average number of stops recorded from mice of each genotype in each day, during the 30 min running, is shown in the left (WT, black line; *mdx*, dotted line; *mdx*/$\theta^{-/-}$, grey line). The extreme time points (1st and 6th days) performance were not included. Same results are expressed as percentage in respect to WT mice (assumed as 100%; *mdx*/$\theta^{-/-}$, grey bars; *mdx*, black bars) in the right, for each day performance.

analyses are required, no alterations in the level of metabolic enzymes were observed in *mdx*/$\theta^{-/-}$ muscle, as compared to *mdx* (Madaro et al. unpublished observation). The observed myogenin up-regulation should thus primarily reflect the increase in differentiating satellite cells, as a feature of regenerating muscle, rather than a sustained myogenin expression. These results may appear in contrast with our previous observation showing that PKC$\theta^{-/-}$ regenerating muscle displayed the characteristics of delayed regeneration, as compared to time-matching regenerating WT muscle. However, those effects were not due to an impairment in muscle precursors activation/differentiation, rather to an impairment of the late phases of growth/regeneration process, delaying further addition of myonuclei to newly formed myofibers, due to reduction in the expression of the so-called pro-fusion genes. Indeed, we show in this paper that primary cells derived from *mdx*/$\theta^{-/-}$ muscle formed thinner myotubes with reduced number of myonuclei *in vitro*, as we observed in PKC$\theta^{-/-}$ muscle cells, when compared to those derived from *mdx* muscle. It

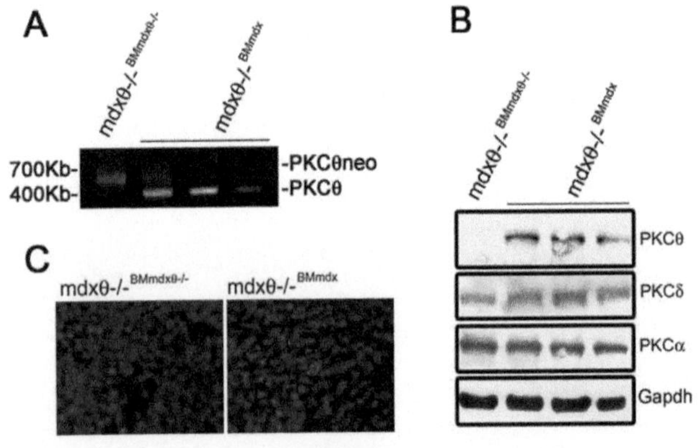

Figure 6. Rescue of PKCθ expression in hematopoietic cells in *mdx*/$\theta^{-/-}$ mice, by bone marrow transplantation. (**A**) PCR analysis of BM cell suspension genomic DNA derived from *mdx*/$\theta^{-/-BMmdx/\theta-/-}$ and from *mdx*/$\theta^{-/-BMmdx}$, as indicated. The reappearance of the 400 kb PCR product from the PKCθ gene in *mdx*/$\theta^{-/-BMmdx}$ is shown. (**B**) Western Blot analysis of PKCθ, PKCδ and PKCα expression in total protein extract from thymi derived from *mdx*/$\theta^{-/-BM\ mdx/\theta-/-}$ and from *mdx*/$\theta^{-/-BMmdx}$. (**C**) Immunofluorescence analysis of PKCθ-expressing cells (green) in the spleens derived from *mdx*/$\theta^{-/-BMmdx/\theta-/-}$ and from *mdx*/$\theta^{-/-BMmdx}$. Hoechst was used to counterstain the nuclei (blue).

is thus conceivable that in *mdx* chronic inflammation contributes significantly in preventing activation of muscle precursor cells, which however are able to efficiently differentiate, as they do *ex vivo*. By contrast, lack of PKCθ, though delaying the fusion process itself, makes a more favourable environment for muscle precursors cells to differentiate by reducing inflammation, thus allowing muscle repair. Indeed, *mdx*/$\theta^{-/-}$ regenerating myofiber CSA was reduced, as compared to *mdx*. As a result, muscle tissue structure is preserved, though the myofiber CSA variability and the percentage of centrally-nucleated myofibers were similar to those in *mdx*. Importantly, lack of PKCθ preserved exercise performance in *mdx*, suggesting that the improved phenotype correlates to improved functionality of the muscle. Taken together, these results demonstrate that lack of PKCθ in *mdx* mice dramatically reduces cell infiltration and inflammation, improving muscle regeneration and performance, suggesting a hitherto unrecognized crucial role of PKCθ in promoting immune response in muscular dystrophy. Consistent with this hypothesis, PKCθ expressing BM derived cells partly counteracted improvement in inflammation, muscle wasting and regeneration in *mdx*/$\theta^{-/-}$ mice, as shown by BM transplantation experiments. Intriguingly, exercise performance was not impaired, instead. This fact may depend on the time required for muscle to decline its functionality: it is conceivable that 6 weeks upon BM transplantation were sufficient to partly restore immune response, which however was still less than in age-matching *mdx*, but not to worsen muscle performance. Although the specific cell population(s) involved is not clear yet, an attractive hypothesis would be that lack of PKCθ prevents T_{eff} cells activity, while maintaining T_{reg} cells activity, which in turn may modulate macrophages phenotype and activity. However, as PKCθ lacks in all cells in the *mdx*/$\theta^{-/-}$ model, the possible contribution of other tissue components, in particular skeletal muscle, in the observed phenotype cannot be ruled out. In any case, the results obtained demonstrate that PKCθ expression/activity in immune cells is required for the robust inflammatory response in *mdx*, which, in turn, exacerbates the muscle pathology. Further studies are needed to verify long term efficacy, as well as the effect of targeting PKCθ in older animals, when pathological features are already established. Moreover, as being muscular dystrophy a multi-factorial disease, in which inflammation plays a crucial role, the possibility to combine PKCθ targeting as anti-inflammatory therapy, to other gene- or cell- based therapeutic approaches to significantly improve and optimize the therapeutic efficacy, should be explored [5]. The availability, and the ongoing development, of

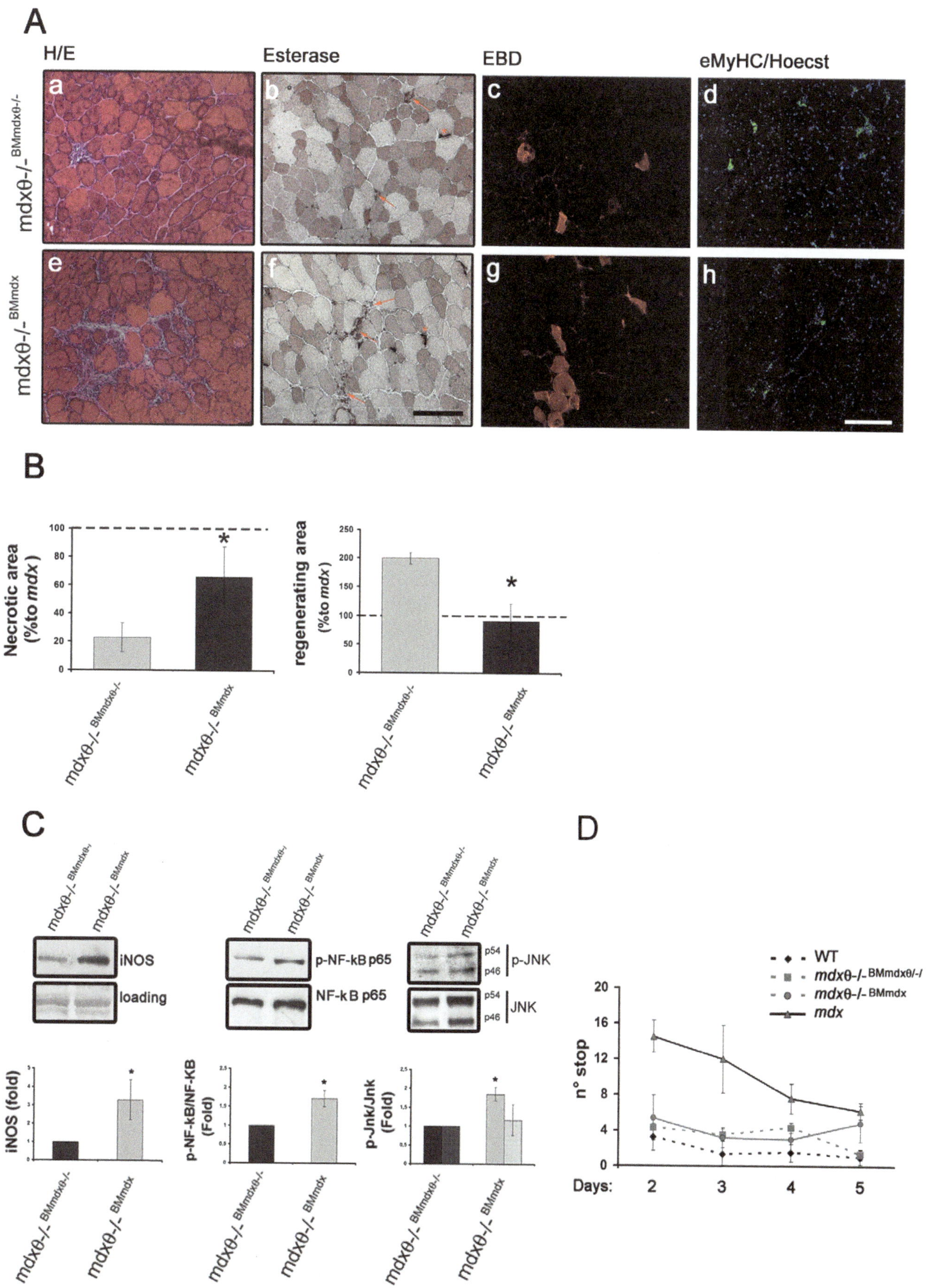
A
H/E
Esterase
EBD
eMyHC/Hoecst
mdxθ-/- BMmdxθ-/-
mdxθ-/- BMmdx
B
Necrotic area (%to mdx)
regenerating area (%to mdx)
C
iNOS
loading
p-NF-kB p65
NF-kB p65
p-JNK
JNK
p54
p46
iNOS (fold)
p-NF-kB/NF-KB (Fold)
p-Jnk/Jnk (Fold)
D
WT
mdxθ-/-BMmdxθ/-/
mdxθ-/-BMmdx
mdx
n° stop
Days:

Figure 7. Rescue of PKCθ expression in hematopoietic cells in *mdx/θ*$^{-/-}$ mice restores *mdx* mice phenotype. (**A**) H/E (**a** and **e**) and esterase (**b** and **f**, arrows indicate cell infiltrates, asterisks indicate neuromuscular junctions) as well as EBD uptake (**c** and **g**) and eMyHC immunofluorescence (**d** and **h**, Hoechst was used to counterstain the nuclei) of TA cryosections derived from $mdx/\theta^{-/-BMmdx/\theta-/-}$ (**a–d**) and from $mdx/\theta^{-/-BMmdx}$ (**e–h**) mice. Bar = 200 μm. (**B**) Extension of regenerating, eMyHC^{+ve}, area and of necrotic, EBD^{+ve}, area in $mdx/\theta^{-/-BMmdx/\theta-/-}$ and in $mdx/\theta^{-/-BMmdx}$ expressed as the percentage in respect to the respective areas in *mdx* (assumed as 1), *$p<0.05$, n = 3/genotype; (**C**) Western Blot analysis of iNOS expression, and of NFkB and JNK expression and activation (phosphorylation) in protein extracts from TA muscle derived from $mdx/\theta^{-/-BMmdx/\theta-/-}$ and from $mdx/\theta^{-/-BMmdx}$; red Ponceau staining of the membrane is shown (loading) for equal loading. Representative blots are shown. The level of expression or of the activation of the above molecules was determined by densitometric analysis and expressed as fold induction in $mdx/\theta^{-/-BMmdx}$ in respect to $mdx/\theta^{-/-BMmdx/\theta-/-}$, assumed as 1, evaluated from 3 separate experiments (bottom). *$p<0.05$ (**D**) Treadmill exercise test performed on $mdx/\theta^{-/-BMmdx/\theta-/-}$ (squares, dotted grey line) and $mdx/\theta^{-/-BMmdx}$ (circles, grey.line) mice for a 30 min running, twice a week, for 3 weeks, as above, starting 6 weeks after transplantation (n = 3/genotype). Performance of *mdx* (triangles, black line) and $mdx/\theta^{-/-}$ (rhombi, dotted black line) mice is also reported for comparison.

specific PKCθ inhibitors, some of which are already in clinical trials for immune disorders, thus opens new perspectives for pharmacological approach to muscular dystrophy [15,19].

Methods

Mice

PKCθ$^{-/-}$ mice were gently provided by Prof Dan Littman [36]. *Mdx* mice were purchased from Charles River. The mice were cross-bread to generate *Mdx/θ*$^{-/-}$ double mutant mice. C57BL10 WT mice (Charles River) were used as control. The animals were housed at the Histology Unit accredited animal facility, in individual cages in an environmentally controlled room (23°C, 12-h light-dark cycle) and provided food and water *ad libitum*. All the procedures were approved by the Italian Ministry for Health and conducted according to the US National Institutes of Health guidelines.

Antibodies

The anti-PKC-θ and the anti-phospho^{Thr538} PKC-θ, the anti-p65NFkB, the anti-phospho p65NFkB, the anti-JNK and the anti-phospho JNK rabbit polyclonal antibodies were purchased from Cell Signalling Inc., Danvers, MA, USA; the anti-iNOS mouse monoclonal antibody was from BD Bioscience, CA, USA, while the anti-dystrophin was from Leica Microsystems, Germany; the anti-myogenin F5D and the anti-embryonic myosin heavy chain F1.652 mouse monoclonal antibodies were from Developmental Studies Hybridoma Bank, Iowa City, Iowa, USA.

Primary muscle cell cultures

Primary cultures were prepared from total limb muscles of *mdx* or *mdx/θ*$^{-/-}$· mice, as previously described [27]. Muscle derived cells were grown on collagen-coated dishes, in growth medium, GM (Dulbecco's modified Eagle's medium, D-MEM containing 20% HS, 3% EE, all from Gibco Invitrogen, Carlsbad, CA, USA) in a humidified 5% CO_2 atmosphere at 37°C. Differentiation was induced by replacing the medium with medium containing lower serum and EE concentration, DM (D-MEM containing 5% HS, 0,75% EE). The cells were fixed after 48 hrs in DM and stained with Wright's solution. The mean number of nuclei contained within each myotube was determined as previously described [27]. Approximately 100 myotubes were counted per dish.

Histological and immunohistochemical analyses

Muscle cryosections were fixed in 4% paraformaldehyde (Sigma-Aldrich, MO, USA) on ice. Myofiber necrosis was evaluated on muscle cryosections prepared from mice intraperitoneally injected with a 1% Evan's Blue Dye solution (EBD, Sigma-Aldrich, MO, USA), at 1% volume relative to body mass, between 16 and 24 h prior to tissue sampling. The samples were analyzed under an epifluorescence Zeiss Axioskop 2 Plus microscope. For histological analysis, muscle cryosections were stained with Hematoxylin/Eosin solution (Sigma-Aldrich, MO, USA). The muscle fiber mean cross sectional area (CSA) was determined by measuring CSA of fibres in the entire section, using Scion Image 4.0.3.2 software (NIH, Bethesda, MA, USA). Myofiber variability was determined by multiplying the standard deviation of all measurements by 1,000 and dividing it by the mean fiber diameter [37]. Esterase localization and activity was evaluated on cryosections by esterase staining (α-naphthyl butyrate/hexazotized pararosaniline).

Immunofluorescence analysis on crysections was performed as previously described [27]. Nuclei were counterstained with Hoechst 33342 (Fluka, WI, USA) and the samples were analyzed under an epifluorescence Zeiss Axioskop 2 Plus microscope.

Gelatin zymography

Muscle extracts were prepared in non-reducing lysis buffer [50 mM Tris-Cl (pH 8.0), 200 mM NaCl, 50 mM NaF, 0.3% IGPAL CA-630 and protease inhibitors], as previously described [28]. Equal amount of proteins was separated on 8% SDS–PAGE containing 1 mg/ml gelatin B (Fisher Scientific) under non-reducing conditions. Gels were washed in 2.5% Triton X-100 for 1 h at room temperature followed by incubation in reaction buffer [50 mM Tris–HCl (pH 8.0), 50 mM NaCl, 5 mM CaCl2 and 0.02% sodium azide] for 48 h at 37°C. To visualize gelatinolytic bands, gels were stained with Coomassie Brilliant Blue dye, followed by extensive washing in destaining buffer (40% methanol/10% acetic acid).

Western blot analysis

Tissue samples were homogenized in ice-cold buffer containing 20 mM Tris (pH 7.5), 2 mM EDTA, 2 mM EGTA, 250 mM sucrose, 5 mM DTT, 200 mg/ml leupeptin, 10 mg/ml Trasylol, 1 mM PMSF, and 0.1% Triton X-100, as previously described [27]. An equal amount of protein from each sample was loaded onto 10% SDS-polyacrylamide gels and transferred to a nitrocellulose membrane (Schleicher and Schuell, Dassel, Germany). The membranes were incubated with the appropriate primary antibodies, and processed as described [27]. Densitometric analysis was performed using the Aida 2.1 Image® software.

Flow Cytometry Analysis. Mononuclear cell population was isolated from dissected muscles by enzymatic digestion. 10^6 cells were incubated on ice with 1 μg of the anti- Mac-3 FITC-labelled and of the anti-CD45 PE-labelled antibodies (BD Biosciences, CA, USA) and analysed with a FacsStar Plus cytofluorimeter. Non-specific fluorescent emission was determined with the specific labelled isotypes.

Treadmill exercise performance

Mice were first acclimated to the treadmill (LE 8710, PanLab S.L.U., Barcelona, Spain) before running by placing them on an unmoving treadmill for 10 min. The treadmill was set at a speed of 15 cm/sec. The test ran for 30 min, and the number of times at which mice failed to keep running, was recorded. Each test was performed twice a week for three weeks, and the results were averaged for each mouse.

Bone Marrow transplantation

1 mo old $mdx/\theta^{-/-}$ mice were X-irradiated (8 Gy per mouse) by a Siemens linear accelerator operating at 10 MV, at a rate of 3 Gy/min. At 2 h postirradiation, the mice were intravenously injected with bone marrow cell suspension derived from age-matching *mdx* or $mdx/\theta^{-/-}$ mice. Mice were sacrificed 6 weeks after transplantation, and bone marrow, thymus and spleen were evaluated for exogenous cells accumulation. Different muscles were dissected for morphological and molecular analyses.

PCR genotyping

Genomic DNA PCR, for $PKC\theta^{-/-}$ mutation, was conducted as previously described [36]. A 700 bp PCR fragment is expected for the mutated PKCθ gene, while a 400 bp PCR fragment for the WT one.

Statistical analysis

Quantitative data are presented as means ± SEM or ± SD (as specified) of at least three independent experiments. Statistical analysis to determine significance was performed using paired Student's t tests. Differences were considered to be statistically significant at the $p<0.05$ level.

Aknowledgments

We thank Prof. DA. Littman, NYU, NY, USA, for providing the PKCθ mutant mice; we also thank Prof. Antonio Musarò, for advice in functional studies, Prof. Mario Molinaro, for continuous discussion and critical reading of the manuscript, and Dr. Fabrizio Padula, for FACS analysis, all from DAHFMO, Histology Unit.

Author Contributions

Conceived and designed the experiments: LM SS MB. Performed the experiments: LM AP GB CN AC VM. Analyzed the data: LM MB. Contributed reagents/materials/analysis tools: MB SS. Wrote the paper: MB LM.

References

1. Evans NP, Misyak SA, Robertson JL, Bassaganya-Riera J, Grange RW (2009) Dysregulated intracellular signaling and inflammatory gene expression during initial disease onset in Duchenne muscular dystrophy. Am J Phys Med Rehabil 88: 502–522.
2. Goyenvalle A, Babbs A, Powell D, Kole R, Fletcher S, et al. (2010) Prevention of dystrophic pathology in severely affected dystrophin/utrophin-deficient mice by morpholino-oligomer-mediated exon-skipping. Mol Ther 18: 198–205.
3. Tedesco FS, Dellavalle A, az-Manera J, Messina G, Cossu G (2010) Repairing skeletal muscle: regenerative potential of skeletal muscle stem cells. J Clin Invest 120: 11–19.
4. Odom GL, Gregorevic P, Allen JM, Chamberlain JS (2011) Gene therapy of mdx mice with large truncated dystrophins generated by recombination using rAAV6. Mol Ther 19: 36–45.
5. Mendell JR, Campbell K, Rodino-Klapac L, Sahenk Z, Shilling C, et al. (2010) Dystrophin immunity in Duchenne's muscular dystrophy. N Engl J Med 363: 1429–1437.
6. Evans NP, Misyak SA, Robertson JL, Bassaganya-Riera J, Grange RW (2009) Immune-mediated mechanisms potentially regulate the disease time-course of duchenne muscular dystrophy and provide targets for therapeutic intervention. PM R 1: 755–768.
7. Balaban B, Matthews DJ, Clayton GH, Carry T (2005) Corticosteroid treatment and functional improvement in Duchenne muscular dystrophy: long-term effect. Am J Phys Med Rehabil 84: 843–850.
8. Keeling RM, Golumbek PT, Streif EM, Connolly AM (2007) Weekly oral prednisolone improves survival and strength in male mdx mice. Muscle Nerve 35: 43–48.
9. Hodgetts S, Radley H, Davies M, Grounds MD (2006) Reduced necrosis of dystrophic muscle by depletion of host neutrophils, or blocking TNFalpha function with Etanercept in mdx mice. Neuromuscul Disord 16: 591–602.
10. Messina S, Bitto A, Aguennouz M, Minutoli L, Monici MC, et al. (2006) Nuclear factor kappa-B blockade reduces skeletal muscle degeneration and enhances muscle function in Mdx mice. Exp Neurol 198: 234–241.
11. Acharyya S, Villalta SA, Bakkar N, Bupha-Intr T, Janssen PM, et al. (2007) Interplay of IKK/NF-kappaB signaling in macrophages and myofibers promotes muscle degeneration in Duchenne muscular dystrophy. J Clin Invest 117: 889–901.
12. Vidal B, Serrano AL, Tjwa M, Suelves M, Ardite E, et al. (2008) Fibrinogen drives dystrophic muscle fibrosis via a TGFbeta/alternative macrophage activation pathway. Genes Dev 22: 1747–1752.
13. Messina S, Bitto A, Aguennouz M, Mazzeo A, Migliorato A, et al. (2009) Flavocoxid counteracts muscle necrosis and improves functional properties in mdx mice: a comparison study with methylprednisolone. Exp Neurol 220: 349–358.
14. Eghtesad S, Jhunjhunwala S, Little SR, Clemens PR (2011) Rapamycin ameliorates dystrophic phenotype in mdx mouse skeletal muscle. Mol Med.
15. Baier G, Wagner J (2009) PKC inhibitors: potential in T cell-dependent immune diseases. Curr Opin Cell Biol 21: 262–267.
16. Baier G (2003) The PKC gene module: molecular biosystematics to resolve its T cell functions. Immunol Rev 192: 64–79.
17. Sutcliffe EL, Bunting KL, He YQ, Li J, Phetsouphanh C, et al. (2011) Chromatin-Associated Protein Kinase C-theta Regulates an Inducible Gene Expression Program and MicroRNAs in Human T Lymphocytes. Mol Cell 41: 704–719.
18. Zanin-Zhorov A, Ding Y, Kumari S, Attur M, Hippen KL, et al. (2010) Protein kinase C-theta mediates negative feedback on regulatory T cell function. Science 328: 372–376.
19. Boschelli DH (2009) Small molecule inhibitors of PKCTheta as potential antiinflammatory therapeutics. Curr Top Med Chem 9: 640–654.
20. Cooles FA, Isaacs JD (2011) Pathophysiology of rheumatoid arthritis. Curr Opin Rheumatol.
21. Kwon MJ, Wang R, Ma J, Sun Z (2010) PKC-theta is a drug target for prevention of T cell-mediated autoimmunity and allograft rejection. Endocr Metab Immune Disord Drug Targets 10: 367–372.
22. Serra C, Federici M, Buongiorno A, Senni MI, Morelli S, et al. (2003) Transgenic mice with dominant negative PKC-theta in skeletal muscle: a new model of insulin resistance and obesity. J Cell Physiol 196: 89–97.
23. D'Andrea M, Pisaniello A, Serra C, Senni MI, Castaldi L, et al. (2006) Protein kinase C theta co-operates with calcineurin in the activation of slow muscle genes in cultured myogenic cells. J Cell Physiol 207: 379–388.
24. Gao Z, Wang Z, Zhang X, Butler AA, Zuberi A, et al. (2007) Inactivation of PKCtheta leads to increased susceptibility to obesity and dietary insulin resistance in mice. Am J Physiol Endocrinol Metab 292: E84–E91.
25. Tokugawa S, Sakuma K, Fujiwara H, Hirata M, Oda R, et al. (2009) The expression pattern of PKCtheta in satellite cells of normal and regenerating muscle in the rat. Neuropathology 29: 211–218.
26. Messina G, Biressi S, Monteverde S, Magli A, Cassano M, et al. (2010) Nfix regulates fetal-specific transcription in developing skeletal muscle. Cell 140: 554–566.
27. Madaro L, Marrocco V, Fiore P, Aulino P, Smeriglio P, et al. (2011) PKC{theta} signaling is required for myoblast fusion by regulating the expression of caveolin-3 and {beta}1D integrin upstream focal adhesion kinase. Mol Biol Cell 22: 1409–1419.
28. Kumar A, Bhatnagar S, Kumar A (2010) Matrix metalloproteinase inhibitor batimastat alleviates pathology and improves skeletal muscle function in dystrophin-deficient mdx mice. Am J Pathol 177: 248–260.
29. Kumar A, Boriek AM (2003) Mechanical stress activates the nuclear factor-kappaB pathway in skeletal muscle fibers: a possible role in Duchenne muscular dystrophy. FASEB J 17: 386–396.
30. Tang Y, Reay DP, Salay MN, Mi MY, Clemens PR, et al. (2010) Inhibition of the IKK/NF-kappaB pathway by AAV gene transfer improves muscle regeneration in older mdx mice. Gene Ther 17: 1476–1483.
31. Kumar A, Khandelwal N, Malya R, Reid MB, Boriek AM (2004) Loss of dystrophin causes aberrant mechanotransduction in skeletal muscle fibers. FASEB J 18: 102–113.
32. Sacco A, Mourkioti F, Tran R, Choi J, Llewellyn M, et al. (2010) Short telomeres and stem cell exhaustion model Duchenne muscular dystrophy in mdx/mTR mice. Cell 143: 1059–1071.
33. Consalvi S, Saccone V, Giordani L, Minetti G, Mozzetta C, et al. (2011) Histone deacetylase inhibitors in the treatment of muscular dystrophies: epigenetic drugs for genetic diseases. Mol Med.
34. Ekmark M, Gronevik E, Schjerling P, Gundersen K (2003) Myogenin induces higher oxidative capacity in pre-existing mouse muscle fibres after somatic DNA transfer. J Physiol 548: 259–269.
35. Siu PM, Donley DA, Bryner RW, Alway SE (2004) Myogenin and oxidative enzyme gene expression levels are elevated in rat soleus muscles after endurance training. J Appl Physiol 97: 277–285.

36. Sun Z, Arendt CW, Ellmeier W, Schaeffer EM, Sunshine MJ, et al. (2000) PKC-theta is required for TCR-induced NF-kappaB activation in mature but not immature T lymphocytes. Nature 404: 402–407.

37. Iwata Y, Katanosaka Y, Arai Y, Shigekawa M, Wakabayashi S (2009) Dominant-negative inhibition of Ca2+ influx via TRPV2 ameliorates muscular dystrophy in animal models. Hum Mol Genet 18: 824–834.

15

Deep Learning of Orthographic Representations in Baboons

Thomas Hannagan*, Johannes C. Ziegler, Stéphane Dufau, Joël Fagot, Jonathan Grainger

Laboratoire de Psychologie Cognitive, Aix-Marseille University, & CNRS (Centre National de la Recherche Scientifique), Marseille, France

Abstract

What is the origin of our ability to learn orthographic knowledge? We use deep convolutional networks to emulate the primate's ventral visual stream and explore the recent finding that baboons can be trained to discriminate English words from nonwords [1]. The networks were exposed to the exact same sequence of stimuli and reinforcement signals as the baboons in the experiment, and learned to map real visual inputs (pixels) of letter strings onto binary word/nonword responses. We show that the networks' highest levels of representations were indeed sensitive to letter combinations as postulated in our previous research. The model also captured the key empirical findings, such as generalization to novel words, along with some intriguing inter-individual differences. The present work shows the merits of deep learning networks that can simulate the whole processing chain all the way from the visual input to the response while allowing researchers to analyze the complex representations that emerge during the learning process.

Editor: Thomas Wennekers, The University of Plymouth, United Kingdom

Funding: This research was supported by a grant from the European Research Council (ERC grant 230313) to Jonathan Grainger. It was carried out within the Labex BLRI (ANR-11-LABX-0036) and has benefited from support from the French government, managed by the French National Agency for Research (ANR), Investments of the Future A*MIDEX (ANR-11-IDEX-0001-02). The funders had no role in study design, data collection and analysis, decision to publish, or preparation of the manuscript.

Competing Interests: The authors have declared that no competing interests exist.

* E-mail: thom.hannagan@gmail.com

Introduction

Baboons can learn to classify strings of letters as real English words or not [1]. Generalization performance to novel words [1] and the existence of transposed letter effects in baboons [2] clearly suggested that the monkeys had learned some kind of orthographic information. However, there is still considerable debate about the precise nature of the information that allows monkeys to achieve such remarkable performance [3,4]. Are they using purely visual information, unordered strings of letters, or higher-level abstract letter combinations? To shed light on these mechanisms, we used state-of-the-art neural networks of invariant object identification that were designed to simulate the primate's visual ventral stream. The networks were trained with an incremental supervised procedure to perform a lexical decision task on realistic, pixel-based strings of letters using the exact same stimuli and training regime as the baboons in the experiment. We first show that the networks can account for several empirical effects observed in baboons as well as for intriguing inter-individual differences. We then establish that the networks developed ordered letter representations, therefore demonstrating the efficiency of such representations for a purely visually driven system that must learn to tell words and nonwords apart. This study allows us to move one step further towards the goal of separating the "visual" from the "linguistic" in human reading, while setting a new standard for more realistic computational models of visual word recognition and learning.

Because reading is acquired late in development compared to speech, visual words must map onto pre-existing linguistic knowledge [5]. Not surprisingly, theories of reading have assumed an important role of phonology and semantics in orthographic development [6]. The key finding of the Grainger et al. [1] experiment is that orthography, that is, information about letter identity and letter order, can be acquired in the total absence of prior linguistic knowledge in the primate (because neither sounds nor meanings were associated with the letter strings processed by the baboons). How baboons actually performed the task is not known. Current computational models of orthographic coding in human adults essentially involve either letter-based schemes, whereby letters are being assigned positions in the word in more or less flexible ways, or schemes based on letter combinations, whereby, for instance, the word "LIFE" would be represented as the set of bigrams LI, LF, LE, IF, IE, FE (see [7,8] for reviews). A consensus has not yet emerged within cognitive psychology as to which scheme captures the data in the most parsimonious way. However cognitive neuroscience has provided direct evidence [9,10] that letter combinations, so-called n-grams, are processed in the Visual Word Form Area - VWFA hereafter, a brain region that specifically responds to orthographic information in humans [11]. Given the solid and long standing evidence that primates represent visual objects by means of combinations of features [12,13], it appears plausible that baboons might use letter combinations or ngrams to represent written words.

Deep learning by convolutional networks

In contrast to our claim that baboons use letter combinations or ngrams to perform word/nonword classifications, several authors have now suggested that baboons in the Grainger et al. experiment used exclusively letter-based schemes. For instance, Frost and Keuleers [4] have argued, without a computational proof, that the task could have been performed simply based on the detection of unordered sets of letters in the stimuli. Others have implemented offline classification systems that can learn to classify the letter strings based on letter positions alone, without invoking any type of ngrams [3]. There is often a host of possible explanations to any given phenomenon, and orthographic coding in baboons is no exception. But none of the explanations described above, when they are actually implemented, brings us any closer to understanding how baboons actually solve the task. This is because they all assume either unlimited computational power, or perfect memory resources, or the availability of all training stimuli at once. In addition, these explanations do not attempt to capture the full range of behavior that has now been documented, but instead, focus on one aspect of the data at best. What is needed is a comprehensive computational account consistent with resource-limited brains, with what we know of the primate's visual system, and with systems being subjected to real visual stimuli in an incremental supervised training procedure. Deep convolutional networks possess many if not all of these properties, and are therefore natural candidates for our purpose.

Figure 1 illustrates our general approach. The top left panel (1A) describes the operant-conditioning experiment simulated in this article, in which baboons were shown strings of letters and were rewarded when they classified them accurately as words or nonwords. The top right panel (1B) shows the main pathway presumed to be involved in the baboon's brain when solving this task [14], constituted by a succession of processing stages within the ventral visual stream from V1 to TEO (posterior inferior temporal cortex) and from there to the orbito-frontal cortex. The lower panel (1C) illustrates the convolutional network architecture of the model.

Convolutional networks were initially inspired by the primate's visual system [15,16]. This is visible in the hierarchical organization of the network Figure 1C, in the restricted connectivity and local sampling of units that captures the increasingly large receptive fields observed from V1 to TEO [17], and in the alternation of convolutional and sampling layers, which mimic the simple and complex cells found in the primary visual cortex. These characteristics have met tremendous success in computer vision and machine learning, and it is fair to say that today convolutional networks display state-of-the-art performance [18], including on tasks such as handwritten character recognition [16], and scene labeling [19]. Closer to our work, the HMAX model, which has been presented in computational neuroscience as the "standard model" of the visual system [20], is very similar in design to convolutional networks.

The model

The model is a convolutional network with a total of 7 layers, a simplified version of the LeNet5 network [16] built using the Theano deep learning libraries [21]. The first layer receives input images (150×30 pixel units) and the output layer produces a binary decision: Word or Nonword. In between, the S1,C1,S2 and C2 layers implement the succession of convolutional (hereafter S, for simple) and sampling (hereafter C, for complex) stages discussed above.

A convolutional layer consists of a number of feature maps (here, 20 for S1 and 50 for S2). Each unit in each feature map is connected to a restricted patch of units lying directly underneath it in all the maps forming the previous layer (for S1, there is only one map in the preceding layer: the input). Although units in the same feature map are connected to different patches of the previous layer, they all share the same connection weight values, according to the principle that what is learned at one location must hold for all locations (i.e. the principle of "weight sharing").

Contrary to convolutional layers, sampling layers reduce the dimensionality of the preceding layer. In a sampling layer, each unit in each feature map represents a larger patch from the preceding layer (in our model, patches of dimensions 2×2 units were used), and together all patches form a partition of the sampled layer. The activation of a sampling unit is obtained in the model by a max pooling operator, which outputs the maximally activated unit within the patch considered. After two series of convolution and sampling, the C2 layer is eventually used as input to a standard multilayer perceptron, which has full connectivity from C2 to the hidden layer, and from the hidden layer to the output layer.

Because baboons in [1] were reinforced on each trial during training, an incremental supervised learning procedure was also needed for the networks and we used a stochastic gradient descent algorithm, with the error gradient being evaluated on each trial and a fixed learning rate used at all epochs (annealing of the learning rate may be introduced in future work but was not found necessary to reproduce the baboons learning curves). The error signal in the case of a binary word/nonword classification task like ours amounts to providing always the same reward in case of success (error of 0) and no reward in case of failure (error of 1), which was precisely the training by reinforcement that the baboons had received. In the experiment, new words were introduced one at a time and interleaved with old words and with nonwords. They were shown repeatedly until the baboon succeeded in categorizing a new word more than 80% of the time, at which point the exemplar became a known word and a new word was introduced. Likewise, each network was exposed to exactly the same sequence of word and nonword stimuli as presented to the corresponding baboon. For instance, the "DAN network" (simulating the baboon of the same name) was exposed to 56 689 strings, half being words and half nonwords, with these categories being balanced for positional letter frequencies but differenciated by positional bigram frequency distributions. It should be noted, however, that the six baboons had received plenty of visual experience before the experiment, whereas our networks started out as a blank slate, with random connection weights. We will return to this limit of our model in the Discussion. For this study, assessment of classification accuracy for both baboons and networks was done every 1000 trials during training.

Six networks were instantiated, one for each of the six baboons (DAN, ART, CAU, DOR, VIO, and ARI) trained in the original study [1]. All networks had two free parameters: learning rate and convolution filter size. We launched a brute force exploration of this parameter space on six possible learning rates (from 10^{-1} to 10^{-6}) and eight possible filter sizes (from 3 to 10 units). Recognition accuracy was collected every 1000 training trials, compared to the corresponding baboon performance, and the parameter values that minimized this distance for each monkey/model pair were selected. Therefore, the models were not fitted to display any generalization, transposition, or letter similarity effect, but only to match the global performance of the baboons during training.

Results

Figure 2A shows the evolution of performance across the first 40 000 trials, averaged over all fitted networks (black crosses) or baboons (gray filled circles). The main feature of a noisy overall increase in correct responses is captured, and a simple similarity index between model and data based on the absolute difference in

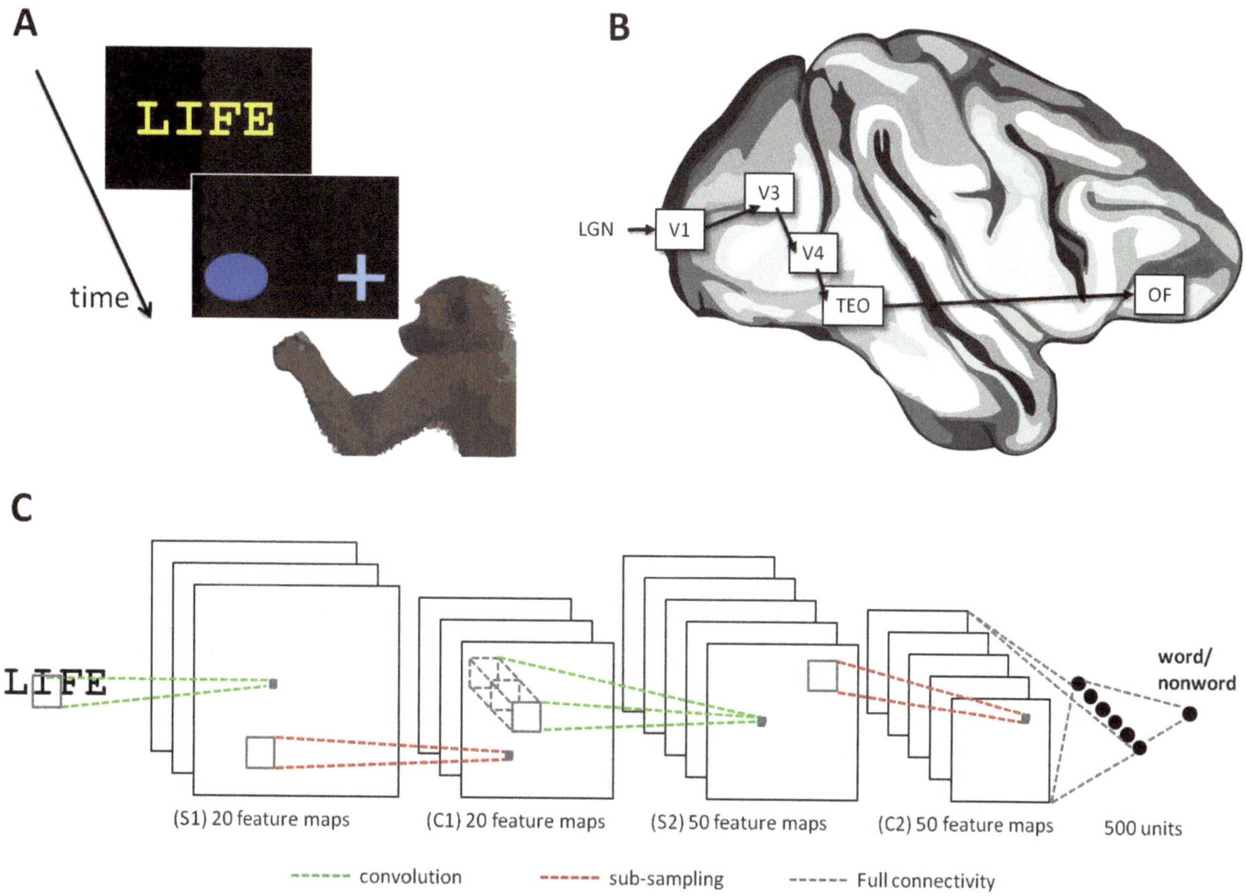

Figure 1. (A) Word-Nonword classification by operant conditioning in baboons [1]. Six baboons were trained to distinguish English words (such as "LIFE") from nonwords (such as "LEFE"), by touching either the cross or the oval shape presented immediately after the word or nonword. A correct response was followed by a blank screen and a food reward, whereas an incorrect response prompted a green screen to appear for 3s. (B) Main cortical network presumed to be involved in distinguishing between words and nonwords, formed by a succession of increasingly selective regions throughout the ventral stream (LGN-V1-V3-V4-TEO) towards the orbito-frontal cortex (OF). (C) Deep learning convolutional network implementing the local and hierarchical connectivity found in the ventral pathway, through a succession of alternating simple convolution and complex pooling levels (cortical regions not mapped to network levels).

performance every 1000 trials approaches the maximal value of 1. By the end of training (not shown in Figure 2A), performance levels were similar though slightly higher for networks (mean accuracy 80.9%) than for baboons (mean accuracy 73.6%).

Figure 2B reports on generalization effects in the networks, that is, the finding that baboons were more likely to classify a real word seen for the first time as a word than a nonword. It can be seen that networks slightly overestimate baboons' performance, but that both are in very close agreement, with much less nonword responses being made upon encountering for the first time a new word (mean baboons 44.8%, mean networks 39.5%) than the next nonword in the sequence (mean baboons 75.6%, mean networks 71.4%). A one-way analysis of variance (ANOVA) on nonword responses for new word stimuli versus nonword stimuli yielded significant differences both for baboons and the model (both ps<.0001). Although a strong similarity in accuracy during training was expected from (indeed the purpose of) our parameter fitting phase, we note that there was no a priori reason why the networks should have displayed quantitatively similar generalization behavior as the baboons, as networks were not fitted to do so.

We then tested whether the networks displayed letter transposition or letter similarity effects, using the exact same stimuli and conditions as in [2]. Letter transposition effects reflect the finding that transposed-letter (TL) nonwords, such as "caniso", are more likely to be misclassified as words than controls like "caviro" [22]. 80 nonwords were constructed from four-letter English words for each monkey: 20 were obtained by substituting two letters (condition DS, e.g. LGAE), 20 by transposing two letters (condition TL, e.g. LFIE), 20 by substituting a single letter by another that was visually dissimilar (condition Dis, e.g. LOFE), and 20 by substituting a single letter by a visually similar one (condition Sim, e.g. LJFE). Figure 2C shows the results of our TL simulations, along with the experimental data. The baboons showed a significant TL effect, making more word responses in the TL condition than in the DS but no Visual similarity effect ($F(1,5) = 5.2$, $p = 0.046$). The networks also made on average more word responses in the TL than in the DS condition, although this did not reach significance ($F(1,5) = 1.56$, $p = 0.24$). Figure 3D reports on the Visual similarity effect: the results show that an equal number of word responses in the Sim and Dis conditions both by baboons ($F(1,5) = 0.04$, $p = 0.84$) and by the networks ($F(1,5) = 0.09$, $p = 0.77$). Table 1 presents the individual parameters of the networks, their performance on accuracy, generalization, transposition effect, and visual similarity effects, along with the corresponding baboon performance.

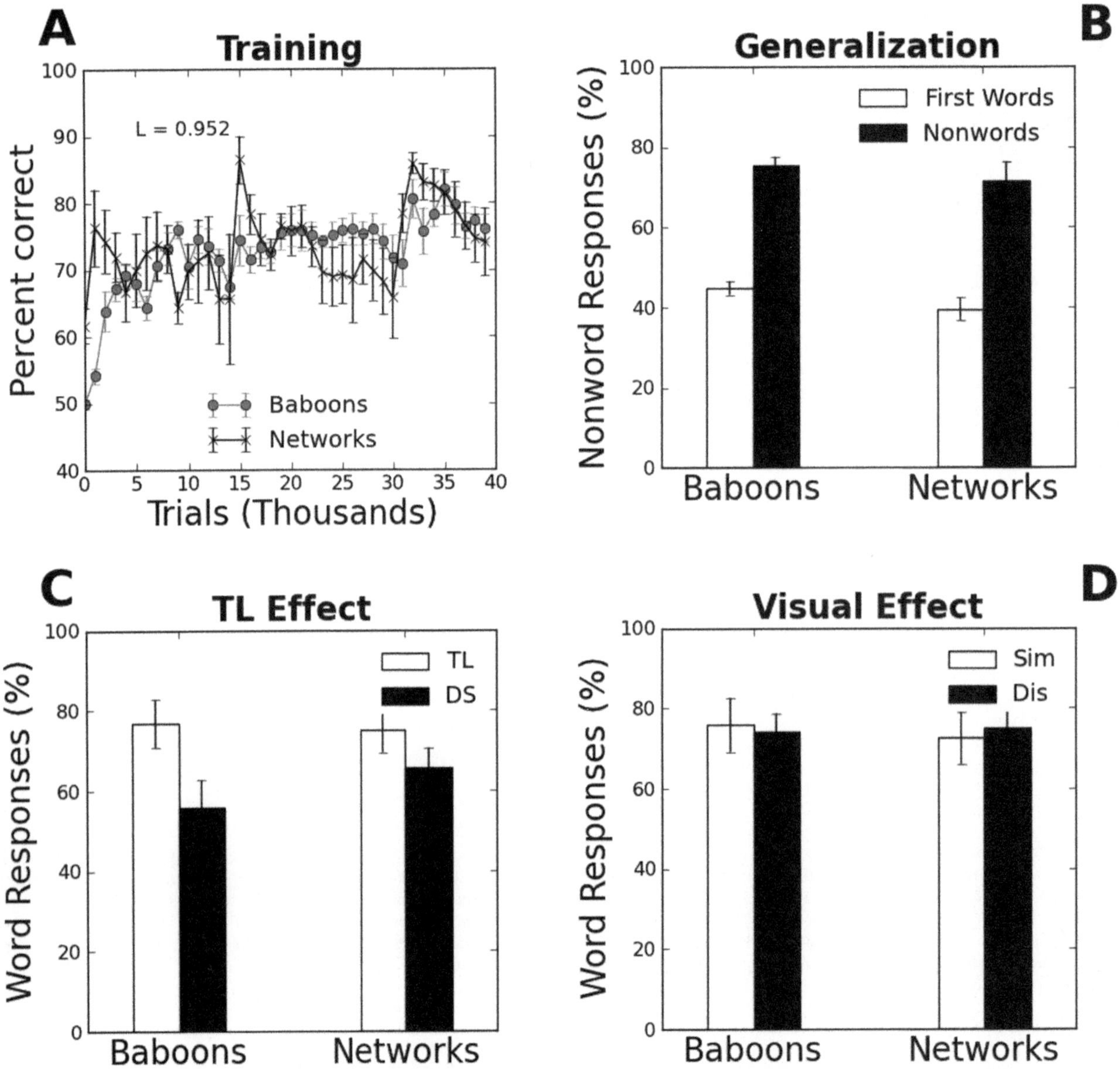

Figure 2. (A) Learning curves in the fitted model, showing the evolution in performance across traning trials in the networks (black crosses) and the baboons (gray dots). Only the first 40 000 trials are shown. (B) Generalization effect: percentage of nonword responses for baboons and networks upon presentation of a new word (white) during training, as opposed to the next nonword (black) in the sequence (experimental data from [1]). (C) Transposition effect: percentage of word responses produced in the networks and in the baboons upon presentation of nonword stimuli that differed from known words by two transposed letters (TL, white bars) or two letter substitutions (DS, black bars) (experimental data from [2]). (D) Visual similarity effect: word responses made on nonword stimuli that were visually similar (Sim, gray bars) or dissimilar (Dis, white bars) to known words (experimental data from [2]).

Analysis

The fact that deep learning convolutional networks can reproduce several of our key experimental findings, without being fitted to do so, forces us to ask how they achieved this feat. There are a number of properties over and above a deep architecture and supervised learning that might explain the success of these networks in capturing the results —namely the use of weight sharing, feature maps, and interleaved convolutional/pooling stages involving in particular the Max operator. Some of these assumptions are not beyond criticism from the point of view of neuroscience. For instance, doubts have been raised about the classification of cells into simple and complex types [23] —a classification which inspired the convolutional and pooling stages— and even assuming such a classification, there is no evidence for layers of simple and complex cells being interleaved throughout the ventral stream [24,25]. Other assumptions of convolutional networks appear to be better motivated. For example, Léveillé and Hannagan [26] recently provided a formal analysis of how connection weights suitable for a max operation could arise in networks trained under a Hebbian learning rule with a temporal window. Yet other assumptions, such as weight sharing, are intriguing and would deserve empirical investigation. But regardless of how well-established or plausible they are however, all of these network characteristics fall short of describing the exact learned mechanism or coding scheme whereby the

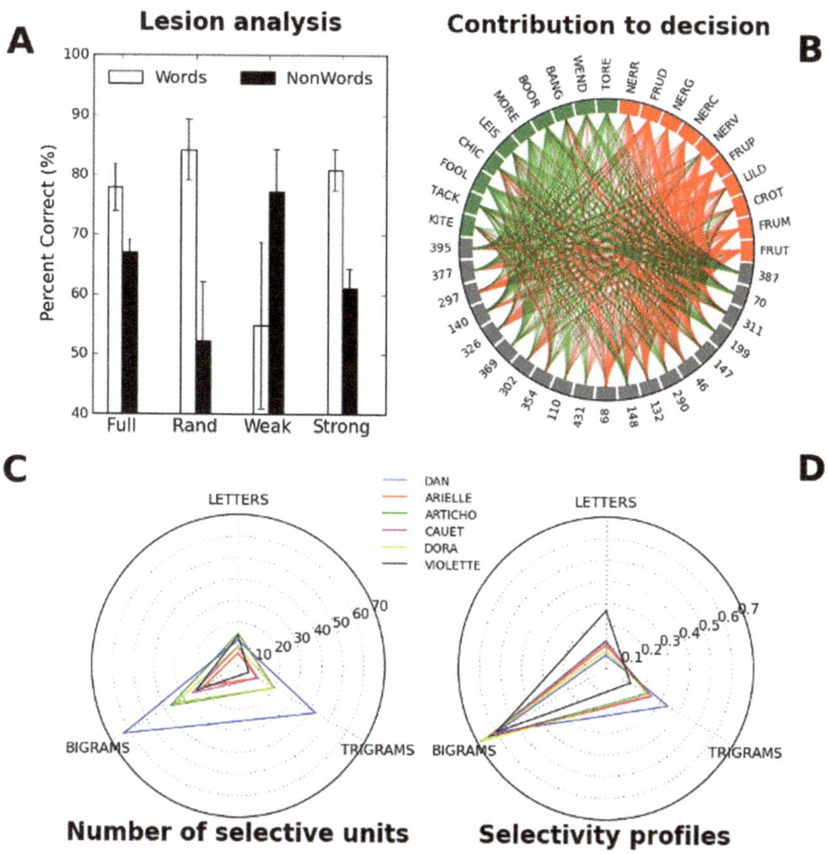

Figure 3. (A) Performance in the normal (Full) and lesioned model, when the 20% intact connections are either chosen at random (Rand), or to be the weakest (Weak) or as the strongest (Strong) connections. (B) Contribution of extremal network units to the lexical decision. Word stimuli occupy the upper left half of the circle, nonwords the upper right half, and the "extremal" units from the last-but-one layer occupy the lower half. A green connection (resp. red) between a unit and a stimulus signifies that this unit contributes a word signal (resp. nonword) to the decision node upon presentation of this stimulus —only 20 units out of 100 are shown. (C) Number of extremal units selective for at least one letter, bigram or trigram, according to a d-prime statistics. A unit was deemed selective for an entity (letter, bigram or trigram) if upon presentation of a word containing or not this entity, the d-prime defined as the difference of z-values for hit rate and false alarm rate reached an arbitrary common threshold. (D) Global proportion of letters, bigrams and trigrams coded for across all extremal units.

model succeeds in solving the task. Hence a dedicated analysis is called for.

In recent years, the representations formed by deep learning networks have been analyzed in a number of ways. Erhan, Courville, and Bengio [27] proposed, among other strategies, to circumscribe the function of any network unit by launching a systematic exploration of input space in search of the patterns that would maximize the unit's activation. Stoianov and Zorzi [28]

Table 1. Summary of networks parameters and performance.

	Train. Base		Network Parameters		Network Selectivity			Network Performance				Baboon Performance			
	Trials	*Voc.*	*CFilter*	*LRate*	*Let.*	*Big.*	*Tri.*	*Acc.*	*Gen.*	*TL*	*VS*	*Acc.*	*Gen.*	*TL*	*VS*
DAN	56 689	308	9	1E-4	12	63	43	91.5	57.09	5	5	79.81	33.7	35	−25
ARIELLE	55 407	87	3	1E-5	6	19	11	68.4	9.20	−10	−4.74	71.14	29.7	30	20
ARTICHO	61 142	112	3	1E-4	15	37	20	92	40.98	20	20	73.41	26.4	−5	−15
CAUET	49 608	121	6	1E-5	9	25	11	70.4	15.18	30	−10	72.43	27.0	0	20
DORA	43 041	81	3	1E-4	9	34	20	82.1	35.54	10	−10	73.15	29.5	40	20
VIOLETTE	50 985	125	3	1E-4	14	23	6	80.8	33.33	0	15	71.55	38.0	25	−10

Trials: number of trials. Voc: number of words in the training base. Conv. Filter: convolutional filter size. Learn. rate: learning rate. Let, Big and Tri: number of units respectively selective to letters, bigrams and trigrams in the penultimate layer. Acc: accuracy. Gen: generalization effect. TL: transposed letter effect. VS: visual similarity effect.

were able to determine the selectivity of network units by regressing their activations against some continuous properties of the stimulus, in their case numerosity and cumulative surface area. In what follows we choose a different approach. We contend that a key step towards understanding the model can be taken by partly destroying it. In neuropsychology and computational modeling, lesion studies have provided important insights into cognition [29–31], and accordingly we start by asking how the model performs when deprived of 80% of its connections in the penultimate layer — that is, the connections coming from the 500 units that entirely and directly subserve lexical decision.

Figure 3A shows the results of our simulated lesion study. We first tested each trained network's performance on all the words in its training base, and on an equal number of nonwords (condition "Full"). Intact networks reached performance levels that were quantitatively very close to observed accuracies for baboons, both for words and for nonwords (Network mean word accuracy 77.9%, Network mean nonword accuracy 67.0%; Baboons mean word accuracy 76.6%, Baboons mean nonword accuracy 70.0%). We then considered three lesioning conditions, "Rand", "Weak", and "Strong", which had in common the removal of 400 connections and only differed in the 100 being left intact. It can be seen that a lesioned model with all but its strongest connections lesioned (condition Strong) retained essentially the same performance on the training base as an intact model. If anything, performance was slightly improved for words and diminished for nonwords (mean word accuracy 80.8%, mean nonword accuracy 61.2%). When all but 100 randomly selected connections were lesioned (condition Rand), network performance on both words and nonwords was impaired (mean word accuracy 84.1%, mean nonword accuracy 52.2%). Finally when all but the 100 weakest connections were lesioned, performance on words approached chance level whereas performance on nonwords increased (condition Weak, mean word accuracy 54.9%, mean nonword accuracy 77.1%). This analysis suggests that the knowledge used by the model in order to tell whether a stimulus is a word or not is essentially embodied in the 100 strongest connections. Therefore, in order to characterize this knowledge, it is sufficient to restrict the analysis to these 100 "extremal" units that correspond to the strongest connections to the output node. The fact that networks restricted to extremal units can still operate very well is less surprising than it may seem at first blush, being in line with at least one deep learning study, which has found that adequate selection of features in the penultimate stage could actually improve network performance on a task of generic visual object classification [32].

We analyze the connectivity pattern of these extremal units in Figure 3B, where extremal units all lie in the lower half of the circle, word stimuli are in the upper left half, and nonword stimuli in the upper right half. The color of the connection between a given unit and stimulus shows the kind of signal sent by this unit to the decision output upon presentation of the stimulus: a green connection for a positive word signal, and a red connection for a negative word signal. The one fact that immediately stands out in Figure 3B is that word and nonword signals coming from this sample of extremal units are very well matched to word and nonword stimuli. The fact that extremal signals are well matched to lexical status explains why these units are so important in decision making.

Next we assessed the selectivity of extremal units. In this analysis, we exposed each network successively to all the words it had been exposed to and asked whether the observed activity patterns in extremal units showed any regularity with the presence of letters, bigrams, and trigrams in the stimuli at specific positions. We restricted our search to position-specific letters, bigrams, and trigrams, and we adopted two operational definitions for the activity and selectivity of a unit. First, we called a unit *active* if upon presentation of a stimulus, the signal it sent forward to the output layer had a magnitude three standard deviations above the mean across all units and stimuli. Under this count, we found that 15 units were active on average upon presentation of any stimulus, and that a unit was active for on average 22 stimuli. On the other hand, a unit was deemed *selective* for a given entity (letter, bigram, or trigram) if it was mostly active for stimuli that contained this entity and inactive for stimuli that did not contain this entity. We formalized this notion using the d-prime statistic, and granted selectivity to a unit if the difference in Z-scores between hit rate and false alarm rate for the considered entity was superior to an arbitrary and constant threshold (see Methods section for more details).

Figure 3C reports the number of units selective for any letters, bigrams, or trigrams, for each network. It can be seen that every simulated baboon had selective units, with some networks having a majority of their extremal units selective (as in the case of DAN). It is also apparent in this case, because the total sums up to more than 100, that the same same unit can be multiplex, in the sense that it can be selective for several different entities simultaneously. Notably enough, the number of bigram and trigram-selective units in a network did not correlate with the size of the training sets ($r_{bi} = 0.29$; $r_{tri} = 0.29$), but it strongly correlated with the number of words in them ($r_{bi} = 0.87$; $r_{tri} = 0.84$), with network accuracy ($r_{bi} = 0.78$; $r_{tri} = 0.68$), and with generalization performance ($r_{bi} = 0.88$; $r_{tri} = 0.79$). These relations extended to the experimentally observed values in baboons as for accuracy ($r_{bi} = 0.98$; $r_{tri} = 0.97$), but not for generalization ($r_{bi} = 0.12$; $r_{tri} = 0.03$). The number of letter-selective units correlated globally in the same way with these variables, only consistently weaker (number of words, $r_{let} = 0.3$; model accuracy, $r_{let} = 0.79$; baboon accuracy, $r_{let} = 0.27$; model generalization $r_{let} = 0.69$; baboon generalization $r_{let} = 0.30$). Finally Figure 3D provides an overview of selectivity for each baboon, by displaying the probability that a given selective unit will code for a letter, a bigram, or a trigram. Unlike the absolute number of selective units, it can be seen that the portfolio of entities selected for in each model is quite similar.

Individual differences: The case of Dan

In our experimental studies, one baboon, Dan, stood out as having acquired the largest vocabulary of all (308 words, more than twice the vocabulary of any other subject), reaching the best accuracy on words (80.0%) and nonwords (79.6%), and showing massive generalization and strong TL effects (33.7 and 35 percentage points, respectively). This is especially puzzling given that Dan was not the baboon who was trained the most (see Table 1). It is striking in Figure 3C and Table 1 that the Dan network is precisely the one that has developed the largest number of bigram and trigram-selective units (resp. 63 and 43 units out of the 100 considered). It appears that the uniqueness of this network can be traced back to the parameter search phase. Although the search yielded a perfectly unremarkable learning rate of 0.0001 for Dan, the size of the convolutional filter was by far the largest, covering 9 units in radius. The straightforward consequence of using a wide convolutional filter is that it allows for a much larger context to be integrated at every convolutional stage, thereby explaining the high number of bigram and trigram selective units that are ultimately available for the Dan network to take decisions upon. Indeed convolutional filter size in the model bore no relationship to the final number of letter-selective units ($r_{let} = 0.03$), but correlated strongly with the number of bigram ($r_{bi} = 0.76$) and trigram ($r_{tri} = 0.75$) selective units.

Discussion

This computational study allowed us to propose a general explanation for how baboons can learn to distinguish words from nonwords. Our results suggest that after tens of thousands of training trials the baboons have developed banks of position-specific letter combination detectors. Because words and nonwords in the training base were explicitly designed so as to maximize the difference in bigram frequencies at specific positions, such a strategy explains the baboons' good performance on the training base as well as their ability to generalize to new stimuli, and is well supported by the strong correlation between both of these abilities and the number of bigram selective units. Likewise, our modeling explains why no visual effect was obtained in baboons: if the final decisions were only based on highly selective letter combination detectors in the top layer of the network, then one would expect that the details pertaining to the specific shapes of the letters would have disappeared at this late stage of integration.

Let us now discuss our findings in the light of other recent computational studies of orthographic coding. Two independent analyses on different word recognition networks have now failed to found any support for the emergence of letter combinations, and have instead reported solid evidence for letter-based schemes [33,34]. Although both networks used artificial letter strings as inputs, the first had a shallow architecture [33], whereas the second used a deep architecture [34]. This suggests that in itself a deep learning design is not sufficient to trigger the emergence of such detectors, and the use of pixel-based visual inputs is unlikely to make a difference in this respect because of one very constraining and critical property shared by these networks: the use of full connectivity between layers. Indeed in both networks, every unit at a given layer was connected to every unit in adjacents layers, effectively forcing every unit to take all the information from the preceding layer into account. In contrast the sparse connectivity used in convolutional networks and in the primate ventral visual stream frees the network from this constraint, and allows each unit to only analyze a limited patch of information. This encourages the specialization of units on different fragments of the pattern from the previous layer, and these fragment detectors in turn get combined to form more complex fragment detectors at the next convolutional step. According to this view, the presence of two critical factors, sparse connectivity and deep architecture, is required in order to produce letter combination detectors.

It coud also be argued that a position-specific letter combination account of orthographic coding in baboons is not entirely compelling in that it does not a priori predict any TL effect, and indeed the networks only showed a non-significant trend towards one, at odds with the observed data. One way to make sense of this conflict is that we may be dealing with a very volatile effect, whose magnitude would depend on minute factors that are not necessarily captured in the model. According to the model, what ultimately determines the presence or absence of a TL effect is not whether the code is position specific or position invariant, but rather, how many more selective units will send a "word" signal to the output layer in the TL condition relative to the DS one. This would depend, first, on the differences in letter, bigram and trigram frequencies between TL and DS stimuli, and second on the exact selectivity profile of the network's units which is a result of the whole training history. It is possible that we made some implausible simplifications during the training phase, for instance by keeping the learning rate and convolution kernel parameters constant throughout training, or by specifying that the perceived reward was always constant. One obvious implausible simplification of this model is that stimuli were always presented in the same position and size on the simplistic simulated retina, whereas actual fixations made by baboons on the letter strings were bound to have varied in locations. Stochastic presentation went beyond the main point of our model, which was essentially to acknowledge the limited resources and hierarchical organization of the primate's visual system. In addition, and due to their built-in weight sharing mechanism for location invariance, deep convolutional networks may not be the best-suited framework for studying the effects of stochastic stimulus presentation. Regardless of this, the distribution of words on the retina may affect orthographic representations, and indeed the finding that orthographic representations in the VWFA are location sensitive has lately been a topic of much interest [35], which some of our computational work has aimed at explaining [36].

We have also trained our networks on a lexical decision task without the benefit of any previous visual expertise, whereas baboons were already experts at recognizing all sorts of visual objects by the time they started experimental training. Introducing previous visual training in this model would be very interesting for a number of reasons, not least because it would allow for the study of mirror invariance, a general characteristic of the expert visual system that has been theorized to be lost during the acquisition of reading [37]. Would the deep learning networks develop mirror-invariance for generic visual stimuli, and would they lose it upon exposure to strings of letters? Regardless of what these answers turn out to be, previous visual training would almost surely impact on the type of features used to perform lexical decision, but we would argue that it is unlikely to threaten our main conclusions here. Indeed, a developmental-like, unsupervised training stage has already been introduced in extant work with deep learning networks on visual categorization, and suggests that such training would result in a dictionary of feature combinations [20]. It is hard to see how the presence of feature combinations at the beginning of training on the lexical task could prevent the emergence of letter combination detectors later on, or could prevent these more developmentally realitistic networks to display any of the effects reported in this study.

We may finally speculate on how the penultimate layer, that was the focus of our analysis, relates to the baboon's brain or to the VWFA. The penultimate layer appears late in the processing stream and its units stand out as the only ones having access to all the information across all feature maps and all locations in the preceeding layer. For that reason, it is perhaps more plausibly related to a region lying between TEO and the orbital frontal cortex. Alternatively, because clear evidence for letter, bigram and quadrigram representations has been found in the VWFA [10], the penultimate layer could be a simplified, structureless homologue of the VWFA itself, whose detailed connectivity is not yet known. In this case, assuming that the same causes produce the same effects in a neural hardware shared across baboons and humans, the model's prediction would be that a region analogue to the VWFA should be observed in baboons that would only be activated for words and legal nonwords but not for false-font letter strings or for illegal, random strings of letters.

Conclusion

We have used deep learning networks to uncover the nature of the orthographic code which allows baboons to perform word-nonword classifications at a high level of accuracy. To do so, we have applied a range of techniques inspired from neuroscience, especially lesion studies and single cell recordings. The understanding of the model's inner workings gained hereby provides new insights into the nature of the orthographic code in humans. Under the dual pressure of visual processing constraints shared by all primates, and of an incremental training procedure, the model has converged towards a code that uses an assortment of position-specific letter combination detectors in order to categorize letter strings. This is in line with a large but still controversial class of theoretical models of visual word recognition proposed over the last decade [8,13,38–40], and weighs in favor of similar letter combinations being involved in the human faculty of reading.

Methods

No new data was collected for this study, this being a computational modeling article that aims at explaining data already published [1,2]. The previous experiments on baboons had the following characteristics:

- Name of the Institutional Animal Care and Use Committee (IACUC) or ethics committee that approved this study: "comité d'éthique de Marseille pour l'expérimentation animale".
- Details of animal welfare: Housing conditions; 650 m^2 outdoor enclosure with various climbing structures connected by tunnels to an indoor enclosure.
- Feeding regimens: Baboons were not food deprived, and received daily food ratio (monkey chow, fruits, vegetables) at 5 pm, water was provided ad libitum.
- Environmental enrichment: Climbing structures, live in social group. Use of the ALDM ("Automated Learning Device for Monkeys") is a form of behavioral enrichment.
- Steps taken to alleviate suffering: no suffering.
- Methods of sacrifice: none.

Summary of the experiments

Here, we provide a summary of the two experiments modeled in this paper, and we refer the interested reader to the original studies for more details about the material and procedure [1,2] Six guinea baboons (Papio papio; 3 females) living in semi-freedom participated in the original word learning experiment of [1]. Baboons were free to access and leave the automated experimental booths whenever they wanted. Self-initiated trials began with the presentation of a stimulus (word or nonword) until the baboon touched the screen, at which point the screen displayed a dark blue cross on the left (correct nonword response) and a light blue oval shape on the right (correct word response). Response was followed by a food reward (dry wheat) when a correct word or nonword response had been given, or followed by a 3 sec delay with a green screen if the response was incorrect. Words and nonwords were four letter strings designed so as to differ in bigram frequencies. Each baboon saw an equal number of words and nonwords, but because of the self-initiated nature of the trials, their total number differed across baboons (DAN: 56 689, ART: 50 985, CAU: 61 142, DOR: 49 608, VIO: 43 041, ARI: 55 407).

Immediately after training, baboons were tested for effects of visual similarity and transposition [2]. The test items were constructed individually for each monkey on the basis of the words that he or she had learned (e.g. LIFE), by substituting two letters (condition DS, e.g. LGAE), transposing two letters (condition TL, e.g. LFIE), by substituting a single letter with another that was visually dissimilar (condition Dis, e.g. LOFE), and by substituting a single letter with a visually similar one (condition Sim, e.g. LJFE). All of these items were reinforced as nonword trials.

Statistical analysis of the models

For each model, accuracy on word/nonword discrimination and generalization performance were calculated as described in [1], with the only difference being that for practical reasons in our paper these calculations were done every 1000 trials, both for networks and for baboons. Generalization scores were computed as the proportion of nonword responses made for first words minus the proportion of nonword responses made for nonwords. Transposed-letter effects were calculated as the number of word responses produced by TL nonwords minus the number of word responses produced by double substitution nonwords. Visual similarity effects were obtained as the number of word responses made on nonword stimuli that were visually similar to a word, minus the number of word responses made on nonwords that were visually dissimilar to known words. Following Ziegler et al. [2], the statistical analysis performed across models was a standard one-way analysis of variance.

Lesion analyses

Networks were lesioned after training was complete, by removing 400 connections out of the 500 arriving at the output node. The 100 connections left intact were either chosen to be the strongest (condition "strong"), the weakest (condition "weak"), or random (condition "random"). Strong and weak connections were defined by the absolute value of their weights: the weight of a strong connection had a large absolute value whereas the weight of a weak connection had a small absolute value (i.e. close to zero).

Sparseness, activity, and selectivity analyses

For each model, we first collected the activation patterns in all 100 extremal units in the last-but-one layer, upon presentation of all the words in the dataset, and of an equal number of nonwords. Because datasets differed across baboons but were exactly identical for a baboon and for its model, the number N of pairs of word/nonword presented varied across models ($N_{DAN}=308$; $N_{ART}=125$; $N_{CAU}=112$; $N_{DOR}=121$; $N_{VIO}=81$; $N_{ARI}=87$). We then computed the mean activation level and standard deviation of all activities across all extremal units and all patterns, and thresholded activation patterns to three standard deviations above the mean - an arbitrary threshold for what counts as an active unit in the model. The resulting Nx100 binary matrix of words-by-units activity served as a base for the computation of sparseness and selectivity in each model (see Figures 3B, 3C and 3D).

Selectivity of a unit to a particular entity (letter, bigram or trigram) was quantified by the d-prime statistic, a signal detection procedure based on the calculation of two quantities: the hit rate (HR), defined as the number of times the unit was active upon presentation of this specific entity divided by the total number of times the entity had been seen, and the false alarm rate (FAR), defined as the number of times the unit was active despite the absence of the entity in the stimulus, divided by this number plus the number of times the unit was not active in the absence of the stimulus (correct rejection). The d-prime was then simply obtained as the difference of z-scores for HR and for FAR, and a unit was deemed selective for an entity if its d-prime was above a common arbitrary value of 2.

All networks were built, trained, and tested using the Theano library [21] running on the Enthought Python Distribution.

Acknowledgments

The first author would like to thank Anne Warlaumont for stimulating discussions.

Author Contributions

Conceived and designed the experiments: TH JZ JG. Performed the experiments: TH. Analyzed the data: TH JZ JF JG. Contributed reagents/materials/analysis tools: TH. Wrote the paper: TH JZ JF SD JG.

References

1. Grainger J, Dufau S, Montant M, Ziegler JC, Fagot J (2012) Orthographic Processing in Baboons (Papio papio). Science 336, 245–248.
2. Ziegler JC, Hannagan T, Dufau S, Montant M, Fagot J, et al. (2013) Transposed Letter Effects Reveal Orthographic Processing in Baboons Psychological Science 24(8), 1609–1611.
3. Bains W (2012) Comment on "Orthographic Processing in Baboons (Papio papio)". Science 337, 1173. Available: www.sciencemag. org/cgi/content/full/337/6099/1173-b.
4. Frost R, Keuleers E (2013) What can we learn from monkeys about orthographic processing in humans? A reply to Ziegler et al. Psychological Science 24(9), 1868–1869.
5. Ziegler JC, Goswami U (2005) Reading acquisition, developmental dyslexia, and skilled reading across languages: a psycholinguistic grain size theory. Psychological Bulletin 131(1), 3–29.
6. Ziegler JC, Perry C, Zorzi M (in press) Modelling reading development through phonological decoding and self-teaching: Implications for dyslexia. Philosophical Transactions of the Royal Society B.
7. Grainger J (2008) Cracking the orthographic code: An introduction. Language and Cognitive Processes 23(1), 1–35.
8. Hannagan T, Grainger J (2012) Protein Analysis Meets Visual Word Recognition: A Case for String Kernels in the Brain. Cognitive Science 36, 575–606.
9. Binder JR, Medler DA, Westbury CF, Liebenthal E, Buchanan L (2006) Tuning of the human left fusiform gyrus to sublexical orthographic structure. Neuro-image 33, 739–748.
10. Vinckier F, Dehaene S, Jobert A, Dubus JP, Sigman M, et al. (2007) Hierarchical coding of letter strings in the ventral stream: dissecting the inner organization of the visual word-form system. Neuron 55(1):143–156.
11. Cohen L, Dehaene S, Naccache L, Lehéricy S, Dehaene-Lambertz G, et al. (2000) The visual word form area: Spatial and temporal characterization of an initial stage of reading in normal subjects and posterior split-brain patients. Brain 123:291–307.
12. Brincat SL, Connor CE (2006) Dynamic Shape Synthesis in Posterior Inferotemporal Cortex. Neuron 49(1), 17–24.
13. Dehaene S, Cohen L, Sigman M, Vinckier F (2005) The neural code for written words: a proposal. Trends Cogn Sci 9:335–341.
14. Blaizot X, Landeau B, Baron JC, Chavoix C (2000) Mapping the visual recognition memory network with PET in the behaving baboon. Journal of Cerebral Blood Flow and Metabolism 20, 213–219.
15. Fukushima K (1980) Neocognitron: A self-organizing neural network for a mechanism of pattern recognition unaffected by shift in position. Biological Cybernetics 36(4), 193–202.
16. LeCun Y, Bottou L, Bengio Y, Haffner P (1998) Gradient-based learning applied to document recognition. Proceedings of the IEEE 86(11), 2278–2324.
17. Van Essen DC, Gallant JL (1994) Neural mechanisms of form and motion processing in the primate visual system. Neuron 13:1–10.
18. Ciresan DC, Meier U, Schmidhuber J (2012) Multi-column Deep Neural Networks for Image Classification. IEEE Conf. on Computer Vision and Pattern Recognition.
19. Farabet C, Couprie C, Najman L, LeCun Y (2013) Learning hierarchical features for scene labeling. IEEE Transactions on Pattern Analysis and Machine Intelligence.
20. Serre T, Oliva A, Poggio T (2007) A Feedforward Architecture Accounts for Rapid Categorization. Proc Natl Acad Sci USA 104(15):6424–6429.
21. Bergstra J, Bastien F, Breuleux O, Lamblin P, Pascanu R, Delalleau O, Desjardins G, Warde-Farley D, Goodfellow I, Bergeron A, Bengio Y (2011) Theano: Deep learning on GPUs with python. In NIPS 2011, BigLearning Workshop, Granada, Spain.
22. Perea M, Lupker SJ (2004) Can CANISO activate CASINO? Transposed-letter similarity effects with nonadjacent letter positions. Journal of Memory and Language 51, 231–246.
23. Mechler F, Ringach DL (2002) On the classification of simple and complex cells. Vision research 42, 1017–1033.
24. Hubel DH, Wiesel TN (1968) Receptive fields and functional architecture of monkey striate cortex. Journal of Physiology-London 195, 215–243.
25. Skottun BC, De Valois RL, Grosof DH, Movshon JA, Albrecht DG, et al. (1991) Classifying simple and complex cells on the basis of response modulation. Vision Research 31, 1079–1086.
26. Léveillé J, Hannagan T (2012) Learning spatial invariance with the trace rule in non-uniform distributions. Neural Computation 25(5), 1261–1276.
27. Erhan D, Courville A, Bengio Y (2010) Understanding Representations Learned in Deep Architectures. Technical Report 1355, Université de Montréal.
28. Stoianov I, Zorzi M (2012) Emergence of a "visual number sense" in hierarchical generative models. Nature neuroscience 15(2):194–6.
29. Plaut DC, Shallice T (1993) Deep dyslexia: A case study of connectionist neuropsychology. Cognitive Neuropsychology 10, 377–500.
30. Cohen G, Johnston RA, Plunkett K (2000) Exploring cognition: Damaged brains and neural networks, readings in cognitive neuropsychology and connectionist modelling. Hove, UK: Psychology Press.
31. Thomas MSC, Purser HRM, Tomlinson S, Mareschal D (2011) Are imaging and lesioning convergent methods for assessing functional specialisation? Investigations using an artificial neural network. Brain and Cognition 78(1): 38–49.
32. Mutch, J. &Lowe, D.G. (2008) Object class recognition and localization using sparse features with limited receptive fields. International Journal of Computer Vision 80(1): 45–57.
33. Hannagan T, Dandurand F, Grainger J (2011). Broken symmetries in a location-invariant word recognition network. Neural Comput. 23: 251–283.
34. Di Bono MG, Zorzi M (2013) Deep generative learning of location-invariant visual word recognition. Front. Psychol. 4:635.
35. Rauschecker AM, Bowen RF, Parvizi J, Wandell BA (2012) Position sensitivity in the visual word form area. Proc Natl Acad Sci USA 109(24): 1568–1577.
36. Hannagan T, Grainger J (2013) The Lazy Visual Word Form Area: Computational Insights into Location-Sensitivity. PLoS Comput Biol 9(10): e1003250.
37. Pegado F, Nakamura K, Cohen L, Dehaene S (2011) Breaking the Symmetry: Mirror discrimination for single letters but not for pictures in the Visual Word Form Area. NeuroImage 55:742–9.
38. Mozer M (1987) Early parallel processing in reading: A connectionist approach. In M. Coltheart (Ed.) Attention and Perfomance XII: The Psychology of Reading. 83–104. Hillsdale, NJ: Lawrence Erlbaum.
39. Whitney CS, Berndt RS (1999) A new model of letter string encoding: Simulating right neglect dyslexia. Progress in Brain Research 121: 143–163.
40. Grainger J, van Heuven W (2003) Modeling letter position coding in printed word perception. In Bonin P. (Ed.), Mental lexicon: "Some words to talk about words" 1–23. New York: Nova Science.

Intranasal Mesenchymal Stem Cell Treatment for Neonatal Brain Damage: Long-Term Cognitive and Sensorimotor Improvement

Vanessa Donega[1], Cindy T. J. van Velthoven[1◔], Cora H. Nijboer[1◔], Frank van Bel[2], Martien J. H. Kas[3], Annemieke Kavelaars[4], Cobi J. Heijnen[1,4]*

1 Laboratory of Neuroimmunology and Developmental Origins of Disease, University Medical Centre Utrecht, Utrecht, The Netherlands, **2** Department of Neonatology, University Medical Center Utrecht, Utrecht, The Netherlands, **3** Rudolf Magnus Institute of Neuroscience, Department of Neuroscience and Pharmacology, University Medical Center Utrecht, Utrecht, The Netherlands, **4** Department of Symptom Research, MD Anderson Cancer Center, Houston, Texas, United States of America

Abstract

Mesenchymal stem cell (MSC) administration via the intranasal route could become an effective therapy to treat neonatal hypoxic-ischemic (HI) brain damage. We analyzed long-term effects of intranasal MSC treatment on lesion size, sensorimotor and cognitive behavior, and determined the therapeutic window and dose response relationships. Furthermore, the appearance of MSCs at the lesion site in relation to the therapeutic window was examined. Nine-day-old mice were subjected to unilateral carotid artery occlusion and hypoxia. MSCs were administered intranasally at 3, 10 or 17 days after hypoxia-ischemia (HI). Motor, cognitive and histological outcome was investigated. PKH-26 labeled cells were used to localize MSCs in the brain. We identified 0.5×10^6 MSCs as the minimal effective dose with a therapeutic window of at least 10 days but less than 17 days post-HI. A single dose was sufficient for a marked beneficial effect. MSCs reach the lesion site within 24 h when given 3 or 10 days after injury. However, no MSCs were detected in the lesion when administered 17 days following HI. We also show for the first time that intranasal MSC treatment after HI improves cognitive function. Improvement of sensorimotor function and histological outcome was maintained until at least 9 weeks post-HI. The capacity of MSCs to reach the lesion site within 24 h after intranasal administration at 10 days but not at 17 days post-HI indicates a therapeutic window of at least 10 days. Our data strongly indicate that intranasal MSC treatment may become a promising non-invasive therapeutic tool to effectively reduce neonatal encephalopathy.

Editor: Olivier Baud, Hôpital Robert Debré, France

Funding: This study was supported by EU-7 Neurobid (HEALTH-F2-2009-241778) from the European Union and Zon-MW Project (no 116002003). The funders had no role in study design, data collection and analysis, decision to publish, or preparation of the manuscript.

Competing Interests: The authors have declared that no competing interests exist.

* E-mail: CJHeijnen@mdanderson.org

◔ These authors contributed equally to this work.

Introduction

Neonatal encephalopathy due to perinatal hypoxia-ischemia (HI) remains a significant cause of neonatal mortality and long-term neurological deficits such as cerebral palsy, mental retardation and seizures in babies born at term [1–6]. Presently, the only available treatment, hypothermia, has limited beneficial effects and is only effective in mildly-affected children born at term [7,8]. Moreover, hypothermia has a narrow therapeutic window of 6 hours. Hence, there is an urgent need to develop therapeutic strategies with a longer therapeutic window.

One emerging strategy with therapeutic potential is mesenchymal stem cell (MSC) treatment. A growing number of studies in rodent models show that MSC treatment significantly improves motor outcome and reduces lesion volume after neonatal brain injury [9–18]. Currently, in most studies, MSCs are administered intracranially, which has serious disadvantages for clinical application. In a previous study, we explored the potential of intranasal MSC administration in a mouse model of neonatal HI brain damage. Our results showed that intranasal MSC treatment improved sensorimotor behaviour and decreased lesion volume 4 weeks after HI, suggesting a therapeutic potential [16].

Neonatal encephalopathy in humans is often associated with cognitive impairment [2,3,5]. Therefore, we investigated for the first time whether intranasal MSC treatment after neonatal HI brain damage restores cognitive function and sensorimotor function 8 weeks after HI. We also determined the dose response relationships and therapeutic window of intranasal treatment in the HI mouse model. Moreover, we explored the early presence of MSCs at the lesion site in relation to the therapeutic window.

Materials and Methods

Ethics statement

Experiments were performed according to the international guidelines and approved by the Experimental Animal Committee Utrecht (University Utrecht, Utrecht, Netherlands).

Animals

Unilateral HI cerebral damage was induced in 9-day old C57Bl/6 mice by permanent occlusion of the right common carotid artery under isoflurane anesthesia followed by 45 min 10% oxygen at 35°C [14]. Our HI induction procedure has a 10% death rate. Sham-controls underwent anesthesia and incision only. Pups from at least five litters were randomly assigned to experimental groups. Analyses were performed in a blinded set-up.

Mesenchymal stem cells from C57/bl6 mice were purchased from Invitrogen (GIBCO mouse C57Bl/6 MSCs, Life Technologies, UK) and cultured according to the manufacturer's instructions. Characterization of cell specific antigens was performed in a previous study from our group [16]. 3 µl of hyaluronidase in PBS (100 U, Sigma-Aldrich, St. Louis, MO) was administered to each nostril. Thirty minutes later animals received 3 µl MSCs or 3 µl PBS (vehicle) twice in each nostril.

Sensorimotor function

Unilateral sensorimotor deficits were evaluated in the cylinder rearing test (CRT). Weight-bearing left (impaired), right (unimpaired) or both paw(s) contacting the wall during full rear were counted. Paw preference was calculated as ((right−left)/(right+left+both))×100%.

Cognitive function

To assess cognitive function, we used the social discrimination test as described [19]. After 10 minutes of habituation to the test environment, the test mouse was allowed to explore a novel conspecific of the same gender (Mouse 1), which is placed in a wire cage for 10 minutes. Five minutes later, the test mouse is exposed to a novel mouse (Mouse 2), which is placed in the empty wire cage while Mouse 1 remains in the other side chamber. This session is repeated 5 min and 3 hours later with a novel unfamiliar mouse (Mouse 3 and Mouse 4) and the now-familiar Mouse 2. Time spent interacting with the familiar or unfamiliar conspecific was scored. Percent time spent with the novel mouse was calculated as ((interaction time novel mouse)/(total interaction time)).

Histology

Coronal paraffin sections (8 µm) of paraformaldehyde (PFA)-fixed brains were incubated with mouse-anti-myelin basic protein (MBP) (Sternberger Monoclonals, Lutherville, MD,) or mouse-anti-microtubuli-associated protein 2 (MAP2) (Sigma-Aldrich) followed by biotinylated horse-anti-mouse antibody (Vector Laboratories, Burliname, CA). Binding was visualized with Vectastain ABC kit (Vector Laboratories) and diaminobenzamidine. Ipsilateral MAP2 and MBP area loss were determined on sections −1.85 mm from bregma [20]. MBP and MAP2 staining were quantified using ImageJ software (http://rsb.info.nih.gov/ij/) and Adobe Photoshop CS5, respectively.

MSC tracking

MSCs were labeled with PKH-26 Red fluorescent cell linker kit (Sigma-Aldrich) and administered intranasally. 24 hours after MSC treatment brains were fixed in 4% PFA. Coronal cryosections (8 µm) were stained with DAPI. Fluorescent images were captured using a EMCCD camera (Leica Microsystems, Benelux) and Softworx software (Applied Precision, Washington).

Statistical analysis

Data were analyzed using (repeated measures) one-way ANOVA followed by Bonferroni post-tests. Significance for social discrimination was analyzed with the one sample t-test. $p<0.05$ was considered statistically significant. Data are presented as mean ± SEM.

Results

Dose of intranasally administered MSC

To determine the MSC dose response relationship, we administered 0.25×10^6, 0.5×10^6, 1×10^6 MSCs or vehicle intranasally at 10 days post-HI. Exposure to HI markedly impaired sensorimotor performance in the CRT (Figure 1A).

The doses of 0.5×10^6 and 1×10^6 MSCs significantly improved sensorimotor function as determined 3, 4 and 5 weeks post-HI (data shown for 5 weeks; Figure 1A). The beneficial effect of MSC treatment on sensorimotor function was lost when the dose was reduced to 0.25×10^6 MSCs.

We analyzed MAP2 loss as a measure of gray matter damage and MBP loss as a parameter of white matter loss at 5 weeks post-insult [14]. HI-mice receiving 0.5×10^6 or 1×10^6 MSCs on day 10 showed a substantial decrease in MAP2 loss (Figure 1B) and MBP loss (Figure 1C). When a lower dose of 0.25×10^6 MSCs was used, we no longer observed a beneficial effect on MAP2 or MBP loss.

Therapeutic window

To determine the therapeutic window of intranasal MSC treatment, we administered 0.5×10^6 MSCs at 3, 10 or 17 days post-HI and assessed sensorimotor performance in the CRT. Our data show that MSC treatment given 3 or 10 days post-insult is effective in improving motor behavior significantly. However, when we postponed MSC treatment to 17 days after HI, we no longer observed improvement of sensorimotor function as analyzed 5 weeks post-insult (Figure 2A).

With respect to lesion volume, MSC treatment at 3 or 10 days, but not 17 days after HI significantly reduced MAP2 and MBP loss (Figure 2B and C).

Effect of two MSC dosages

Previous results from our group showed that two intracranial MSC injections at 3+10 days compared to a single injection 3 days post-HI, further improved motor performance and decreased neuronal and white matter loss [14]. The data in Figure 3A show that one intranasal dose at either 3 or 10 days is as effective in restoring sensorimotor function as two intranasal doses at 3+10 days post-insult. Histological data on MAP2 or MBP staining confirmed the functional CRT analyses, showing that one single intranasal dose at 3 or 10 days is as effective as two doses at 3+10 days analyzed at 5 weeks post-HI (Figure 3B and C). Furthermore, the positive effects of either one or two intranasal doses of MSCs on sensorimotor and histological outcome can still be observed at respectively 8 and 9 weeks after HI, and no additive effect of two doses was observed at this time point either (Figure 4).

Effect of MSC treatment on cognitive function

To determine whether MSC treatment improves HI-induced cognitive impairment, we used the social discrimination test at 7 weeks post-HI. This test uses the preference for social novelty as a measure of cognitive function *i.e.* the ability to discriminate between known or unknown conspecific.

As anticipated, sham-control mice showed preference for the novel mouse at 5 min and 3 h after the training session (Figure 5A and B). Vehicle-treated HI mice lost the capability to discriminate between the novel and familiar mouse at both time points. Interestingly, performance in the social discrimination test normalized in the groups treated with MSCs at 10 days or 3+10

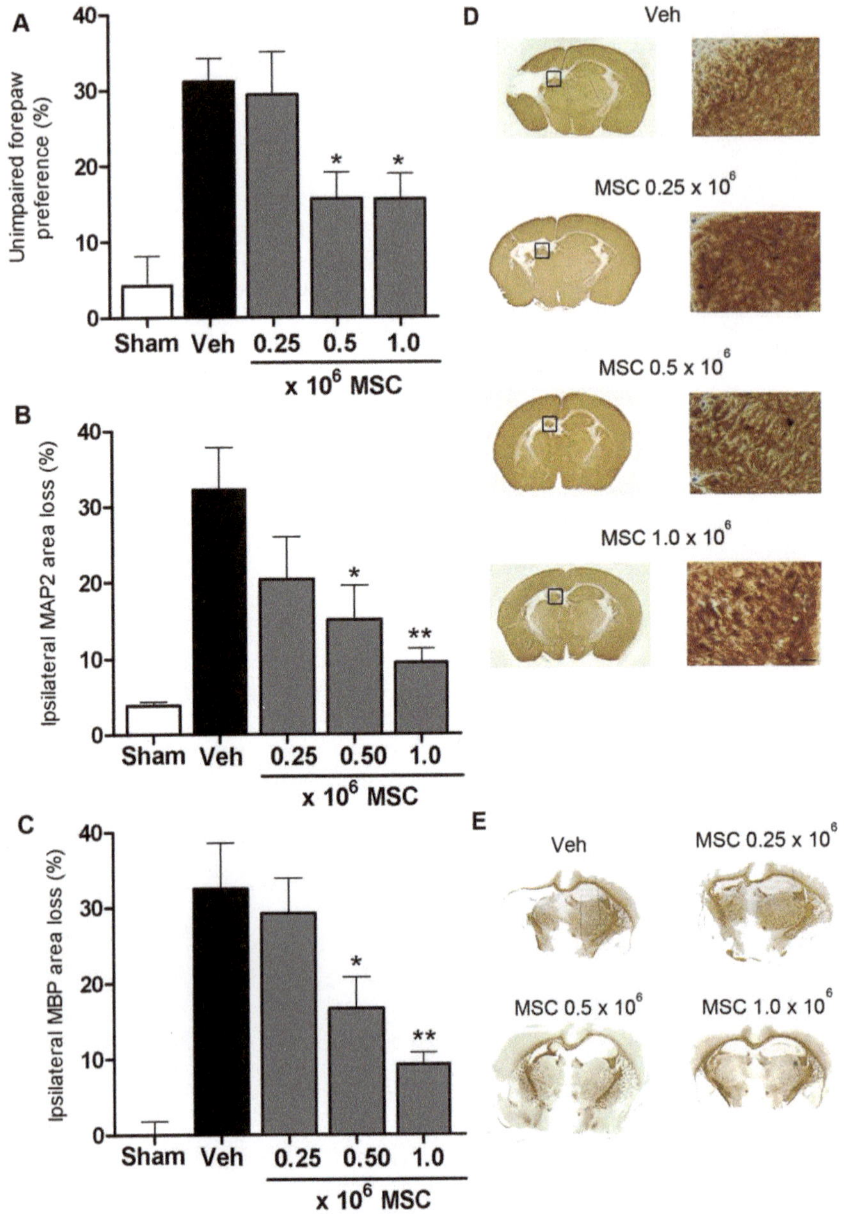

Figure 1. Dose effect of MSCs on motor performance and lesion volume at 35 days post-HI. Mice received 0.25×10^6, 0.5×10^6, 1×10^6 MSCs or Vehicle (Veh) treatment at 10 days after induction of HI. (A) Paw preference to use the unimpaired forepaw in the cylinder rearing test (CRT) was assessed at 5 weeks post-HI. Sham-operated littermates (Sham) were used as controls. Quantification of ipsilateral MAP2 (B) and MBP(C) area loss measured as 1- (ipsi-/contralateral MAP2- or MBP-positive area) at 5 weeks post-HI. (D) Representative sections of MAP2 loss. Insets show higher magnification of corresponding MAP2 sections. Scale bar = 100 µm. (E) Representative sections of MBP area loss. Data represent mean ± SEM. Sham n = 8; Veh n = 10; 0.25×10^6 MSC n = 11; 0.5×10^6 MSC n = 10; 1×10^6 MSC n = 13. *$p<0.05$; **$p<0.01$ vs Veh. Data presented in this figure are results from pooled experiments out of 8 different litters. Treatment groups were randomly distributed between litters.

days following HI. During the training session we did not observe group differences in social interaction times (Figure 5C).

MSC migration towards lesion site

To determine whether MSCs migrate from the nose to the lesion site, we used PKH-26 labeled MSCs and analyzed brain sections 24 h after intranasal administration. PKH-26+ MSCs administered 3 days post-HI are present predominantly in the damaged hippocampus (Figure 6A; ipsilateral).

MSCs given 10 days after HI localize in the area surrounding the lesion site. At this time point, the entire hippocampus was lost and PKH-26+ cells were now present in cortical layers 5 and 6 (Figure 6B; ipsilateral) and in the dorsal and epithalamic regions (data not shown). Interestingly, hardly any MSCs could be detected in the brain when given 17 days post-HI (Figure 6C; ipsilateral).

No labeled MSCs were present in the contralateral hemisphere, although MSCs were given in both nostrils (Figure 6A to C). In addition, no PKH-26+ cells were observed in brains from sham-operated mice (Figure 6D; sham 3 d and 10 d). Furthermore, we found no evidence for tissue autofluorescence after HI either 3 or 10 days following the insult (Figure 6D; vehicle).

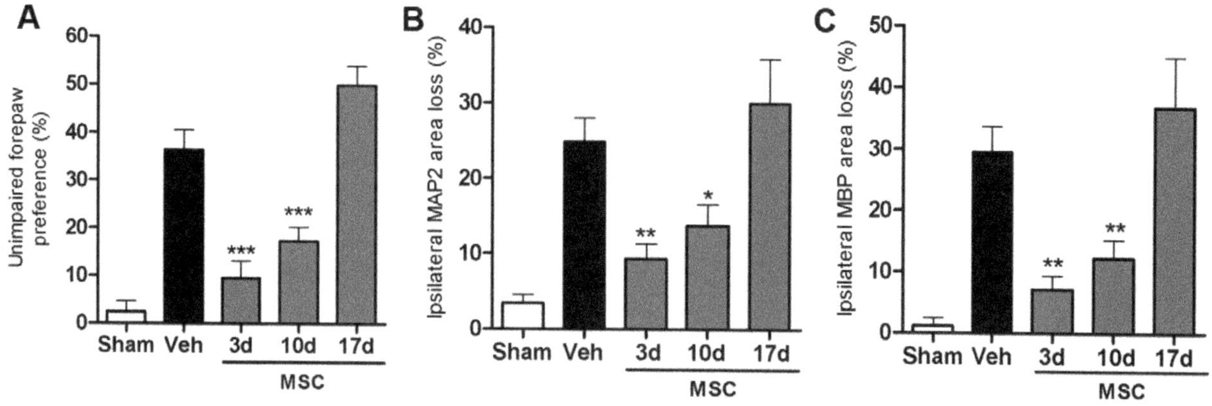

Figure 2. Therapeutic window for MSC treatment. Mice received 0.5×10^6 MSCs or Veh at 3, 10 or 17 days post-HI. Because no significant difference was found between Veh groups treated at different time points we pooled all animals into one group. (A) Unimpaired forepaw preference in the CRT at 5 weeks post-insult. Quantification of ipsilateral MAP2 (B) and MBP (C) area loss measured as 1- (ipsi-/contralateral MAP2- or MBP-positive area) at 5 weeks post-insult. Insets show representative examples of MAP2 or MBP staining. Data represent mean ± SEM. Sham n = 8; Veh n = 19; MSC 3 d n = 12; MSC 10 d n = 17; MSC 17 d n = 9; *p<0.05; **p<0.01; ***p<0.001 vs. Veh. Data presented in this figure are results from pooled experiments out of 12 different litters. Treatment groups were randomly distributed between litters.

Discussion

In the present study we show that intranasally administered MSCs reach the brain within just 24 h after intranasal administration. Interestingly, the MSCs migrate specifically towards the lesion site. We determined the dose response relationship of MSC administration and demonstrate that one single intranasal MSC administration is sufficient to induce maximal improvement of functional outcome and reduction in brain damage. A single intranasal dose of MSC has long-lasting effects on motor, cognitive and histological outcome up to 9 weeks following HI injury. Moreover, we are the first to show that intranasal MSC treatment also improves cognitive function, which holds significant clinical relevance.

To determine whether the beneficial effects of intranasal MSC application are associated with migration of MSCs from the nasal mucosa to the lesion site, we used the cell tracking dye PKH-26. In contrast to other commonly used cell tracking dyes, PKH-26 does not have any adverse effects on cell proliferation or survival. Furthermore, no significant cell to cell transfer or dye leakage from cells was measured [21]. No PKH-26+ MSCs were observed at the hippocampal level of sham-operated mice treated at 3 or 10 days post-HI. Interestingly, we clearly detected PKH-26 labeled MSCs at the HI lesion site, but not in the contralateral hemisphere of mice subjected to HI that were treated with MSCs via both the ipsilateral and contralateral nostril. MSCs given at 3 days post-HI migrate specifically to the damaged hippocampus. At 10 days the damage has evolved into an extensive lesion site, including the entire hippocampus and part of the cortex. When MSCs are administered intranasally at 10 days following injury, 24 h later MSCs are located in the remaining regions of the cortex and thalamus. In contrast, when MSCs were administered at 17 days post-HI, only a sporadic PKH-26+ MSC was detected in the lesion, although there is still substantial injury at 17 days post-insult. Notably, MSC treatment at this time point did not have any effect on lesion size nor on sensorimotor function. Combined, these findings indicate that the loss of effect of MSCs given at 17

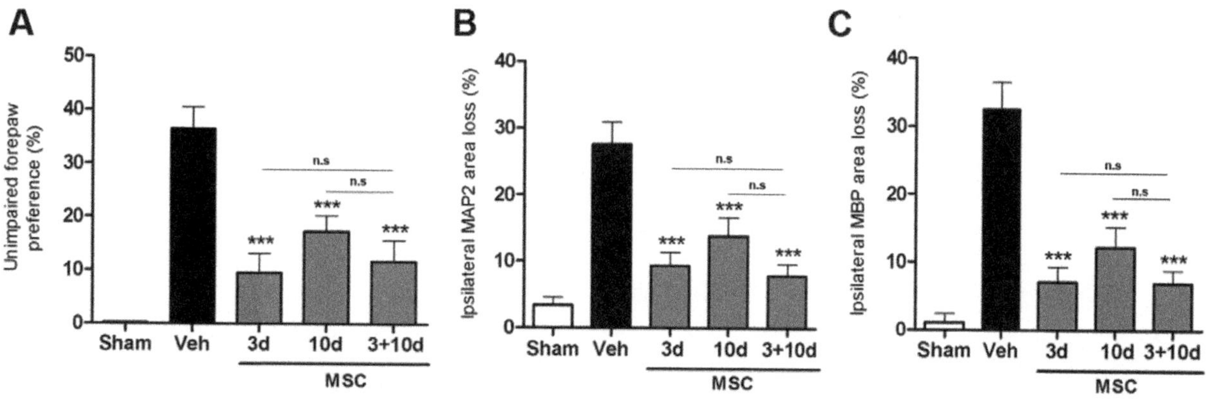

Figure 3. Effect of two MSC treatments on sensorimotor function and lesion size. Mice received 0.5×10^6 MSCs or Veh at 3, 10 or 3+10 days post-insult. Because no significant difference was found between Veh groups treated at different time points we pooled all animals into one group. (A) Unimpaired forepaw preference in the CRT at 5 weeks post-HI. Quantification of ipsilateral MAP2 (B) and MBP (C) area loss measured as 1- (ipsi-/contralateral MAP2- or MBP-positive area) at 5 weeks post-HI. Insets show representative examples of MAP2 or MBP staining. Data represent mean ± SEM. Sham n = 7; Veh n = 19; MSC 3+10 d n = 13; MSC 10+17 d n = 11; MSC 10 d n = 17. ***p<0.001 vs. Veh., n.s. = non-significant. Data presented in this figure are results from pooled experiments out of 14 different litters. Treatment groups were randomly distributed between litters.

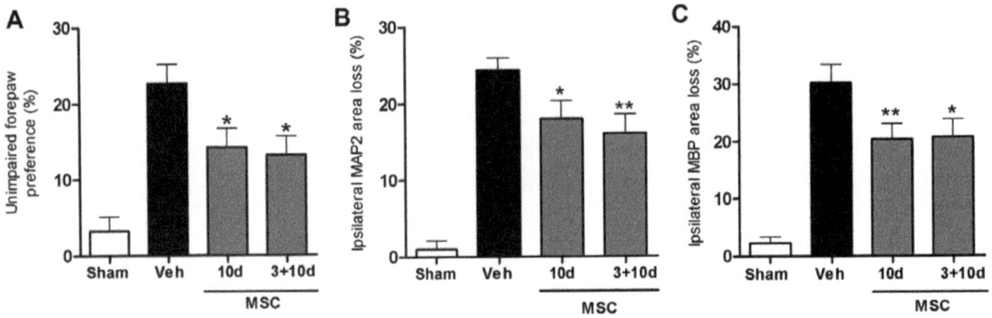

Figure 4. Long-term effect of MSC treatment on sensorimotor function and lesion volume. Mice received 0.5×10^6 MSCs or Veh at 3, 10 or 3+10 days post-HI. Because no significant difference was found between Veh groups treated at different time points we pooled all animals into one group. (A) Unimpaired forepaw preference in the CRT at 8 weeks post-HI. Quantification of ipsilateral MAP2 (B) and MBP (C) area loss measured as 1-(ipsi-/contralateral MAP2- or MBP-positive area). Insets show representative examples of MAP2 or MBP staining at 9 weeks post-insult. Data represent mean ± SEM. Sham n = 23; Veh n = 23; MSC 10 d n = 23; MSC 3+10 d n = 12. *$p<0.05$; **$p<0.01$ vs. Veh. Data presented in this figure are results from pooled experiments out of 13 different litters. Treatment groups were randomly distributed between litters.

days after the insult will be due to reduced migration of MSC to the lesion site when cells are given at this time point.

The social discrimination test was used to assess cognitive performance following HI. This test was chosen as it is a very sensitive measure for short-term and long-term recognition memory function [19]. This type of memory describes the ability to discriminate between familiar and unfamiliar stimuli. The social discrimination test exploits the natural interest of an adult mouse to explore an unknown conspecific over a known conspecific. The molecular mechanisms and neuro-anatomical structures underlying this behavioural paradigm remain elusive. Studies have shown that performance on the social discrimination test requires functional olfactory bulb, bed nucleus of the stria terminalis and amygdalo-hippocampal network [22,23]. The hippocampus is known to be important in memory consolidation, which is essential for long-term memory storage in other cortical areas [24]. We are the first to establish that intranasal MSC treatment not only improves sensorimotor but also cognitive behavior.

The exact mechanisms underlying MSC-induced improvement of short-term memory after neonatal HI brain damage have yet to be clarified. HI injury leads to extensive unilateral loss in cortical and hippocampal areas, which explains the inability of vehicle-treated group to form and/or store new information. Intranasal MSC treatment significantly reduces loss of cortical and hippocampal areas, which are both important in memory formation and storage.

Our present and previous findings [14] indicate that there is no further increase in lesion size after day 10 post-HI. Therefore, we propose that the major effect of MSC treatment is not mediated by inhibition of injurious processes, but rather by stimulating repair. We indeed have evidence from earlier studies in which we applied MSC intracranially that formation of new cells and differentiation of these new cells into neurons and oligodendrocytes is promoted by MSC treatment [14,15]. Moreover, we showed that only a very small proportion of cells of donor origin survived in the brain [16]. Therefore, it is highly likely that MSC treatment restores cognitive as well as motor circuitries in the brain through stimulation of endogenous regenerative processes.. However, it may well be possible that inhibition of injurious processes contributes to the observed beneficial effects of MSC treatment. If so, this will especially be the case when MSC are administered early after the insult. In our experiments this is 3 days after the insult as at this time point the damage has not yet fully developed.

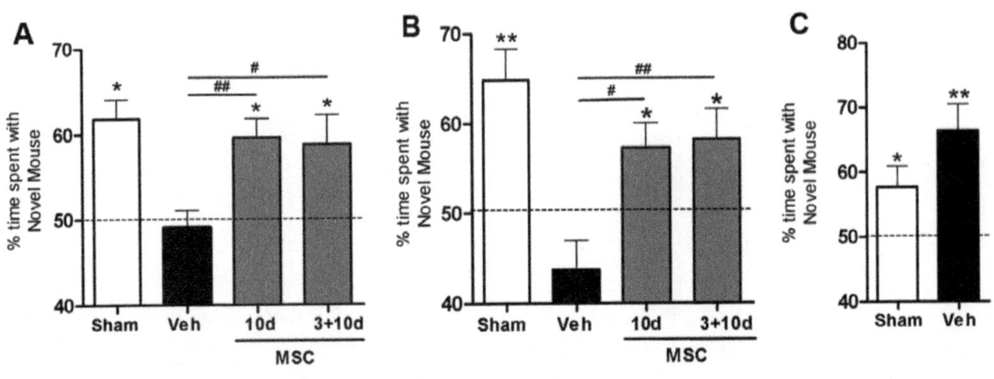

Figure 5. Effect of MSC treatment on cognitive behavior. Mice received 0.5×10^6 MSCs or Veh at 10 or 3+10 days post-HI. Because no significant difference was found between Veh groups treated at different time points we pooled all animals into one group. Animals were tested for cognitive function using the social discrimination test at 7 weeks post-HI. Preference for novel conspecific is expressed as exploratory ratio. The total social interaction time did not differ between groups. (A) Preference for novel mouse after a 5 min interval. (B) Preference for novel mouse after a 3 hour interval. (C)Training session to measure preference for social novelty as an indication for social avoidance. Sham n = 23; Veh n = 23; MSC 10 d n = 23; MSC 3+10 d n = 12. *$p<0.05$; **$p<0.01$ in relation to 50% (no discrimination); $^{\#}p<0.05$; $^{\#\#}p<0.01$ vs. Veh. Data presented in this figure are results from pooled experiments out of 13 different litters. Treatment groups were randomly distributed between litters.

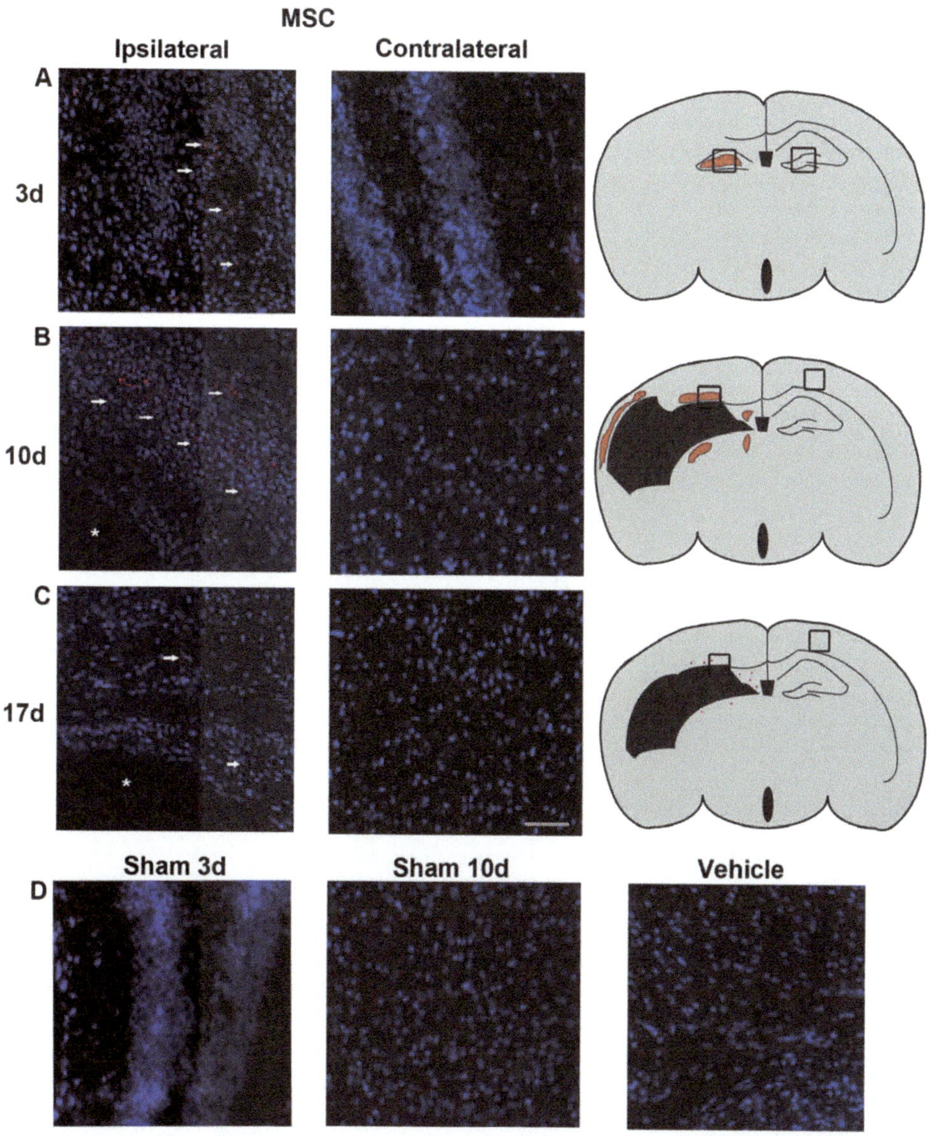

Figure 6. Presence of MSCs in the brain. PKH-26 labeled 1.0×10^6 MSCs were administered intranasally at 3, 10 and 17 days post-HI. Because no significant difference was found between Veh groups treated at different time points we pooled all animals into one group. (A, B, C) Notice the severe HI-induced damage, as the layer structure of the ipsilateral cortex and hippocampus are lost. (A) MSCs (red) in the ipsilateral hippocampus (see arrow heads) 24 h after administration at 3 days post-HI. (B) MSCs (see arrow heads) in the ipsilateral damaged cortex 24 h after administration at 10 days post-HI. (C) Lack of MSCs (see arrow heads) in the ipsilateral cortical areas surrounding the lesion site when MSCs are given at 17 days post-HI. Contralateral pictures depict hippocampal area (in A) and cortical area (in B and C). (D) Control groups showing lack of MSCs in the hippocampal area and cortical area at 3 and 10 days, respectively, after MSC administration in sham-operated animals and HI-Vehicle treated brain without MSC treatment. Asterisk = lesion site. Blue = Dapi staining. Scale bar 50 µm. Data presented in this figure are results from pooled experiments out of 10 different litters. Treatment groups were randomly distributed between litters.

We also propose that beneficial effects of MSC transplantation are cell-specific, because intracranial fibroblast (3T3) administration did not affect performance in the CRT as well as on lesion size (unpublished observations). Although we now studied the effect of intranasal administration, we anticipate that the same cellular specificity will apply for intranasal administration, give an effect on repair, it should anyhow occur when cells are given directly at the lesion site.

In preparation for clinical translation of our finding that nasally administered MSCs improve cognitive as well as sensorimotor outcome and reduce lesion size, we investigated the therapeutic window, the treatment dose and frequency. We report here that sensorimotor function improves and lesion size reduces significantly when 0.5×10^6 cells are administered via the nasal route. Decreasing the dose to 0.25×10^6 MSCs had no significant effect on either the CRT performance or MAP2 and MBP outcome, while increasing the dose to 1.0×10^6 did not further improve sensorimotor function or reduce lesion size. Thus, a minimum of 0.5×10^6 MSCs per mouse is required to have a long-lasting beneficial effect on behavior and lesion size in our model of HI-induced brain damage.

In the present study, we demonstrate that multiple doses at 3+10 days via the intranasal route did not have an additional effect on sensorimotor performance and brain damage compared to a single dose. This finding is clinically important as it would mean that only one intranasal dose of MSCs will be sufficient for optimal therapeutic benefit in the neonate. In contrast, we showed earlier that when MSCs are administered intracranially, two gifts of MSC have a stronger beneficial effect on outcome than one gift. One possible explanation for the finding that one intranasal MSC treatment already provides optimal effects is that intranasally administered MSCs may perform better as migration from the nasal mucosa towards the lesion may allow adaptation to the detrimental milieu in the brain. In contrast, intracranially administered cells were directly injected adjacent to the lesion size, consisting of an apoptotic/inflammatory milieu which may partially impair the functionality of MSCs.

Conclusions

In conclusion, this study shows that intranasal MSC treatment has a wide therapeutic window, leads to long-term improvement in sensorimotor and cognitive function and decreases gray and white matter damage after HI. MSC migrate specifically to the site of injury, despite contralateral administration. Our results clearly establish that intranasal MSC treatment has the potential to become a novel therapeutic strategy for neonatal encephalopathy.

Acknowledgments

The authors are greatly indebted to Dr. Marijke Laarakker, Jitske Zijlstra, Marijke Tersteeg-Kamperman and Karima Amarouchi for technical assistance.

Author Contributions

Conceived and designed the experiments: VD CTJV CN FB AK CJH. Performed the experiments: VD. Analyzed the data: VD CTJV CN AK CJH. Contributed reagents/materials/analysis tools: MJHK. Wrote the paper: VD CTJV CN MJHK FB AK CJH.

References

1. Dammann O, Ferriero D, Gressens P (2011) Neonatal encephalopathy or hypoxic ischemic encephalopathy? Appropriate terminology matters. Pediatr Res 70: 1–2.
2. De Haan M, Wyatt JS, Roth S, Vargha-Khadem F, Gadian D, et al. (2006) Brain and cognitive-behavioural development after asphyxia at term birth. Dev Sci 9: 350–358.
3. Ferriero DM (2004) Neonatal brain injury. N Engl J Med 351: 1985–1995.
4. Graham EM, Ruis KA, Hartman AL, Northington FJ, Fox HE (2008) A systematic review of the role of intrapartum hypoxia-ischemia in the causation of neonatal encephalopathy. Am J Obstet Gynecol 199: 587–595.
5. van Handel M, Swaab H, de Vries LS, Jongmans MJ (2007) Long-term cognitive and behavioural consequences of neonatal encephalopathy following perinatal asphyxia: a review. Eur J Pediatr 166: 645–654.
6. Volpe JJ (2001) Perinatal brain injury: from pathogenesis to neuroprotection. Ment Retard Dev Disabil Res Rev 7: 56–64.
7. Edwards AD, Brocklehurst P, Gunn AJ, Halliday H, Juszczak E, et al. (2010) Neurological outcomes at 18 months of age after moderate hypothermia for perinatal hypoxic ischaemic encephalopathy: synthesis and meta-analysis of trial data. BMJ 340: c363–370.
8. Azzopardi DV, Strohm B, Edwards AD, Dyet L, Halliday HL, et al. (2009) Moderate hypothermia to treat perinatal asphyxial encephalopathy. N Engl J Med 361: 1349–1358.
9. Borlongan CV, Weiss MD (2011) Baby STEPS: A giant leap for cell therapy in neonatal brain injury. Pediatr Res 70: 3–9.
10. Lee JA, Kim BI, Jo CH, Choi CW, Kim EK, et al. (2010) Mesenchymal stem-cell transplantation for hypoxic-ischemic brain injury in neonatal rat model. Pediatr Res 67: 42–46.
11. Pimentel-Coelho PM, Mendez-Otero R (2010) Cell therapy for neonatal hypoxic-ischemic encephalopathy. Stem Cells Dev 19: 299–310.
12. Titomanlio L, Kavelaars A, Dalous J, Heijnen C, Baud O, et al. (2011) Stem cell therapy for neonatal brain injury: perspectives and challenges. Ann of Neurol 70: 698–712.
13. van Velthoven CT, Kavelaars A, Heijnen CJ (2012) Mesenchymal stem cells as a treatment for neonatal ischemic brain damage. Pediatr Res 71: 474–481.
14. van Velthoven CT, Kavelaars A, van Bel F, Heijnen CJ (2010) Mesenchymal stem cell treatment after neonatal hypoxic-ischemic brain injury improves behavioral outcome and induces neuronal and oligodendrocyte regeneration. Brain Behav Immun 24: 387–393.
15. van Velthoven CT, Kavelaars A, van Bel F, Heijnen CJ (2010) Repeated mesenchymal stem cell treatment after neonatal hypoxia-ischemia has distinct effects on formation and maturation of new neurons and oligodendrocytes leading to restoration of damage, corticospinal motor tract activity, and sensorimotor function. J Neurosci 30: 9603–9611.
16. van Velthoven CT, Kavelaars A, van Bel F, Heijnen CJ (2010) Nasal administration of stem cells: a promising novel route to treat neonatal ischemic brain damage. Pediatr Res 68: 419–422.
17. Yasuhara T, Matsukawa N, Yu G, Xu L, Mays RW, et al. (2006) Behavioral and histological characterization of intrahippocampal grafts of human bone marrow-derived multipotent progenitor cells in neonatal rats with hypoxic-ischemic injury. Cell Transplant 15: 231–238.
18. Yasuhara T, Hara K, Maki M, Mays RW, Deans RJ, et al. (2008) Intravenous grafts recapitulate the neurorestoration afforded by intracerebrally delivered multipotent adult progenitor cells in neonatal hypoxic-ischemic rats. J Cereb Blood Flow Metab 28: 1804–1810.
19. Nadler JJ, Moy SS, Dold G, Trang D, Simmons N, et al. (2004) Automated apparatus for quantitation of social approach behaviors in mice. Genes Brain Behav 3: 303–314.
20. Nijboer CH, Kavelaars A, Vroon A, Groenendaal F, van Bel F, et al. (2008) Low endogenous G-protein-coupled receptor kinase 2 sensitizes the immature brain to hypoxia-ischemia-induced gray and white matter damage. J Neurosci 28: 3324–3332.
21. Ashley DM, Bol SJ, Waugh C, Kannourakis G (1993) A novel approach to the measurement of different in vitro leukaemic cell growth parameters: The use of PKH GL fluorescent probes. Leuk Res 17: 873–882.
22. Auger CJ, Vanzo RJ (2005) Progesterone Treatment of Adult Male Rats Suppresses Arginine Vasopressin Expression in the Bed Nucleus of the Stria Terminalis and the Centromedial Amygdala. J of Neuroendoc 18: 187–194.
23. Ferguson JN, Young LJ, Insel TR (2002) The neuroendocrine basis of social recognition. Front Neuroendocrinol 23: 200–224.
24. Morgado-Bernal I (2011) Learning and memory consolidation: Linking molecular and behavioural data. Neurosci 176: 12–19.

17

Combined Effects of Ventilation Mode and Positive End-Expiratory Pressure on Mechanics, Gas Exchange and the Epithelium in Mice with Acute Lung Injury

Apiradee Thammanomai, Hiroshi Hamakawa, Erzsébet Bartolák-Suki, Béla Suki*

Department of Biomedical Engineering, Boston University, Boston, Massachusetts, United States of America

Abstract

The accepted protocol to ventilate patients with acute lung injury is to use low tidal volume (V_T) in combination with recruitment maneuvers or positive end-expiratory pressure (PEEP). However, an important aspect of mechanical ventilation has not been considered: the combined effects of PEEP and ventilation modes on the integrity of the epithelium. Additionally, it is implicitly assumed that the best PEEP-V_T combination also protects the epithelium. We aimed to investigate the effects of ventilation mode and PEEP on respiratory mechanics, peak airway pressures and gas exchange as well as on lung surfactant and epithelial cell integrity in mice with acute lung injury. HCl-injured mice were ventilated at PEEPs of 3 and 6 cmH_2O with conventional ventilation (CV), CV with intermittent large breaths (CV_{LB}) to promote recruitment, and a new mode, variable ventilation, optimized for mice (VV_N). Mechanics and gas exchange were measured during ventilation and surfactant protein (SP)-B, proSP-B and E-cadherin levels were determined from lavage and lung homogenate. PEEP had a significant effect on mechanics, gas exchange and the epithelium. The higher PEEP reduced lung collapse and improved mechanics and gas exchange but it also down regulated surfactant release and production and increased epithelial cell injury. While CV_{LB} was better than CV, VV_N outperformed CV_{LB} in recruitment, reduced epithelial injury and, via a dynamic mechanotransduction, it also triggered increased release and production of surfactant. For long-term outcome, selection of optimal PEEP and ventilation mode may be based on balancing lung physiology with epithelial injury.

Editor: Derek Abbott, University of Adelaide, Australia

Funding: This study was funded by National Institutes of Health (NIH) HL-098976. The funders had no role in study design, data collection and analysis, decision to publish, or preparation of the manuscript.

Competing Interests: The authors have declared that no competing interests exist.

* E-mail: bsuki@bu.edu

Introduction

In acute respiratory distress syndrome (ARDS), the currently accepted protocol is to use low tidal volume (V_T) in combination with a procedure that helps keep the lung open such as a recruitment maneuver (RM) or adding positive end-expiratory pressure (PEEP) [1]. PEEP stabilizes the injured alveoli and prolongs the effects of RM such as sustained high airway pressure [2]. However, PEEP can also decrease cardiac output [3], or increase pulmonary edema [4]. Furthermore, ventilation superimposed on a high PEEP can lead to ventilator induced lung injury (VILI) due to barotrauma and/or volutrauma [5].

The development of VILI is related to the heterogeneous nature of ARDS lungs [6]. When a high PEEP is required to maintain an open lung, normal lung regions will be overinflated. In contrast, an inadequate PEEP can result in cyclic recruitment/derecruitment during ventilation with high non-physiologic shear and normal stresses on the epithelium which can generate epithelial injury [7,8]. Consequently, the selection of an optimal PEEP is a highly debated topic [9–13]. A moderate PEEP together with regular delivery of large breaths as RMs might be able to keep the lung open while minimizing the risk of VILI. This approach works in mice [14–16] but clinical studies showed mixed results [10,17,18].

Variable ventilation (VV) introduced by Lefevre et al. [19] better maintains an open lung than conventional ventilation (CV) [20–27]. Recently, we introduced a new optimized VV (VV_N) for both normal and HCl-injured mice which significantly improved respiratory mechanics and oxygenation over several other methods without causing additional injury [27]. Since both PEEP and RM have a significant impact on ventilator performance, the aim of the current study was to investigate the combined effects of ventilation mode and PEEP on physiology including mechanics and gas exchange and on lung biology including surfactant and epithelial cell integrity in mice with acute lung injury (ALI) at two PEEP levels. To this end, HCl-injured mice were ventilated with CV, CV with intermittent large breaths (CV_{LB}) to promote recruitment, and VV_N at PEEPs of 3 and 6 cmH_2O. While the raw physiological data including mechanics and gas exchange at 3 cmH_2O PEEP were reported in our previous paper [27], the mechanics data were reanalyzed using an advanced model. Additionally, all the biochemical data reported here are novel.

Methods

Animal Preparation

The protocol was approved by the Animal Care and Use Committee of Boston University (approval number: 04–033). Male

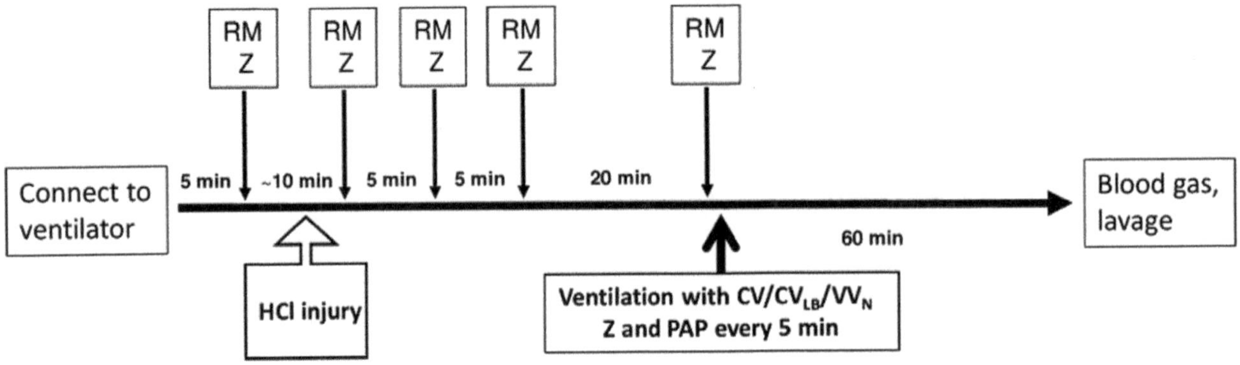

Figure 1. Time course of the protocol. Once the animal was connected to the ventilator, a recruitment maneuver (RM) was followed by impedance measurement (Z) and HCl treatment. After a stabilization period and several RM and Z, a 60 min ventilation period (thick arrow) was started. During the ventilation period, Z and peak airway pressure (PAP) were recorded at 5 min intervals. At the end of the protocol, blood gases and a lavage sample were obtained and the lung was isolated for further processing.

C57BL/6 mice (weight: 22–26 g, Charles River Laboratories, Wilmington, MA) were used throughout the studies. The mice were anesthetized with an intraperitoneal injection of 70 mg/kg of pentobarbital sodium, tracheostomized with an 18-guage metal cannula and placed on a heated pad to maintain a constant body temperature (37°C) throughout the experiment. Extra doses of pentobarbital sodium (20 mg/kg) were administered every 20 minutes to keep the animal in a deeply anesthetized state and minimize any suffering. The tracheal cannula was later connected to the outlet of a small animal ventilator-oscillator system (flexiVent ventilator, SCIREQ, CA).

Acid Aspiration Lung Injury

In order to obtain a mouse model of lung injury, hydrochloric acid (HCl, 0.1 M, pH = 1.25) was introduced intratracheally in 1 μL/g increments with a bolus of air in between for a total of 3 μL/g. To prevent lung collapse, the mice were connected to the ventilator, immediately given 2 RMs defined as a ramp increase in volume to 1 ml in 4 sec, and subsequently 1 RM after 5 and 10 minutes. The animals were then ventilated using a constant V_T of 8 ml/kg for an additional period of 20 minutes at a PEEP of 3 cmH_2O. The dose and the delivery method were developed and tested in a pilot study using 6 mice.

Ventilation Protocols

We aimed to investigate the physiological effects of each ventilation mode at 2 levels of PEEP using 3 ventilation modes that were identical to those in our previous study in which actual time series of V_T are also shown [27]:

CV group. Mice were ventilated with a constant V_T of 8 ml/kg and a breathing frequency (f) of 240 breaths per minute.

VV_N group. Mice were ventilated with a ventilation mode that was designed previously [27]. Briefly, the shape of the V_T distribution is flat for small V_Ts followed by a power law decrease for larger V_Ts. The mean V_T (V_{MEAN}) was set to 8 ml/kg and f was also adjusted to obtain constant minute ventilation on a cycle-by-cycle basis. The parameters specifying the distribution of V_T were as follows: the smallest V_T was $V_{MIN} = 0.7\ V_{MEAN}$; the V_T at which the distribution changes to a power law (with an exponent $\alpha = 5.1$) was $V_P = 0.9V_{MEAN}$. The maximum delivered V_T (V_{MAX}) was set to 2.25 times the V_{MEAN}. The median V_T for VV_N was 7.3 ml/kg.

CV_{LB} group. Mice were ventilated with a V_T of 8 ml/kg using a constant frequency at 240 breaths per minute. Twice in every minute, the animals received a large breath. Similar to VV_N, f was reduced for each large breath to maintain constant minute ventilation. The size of the two large breaths was matched with the 2 largest V_T values of the VV_N mode in 1 minute.

The time course of treatment and ventilation is illustrated in Fig. 1. A total of 48 animals were used with 8 mice in each ventilation group, 3 ventilation and 2 PEEP groups. At the beginning of the ventilation protocol, animals received 2 RMs to standardize volume history and then ventilated with room air for 60 minutes with CV mode, VV_N mode or CV_{LB} mode. The desired PEEP, 3 or 6 cmH_2O, was maintained by placing the expiratory port of the ventilator in a water trap. For the animals to be ventilated at a PEEP of 6 cmH_2O, the PEEP level was adjusted immediately after 2 RMs at the beginning of the protocol and the baseline mechanics were measured.

Unventilated group. Additionally, 4 animals received HCl injury and the initial treatment as shown in Fig. 1. However, these animals were sacrificed just before the 60 min ventilation protocol started. The lungs of these animals were lavaged and homogenized for biochemical analysis (see below) and represent the state of the lung at the beginning of the 60 min ventilation protocol.

Impedance Measurement

Respiratory mechanics were determined by the force oscillation technique by using the Optimum Ventilation Waveform [28]. Airway opening pressure and flow were obtained from the flexiVent ventilator and the data were processed offline using Fourier analysis to obtain respiratory input impedance (Z_{RS}) every 5 minutes throughout the ventilation protocol.

Model Description and Parameter Estimation

Respiratory input impedance Z_{RS} was fitted by the single compartment constant-phase model [29] and a heterogeneous model [30], which characterizes heterogeneity of tissue elasticity of the lung that invariably occurs after lung injury [31]. Briefly, the constant-phase model characterizes tissue impedance by a tissue damping coefficient (G) and a tissue elastance coefficient (H) while the airway structure is partitioned to airway resistance (R_{aw}) and inertance (I_{aw}). For the total respiratory system, the chest wall also contributes to the resistance and the model provides an estimate of the total Newtonian resistance (R). The heterogeneous tissue model represents the airway tree by a set of parallel pathways, each composed of R_{aw} and I_{aw} and a tissue compartment. The values of R_{aw} and I_{aw} are assumed to be the same for every

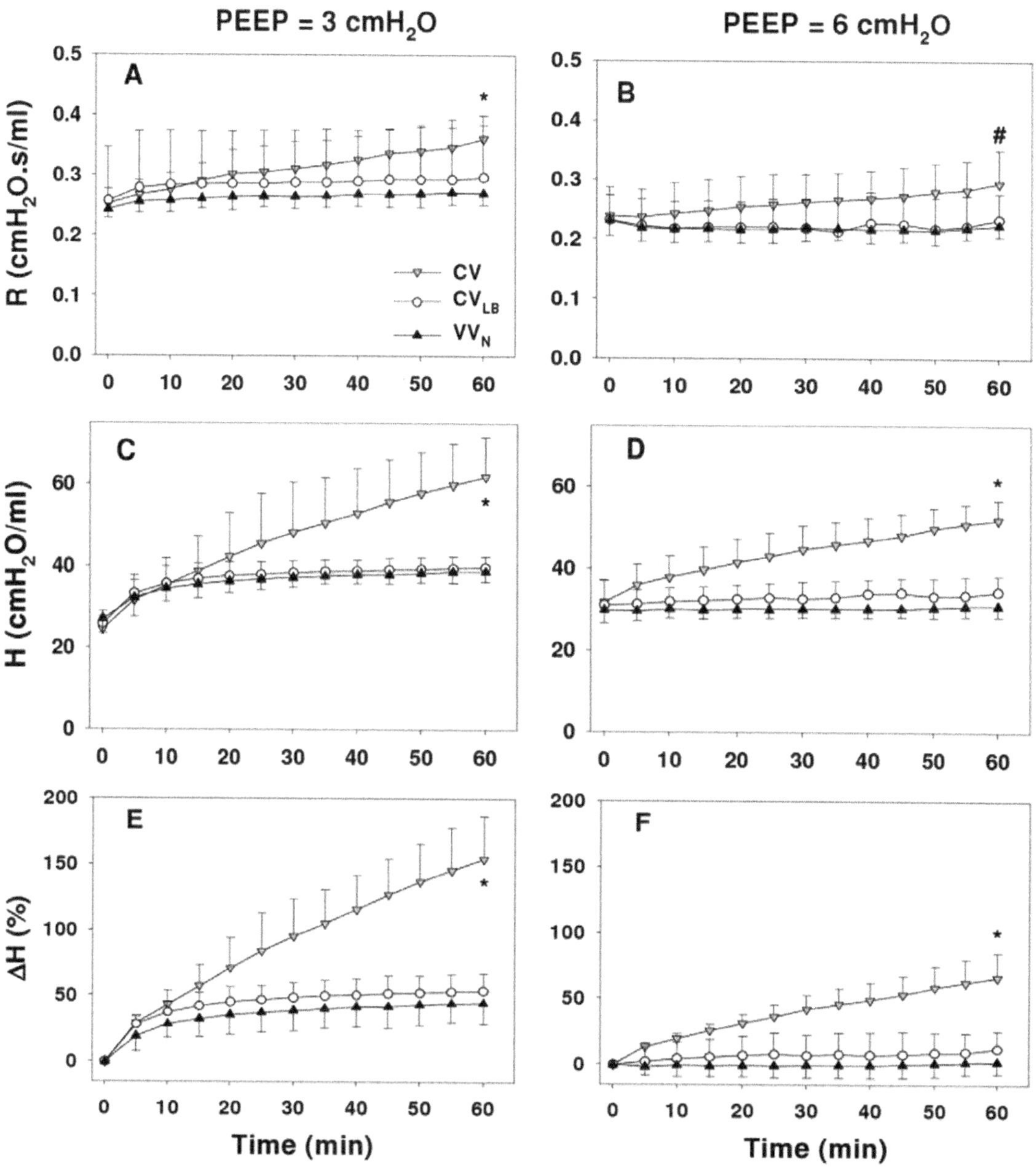

Figure 2. Mechanical parameters from the single compartment model. The graphs compare the time courses of Newtonian resistance (*R*, panels A and B), tissue elastance (*H*, panels C and D) and the change in H (*ΔH*, panels E and F) during 60 min of ventilation using conventional ventilation (CV), conventional ventilation with large breaths (CV_{LB}), or variable ventilation (VV_N) in HCl-injured mice at PEEPs of 3 (left panels) and 6 cmH_2O (right panels). * denotes significant difference between CV and CV_{LB} as well as CV and VV_N at 60 min; # denote significant difference between CV and VV_N at 60 min. Additional significance levels are given in the text.

pathway whereas each tissue compartment is described by the constant-phase model. The *H* is assumed to be distributed according to a probability density function *n(H)* which is proportional to *1/H* between a minimum (H_{min}) and a maximum (H_{max}) value of H. In this study, only the values of *R*, *H*, H_{min}, H_{max} and the standard deviation (SD) of *n(H)* are reported. All parameters were estimated using a global optimization algorithm which minimized the root-mean squared (RMS) error between data and model [32].

Airway Pressure Measurement

The airway pressure was monitored and recorded throughout the experiments using a separate pressure transducer (World Precision Instruments, Sarasota, FL) attached to the tracheal cannula. The mean peak airway pressure was calculated for every 5 minute period.

Sample Processing

At the conclusion of the 60 min ventilation, arterial blood was collected from the carotid artery by clamping both the upper and lower parts with small clamps. A small incision was made between the clamps. The lower clamp was then released and the arterial blood was collected using capillary tubes and immediately analyzed for partial pressure of oxygen (PaO_2), partial pressure of carbon dioxide ($PaCO_2$), pH, and percent oxygen saturation

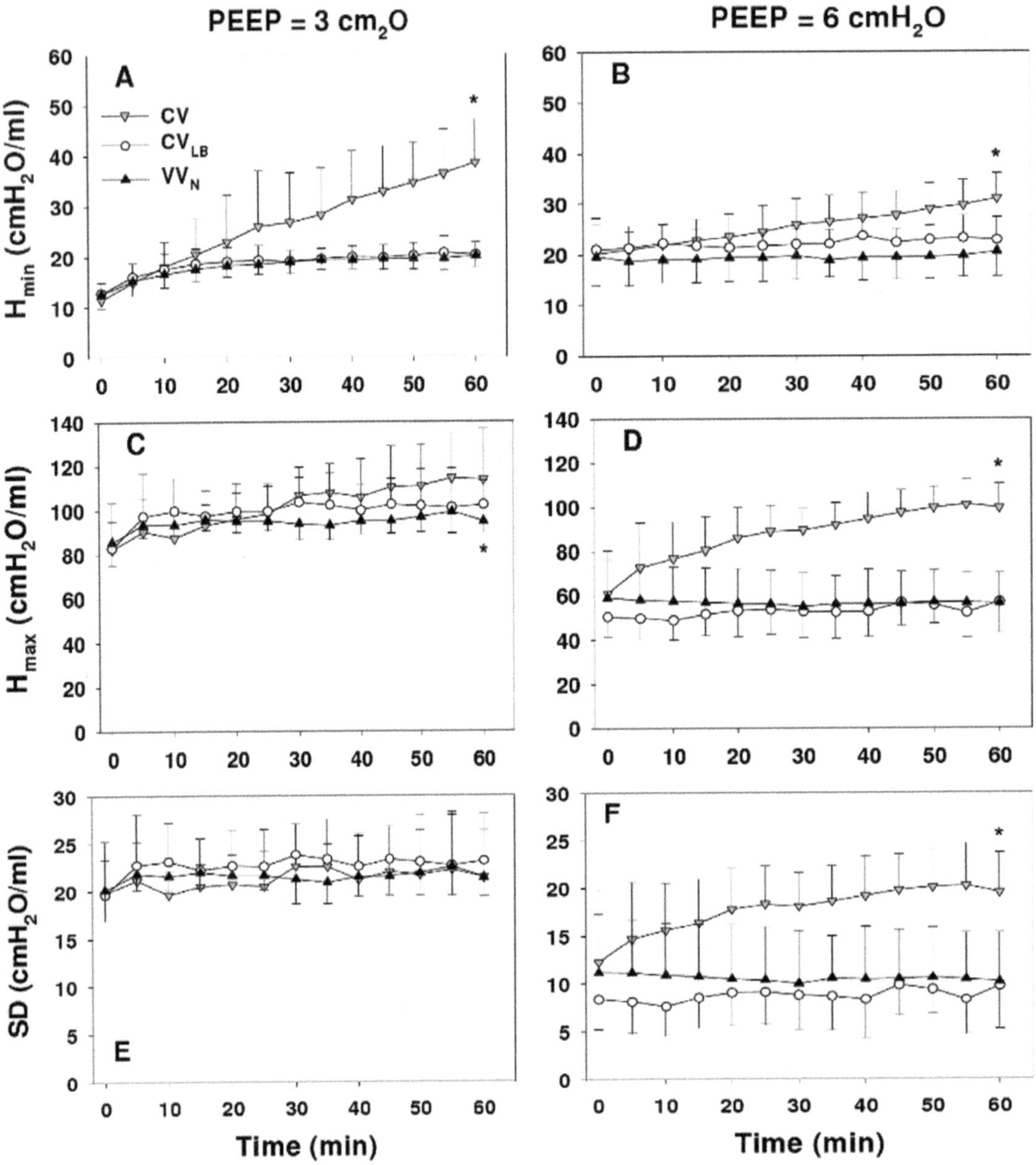

Figure 3. Mechanical parameters from the distributed model. The graphs compare the time courses of minimum (H_{min}, panels A and B), maximum (H_{max}, panels C and D) and standard deviation (SD, panels E and F) of the distribution of elastance in the heterogeneous tissue model during 60 min of ventilation using CV, CV_{LB} or VV_N in HCl-injured mice at PEEPs of 3 (left panels) and 6 cmH$_2$O (right panels). * denotes significant difference ($p<0.001$) between CV and CV_{LB} as well as CV and VV_N at 60 min.

(%sO_2) using an I-STAT® blood gas analyzer (Abbott Laboratories, Abbott Park, Illinois, USA).

Lavage samples were also collected at the conclusion of the ventilation by instilling 1 ml of warm saline (37°C) via the tracheal cannula and slowly retrieving approximately 0.9 ml. The lavage sample was centrifuged and the cell-free supernatant was frozen until further analysis. The lung was then removed and homogenized in 2 ml of PBS at a pH of 7.2 and centrifuged. Supernatant was transferred to microcentrifuge tubes and stored frozen until further analysis.

Total Protein and Western Blot Analysis

The amount of protein in the lavage and homogenate samples was measured using BCA protein assay reagent kit (Pierce, Rockford, IL). Equal amounts of total protein (7.8 μg) or equal volume (15 μl) of samples were separated using 4–20% SDS-polyacrylamide gels and transferred onto polyvinylidene fluoride membranes (Millipore, Bedford, MA). To test whether the ventilation mode had an effect on epithelial injury, Western blot analysis was carried out for E-cadherin from the lavage samples using a primary antibody for E-cadherin (CHEMICON, Temecula, CA). To test whether the ventilation mode had an effect on surfactant release and production, Western blot analysis was carried out for surfactant protein B (SP-B) from lavage and its proprotein form (proSP-B) (CHEMICON, Temecula, CA) from the homogenate samples. Densitometry was performed after chemiluminescence detection.

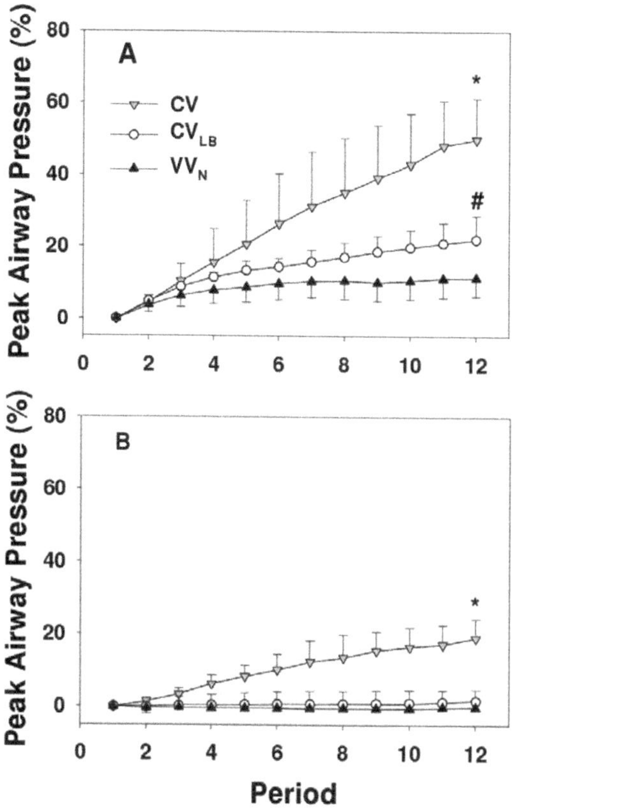

Figure 4. Peak airway pressures as a function of ventilation mode. The graphs compare the time courses of the relative percentage change in mean peak airway pressure during 60 min of ventilation using CV, CV_{LB} or VV_N in HCl-injured mice at PEEPs of 3 (panel A) and 6 cmH_2O (panel B). Each point is calculated from the average of the peak airway pressure in a 5-minute ventilation period compared to its value at time 0. *denotes significant difference ($p<0.001$) between CV and CV_{LB} as well as CV and VV_N at 60 min; # denote significant difference ($p<0.02$) between CV_{LB} and VV_N at 60 min.

Statistical Analysis

The statistical differences among parameters at different conditions were tested using t-tests and analyses of variance. For example, we used one-way ANOVA to compare parameters at 60 min, one-way repeated measure ANOVA to analyze time courses, two-way ANOVA to test the effects of PEEP and ventilation mode and their interactions. For post-hoc analyses, we used Tukey pairwise comparison. When data were not normally distributed, one way-ANOVA on ranks was used to analyze statistical differences. A value of $p<0.05$ was used to establish statistical significance. (SigmaStat, SPSS Inc, Chicago, IL).

Results

The time course of mechanical parameters in the HCl-injured mice ventilated at PEEPs of 3 cmH_2O (PEEP3) and 6 cmH_2O (PEEP6) are shown in Fig. 2. All parameters, R and H and the percent change in H (ΔH), depended on time ($p<0.001$). The ΔH during CV increased linearly while it reached a plateau during VV_N and CV_{LB}. At 60 min, all parameters, R, H and ΔH, depended on both the ventilation mode ($p<0.001$) and PEEP ($p<0.001$). At PEEP3, R in the CV group was significantly higher than in the VV_N and CV_{LB} groups ($p=0.002$ and $p=0.028$, respectively) whereas at PEEP6, R during CV was significantly higher than during VV_N ($p=0.012$), but not during CV_{LB}. Both H and ΔH during CV was higher than during VV_N and CV_{LB} ($p<0.001$) at both PEEPs while there was no difference between ΔH during VV_N and CV_{LB}.

The H_{min} (Fig. 3) describing the lowest regional stiffness increased with all ventilation modes at both PEEP3 ($p<0.001$) and PEEP6 ($p<0.001$ for CV, and $p<0.05$ for CV_{LB} and VV_N). The H_{max} describing the stiffest region increased only during CV and CV_{LB} ($p<0.001$) at PEEP3 whereas at PEEP6, it increased only during CV ($p<0.001$). The SD of H representing heterogeneity of regional lung stiffness increased only during CV at PEEP6 ($p<0.001$). At 60 min, both H_{min} and H_{max} were higher during CV than during CV_{LB} and VV_N ($p<0.001$). At PEEP6, the SD of H during CV at 60 min was also different from that during CV_{LB} and VV_N ($p<0.001$).

At time 0, the mean peak airway pressure ($_{PAP}$) started at 10 and 14 cmH_2O at PEEP3 and PEEP6, respectively. By 60 min during CV, PAP reached 14 and 18 cmH_2O at PEEP3 and PEEP6, respectively. The corresponding percent increases in PAP (ΔPAP) during CV_{LB} and VV_N were much smaller than during CV (Fig. 4). For CV and CV_{LB} at both PEEPs, ΔPAP was significantly affected by time ($p<0.001$) whereas during VV_N, ΔPAP depended on time only at PEEP3 ($p<0.001$). The PAP at 60 min was significantly lower during VV_N and CV_{LB} than during CV ($p<0.005$). At 60 min, ΔPAP was significantly affected by both ventilation mode and PEEP ($p<0.001$) and it was higher during CV than VV_N and CV_{LB} ($p<0.001$). At 60 minutes, ΔPAP at PEEP3 was higher during CV_{LB} than during VV_N ($p<0.02$).

The PaO_2, $\%sO_2$ as well as A-a gradient were affected by both ventilation ($p<0.001$) and PEEP ($p<0.03$) (Fig. 5). At PEEP3, PaO_2 and $\%sO_2$ were higher during VV_N and CV_{LB} compared to CV ($p<0.005$ and $p<0.02$, respectively). At PEEP6, PaO_2 during VV_N was higher than during CV ($p=0.012$). Furthermore, at PEEP3, A-a gradient during CV was higher than during VV_N and CV_{LB} ($p<0.05$), while at PEEP6, the A-a gradient was higher during CV than VV_N ($p=0.019$).

Example Western blots for SP-B, proSP-B and E-cadherin are demonstrated in Fig. 6 and group means normalized to the mean of the unventilated groups are given in Fig. 7. At PEEP3, SP-B was significantly higher than at PEEP6 ($p<0.001$) and while CV and CV_{LB} had lower levels than the unventilated group ($p<0.001$), VV_N had a higher level than any other group ($p<0.001$). The proSP-B also significantly depended on PEEP ($p=0.004$). At PEEP3, animals during CV and VV_N, but not CV_{LB}, had higher proSP-B than in the unventilated group ($p=0.006$ and $p=0.001$, respectively). For E-cadherin, we detected its 85 kDa soluble form in the lavage with significant interaction between PEEP and ventilation ($p<0.03$). During CV and VV_N, the E-cadherin was higher at PEEP6 than at PEEP3 ($p<0.05$) whereas at PEEP3, E-cadherin during CV_{LB} was higher than during CV, VV_N ($p<0.05$) and the unventilated ($p<0.003$) groups. Excluding the unventilated groups, there was no difference in any of these expressions among the ventilation groups at PEEP6.

Discussion

The lower inflection point (LIP) on the pressure-volume (P-V) curve is often thought to represent alveolar recruitment and setting the PEEP just above LIP can improve lung function in patients with ARDS and increase their chances of early weaning [33]. However, Mergoni et al. showed that LIP is not a good indicator of alveolar recruitment [34]. Indeed, while the regional P-V curves can exhibit behavior consistent with recruitment, the total P-V curve may not show an LIP [9]. It is generally accepted that high

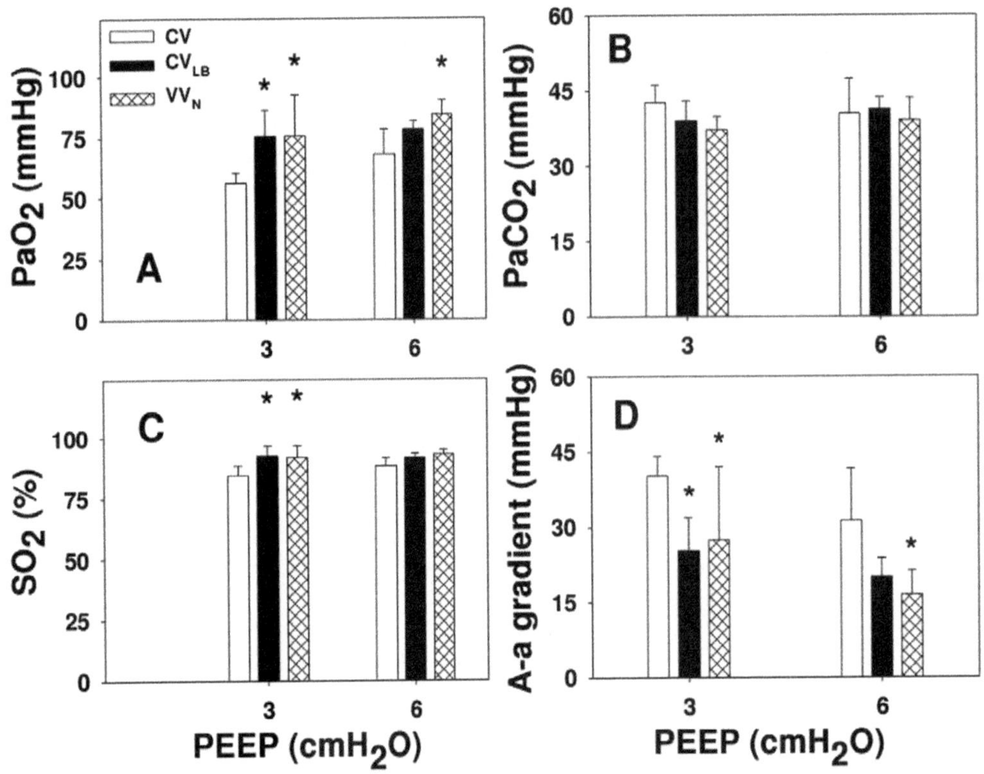

Figure 5. Gas exchange as a function of ventilation mode. The graphs show the partial pressures of oxygen and carbon dioxide (PaO_2 and $PaCO_2$, respectively, in panels A and B), percent oxygen saturation (SO_2, panel C) and Alveolar-arterial gradient (A–a gradient, panel D) obtained at the end of 60 min ventilation using CV, CV_{LB} or VV_N in HCl-injured mice at PEEPs of 3 and 6 cmH_2O. * denotes significant difference compared to CV ($p<0.05$).

V_T and plateau airway pressure lead to lung injury [35] via exposing the epithelium to large stresses and strains [36]. Recently, Chiumello et al. [37] showed that V_T and airway pressure are not adequate surrogates of lung stress and strain due to the large heterogeneity of lung stiffness in ARDS patients in agreement with the above notion that regional heterogeneity is not reflected in the total P-V curve. A mathematical index of stress and strain was introduced by Brunner and Wysocki [38] to predict the best combination of V_T and f as a function of PEEP. Other methods include titrating the PEEP for best gas exchange [39] or using

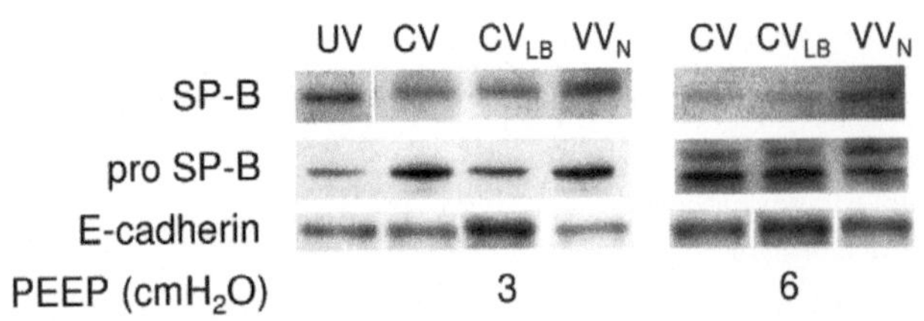

Figure 6. Example Western blots. Representative blots are shown for surfactant protein (SP)-B and its pro form (proSP-B) as well as E-cadherin obtained from lung homogenates and lavage fluid at the conclusion of 60 min of ventilation using CV, CV_{LB} or VV_N at PEEPs of 3 (left) and 6 cmH_2O (right). UV denotes blots from HCl-injured but unventilated group of animals. Note the small gaps between several images for a given protein. These blots are from the same film, but not in the same order as the rest of the blots. The original image was first cut into pieces, without changing the image, and then reassembled in the desired order.

dynamic features derived from respiratory mechanics [9,11,31].The general findings are that within some limits, a higher PEEP is better in terms of lung function as it maintains the lung open.

The above studies do not address two aspects of mechanical ventilation. The first is the combined effects of PEEP and less conventional modes of ventilation such as CV with RM and VV on physiology. The second is that these studies do not consider epithelial mechanobiology. It is implicitly assumed that the PEEP-V_T combination best for physiological outcome also protects the epithelium. In this study, we compared the effects of three ventilation modes on physiology in HCL-injured mice at two PEEP levels. We also investigated the effects of ventilation mode on SP-B that is important in maintaining low surface tension as well as epithelial cell injury characterized by E-cadherin. The main results are that *1)* ventilation performance was PEEP dependent; *2)* the performance of VV_N and CV_{LB} was significantly better than that of CV, the current clinical method; *3)* there were small, but consistent improvements in lung physiology during VV_N compared to CV_{LB}; 4) despite better physiology including mechanical parameters, airway pressures and gas exchange at the higher PEEP, the lower PEEP was advantageous in terms of surfactant and epithelial cell integrity; and 5) VV_N was superior to CV_{LB} in terms of epithelial mechanobiology.

Effects of PEEP on Ventilation Performance

The effects of PEEP on lung physiology and biology have been studied in numerous animal models of lung injury [8,11,15,16,40–43]. In our study, we compared the performance of three

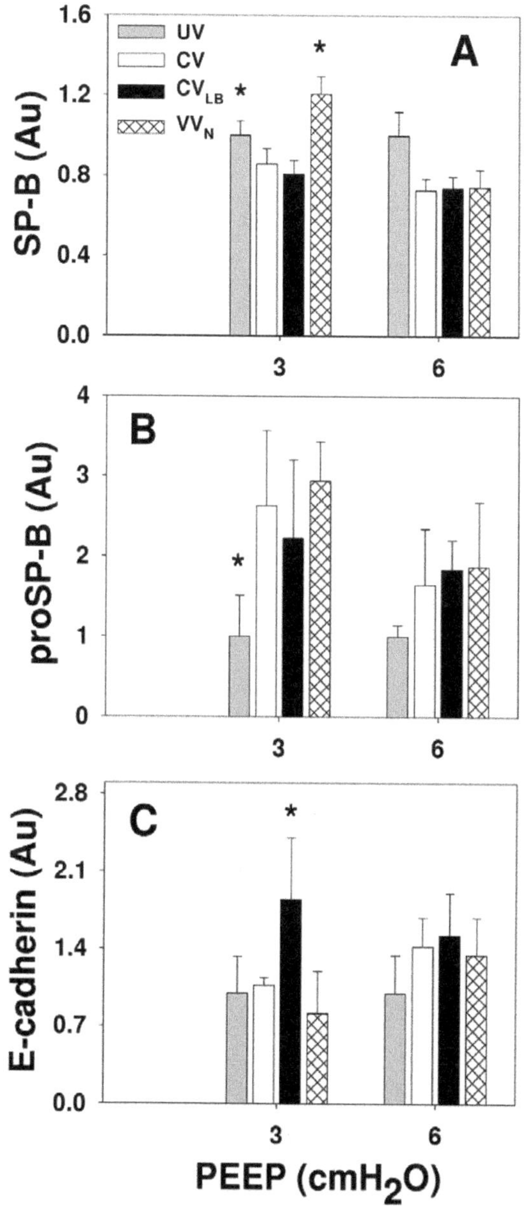

Figure 7. Analysis of Western blots. Graphs show the relative amounts of SP-B, proSP-B and soluble E-cadherin in HCl-injured mice at the conclusion of 60 min of ventilation using CV, CV_{LB} or VV_N at PEEPs of 3 and 6 cmH_2O. The data are normalized with the corresponding mean values of the unventilated group with baseline injury. The two bars in the unventilated group at the two PEEP levels correspond to samples from the same lungs on two separate Western blots. * denotes significant difference compared to CV ($p<0.05$).

ventilation modes at PEEP3 and PEEP6 that are well tolerated by the mouse during injury [14,15]. The PEEP3 was a necessary minimum PEEP because in a preliminary study, we found that mice with HCl injury did not tolerate the 60 min ventilation at a PEEP of 1 cmH_2O. The PEEP6 in mice may correspond to a PEEP of about 15 cmH_2O in humans. The reason is as follows. The *in situ* transpulmonary pressure of the mouse lung is between 1.5 and 2 cmH_2O and PEEP6 is about 3–4 times higher than this range. In humans, the average transpulmonary pressure is around 5 cmH_2O. Thus, our PEEP6 in the mouse might correspond to

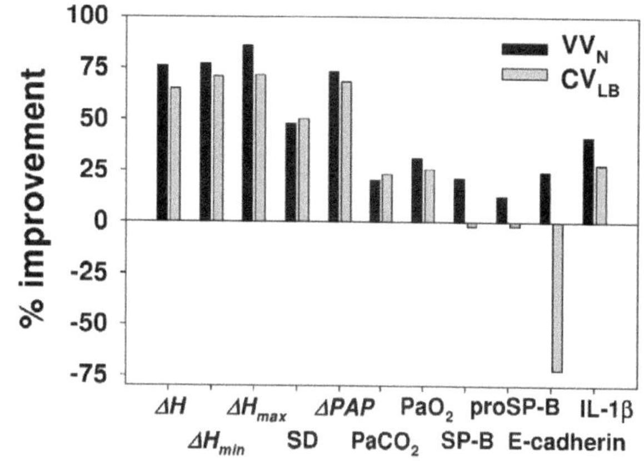

Figure 8. Comparison of ventilation modes. The graph compares the average improvements of VV_N and CV_{LB} over CV. The SD bar represents different experimental conditions including 2 PEEP levels and normal lung from our previous study [27]. Only those parameters are included for which either CV_{LB}, VV_N or both significantly improved over CV. See text for more explanation.

a PEEP of 15–20 cmH_2O PEEP in human patients, not uncommon in the ICU. However, the peak airway pressures at PEEP6 barely reached 20 cmH_2O even during CV. Allen et al. [14] investigated the effects of deep inspiration on lung mechanics and found a strong PEEP dependence albeit during short term ventilation. Since the effects of PEEP were studied in the same animals, it is possible that the results at different PEEP levels were not independent of each other. This was ruled out here because each mouse was ventilated for an hour at a single PEEP.

Respiratory mechanics. Using the constant phase model (Fig. 2), all parameters were highly PEEP dependent consistent with the data of Allen et al. in that the recovery in H over a 7-min period after a large breath was inversely related to PEEP between 1 and 6 cmH_2O in both acid and endotoxin models of ALI [14]. All parameters during CV substantially increased with time suggesting massive derecruitment. In contrast, respiratory mechanics at PEEP3 during VV_N or CV_{LB} reached their respective plateau levels after 30 min implying that CV_{LB} and VV_N maintained the lung open albeit at different levels.

In the heterogeneous model (Fig. 3), only H_{max} and SD were PEEP dependent. Although H_{min} was not PEEP dependent, its percent increase highly depended on PEEP with a rate much slower at the higher PEEP. This is perhaps not surprising, since a higher PEEP prevents or slows down the collapsing process [2]. At PEEP6, there was no change in ΔPAP during VV_N. Thus, PEEP significantly affected the rate of lung collapse in a ventilation mode dependent manner. Further, in agreement with Chiumello et al. [37], the SD of H increased over time during CV becoming significantly larger than during CV_{LB} or VV_N. Such increase in regional lung stiffness is directly related to the heterogeneity of regional strain and hence is at the heart of inducing VILI during CV.

Gas exchange. When examined separately for each ventilation mode, only PaO_2 of the CV group was PEEP dependent. Thus, even though raising the PEEP opened more regions and resulted in lower stiffness in the VV_N and CV_{LB} groups, it did not significantly improve PaO_2. Nevertheless, the A-a gradient significantly improved by increasing the PEEP during VV_N implying that the higher PEEP also reduced ventilation/perfusion mismatch.

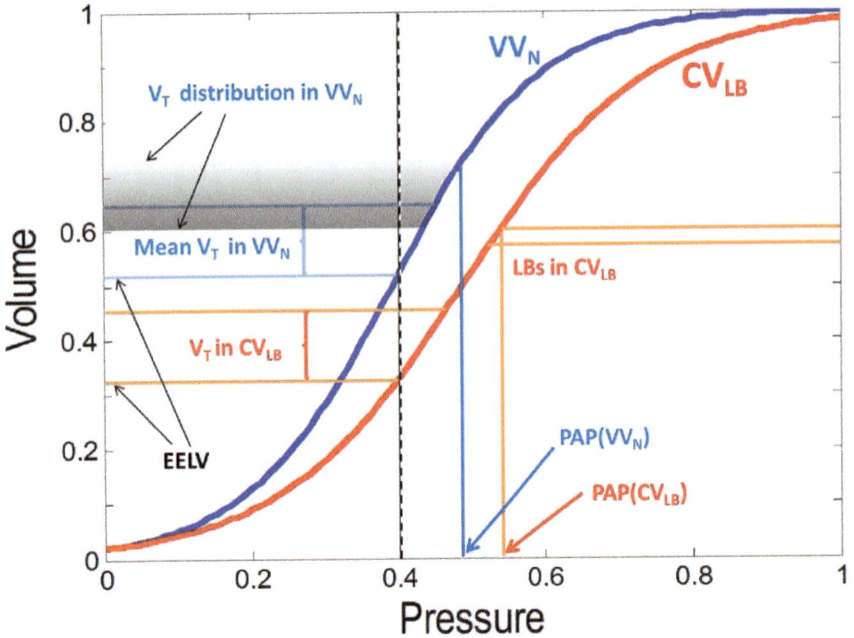

Figure 9. Schematic representation of ventilation along the normalized pressure-volume curve during CV_{LB} and VV_N. The vertical dashed black line at 0.4 represents PEEP. The intersections of PEEP and the pressure-volume curves mark the end-expiratory lung volumes (EELV) during the two ventilation modes upon which V_T is superimposed. For CV_{LB} (red), we also show the large breaths (LB) and the corresponding peak airway pressure (PAP). For VV_N (blue), there is a range of V_Ts superimposed on EELV. The corresponding end-inspiratory volumes have a distribution shown by the shaded area. The probability of a given tidal volume is proportional to the gray scale. Also notice that the mean V_T in VV_N is the same as in CV_{LB}, but the distribution of V_Ts goes below the V_T of CV_{LB} and stretches up to the LBs in CV_{LB}.

Surfactant protein and epithelial integrity. The levels of both SP-B and proSP-B highly depended on PEEP (Fig. 7). At PEEP6, SP-B in all ventilation modes dropped compared to unventilated mice suggesting that a static stretch higher than the physiological level corresponding to FRC hinders SP-B release. At PEEP3, SP-B during CV and CV_{LB} was lower while during VV_N, it was higher than the unventilated group in agreement with findings in normal guinea pigs [44]. Because proSP-B also increased at PEEP3 with VV_N, this is evidence that variability in stretch amplitude delivered to the epithelium during VV_N triggers a dynamic mechanotransduction in alveolar epithelial type II cells that leads to the upregulation of both the release and production of SP-B even in severely injured lungs. Interestingly, while the proSP-B increased during CV, CV_{LB} was not able to increase proSP-B above the level of the unventilated group at PEEP3.

With regard to the integrity of the epithelium, we examined E-cadherin in the lavage fluid. The E-cadherin is a 120-kDa transmembrane glycoprotein localized to the lateral sides of epithelial cells linking them together and plays an important role in cell-cell mechanical signaling [45]. It is expressed in epithelial tissues but in the normal lung, it is not present in the alveolar liquid lining. Hence, appearance of the soluble fragments of E-cadherin in the lavage fluid is a clear indication of epithelial cell injury [42,46,47]. We found the 85 kDa soluble fragments of E-cadherin in the lavage fluid of unventilated animals which is likely due to the combined effects of HCl injury and the lavage process itself (Fig. 7C). Interestingly, however, while the higher PEEP was advantageous in terms of mechanics, it increased epithelial cell injury during CV and VV_N. Unexpectedly, E-cadherin was not increased following CV at PEEP3. Perhaps the 60 min ventilation at PEEP3 was not sufficient in this mouse model of ALI to damage the epithelium beyond the unventilated group. On the other hand, stretch due to the intermittent large breaths in CV_{LB} without the compensating effects of variability started to break up the link between epithelial cells that showed up in the bulk biochemical assays even at PEEP3. Comparing the patterns of proSP-B and E-cadherin in Fig. 7B and C, we can see an apparent inverse relation at PEEP3: the highest level of proSP-B was associated with the lowest level of E-cadherin and vice versa. We can interpret this as follows. Increasing E-cadherin fragments in the lavage is likely due to overstretching the HCl-injured epithelium at least regionally. This in turn results in a breakup of E-cadherin anchoring epithelial cells. If the regional signal is strong enough, it will also show up in the bulk biochemical assays from the whole lung lavage. Furthermore, this breakup also leads to a loss of mechanical coupling among the cells. Since type II epithelial cells are stiffer than type I cells [48], weakening this mechanical interaction reduces the deformation of type II cells during lung inflation with subsequent reduction in surfactant secretion. We conclude that at least in this mouse model of ALI, increasing PEEP has a detrimental effect on surfactant and it does not protect the epithelium with likely consequences on longer term organ level physiology.

The implications of our results cannot be easily extrapolated to ventilating humans. We nevertheless note that a recent reanalysis of data from trials demonstrated that ventilating ARDS patients at a higher PEEP increased hospital survival whereas in patients without ARDS a lower PEEP tended to be more beneficial [49]. Regarding the gas exchange results (Fig. 5), the HCl treatment in this study is closer to a mild form of ARDS. However, ventilation in the human studies was carried out with CV and no epithelial markers were reported. While our results suggest that ventilation at PEEP6 leads to better organ level physiology, ventilation at PEEP3 leads to better epithelial protection. Whether or not these findings are specific to the mouse will need to be tested in future studies.

Comparing the Performance of Ventilation Modes

Conventional ventilation vs alternative modes of ventilation. To better understand how alternative approaches improve lung physiology compared to CV, we computed the percent improvement at 60 min in some of the parameters during VV_N and CV_{LB} compared to CV. Briefly, at PEEP6, the improvements during VV_N and CV_{LB} were higher than at PEEP3. Moreover, at both PEEP levels, H_{min} from the heterogeneous model and its percent increase were significantly improved during VV_N and CV_{LB} compared to CV. Therefore, VV_N and CV_{LB} could keep the most compliant regions of the lung open longer while those regions during CV gradually collapsed which resulted in higher peak airway pressures during CV. Both VV_N and CV_{LB} improved PaO_2, $\%sO_2$ and A-a gradient at PEEP3 while at PEEP6, only VV_N was able significantly improve PaO_2 and A-a gradient over CV. This might be due to a nonlinear synergistic effect between variability in VV_N and PEEP. Finally, the only parameter that was better during CV than CV_{LB} (but not VV_N) was E-cadherin at PEEP3 (Fig. 7C). While the reason is unclear, we note that recruitment maneuvers were recently reported to increase inflammation and fibrogenic response with worsening lung function [50]. It is important to further investigate this phenomenon.

Comparing the performance of CV_{LB} and VV_N. Funk et al. [23] has shown that the biologically variable ventilation improves oxygenation and lung compliance over conventional ventilation with recruitment maneuvers in a porcine model of lung injury. Nevertheless, in that study, the recruitment was delivered once every hour over a 5 hour ventilation period and it is unclear whether the RMs were matched to the largest breaths in VV. Since the effect of RM is transient in lung injury [21], it is likely that this frequency of RM is not enough to sustain an open lung. Indeed, Allen et al. [51] demonstrated that frequently delivered RM's can improve gas exchange and lung mechanics over CV in normal mice. We constructed the large breaths in CV_{LB} so that they matched the 10 largest breaths of VV_N in 5 minutes both in size and timing with 1 large breath every 30 second, the same rate as in the study of Allen et al. [51]. Thus, our study constitutes a fair comparison of CV_{LB} and VV_N.

We also examined the mean airway pressure at PEEP3 and found that at 60 min, there was no difference between CV_{LB} and VV_N (data not shown). However, at 60 min, H (Fig. 2) and ΔPAP (Fig. 4) at PEEP6 were lower during VV_N than CV_{LB}. Additionally, PaO_2 and A-a gradient at PEEP6 was better during VV_N but not CV_{LB} than CV. The likely reason is that CV_{LB} could not improve over CV due to inter-animal variability which was overcome by VV_N suggesting that VV_N is more robust. In order to further explore the differences between these two modes of ventilation, we compiled data in all parameters for which there was a statistically significant improvement of CV_{LB} and/or VV_N over CV. The results are summarized in Fig. 8. Since the VV_N and CV_{LB} were identical in our previous study [27], we also included physiology data from normal mice as well as IL-1β as an indicator of lung injury. In Fig. 8, each bar represents a single number, the percent improvement in that parameter. For example, the first bar was calculated by taking the relative difference of the means of ΔH at 60 min during VV_N and CV for a given condition such as the data at PEEP3. Next, this improvement was calculated and averaged for all available conditions including PEEP3, PEEP6 as well as PEEP3 in normal mice from our previous study [27]. The second bar was obtained similarly but now for CV_{LB}. Thus, the corresponding bars for each parameter represent paired data and a comprehensive comparison of VV_N and CV_{LB} can be obtained by using a non-parametric paired t-test. The median improvements of CV_{LB} and VV_N over CV were 28% and 41%, respectively ($p=0.005$). It is also noteworthy that VV_N but not CV_{LB} caused significantly less inflammatory response than CV as indicated by significantly lower level of IL-1β and CV_{LB} also caused more epithelial injury than CV ($p<0.01$).

Since both VV_N and CV_{LB} had already reached their plateau in every mechanical parameter by 30 minutes of ventilation, we combined all the data points for each parameter from 30 minutes to 60 minutes in the CV_{LB} and VV_N groups. At both PEEPs, the plateau levels of H, ΔH and ΔPAP were significantly higher during CV_{LB} than VV_n. This suggests that VV_N better recruits the lung independent of PEEP which we attribute to the presence of intermediate V_Ts in agreement with previous computer model-based predictions [52]. The pressure-volume curve of the lung during VV_N must therefore be different than during CV_{LB} and ventilation is superimposed on different lung volumes with different compliance. This scenario is summarized schematically in Fig. 9. We can see that the V_Ts are superimposed on the same PEEP (dashed line) but different end-expiratory lung volumes (EELV) which results in a higher compliance (slope of the pressure-volume curve) and a lower peak airway pressure (PAP) during VV_N. At the start of ventilation (time 0), the pressure-volume curve would be somewhat to the left of the VV_N curve (blue) whereas after 60 min of ventilation with CV, the pressure-volume curve would be to the far right of the CV_{LB} curve (red). Thus, better and safer stretching of the lung and especially the epithelium with reduced repetitive opening and closing during VV_N eventually results in molecular changes including increased surfactant and less injury such as lower level of soluble E-cadherin.

To summarize, our first conclusion is that to the extent that our results are not specific to mice, there is room for improvement on current clinical ventilation approaches. Second, PEEP had a significant effect on the performance of all ventilation methods. With regard to physiology, the higher PEEP protected the lung from collapse and reduced tissue heterogeneity. However, the lower PEEP better protected the epithelium and had a positive effect on surfactant especially during VV. Thus, selection of optimal PEEP should be based on balancing organ level physiology with epithelial injury. Third, VV better maintains an open lung, which allows ventilation at a lower PEEP with lower maximum lung stretch reducing epithelial injury. Additionally, the dynamic mechanotransduction induced by VV has a beneficial effect on surfactant metabolism.

Author Contributions

Conceived and designed the experiments: BS AT. Performed the experiments: AT HH EB. Analyzed the data: AT HH EB BS. Contributed reagents/materials/analysis tools: AT EB. Wrote the paper: AT BS.

References

1. Amato MB, Barbas CS, Medeiros DM, Magaldi RB, Schettino GP, et al. (1998) Effect of a protective-ventilation strategy on mortality in the acute respiratory distress syndrome. N Engl J Med 338: 347–354.
2. Halter JM, Steinberg JM, Schiller HJ, DaSilva M, Gatto LA, et al. (2003) Positive end-expiratory pressure after a recruitment maneuver prevents both

alveolar collapse and recruitment/derecruitment. Am J Respir Crit Care Med 167: 1620–1626.
3. Jardin F, Farcot JC, Boisante L, Curien N, Margairaz A, et al. (1981) Influence of positive end-expiratory pressure on left ventricular performance. N Engl J Med 304: 387–392.
4. Bo G, Hauge A, Nicolaysen G (1977) Alveolar pressure and lung volume as determinants of net transvascular fluid filtration. J Appl Physiol 42: 476–482.
5. Tremblay LN, Slutsky AS (1998) Ventilator-induced injury: from barotrauma to biotrauma. Proc Assoc Am Physicians 110: 482–488.
6. Tomashefski JF Jr (2000) Pulmonary pathology of acute respiratory distress syndrome. Clin Chest Med 21: 435–466.
7. Bilek AM, Dee KC, Gaver DP 3rd (2003) Mechanisms of surface-tension-induced epithelial cell damage in a model of pulmonary airway reopening. J Appl Physiol 94: 770–783.
8. Muscedere JG, Mullen JB, Gan K, Slutsky AS (1994) Tidal ventilation at low airway pressures can augment lung injury. Am J Respir Crit Care Med 149: 1327–1334.
9. Bellardine Black CL, Hoffman AM, Tsai LW, Ingenito EP, Suki B, et al. (2007) Relationship between dynamic respiratory mechanics and disease heterogeneity in sheep lavage injury. Crit Care Med 35: 870–878.
10. Hodgson CL, Tuxen DV, Bailey MJ, Holland AE, Keating JL, et al. (2011) A positive response to a recruitment maneuver with PEEP titration in patients with ARDS, regardless of transient oxygen desaturation during the maneuver. J Intensive Care Med 26: 41–49.
11. Kostic P, Zannin E, Andersson Olerud M, Pompilio PP, Hedenstierna G, et al. (2011) Positive end-expiratory pressure optimization with forced oscillation technique reduces ventilator induced lung injury: a controlled experimental study in pigs with saline lavage lung injury. Crit Care 15: R126.
12. Krebs J, Pelosi P, Tsagogiorgas C, Alb M, Luecke T (2009) Effects of positive end-expiratory pressure on respiratory function and hemodynamics in patients with acute respiratory failure with and without intra-abdominal hypertension: a pilot study. Crit Care 13: R160.
13. Zhao Z, Steinmann D, Frerichs I, Guttmann J, Moller K (2010) PEEP titration guided by ventilation homogeneity: a feasibility study using electrical impedance tomography. Crit Care 14: R8.
14. Allen G, Bates JH (2004) Dynamic mechanical consequences of deep inflation in mice depend on type and degree of lung injury. J Appl Physiol 96: 293–300.
15. Allen G, Lundblad LK, Parsons P, Bates JH (2002) Transient mechanical benefits of a deep inflation in the injured mouse lung. J Appl Physiol 93: 1709–1715.
16. Allen GB, Leclair T, Cloutier M, Thompson-Figueroa J, Bates JH (2007) The response to recruitment worsens with progression of lung injury and fibrin accumulation in a mouse model of acid aspiration. Am J Physiol Lung Cell Mol Physiol 292: L1580–1589.
17. Meade MO, Cook DJ, Griffith LE, Hand LE, Lapinsky SE, et al. (2008) A study of the physiologic responses to a lung recruitment maneuver in acute lung injury and acute respiratory distress syndrome. Respir Care 53: 1441–1449.
18. Meade MO, Cook DJ, Guyatt GH, Slutsky AS, Arabi YM, et al. (2008) Ventilation strategy using low tidal volumes, recruitment maneuvers, and high positive end-expiratory pressure for acute lung injury and acute respiratory distress syndrome: a randomized controlled trial. JAMA 299: 637–645.
19. Lefevre GR, Kowalski SE, Girling LG, Thiessen DB, Mutch WA (1996) Improved arterial oxygenation after oleic acid lung injury in the pig using a computer-controlled mechanical ventilator. Am J Respir Crit Care Med 154: 1567–1572.
20. Arold SP, Mora R, Lutchen KR, Ingenito EP, Suki B (2002) Variable tidal volume ventilation improves lung mechanics and gas exchange in a rodent model of acute lung injury. Am J Respir Crit Care Med 165: 366–371.
21. Bellardine CL, Hoffman AM, Tsai L, Ingenito EP, Arold SP, et al. (2006) Comparison of variable and conventional ventilation in a sheep saline lavage lung injury model. Crit Care Med 34: 439–445.
22. Carvalho AR, Spieth PM, Guldner A, Cuevas M, Carvalho NC, et al. (2011) Distribution of regional lung aeration and perfusion during conventional and noisy pressure support ventilation in experimental lung injury. J Appl Physiol 110: 1083–1092.
23. Funk DJ, Graham MR, Girling LG, Thliveris JA, McManus BM, et al. (2004) A comparison of biologically variable ventilation to recruitment manoeuvres in a porcine model of acute lung injury. Respir Res 5: 22.
24. Mutch WA, Harms S, Lefevre GR, Graham MR, Girling LG, et al. (2000) Biologically variable ventilation increases arterial oxygenation over that seen with positive end-expiratory pressure alone in a porcine model of acute respiratory distress syndrome. Crit Care Med 28: 2457–2464.
25. Mutch WA, Harms S, Ruth Graham M, Kowalski SE, Girling LG, et al. (2000) Biologically variable or naturally noisy mechanical ventilation recruits atelectatic lung. Am J Respir Crit Care Med 162: 319–323.
26. Spieth PM, Carvalho AR, Pelosi P, Hoehn C, Meissner C, et al. (2009) Variable tidal volumes improve lung protective ventilation strategies in experimental lung injury. Am J Respir Crit Care Med 179: 684–693.
27. Thammanomai A, Hueser LE, Majumdar A, Bartolak-Suki E, Suki B (2008) Design of a new variable-ventilation method optimized for lung recruitment in mice. J Appl Physiol 104: 1329–1340.
28. Lutchen KR, Yang K, Kaczka DW, Suki B (1993) Optimal ventilation waveforms for estimating low-frequency respiratory impedance. J Appl Physiol 75: 478–488.
29. Hantos Z, Daroczy B, Suki B, Nagy S, Fredberg JJ (1992) Input impedance and peripheral inhomogeneity of dog lungs. J Appl Physiol 72: 168–178.
30. Ito S, Ingenito EP, Arold SP, Parameswaran H, Tgavalekos NT, et al. (2004) Tissue heterogeneity in the mouse lung: effects of elastase treatment. J Appl Physiol 97: 204–212.
31. Kaczka DW, Hager DN, Hawley ML, Simon BA (2005) Quantifying mechanical heterogeneity in canine acute lung injury: impact of mean airway pressure. Anesthesiology 103: 306–317.
32. Csendes T (1988) Nonlinear parameter estimation by global optimization-efficiency and reliability. Acta Cybernet 8: 361–370.
33. Amato MB, Barbas CS, Medeiros DM, Schettino Gde P, Lorenzi Filho G, et al. (1995) Beneficial effects of the "open lung approach" with low distending pressures in acute respiratory distress syndrome. A prospective randomized study on mechanical ventilation. Am J Respir Crit Care Med 152: 1835–1846.
34. Mergoni M, Volpi A, Bricchi C, Rossi A (2001) Lower inflection point and recruitment with PEEP in ventilated patients with acute respiratory failure. J Appl Physiol 91: 441–450.
35. (2000) Ventilation with lower tidal volumes as compared with traditional tidal volumes for acute lung injury and the acute respiratory distress syndrome. The Acute Respiratory Distress Syndrome Network. N Engl J Med 342: 1301–1308.
36. Gattinoni L, Caironi P, Carlesso E (2005) How to ventilate patients with acute lung injury and acute respiratory distress syndrome. Curr Opin Crit Care 11: 69–76.
37. Chiumello D, Carlesso E, Cadringher P, Caironi P, Valenza F, et al. (2008) Lung stress and strain during mechanical ventilation for acute respiratory distress syndrome. Am J Respir Crit Care Med 178: 346–355.
38. Brunner JX, Wysocki M (2009) Is there an optimal breath pattern to minimize stress and strain during mechanical ventilation? Intensive Care Med 35: 1479–1483.
39. Badet M, Bayle F, Richard JC, Guerin C (2009) Comparison of optimal positive end-expiratory pressure and recruitment maneuvers during lung-protective mechanical ventilation in patients with acute lung injury/acute respiratory distress syndrome. Respir Care 54: 847–854.
40. Chiumello D, Pristine G, Slutsky AS (1999) Mechanical ventilation affects local and systemic cytokines in an animal model of acute respiratory distress syndrome. Am J Respir Crit Care Med 160: 109–116.
41. Martynowicz MA, Walters BJ, Hubmayr RD (2001) Mechanisms of recruitment in oleic acid-injured lungs. J Appl Physiol 90: 1744–1753.
42. Thammanomai A, Majumdar A, Bartolak-Suki E, Suki B (2007) Effects of reduced tidal volume ventilation on pulmonary function in mice before and after acute lung injury. J Appl Physiol 103: 1551–1559.
43. Tremblay L, Valenza F, Ribeiro SP, Li J, Slutsky AS (1997) Injurious ventilatory strategies increase cytokines and c-fos m-RNA expression in an isolated rat lung model. J Clin Invest 99: 944–952.
44. Arold SP, Suki B, Alencar AM, Lutchen KR, Ingenito EP (2003) Variable ventilation induces endogenous surfactant release in normal guinea pigs. Am J Physiol Lung Cell Mol Physiol 285: L370–375.
45. le Duc Q, Shi Q, Blonk I, Sonnenberg A, Wang N, et al. (2010) Vinculin potentiates E-cadherin mechanosensing and is recruited to actin-anchored sites within adherens junctions in a myosin II-dependent manner. J Cell Biol 189: 1107–1115.
46. Goto T, Ishizaka A, Katayama M, Kohno M, Tasaka S, et al. (2010) Involvement of E-cadherin cleavage in reperfusion injury. Eur J Cardiothorac Surg 37: 426–431.
47. Goto Y, Uchida Y, Nomura A, Sakamoto T, Ishii Y, et al. (2000) Dislocation of E-cadherin in the airway epithelium during an antigen-induced asthmatic response. Am J Respir Cell Mol Biol 23: 712–718.
48. Azeloglu EU, Bhattacharya J, Costa KD (2008) Atomic force microscope elastography reveals phenotypic differences in alveolar cell stiffness. J Appl Physiol 105: 652–661.
49. Briel M, Meade M, Mercat A, Brower RG, Talmor D, et al. (2010) Higher vs lower positive end-expiratory pressure in patients with acute lung injury and acute respiratory distress syndrome: systematic review and meta-analysis. JAMA 303: 865–873.
50. Santiago VR, Rzezinski AF, Nardelli LM, Silva JD, Garcia CS, et al. (2010) Recruitment maneuver in experimental acute lung injury: the role of alveolar collapse and edema. Crit Care Med 38: 2207–2214.
51. Allen GB, Suratt BT, Rinaldi L, Petty JM, Bates JH (2006) Choosing the frequency of deep inflation in mice: balancing recruitment against ventilator-induced lung injury. Am J Physiol Lung Cell Mol Physiol 291: L710–717.
52. Suki B, Alencar AM, Sujeer MK, Lutchen KR, Collins JJ, et al. (1998) Life-support system benefits from noise. Nature 393: 127–128.

18

Discrimination of Ultrasonic Vocalizations by CBA/CaJ Mice (*Mus musculus*) Is Related to Spectrotemporal Dissimilarity of Vocalizations

Erikson G. Neilans[1], David P. Holfoth[1], Kelly E. Radziwon[1], Christine V. Portfors[2], Micheal L. Dent[1]*

1 Department of Psychology, University at Buffalo, the State University of New York, Buffalo, New York, United States of America, **2** School of Biological Sciences, Washington State University-Vancouver, Vancouver, Washington, United States of America

Abstract

The function of ultrasonic vocalizations (USVs) produced by mice (*Mus musculus*) is a topic of broad interest to many researchers. These USVs differ widely in spectrotemporal characteristics, suggesting different categories of vocalizations, although this has never been behaviorally demonstrated. Although electrophysiological studies indicate that neurons can discriminate among vocalizations at the level of the auditory midbrain, perceptual acuity for vocalizations has yet to be determined. Here, we trained CBA/CaJ mice using operant conditioning to discriminate between different vocalizations and between a spectrotemporally modified vocalization and its original version. Mice were able to discriminate between vocalization types and between manipulated vocalizations, with performance negatively correlating with spectrotemporal similarity. That is, discrimination performance was higher for dissimilar vocalizations and much lower for similar vocalizations. The behavioral data match previous neurophysiological results in the inferior colliculus (IC), using the same stimuli. These findings suggest that the different vocalizations *could* carry different meanings for the mice. Furthermore, the finding that behavioral discrimination matched neural discrimination in the IC suggests that the IC plays an important role in the perceptual discrimination of vocalizations.

Editor: Bernd Sokolowski, University of South Florida, United States of America

Funding: This work was supported by NIH DC009483 and DC012302 to MLD. The funders had no role in study design, data collection and analysis, decision to publish, or preparation of the manuscript.

Competing Interests: The authors have declared that no competing interests exist.

* E-mail: mdent@buffalo.edu

Introduction

Many animals, including humans, use sound to communicate, and the vocal repertoire of a species is often quite large (e.g. [1]). This suggests that different vocalizations convey different information to receivers. The vervet monkey (*Chlorocebus pygerythrus*) is a classic example of this; scouts emit one vocalization for a flying predator and another for a terrestrial predator [2]. Moreover, the finding that vervet monkeys have specific behaviors for each vocalization indicates that they can discriminate between the vocalizations and interpret their meaning. In many other animal species, however, the meaning of different vocalizations is not known and it is not clear whether different vocalizations can be discriminated.

Mice (*Mus musculus*) produce an array of ultrasonic vocalizations (USVs) in a variety of behavioral contexts. Males emit USVs in the presence of females, females produce them in the presence of other females, and infants produce them when separated from their mothers (reviewed in [3]). The vocalizations emitted by mice have been classified into syllable types based on spectrotemporal characteristics, but the numbers and types of categories vary widely [3], [4], [5], [6], [7]. Moreover, these categories have been determined by visual or statistical analyses of spectrograms rather than assessing whether mice can discriminate between the different syllables. The discrimination ability of mice for ultrasonic vocalizations is not yet known.

Playback studies have been useful for establishing the functional importance of mouse USVs, but small effect sizes, habituation, and difficulty interpreting results all hinder their usefulness in understanding mouse communication. Psychophysical experiments, on the other hand, allow us to test the perceptual limits of discriminating between acoustic stimuli [8]. Here, we used psychophysical methods to test the ability of mice to discriminate USVs.

We used the same stimuli that were used to examine neural selectivity to vocalizations in the inferior colliculus (IC) of mice [9]. Holmstrom and colleagues presented awake and restrained mice with four different vocalizations and spectrotemporal manipulations of those vocalizations. These USVs elicited heterogeneous patterns of responses across the neural population and within individual neurons. Neurons also responded differently to unaltered calls compared to those that had modified spectrotemporal properties, suggesting a distinct neural representation of each vocalization. These results suggested that the IC is critical for the encoding of behaviorally relevant sounds, even though the behavioral relevancy of those sounds has not yet been determined. As a first attempt to understand how trained, reliable mouse observers perceive ultrasonic calls, we used psychoacoustic techniques to test the discrimination ability for USVs. We found that the mice were able to discriminate between vocalizations with performance that was always better than chance and that performance was best for USVs that were spectrotemporally dissimilar. These behavioral results are similar to the neural results obtained in the IC, using the same stimuli, which suggests that the IC plays an important role in the perceptual discrimination of vocalizations.

Materials and Methods

Ethics Statement

All procedures were approved by the University at Buffalo, SUNY's Institutional Animal Care and Use Committee under protocol PSY13056N.

Subjects

We tested how well five mice with normal hearing abilities (CBA/CaJ strain; 3 males, 2 females) could discriminate different mouse USVs. Training began when the mice were approximately 6 months old and the experiments lasted approximately 12 months. The mice were housed separately and kept on a reversed day/night cycle (lights off at 6 am and on at 6 pm). The mice were tested during the dark portion of their cycle. They were water restricted and kept at approximately 85% of their free-drinking weight during the course of the experiment. The animals had unrestricted access to food, except while they were participating in the experiments. The mice were bred at the University at Buffalo, SUNY and all procedures were approved by the University at Buffalo, SUNY's Institutional Animal Care and use Committee.

Apparatus

The mice were tested in a wire cage (23×39×15.5 cm) placed in a sound-attenuated chamber (53.5×54.5×57 cm) lined with 4-cm thick Sonex sound attenuating foam (Illbruck Inc., Minneapolis, MN). The chamber was illuminated at all times by a small lamp with a 25-W white light bulb and the behavior of the animals during test sessions was monitored by an overhead web camera (Logitech QuickCam Pro, Model 4000). The test cage consisted of an electrostatic speaker (Tucker-Davis Technologies (TDT), Gainesville, FL, Model ES1), a response dipper (Med Associated Model ENV-302M-UP), and two nose poke holes surrounded by infrared sensors (Med Associates Model ENV-254).

The experiments were controlled by Dell Dimension 3100 computers operating TDT modules and software. Stimuli were sent through an RP2 signal processor, an SA1 power amplifier, a PA5 programmable attenuator, and finally to the speaker. Inputs to and outputs from the testing cages were controlled via RP2 and RX6 processors. Power supplies were used to drive the dipper (Elenco Precision, Wheeling, IL, Model XP-603) and infrared sensors (Elenco Precision, Model XP-650). Custom MATLAB and TDT RPvds software programs were used to control the hardware.

Stimuli

We used the same vocalizations used by [9], plus one additional vocalization not included in that study but acquired in their laboratory. Using the same stimuli allows us to more closely compare the neural correlates of call discrimination in mice and the behavioral data collected from this experiment. The vocalizations were recorded from CBA/CaJ mice during social interactions and analyzed with custom-written MATLAB code implementing a harmonic state-space signal model and the extended Kalman smoother [10]. The stimuli were synthesized in this way to reduce background noise and allow for manipulation of individual parameters in each of the vocalizations. We used five vocalization types, named based on the presence/absence of harmonics and jumps in frequency (all produced by males), or simpler sweep shapes produced by a male or a female: F Upsweep, M Upsweep, 30 kHz Harm/0 Jump, 30 kHz Harm/1 Jump, and 40 kHz Harm/2 Jump. Each of these five vocalizations was also manipulated in eight ways: the fundamental frequency was raised by 10 and 20% and lowered by 10 and 20%, the frequency modulation was removed, the entire vocalization was reversed, and the vocalizations were doubled and halved in duration (while maintaining the original frequency). We manipulated these acoustic parameters because they known to be important cues in tasks such as auditory scene segregation (e.g. [11]; [12]). Stimuli were presented at approximately 65 dB SPL, measured at the position where the mouse's head would normally be during testing. Two of the original vocalizations with several of these manipulations are shown in Figure 1. Sound pressure levels were calculated using an ultrasound recording system (Avisoft Model USG116-200) and Raven Pro (v 1.3, Cornell University) software.

Procedures

Mice were trained using a go-no-go operant conditioning procedure on a discrimination task. Subjects listened to one vocalization presented repeatedly and were required to indicate when they heard any other stimulus type. The first stage in the training process was to shape the mice to nose poke to the observation hole and then approach the dipper for the chocolate Ensure/water reinforcer. The animals were then trained to repeatedly poke to the observation hole until they heard a vocalization, after which they would nose poke to the report hole for the reinforcement. Next, catch trials were phased into the training and the waiting interval was systematically increased. Finally, a repeating background vocalization (different from the target vocalization) was phased into the training in small intensity increments from session to session.

During testing, the mouse began a trial by nose poking through the observation nose-poke hole two times, which initiated a variable waiting interval ranging from 1 to 4 s. During this time, a repeating background of one vocalization alternating with silence was presented repeatedly at a rate of once every 200 ms. After the waiting interval, a single test stimulus was presented, alternating with the background stimulus vocalization two times. In the "go" condition, the target stimulus was either a different vocalization or a manipulated version of the same vocalization. If the mouse discriminated this change between background and target, it was required to nose poke through the report nose-poke hole within 2 s of the onset of the target. In this trial type, a "hit" was recorded if the mouse correctly responded within the response window and the animal received 0.01 ml of Ensure or water as a reinforcement. A "miss" was recorded if the mouse failed to nose poke through the report hole within 2 s. If the mouse responded to the report nose-poke hole during the waiting interval, the trial was aborted and the mouse received a 3-5-s timeout, during which no stimuli would play.

Approximately 30% of all trials were "no go", or catch trials. Here, the repeating background continued to be presented during the response phase. These trials were required to measure the false alarm rate and calculate the animal's response bias. If the subject nose poked to the report hole during a catch trial, a "false alarm" was recorded and the mouse was punished with a 3-s timeout interval. However, if the subject continued to nose poke to the observation hole, a "correct rejection" was recorded and the next trial would begin immediately. In either case, no reinforcement was given. Chance performance was represented by the animal's false alarm rate. Sessions were excluded from analysis if the percentage of false alarms was greater than 20%. Using the

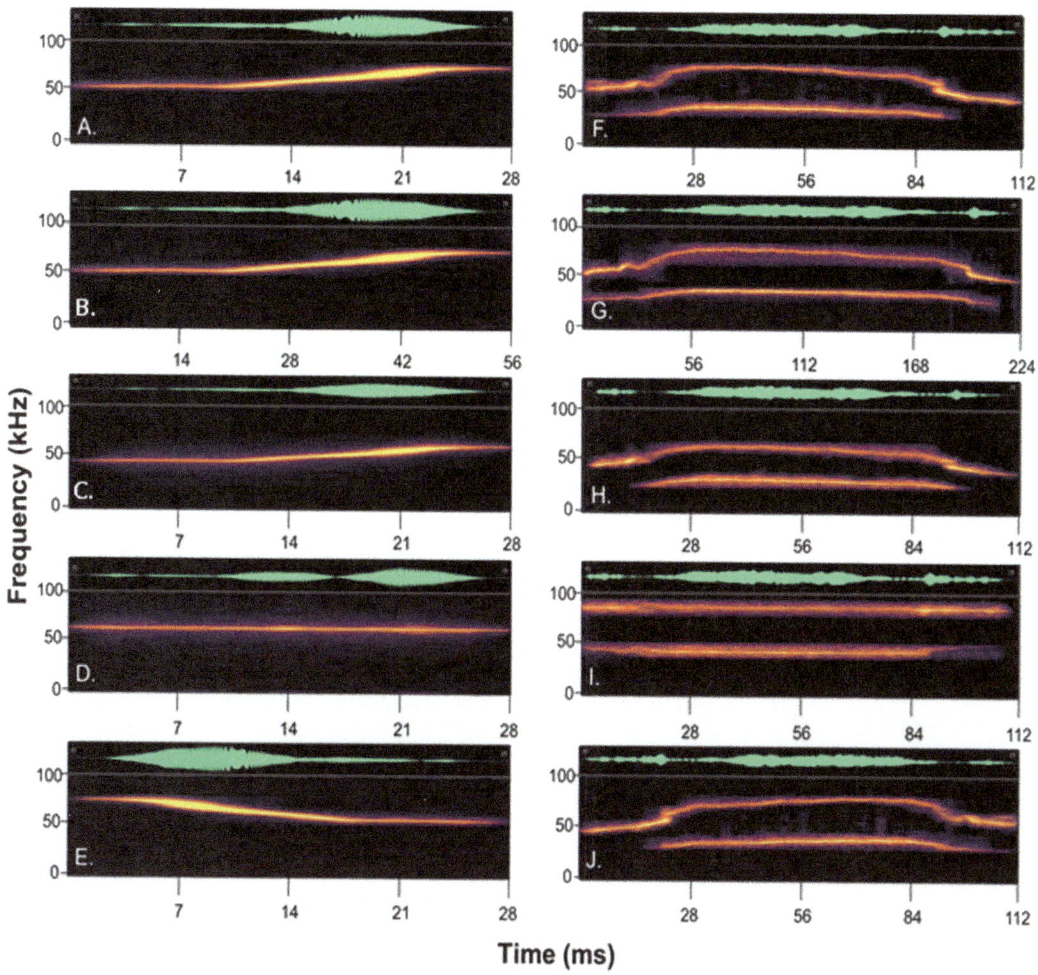

Figure 1. Vocalization spectrograms. Spectrograms of two unmodified vocalizations (A = Male Upsweep and F = 40 kHz Harm/2 Jump), and those same two vocalizations doubled in duration (B and G), lowered in frequency by 20% (C and H), with frequency modulation removed (D and I), and reversed in time (E and J).

criterion of at least 80% hit rate on target trials and below 20% false alarm rate assures that the mice are under stimulus control [8]. Approximately 30% of all sessions were thrown out due to high false alarm rate. There are two reasons to not include these data: First, including these sessions would lead to a false impression of the discrimination acuity found in this strain of mouse. Second, when the false alarm rates are above 20%, it is difficult to determine whether the animals are responding to the stimuli or just randomly nose-poking due to some motivational or attentional bias.

The mice were tested on two 30-min sessions/day, five to six days per week. Typically, the mice ran between 50 and 100 trials per session. All mice were tested on all stimuli in a random order, and a different random order was used for each subject. Testing on each vocalization background continued until results from at least 20 trials of each target type comparison were collected. Those results were used to calculate the percent correct discrimination performance for each vocalization versus every other vocalization and versus every vocalization manipulation.

Data Analysis

Signal detection analysis was performed to factor out the animals' motivational biases. At least 40 trials were obtained for each stimulus comparison for each mouse, where one stimulus was the background and the other was the target for 20 trials, and the reverse was true for another 20 trials. Mean hit rates were calculated for each animal on each stimulus comparison, and false alarms were tracked from session to session to ensure that the animals remained reliable observers under stimulus control. Repeated-measures ANOVAs were used to compare performance across all stimulus types, and Holm-Sidak post-hoc analyses were conducted for pairwise comparisons.

Once discrimination performance for all vocalization combinations (each of the five vocalizations against every other vocalization) was complete, a multidimensional analysis (Proxscal) was conducted. Each cell in a discrimination performance matrix for each mouse contained the responses from a single session involving the discrimination percentage from the first 20 presentations of a single pair of USVs. Once testing was complete, similar to the techniques used by [13], a single matrix representing the mean results across the five different USVs was calculated. The values in the diagonal (i.e., performances from "same" trials) were discarded. Asymmetrical discrimination responses were found across all mice and USV conditions. That is, the discrimination responses varied when a call was the background or target in the USV pairing, although no systematic explanation for this could be found, and asymmetries differed in direction across animals. This required using a full-matrix analysis of discrimination performances, which can be done using Proxscal in SPSS. From this behaviorally derived matrix, the Euclidean distances between the vocalizations were calculated.

Spectrotemporal similarities between the vocalization types and between the vocalizations and their manipulated versions were measured using the cross correlation tool from RAVEN software (Cornell Lab of Ornithology, Ithaca, NY, v. 1.3, biased setting). Those spectrotemporal similarity calculations, along with the behavioral discrimination performance measures, were used for a nonlinear regression analysis.

Results

We found that the mice generally were able to discriminate between vocalizations (Figure 2). Mean discrimination performance ranged from 56–95% across all vocalizations, while the false alarm rate averaged 9.8% (SE = 0.1%) across all animals and sessions. A one-way repeated measures ANOVA revealed significant differences in discrimination abilities across the five stimuli ($F_{9,36} = 8.82$, $p < 0.001$). A Holm-Sidak test revealed several significant pair-wise comparisons (Figure 2). Generally, the two upsweep vocalizations were difficult to discriminate from one another ($p > 0.05$) but were easily discriminated from the harmonic vocalizations ($p < 0.05$), while the harmonic vocalizations were easily discriminated from the upsweeps ($p < 0.05$), but the mice had some difficulty discriminating among the different harmonic vocalizations ($p > 0.05$). These discrimination performances do not appear to be due to an experience-dependent confound. The mice took an average of 5 sessions (~200 trials) to reach criterion and this number did not systematically change as the experiment continued. Specifically, discrimination performance during training and testing were nearly identical for all mice at the beginning of the experiment and at the end of the experiment.

The mean percent correct discrimination performance matrices for every vocalization combination were used to calculate a multidimensional map of vocalization perception in the mice (Figure 3). A two-dimensional map of similarity was conducted from this analysis, accounting for 78% of the dispersion, with the first and second dimensions accounting for 45% and 33% respectively. Along dimension 1 (accounting for slightly more variance), the USVs appear to be divided along the presence of transient frequency "jumps" in the vocalization. Vocalizations with no harmonics or jumps were in the negative portion of dimension 1, while the vocalizations with harmonics and jumps were placed in the positive portion of dimension 1. The one harmonic vocalization with no jumps was in-between the other two sets of vocalizations. It is less clear which acoustic cues the mice are using along the second dimension. It would appear that along dimension 2, the USVs are separated by the spectrotemporal differences found between the start and end of the call. The positive portion of the dimension contains the only USV with the same beginning and end frequency (30k Harm, 0 Jump; Δ = 0 Hz). The calls in the middle of the second dimension (F Upsweep and 30k Harm, 1 Jump) are found to have ~5–10 kHz of frequency separation, whereas the calls in the negative portion of the dimension (M Upsweep and 40k Harm, 2 Jump) were found to have the largest frequency separation, about a ~15–20 kHz change in frequency. These results support the hypotheses stated from [9], that the auditory system may process USVs by the distortion products generated from the beginning and end frequencies of the call. Vocalizations separated by the two smallest Euclidian distances are circled. Based on behavioral discrimination performance alone, the pairs of M Upsweep and F Upsweep and 30kHarm/1Jump and 40kHarm/2Jump were deemed as more similar to each other than the other vocalization comparisons.

Discrimination performance for the manipulated vocalizations also varied across conditions (Figure 4), with percent correct ranging from 32–86%, averaged across the five vocalization types. There was a significant difference between the five background vocalization types ($F_{4,28} = 36.57$, $p<0.001$), the eight vocalization manipulations ($F_{7,28} = 26.04$, $p<0.001$), and a significant interaction between the two variables ($F_{28,112} = 8.16$, $p<0.001$). In general, discrimination between the unaltered and manipulated vocalizations was most difficult for the upsweeps ($p > 0.05$) and least difficult for the harmonic vocalizations ($p < 0.05$). The manipulations that were most difficult to discriminate from the unaltered vocalizations were reversing and compressing the vocalizations, followed by removing the frequency modulation and increasing the frequency by 10%. The easiest manipulations to discriminate from the unaltered vocalizations were lowering the frequency of the vocalizations by 20%, lowering the frequency by 10%, doubling the duration, and increasing the frequency by 20%. Regardless of the manipulation, all discriminations were well above chance performance (false alarm rate = 9.8%). Furthermore, discrimination performance was quite consistent across mice, indicated by the relatively low standard error values across conditions ($SE_{mean} = 4.55$).

Spectrotemporal similarity between the different vocalizations and between the vocalizations and their manipulations were significantly correlated with discrimination performance (Figure 5, $r^2 = 0.13$, $p<0.01$). For example, discrimination performance was >80% for the vocalizations that were most dissimilar. In general, as spectrotemporal similarity between vocalizations increased, discrimination performance decreased. Thus, the mice have the behavioral ability to discriminate between vocalizations that are spectrotemporally dissimilar from one another, and this ability declines for more similar vocalizations.

Discussion

We tested the ability of mice to discriminate between regularly emitted USVs using operant conditioning procedures and the psychophysical Method of Constant Stimuli. We found that mice can reliably discriminate between vocalizations, particularly those with dissimilar spectrotemporal properties. Thus, although the functional importance of different types of mouse vocalizations is not yet known, our findings show that mice are able to behaviorally discriminate vocalizations, suggesting that the different types of vocalizations are perceptually meaningful.

Until now, the only discrimination studies using mouse USVs have been conducted using preference studies. These studies have shown that female mice spend less time with a devocalized male than one that is vocalizing [14], females pair-housed with males later preferred cage areas projecting male vocalizations [15], female Neotropical mice (genus *Scotinomys*) prefer signals produced at a fast rate over a slow rate [16], female mice prefer synthetic multiharmonic calls that are similar to natural pup calls [17], female mice respond to ultrasonic 'songs' with approach behavior [18] and finally, female wild house mice can distinguish between ultrasonic vocalizations produced by their brothers and those from unfamiliar non-kin [19]. However, these types of preference experiments are limited in their ability to tell us about acoustic communication. These studies show large variations in results across laboratories, animals can only be tested once or twice and only for a minute or two before they habituate, and actual

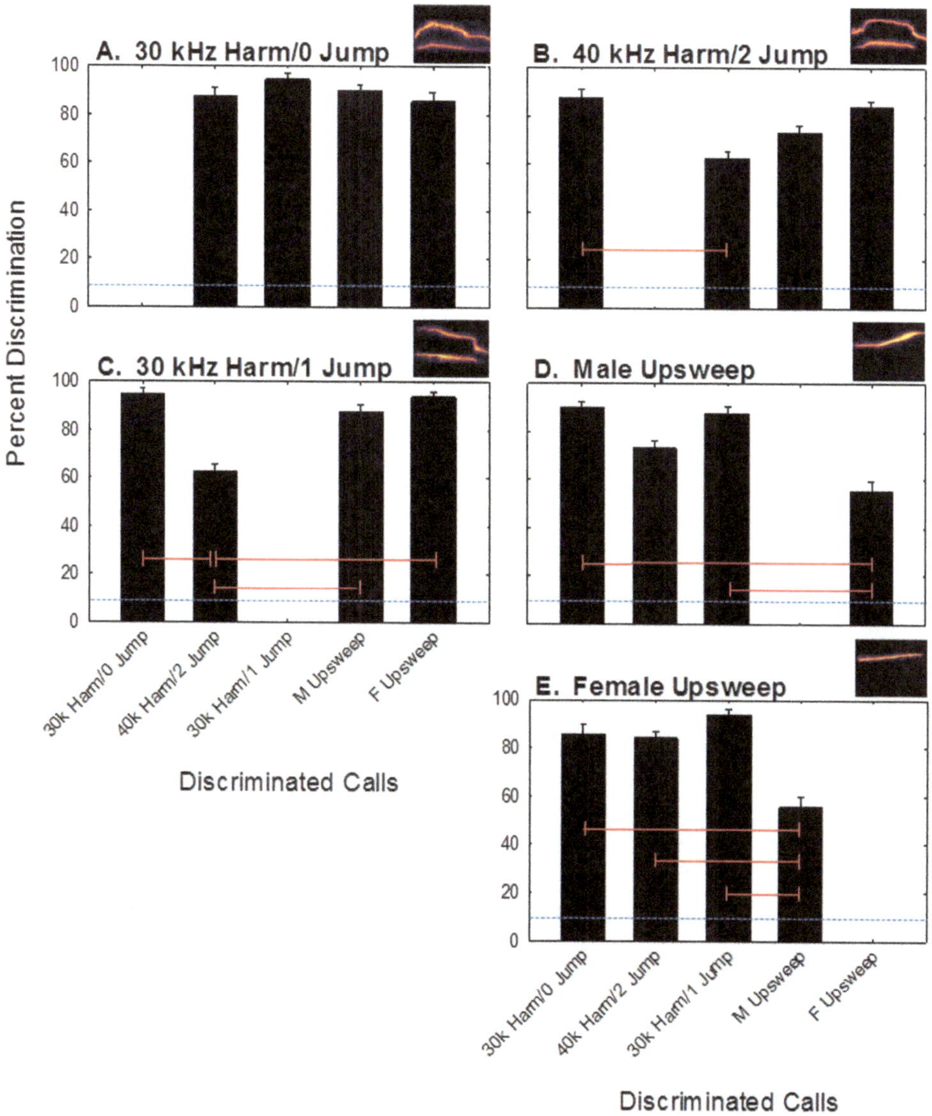

Figure 2. Discrimination of vocalizations. Mean discrimination performance across subjects for the five vocalizations types (A-E) against all other vocalizations. The blue horizontal dashed lines represent chance level performance. Error bars are between-subject standard errors. The red horizontal lines connect significantly different bars. The missing bar in each of the graphs is when the stimulus was used as the background.

amounts of time spent in one compartment compared to another are often quite small (reviewed in [15]). Further, when an animal spends more time in proximity of one stream of vocalizations over another, we know nothing about why it is doing this. Similarly, the finding of no preferences may be because the animal cannot discriminate between two stimuli. For these reasons, psychophysical experiments using reliable observers are much more informative for understanding rodent communication.

It is well known that mice emit ultrasonic vocalizations in a variety of social contexts (reviewed in [3]), and that they emit a number of different vocalization types that have been referred to as syllables [20]. The classification of syllables has been based on spectrotemporal properties of the vocalizations [4], [5], [7] or statistical clustering algorithms [6], [21], rather than on the ability of mice to behaviorally discriminate between different vocalizations. To the best of our knowledge, this is the first study to show that mice can behaviorally discriminate among a small number of commonly emitted vocalizations. This has been an open question since Ehret's early studies [22] on how lactating female mice do not behaviorally differentiate between narrowband noise models and natural calls in a phonotaxis task. Until now, it was uncertain if mice were able to discriminate between these artificial models and natural calls. The current study suggests that if sufficiently motivated, mice can in fact make subtle acoustic discriminations.

In general, the mice were best able to discriminate between vocalizations that had dissimilar spectrotemporal properties. The mice were able to accurately discriminate between the upsweep vocalizations and the harmonic and jump vocalizations, but were not able to accurately discriminate between the two upsweep vocalizations or among the harmonic and jump vocalizations. This is illustrated by the multidimensional map of perceptual space for the five vocalizations. The upsweep vocalizations occupied the negative space while the harmonic vocalizations occupied the positive space along dimension one. This analysis also suggests that peak frequency was of less importance for discrimination behavior than the presence or absence of jumps in the harmonic

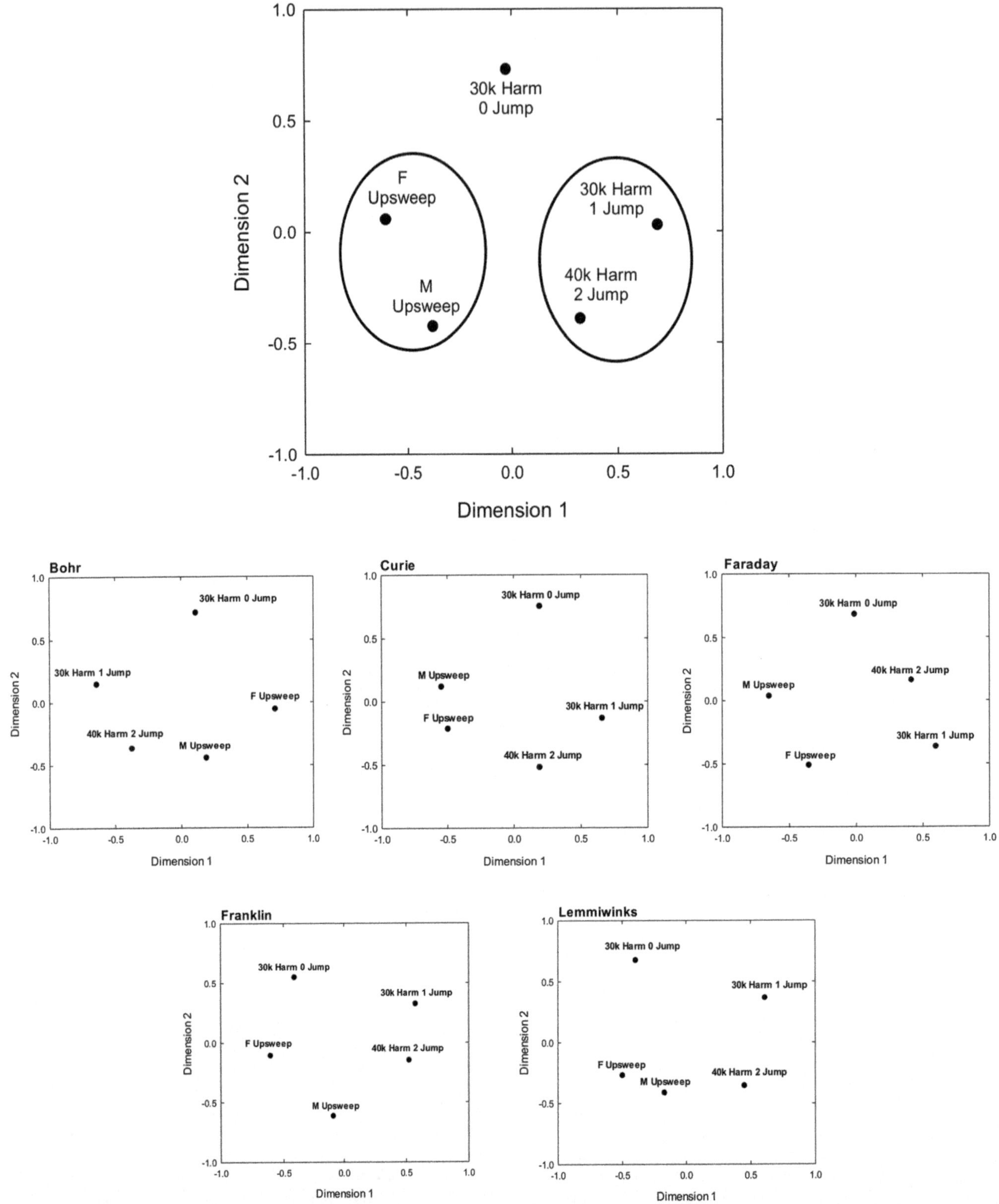

Figure 3. Multidimensional analysis of vocalization discrimination. At the top, multidimensional map of mean vocalization discrimination obtained from matrices of discrimination performance for the five vocalization types. Circles connect vocalizations with the two smallest Euclidian distances. At the bottom are the five individual perceptual maps generated for each mouse across the five vocalization types.

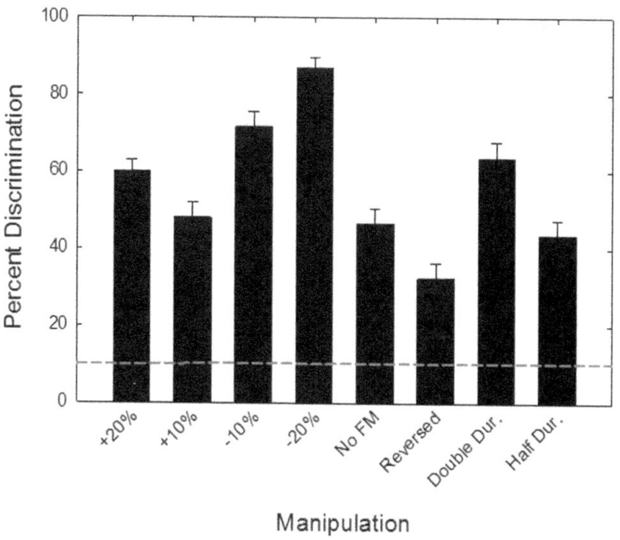

Figure 4. Discrimination of manipulated vocalizations. Mean discrimination performance across subjects and across the five background vocalization types for the eight vocalization manipulations. Error bars are between-subjects standard errors. The blue horizontal dashed lines represent chance level performance.

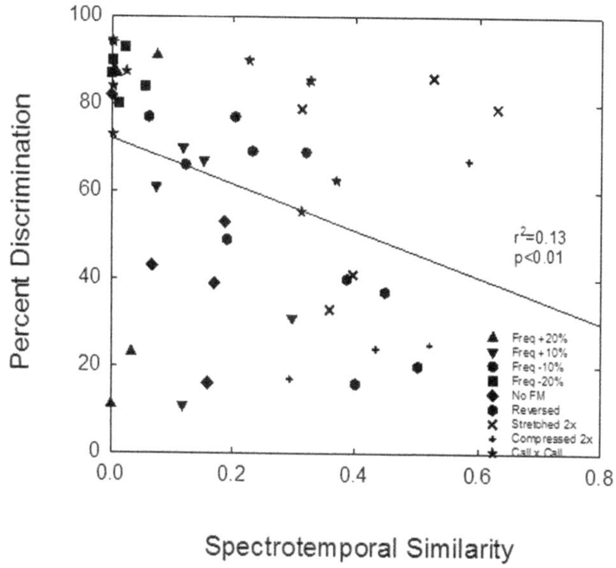

Figure 5. Correlation between spectrotemporal similarity and discrimination performance. Percent correct discrimination as a function of spectrotemporal similarity for all vocalization discriminations in the first experiment and all vocalizations versus their manipulations in the second experiment.

vocalizations, as the two harmonic vocalizations with jumps were perceived as more similar to each other than they were to the harmonic vocalization without jumps. However, frequency does have some relevance, as the two 30 kHz vocalizations occupy positive space in dimension 2, while the 40 kHz harmonic vocalization occupies negative space in the same dimension. These multidimensional scaling results from mice are similar to those from similar studies conducted in budgerigars (*Melopsittacus undulatus*, [13]), where different vocalization "types" occupy different perceptual space, simply measured by discrimination performance. Thus, this type of study provides access to the animals' perceptual capabilities, information not obtained by other experimental methodologies.

Our behavioral discrimination results match well with neural discrimination results obtained in the IC of CBA/CaJ mice [9]. Because the same vocalization stimuli were used in the behavioral and neurophysiological studies, we can gain a better understanding of the neural correlates of behavioral discrimination of vocalizations in mice. Our finding that mice could accurately discriminate between the upsweeps and the harmonic vocalizations corresponds well to the findings that more IC neurons fire to the harmonic vocalizations compared to the upsweeps and that there are differences in the reliability of the temporal firing patterns between the stimuli. Thus, IC neurons can discriminate between the upsweep and harmonic vocalizations, just like the awake behaving mice can. While the mice were less accurate at discriminating between the two upsweep vocalizations, their performance was greater than chance (56% versus 10%), indicating some level of discrimination ability. This result matches somewhat with the finding that more IC neurons responded to the female-emitted upsweep vocalization than to the male-emitted upsweep vocalization [9]. Behavioral performance may have been less accurate than the neural performance because the upsweeps were both short in duration and very high frequency, and it is well known that behavioral discrimination is better with longer duration stimuli (e.g. [23]) and there is little neural representation of vocalizations above 60 kHz in the mouse auditory system [24].

To better understand what spectrotemporal properties in vocalizations are important for behavioral discrimination ability, we independently manipulated parameters within each of the vocalization stimuli and tested how well the mice could discriminate the original vocalization from the modified one. The mice were able to discriminate all of the manipulations with accuracy above chance, although the mice were better able to discriminate between the original and modified vocalization when frequency content was altered. These behavioral results are very similar to how neurons in the IC responded to the same original and modified vocalizations we used in this study [9]. Neurons changed their discharge rate, discharge pattern, or both when the vocalizations were shifted in frequency, when frequency modulation was removed, or when duration was altered. In general, shifting frequency content led to the greatest change in neural response patterns.

Overall, behavioral performance for the five vocalizations and their spectrotemporal manipulations shows that vocalization shape, frequency content, and the presence or absence of frequency jumps all affect the mouse's ability to discriminate among different vocalizations. Discrimination performance was lower for spectrotemporally similar vocalizations and higher for spectrotemporally dissimilar vocalizations. Thus, our results provide evidence that mice are capable of perceptually discriminating among commonly emitted vocalizations that differ in spectrotemporal properties. Furthermore, our behavioral results closely match neural responses in the IC, suggesting that the neural mechanisms underlying selectivity to vocalizations in the IC play a fundamental role in the perception of vocalizations.

Acknowledgments

Thanks to Drs. Richard Salvi, Matthew Xu-Friedman, and Thomas Welch for assistance.

Author Contributions

Conceived and designed the experiments: EN KR CP MD. Performed the experiments: EN DH KR. Analyzed the data: EN MD. Wrote the paper: MD CP.

References

1. Searcy WA, Nowicki S (2005) The evolution of animal communication. Princeton, NJ: Princeton University Press.
2. Seyfarth RM, Cheney DL, Marler P (1980) Vervet monkey alarm calls: semantic communication in a free-ranging primate. Anim Behav 28: 1070–1094.
3. Portfors CV (2007) Types and functions of ultrasonic vocalizations in laboratory rats and mice. J Am Assoc Lab Anim Sci 46: 28–34.
4. Scattoni ML, McFarlane HG, Zhodzishky V, Caldwell HK, Young WS, et al. (2008) Reduced ultrasonic vocalizations in vasopressin 1b knockout mice. Behav Brain Res 187: 371–378.
5. Grimsley JMS, Monaghan JJM, Wenstrup JJ (2011) Development of social vocalizations in mice. PLoS One 6: e17460.
6. Grimsley JM, Gadziola MA, Wenstrup JJ (2012) Automated classification of mouse pup isolation syllables: from cluster analysis to an Excel-based "mouse pup syllable classification calculator". Frontiers Behav Neurosci 6: 89.
7. Mahrt EJ, Perkel DJ, Tong L, Rubel EW, Portfors CV (2013) Engineered deafness reveals that mouse courtship vocalizations do not require auditory experience. J Neurosci 33: 5573–5583.
8. Radziwon KE, June KM, Stolzberg DJ, Xu-Friedman MA, Salvi RJ, et al. (2009) Behaviorally measured audiograms and gap detection thresholds in CBA/CaJ mice. J Compar Physiol A 195: 961–969.
9. Holmstrom LA, Eeuwes LBM, Roberts PD, Portfors CV (2010) Efficient encoding of vocalizations in the auditory midbrain. J Neurosci 30: 802–819.
10. Holmstrom LA, Kim S, McNames J, Portfors C (2009) Stimulus design for auditory neuroethology using state space modeling and the extended kalman smoother. Hear Res 247: 1–16.
11. Bregman A (1990) Auditory Scene Analysis. MIT Press, Cambridge, MA.
12. Bee MA, Micheyl C (2008) The cocktail party problem: what is it? How can it be solved? And why should animal behaviorists study it? J Compar Psychol 122: 235–251.
13. Dooling RJ, Park TJ, Brown SD, Okanoya K, Soli SD (1987) Perceptual organization of acoustic stimuli by budgerigars (*Melopsittacus undulatus*): II. Vocal signals. J Compar Psychol 101: 367–381.
14. Pomerantz SM, Nunez AA, Bean NJ (1983) Female behavior is affected by male ultrasonic vocalizations in house mice. Physiol Behav 31: 91–96.
15. Shepard KN, Liu RC (2011) Experience restores innate female preference for male ultrasonic vocalizations. Genes Brain Behav 10: 28–34.
16. Pasch B, George AS, Campbell P, Phelps SM (2011) Androgen-dependent male vocal performance influences female preference in Neotropical singing mice. Anim Behav 82: 177–183.
17. Ehret G, Riecke S (2002) Mice and humans perceive multiharmonic communication sounds in the same way. Proc Nat Acad Sci 99: 479–482.
18. Hammerschmidt K, Radyushin K, Ehrenreich H, Fischer J (2009) Female mice respond to male ultrasonic 'songs' with approach behavior. Biol Letters 5: 589–592.
19. Musolf K, Hoffman F, Penn DJ (2010) Ultrasonic courtship vocalizations in wild house mice, *Mus musculus*. Anim Behav 79: 757–764.
20. Holy TE, Guo Z (2005) Ultrasonic songs of male mice. PLoS Biol 3: e386.
21. Hammerschmidt K, Radyushin K, Ehrenreich H, Fischer J (2012) The structure and usage of female and male mouse ultrasonic vocalizations reveal only minor differences. PLoS One 7: e41133.
22. Ehret G, Haack B (1982) Ultrasonic recognition in house mice: Key-stimulus configuration and recognition mechanism. J Comp Physiol 148:245–251
23. Ehret G (1976) Temporal auditory summation for pure tones and white noise in the house mouse (*Mus musculus*). J Acoust Soc Am 59: 1421–1427.
24. Portfors CV, Mayko ZM, Jonson K, Cha GF, Roberts PD (2011) Spatial organization of receptive fields in the auditory midbrain of awake mouse. Neurosci 193: 429–439.

19

Social Investigation and Long-Term Recognition Memory Performance in 129S1/SvImJ and C57BL/6JOlaHsd Mice and Their Hybrids

Jana Hädicke[1], **Mario Engelmann**[1,2]*

1 Otto-von-Guericke-Universität, Institut für Biochemie und Zellbiologie, Magdeburg, Germany, **2** Center for Behavioral Brain Sciences, Magdeburg, Germany

Abstract

When tested for their behavioural performance, the mixed genetic background of transgenic mice is a critical, but often ignored, issue. Such issues can arise because of the significant differences in defined behavioural parameters between embryonic stem cell donor and recipient strains. In this context, the commonly used stem cell donor strain '129' shows 'deficits' in different paradigms for learning and long-term memory. We investigated the long-term social recognition memory performance and the investigative behaviour in commercially available 129S1/SvImJ and C57BL/6JOlaHsd mice and two F1-hybrids (129S1/SvImJ×C57BL/6JOlaHsd) by using the social discrimination procedure and its modification, the volatile fraction cage (VFC). Our data revealed an unimpaired olfactory long-term recognition memory not only in female and male 129S1/SvImJ and C57BL/6JOlaHsd mice but also in the two hybrid lines (129S1/SvImJxC57BL/6JOlaHsd) when the full 'olfactory signature' of the 'to-be-recognized' conspecific was presented. Under these conditions we also failed to detect differences in the long-term recognition memory between male and female mice of the tested strains and revealed that the oestrus cycle did not affect the performance in this memory task. The performance in the VFC, based only on the volatile components of the 'olfactory signature' of the 'to-be-recognized' conspecific, was similar to that observed under direct exposure except that females of one F1 hybrid group failed to show an intact long-term memory. Thus, the social discrimination procedure allowing direct access between the experimental subject and the stimulus animal(s) is highly suitable to investigate the impact of genetic manipulations on long-term memory in male and female mice of the strain 129S1/SvImJ, C57BL/6JOlaHsd and 129S1/SvImJxC57BL/6JOlaHsd hybrids.

Editor: Neil R. Smalheiser, University of Illinois-Chicago, United States of America

Funding: ME was supported by the German Research Foundation (DFG) (EN 366/6-1 and 366/8-1). The funders had no role in study design, data collection and analysis, decision to publish, or preparation of the manuscript.

Competing Interests: The authors have declared that no competing interests exist.

* E-mail: mario.engelmann@med.ovgu.de

Introduction

There is increasing evidence that inbred mouse strains differ in their behavioural profile. Indeed, various studies have reported significant differences in defined behavioural parameters between different commercially available mouse strains including the C57BL/6J strain and inbred '129' substrains [1–5]. In particular substrains of the latter, the commonly used embryonic stem cell donor strain, show impaired learning and long-term memory performance [5–7], that is paralleled by alterations in brain circuitry [8–9]. Thus, although the design of molecular tools for manipulating the mouse genome is continually progressing, the mixture of genetic material from different mouse strains remains critical [10–13]. To cope with these mouse strain-differences, guidelines were proposed to enable the use of mutant mice in tasks in which 129 mice show a very poor learning and memory performance (e.g. numerous backcrosses to reduce the genetic material of the 129 strain) and – very recently – a report suggested using animals of the C57BL/6N strain as stem cell donors [14].

The alternative approach would employ a learning and memory task in which animals of the 129 strain show an unimpaired behavioural performance similar to other commonly used mouse strains. Tests that investigate the olfactory cued "social recognition memory" may provide such an alternative. Typically, the olfactory cues are provided by the odour emitted from a given animal, often called 'olfactory signature', which includes both volatile (which are detectable over a distance) and non-volatile components (detectable by direct body contact only) [15–17]. Such tests rely on the animals' intrinsic motivation to acquire the 'olfactory signature' of a conspecific and that unfamiliar items will be investigated over familiar items [18]. Therefore, a significantly longer investigation duration towards a novel versus a previously encountered conspecific is taken as an evidence for successful recognition. However, a recent study reports impaired social memories in 129P2 inbred mice [19] when tested in the "social recognition procedure" originally established in 1987 for testing social memory in rats [20].

The social discrimination test [21–22] was introduced to overcome some technical limitations of the social recognition procedure and offers another possibility to analyse olfactory-cued social recognition memory in mice. The main modification concerns the simultaneous presentation of the previously encountered with a new stimulus animal which enables a direct judgement of the recognition/discrimination abilities of the experimental subject in a single session. In the original version of this test the experimental subject gets direct access to the stimulus animals, thereby enabling the acquisition of the full 'olfactory signature' (including both the volatile and non-volatile components). In a recently published modification of this procedure the access can be restricted to the volatile components only by using the volatile fraction cage (VFC) [23–24].

The present study was designed to examine the investigative behaviour and long-term social recognition memory performance in 129S1/SvImJ mice compared to C57BL/6JOlaHsd animals using the social discrimination task and the VFC. We also established and tested two F1-hybrid lines by mating females or males of the 129S1/SvImJ strain with male and female, respectively, C57BL/6JOlaHsd mice. The use of mothers from the different strains was designed to characterize possible influences of maternal behaviour on the behavioural performance of the heterozygous offspring when tested as adults for their social memory. F1 hybrids provide a model of a mixed genetic background which may easily reveal a potential memory distracting impact in this model. Other authors reported that F1 hybrids of 129 and C57B6 mice show dramatic changes in the ataxia index [25]. The present study extended these findings by including analysis of the olfactory recognition memory performance in female mice of all lines listed.

Materials and Methods

Animals

Adult male and female 129S1/SvImJ (male: n = 23, female: n = 20; The Jackson Laboratory, Charles River, Sulzfeld, Germany) and C57BL/6JOlaHsd mice (male: n = 24, female: n = 26; Harlan-Winkelmann, Borchern, Germany;) and two F1-hybrid lines were used as experimental subjects. The F1-hybrids were created by mating female 129S1/SvImJ with male C57BL/6JOlaHsd (**Hyb1;** male: n = 20, female: n = 18) and female C57BL/6JOlaHsd with male 129S1/SvImJ (**Hyb2;** male: n = 20, female: n = 20). All animals were housed in groups of three to six per cage (size 20×37×15 cm) under standard laboratory conditions (temperature: 23±1°C; humidity: 60±5%; 12 h light/12 h dark cycle with lights on at 07:00) with free access to food and water. All tested mice were at an age of 8–16 weeks during testing and sexually naive. Juvenile mice of both sexes (25–35 days old; C57BL/6JOlaHsd strain) were used as olfactory stimuli. Previous extensive studies in our laboratory have shown that neither the sex nor the defined age of the juveniles significantly affects social recognition memory performance in C57BL/6JOlaHsd mice [26]. If not stated otherwise, stimulus animals were kept in groups of 3–5.

Ethics statement. All experimental manipulations were approved according to German and European legislation by the *Landesverwaltungsamt Sachsen-Anhalt, Referat Verbraucherschutz, Veterinär-angelegenheiten, Halle,* Germany (approval number: 203.k-42502-2-992UniMDG).

Determination of the Stage of the Oestrus Cycle

As vaginal stimulation prior to a behavioural experiment has been shown to influence learning and memory performance in female rodents [27–30], vaginal smears were obtained after the end of choice in each experiment. Sterile Q-tips were immersed in physiological saline and gently inserted into the vaginal tract to remove loose cells. Smears were transferred onto a glass slide, air dried and Nissl stained. Based on the vaginal smear cytology mice were assigned to two groups: oestrus and non-oestrus during choice. According to the previously described impact of the oestrus cycle primarily on memory acquisition we extrapolated the cycle phase during sampling from the samples collected after choice by estimating a duration of the oestrus cycle of 24 h [31]. Thus, females showing an oestrus smear during choice were estimated as being tested in prooestrus during sampling (24 h exposure interval). In contrast, smears not showing typical oestrus signs suggested that the females were tested in non-prooestrus (including dioestrus, metoestrus, oestrus) during sampling.

Social Discrimination Test and the Volatile Fraction Cage (VFC)

A detailed description of both procedures is provided by Engelmann et al. (2011) including the detailed description of the sequence in which the stimulus animals are used during testing. Briefly, the tests consist of two sessions (sampling and choice) during which given juveniles are exposed as stimulus animals to the experimental subjects for 4 min during each session. Previous testing revealed that both male and female juveniles are suitable to serve as stimulus animals [26]. Two hours before starting the test experimental subjects were separated by transferring them into small cages (home cage; 14×20×15 cm). When using the VFC there was an additional separation time of the experimental subject in the VFC (10 min) before starting the experiment. During the first encounter (sampling; performed during the light phase between 8.00 h and 13.00 h), the 'to-be-recognized' juvenile was exposed to the experimental subject for acquiring the stimulus animals' 'olfactory signature'. Then, the juvenile was removed, kept individually in a fresh cage with food and water *ad libitum* for the exposure interval of 24 h. During the second encounter (choice), the familiar juvenile was re-exposed to the experimental subject together with an additional, previously not encountered (novel) juvenile. The monitored olfactory investigatory behaviour of the experimental subject toward the stimulus animals during choice serves as an index for recognition: a significantly longer investigation duration of the novel versus the previously encountered conspecific is taken as an evidence for successful recognition. To allow for easy identification of the juveniles by the observer during the choice, tail marking was used (Edding 30 permanent marker, red or green, Edding AG, Germany), which does not affect social discrimination [26]. After conclusion of each test the animals were put back in the original groups until the next test.

Two series of experiments were performed. All experimental subjects (the number of animals per group is given in the respective figures) were first tested in the first experimental series and one week later in the second experimental series. Another week later the animals were tested in the open field (for the results see **Fig. S1** and **Table S1**). During the **first series,** the social recognition performance was monitored using the social discrimination procedure with direct access to the stimulus animals during both sampling and choice. During the **second series,** the social discrimination procedure was modified by employing the VFC (indirect contact) [26]. Because of computer failures during the measurement of choice one male mouse was excluded in the first experimental series, and five male mice in the second series.

First series: Direct stimulus animal presentation during both sampling and choice. Sampling and choice took place in the home cage of the experimental subject under unrestricted conditions **(Fig. 1A)**. Thus, the experimental subject had direct contact to the juvenile(s) and access to the full juveniles' 'olfactory signature' in both sessions. During each session, the duration of investigatory behaviour of the experimental subject towards each juvenile (licking, sniffing) was monitored separately by a trained observer unaware of the animals genotype and sex by pressing pre-set keys on a keyboard and a computer program.

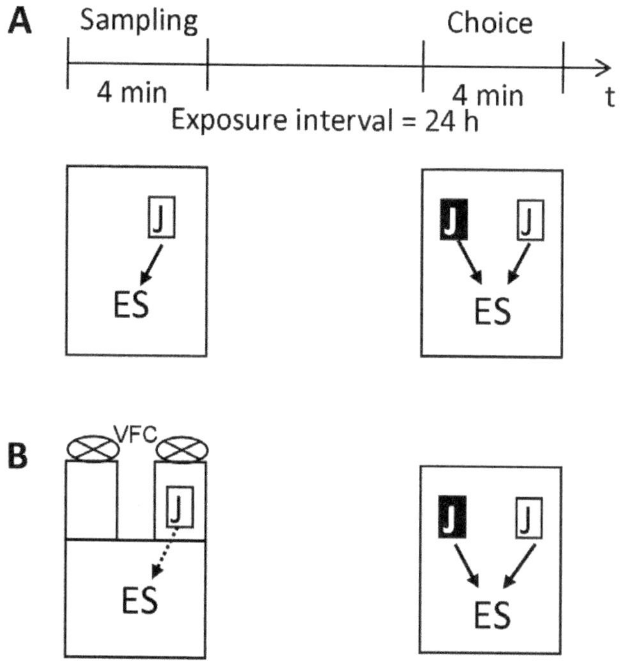

Figure 1. Design of the two series of experiments in which the social discrimination performance was tested. A given juvenile (J in white box) was exposed to the experimental subject (ES) during a 4-min sampling session. This juvenile was removed and, after the exposure interval of 24 h, re-exposed to the experimental subject during the choice session, together with a different, not previously encountered juvenile (J in black box). **A:** Direct exposure of the juvenile(s) in the home cage of the experimental subject with direct interaction between experimental subject and stimulus animal(s) during both sampling and choice allowing the acquisition of the full olfactory signature of the juvenile(s) (black arrow(s)). **B:** Sampling was performed in the volatile fraction cage (VFC, presentation of only the juvenile's volatile components of the 'olfactory signature' = hatched arrows). Choice was performed in the home cage of the experimental subject allowing direct interaction between the animals and, thus, the acquisition of the full olfactory signature (black arrows).

Second series: Sampling performed in the VFC, direct stimulus animal presentation during choice. During the second series, choice was performed in the experimental subject's home cage as indicated above. Sampling was performed in the VFC that separated the experimental subject from the stimulus juveniles. As described in detail elsewhere [26], a fan behind the juvenile produced a constant air flow that carried only the volatile components of the juvenile's 'olfactory signature' to the experimental subject (**Fig. 1B**). Thus, the experimental subject gained access to only the volatile components of the juvenile's 'olfactory signature' during sampling, but was allowed to acquire the full 'olfactory signatures' of the familiar and the novel juvenile during choice. In the VFC, active sniffing of the experimental subject toward the air streams containing the volatile information of the respective juvenile was measured as indicated above. Acquisition of the volatile component of the olfactory signature in the VFC is subsequently termed 'indirect sampling' whereas direct exposure of the stimulus animal to the experimental subject 'direct sampling'.

Statistical Analyses

Statistical analysis was performed using GB-STAT 6.0 (Dynamic Microsystems, Silver Springs, MD, U.S.A.) or GraphPad Prism 4.01 (GraphPad Software, San Diego, CA, U.S.A.). Data obtained from the social discrimination test are presented as means +/− SEM and were used to calculate the recognition index (RI), which is a measure of the recognition ability [32]. The RI was calculated on the parameters measured during choice and is the quotient of the investigation duration of the novel juvenile divided by the sum of the investigation durations of familiar and novel juvenile × 100. Chance performance is 50. A RI significantly above 50 indicates successful recognition of the familiar juvenile [26]. The RIs were analyzed using a one sample t-test against a theoretical mean of 50 and two-way ANOVA (genotype × sex). Additionally, investigation durations towards the novel and the previously encountered juveniles during choice were compared using the paired t-test [26]. For analysing the influence of genotype and sex on the sampling investigation two-way ANOVAs (genotype × sex) were carried out for the two experimental series. For analyzing the influence of juvenile's sex, a three-way ANOVA was used (genotype × experimental subject's sex × juvenile's sex). To analyse the impact of the oestrus cycle on the performance of the female experimental subjects RIs were submitted to a two-way ANOVA (experimental series × oestrus stage (i.e. prooestrus versus non-prooestrus)). ANOVAs were followed by Scheffé's post-hoc test, if appropriate. A $p<0.05$ was considered to be statistically significant.

Results

Influence of Presentation Mode, Genotype and Sex on the Investigation Duration

Investigation durations during sampling measured in the second experimental series with indirect sampling were significantly shorter than those measured in the first experimental series (direct sampling; **Tables 1 and 2**). The two-way ANOVA showed a significant effect of 'genotype' on the investigation duration during sampling for the two experimental series whereas a significant effect of 'sex' was revealed for the first experimental series only (first series: genotype: $F(3,162)=56.17$, $p<0.001$; sex: $F(1,162)=40.95$, $p<0.001$; interaction $F(3,162)=7.22$, $p<0.001$; second series: genotype: $F(3,158)=33.64$, $p<0.001$; sex: $F(1,158)=0.27$, $p=0.60$; interaction $F(3,158)=1.16$, $p=0.33$; **Tables 1 and 2**). Hyb2 mice investigated in both experimental series the juveniles significantly longer than all other genotypes (**Tables 1 and 2**). An interaction between genotype and sex was seen in the investigation duration in the first series only, where males investigated the juvenile significantly longer than females, except for C57BL/6JOlaHsd males (**Table 1**).

Olfactory Recognition Performance

First series: Direct stimulus animal presentation during both sampling and choice. A two-way ANOVA did not show any differences in the RIs of any groups tested (genotype: $F(3,162)=0.09$, $p=0.97$; sex: $F(1,162)=0.82$, $p=0.37$; interaction: $F(3,162)=2.52$, $p=0.06$). As shown in **Figure 2A**, detailed analyses of the RIs using the one sample t-test revealed that mice from all genotypes and sex reached an RI significantly higher than chance level (= 50) and, thus, showed an intact social recognition. This was confirmed by analysing the investigation durations during choice showing that all groups investigated the novel juvenile significantly longer than the previously encountered juvenile (paired t-test, **Table 1**).

Second series: Sampling performed in the VFC, direct stimulus animal presentation during choice. The two-way ANOVA did not show any statistically significant differences between the animals testing the RIs (genotype: $F(3,158)=1.44$,

Table 1. Raw investigation durations during sampling and choice for the first experimental series (sampling direct - choice direct), separately shown for all genotypes and sex of the experimental subjects (ES).

ES genotype	ES sex	Sampling	Choice (familiar juvenile)	Choice (unfamiliar juvenile)
129S1/SvImJ	Male	54.8±7.9 [b]	25.4±2.8	40.1±4.4 ***
	Female	24.5±4.1 [b']	26.5±3.9	44.9±6.4 **
C57BL/6JOlaHsd	Male	41.6±3.8 [b]	21.3±1.9	33.8±2.2 ***
	Female	45.8±3.0 [b']	17.6±1.9	25.3±2.4 **
Hyb1	Male	69.3±6.1 [b]	28.6±3.5	44.9±4.3 **
	Female	39.5±5.1 [b']	24.2±2.7	37.3±4.2 **
Hyb2	Male	129.8±6.8 [a]	56.5±2.7	74.7±5.0 **
	Female	81.5±7.5 [a']	34.7±4.8	65.4±6.1 ***

Data are means ± SEM; for animals/group see Fig. 2A.
**p<0.01;
***p<0.001 versus familiar juvenile in the same genotype and sex, t-test for repeated measures;
a: p<0.01 vs. **b**, **a'**: p<0.01 vs. **b'**, two-way ANOVA (genotype×sex) and Scheffé's post-hoc test; significances shown for within sex comparism only.

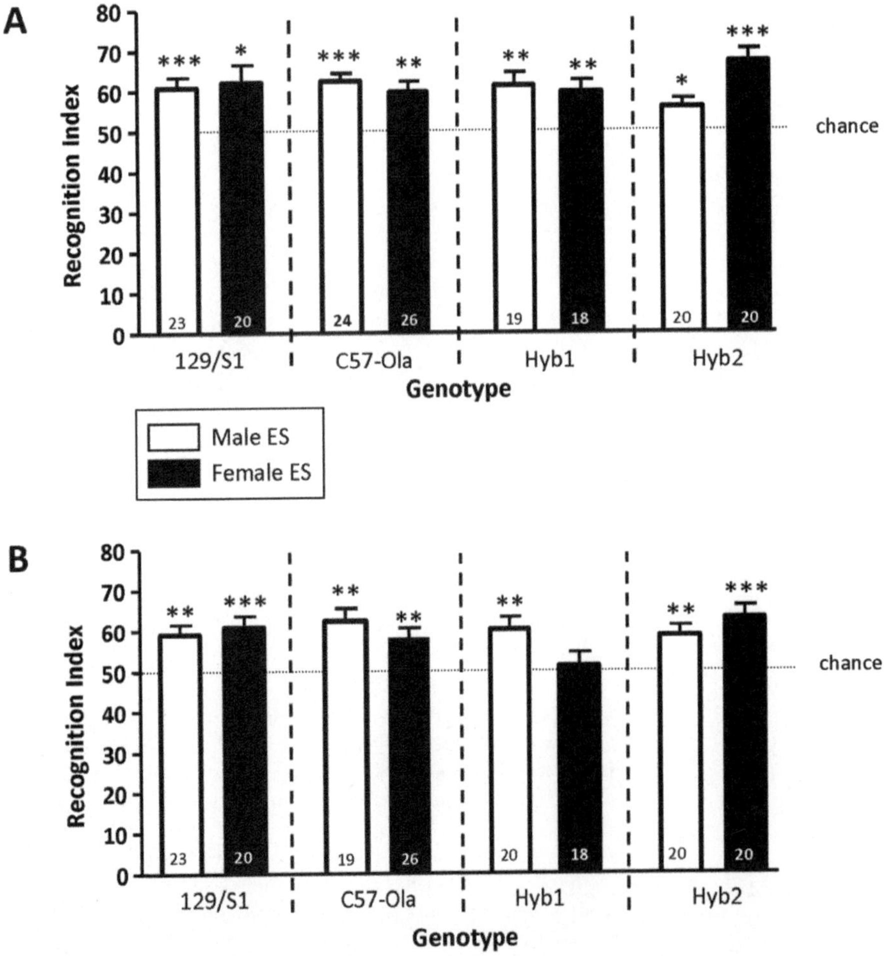

Figure 2. Recognition indices (RI; means+SEM) calculated from investigation durations during choice for the two experimental series. A shows that In the first experimental series the RIs of all tested groups [129S1/SvImJ (129/S1), Hyb1, Hyb2, C57BL/6JOlaHsd (C57-Ola)] of both male (white bars) and female ES (black bars) were significantly higher than 50 (chance) indicating an intact long-term recognition memory. **B** illustrates that in the second experimental series genotypes 129S1/SvImJ (both sex), Hyb2 (both sex), C57BL/6JOlaHsd (both sex), and Hyb1 males showed a significantly higher RI than 50 indicating an intact recognition memory. Hyb1 = F1-hybrid line (♀129S1/SvImJ×♂C57BL/6JOlaHsd), Hyb2 = F1-hybrid line (♀C57BL/6JOlaHsd×♂129S1/SvImJ), C57-Ola = C57BL/6JOlaHsd, 129/S1 = 129S1/SvImJ; numbers in bars = animals/group. * p<0.05, ** P<0.01 and *** P<0.001, one sample t-test *versus* 50.

$p=0.23$; sex: $F(1,158)=0.72$, $p=0.40$; interaction: $F(3,158)=2.49$, $p=0.06$). As shown in **Figure 2B**, only Hyb1 females failed to recognize the previously exposed juvenile; all other groups showed a RI significantly higher than the chance level. This was confirmed by the statistical analyses of the investigation durations during choice. All – except female Hyb1– other groups tested spent significantly more time investigating the novel juvenile than the previously encountered juvenile (**Table 2**).

Influence of Oestrus Cycle and Juvenile Sex on Recognition Ability

A two-way ANOVA testing the impact of prooestrus *versus* non-prooestrus on the memory performance in female mice revealed no significant differences in the RI in 129 ('experimental series': $F(1,36)=0.01$, $p=0.97$; 'oestrus stage': $F(1,36)=0.02$, $p=0.87$; interaction: $F(1,36)=0.21$, $p=0.65$) and Hyb1 females ('experimental series': $F(1,28)=0.10$, $p=0.75$; 'oestrus stage': $F(1,28)=0.78$, $p=0.38$; interaction: $F(1,28)=2.33$, $p=0.14$). Also for female C57BL/6JOlaHsd mice there was no significant impact of the oestrus cycle on the RI ('experimental series': $F(1,48)=0.93$, $p=0.34$; 'oestrus stage': $F(1,48)=0.39$, $p=0.54$; interaction: $F(1,48)=1.23$, $p=0.27$). Statistical analysis in Hyb2 females could not be performed because only one female was tested in prooestrus during both experimental series 1 and 2.

The results of the statistical analysis also indicated that the sex of the juvenile had no influence on the recognition performance in both experimental series (three-way ANOVA, first series: juveniles' sex: $F(1,95)=0.81$, $p=0,37$; interaction between juveniles' sex and genotype: $F(2,95)=1.65$, $p=0.20$; interaction between juveniles' sex and experimental subjects' sex: $F(1,95)=1.23$, $p=0.27$; second series: juveniles' sex: $F(1,109)=0.16$, $p=0,69$; interaction between juveniles' sex and genotype: $F(2,109)=0.32$, $p=0.73$; interaction between juveniles' sex and experimental subjects' sex: $F(1,109)=3.4$, $p=0.07$).

Discussion

Here we investigated the social investigation and olfactory cued social recognition memory of 129S1/SvImJ and C57BL/6JOlaHsd mice and two hybrid lines obtained from the parental strains.

Compared to all other groups, the duration spent in direct social investigation was significantly increased in Hyb2, but not Hyb1 animals (**Table 1**), the former reared by female C57BL/6JOlaHSD, the latter by 129S1/SvImJ mice. At present, it remains purely speculative whether phenomena such as imprinting [34–36], maternal behaviour [37–39] and hybrid vigor (=heterosis) [33] can explain the differences between our two hybrid strains. However, it is important to note that the situation seems very complex, since it appears to play a role which strain was used as the maternal and which other one as the paternal parental strain of the hybrids. Thus, further studies are needed to investigate the interaction between the possible contribution of these parameters on the expression of social investigation during adulthood towards conspecific juveniles in more detail.

All, but C57BL/6JOlaHsd, female mice spend less time investigating the juvenile during sampling than the males of the respective strain (**Table 1**). The fact that female animals show shorter investigation durations than males during direct encounters is well known from rats [40] and is – at least partially – here also confirmed for mice. Further studies have to investigate whether the genetic background plays indeed a role for the observed differences of the different sexes to engage in direct contact with the C57BL/6JOlaHsd juveniles. In any case, a reduced investigation duration during sampling (as seen in females of the three other strains) is not an indicative for a possibly impaired memory performance.

We measured in the VFC a significantly reduced investigation duration during sampling compared to the direct exposure. This implies that the stimulus animals were of less interest if only the volatile fraction of their olfactory signature was accessible in the VFC. It has been suggested that the non-volatile components of the olfactory signature's trigger the intensive investigation of conspecifics in laboratory rodents [41]. Indeed, Keller et al. [42] found a decreased investigatory behaviour in animals with impaired vomeronasal organ resulting in an inability to acquire and process the non-volatile components of the 'olfactory signature'. Therefore, the differences seen in the investigation durations in our study are likely to result from the inaccessibility of the non-volatile components of the conspecifics odour in the VFC apparatus. In addition, investigation durations during sampling were also affected by the genotype and sex of the experimental subjects. Hyb2 mice showed a significantly longer investigation of the juvenile compared to all other genotypes in both

Table 2. Raw investigation durations during sampling and choice for the second experimental series (sampling indirect (VFC) – choice direct), separately shown for all genotypes and sex of the experimental subjects (ES).

ES genotype	ES sex	Sampling	Choice (familiar juvenile)	Choice (unfamiliar juvenile)
129S1/SvImJ	Male	22.6±3.2 [b]	24.1±2.3	33.7±2.8 **
	Female	18.9±3.1 [d]	14.1±1.8	22.8±2.9 **
C57BL/6JOlaHsd	Male	21.3±2.7 [b]	17.6±2.0	28.0±2.2 **
	Female	18.8±1.9 [d]	20.1±1.52	29.0±3.1 **
Hyb1	Male	21.3±4.4 [b]	20.5±1.9	31.4±3.5 **
	Female	30.2±5.2 [e]	39.5±4.6	41.3±5.5
Hyb2	Male	55.0±3.5 [a]	49.7±4.0	68.2±3.5 **
	Female	52.5±4.4 [c]	20.0±3.4	31.0±5.1 ***

Data are means ± SEM; for animals/group see Fig. 2B.
*p<0.05;
**p<0.01;
***p<0.001 versus familiar juvenile, t-test for repeated measures; **a**: p<0.01 vs. **b**, **c**: p<0.01 vs. **d**, **c**: p<0.05 vs. **e**, two-way ANOVA (genotype×sex) and Scheffé's post-hoc test.

experimental series (see **Tables 1** and **2**). These data suggest that animals of mixed genetic background may show a higher social curiosity than those of their parental strains. This increased curiosity, however, has no predictive value of whether social recognition occurs or not. Indeed, animals that show a relatively low investigation duration during sampling successfully acquired for subsequent recognition (see **Table 1**, cf. [43]).

Authors of a recent study report an impaired short- and long-term social memory performance of male 129P2 mice, that was "rescued" by a (genetically engineered) reduced expression of the C-terminal Src kinase, reported to be causally involved in the proper development of the central nervous system and in the glutamatergic neurotransmission in the adult brain [19]. The authors conclude that social recognition is impaired in 129 animals [19]. In contrast, the data obtained in the first experimental series provide evidence for an unimpaired olfactory long-term recognition memory in both sexes and all genotypes tested in the social discrimination test **(Fig. 1A)**. The main difference between our and the previously published data seems to be the experimental design of the behavioural test: The 129P2 mice were tested for their memory performance using the *social recognition procedure* in which during retrieval ('choice') the familiar and novel juvenile are not presented simultaneously, but in separate sessions. We tested using the *social discrimination task*, which provides a direct comparison of the investigation duration of the familiar and novel stimulus animal in the choice session. It seems that the social recognition procedure detected false negative results concerning the social recognition performance of animals of the 129 strain. Indeed, the reported 'abnormality' of the 129 strain in various behavioural parameters [5–6,44–45], which may have also contributed to the deficits reported for this strain in defined learning and memory tasks [1,46], did not interfere in our paradigm of long-term social discrimination memory in either experimental series. To confirm that the 129S1/SvImJ animals used show a behavioural profile similar to that reported for 129 strains in defined behavioural paradigms, we submitted our animals to an open field test. Compared to C57BL/6JOlaHsd mice both males and females of the 129 strain showed the typical decreased motor activity, including a reduced time spent in running, rearing and a reduced number of lines crossed, and an increase in parameters indicating anxiety-like behaviour (see **Fig. S1** and **Table S1**). These observations match with the reported behavioural profile of the 129 strain in this test observed by some authors [2,47]. Therefore, we hypothesize that the unimpaired long-term olfactory recognition memory monitored in our sample of 129S1/SvImJ mice represents the parent population of this strain concerning the performance in the social discrimination test.

The literature is inconsistent concerning the influence of the oestrus cycle on learning and memory performance: Some authors reported an improvement of memory performance during the prooestrus [30,48–49], where others do not observe such an impact [50–51]. The differences were explained by oestrogen effects on different neuronal structures (and processes), [52–53] and a task specific impact of oestrogen [50]. Here, we failed to detect a statistically significant impact of the oestrus cycle on the social recognition performance in female mice of the two parental strains and Hyb1 during both experimental series. Although an analysis of the Hyb2 females could not be performed, the present data extend previous observations in rats [43] and suggest the robustness of the behavioural paradigm under study.

The data obtained in the second experimental series showed only for Hyb1 females an impaired social long-term recognition memory (see **Fig. 2B**). The observation that females and males of parental strains, 129S1/SvImJ and C57Bl/6OlaHsd, as well as the heterozygote Hyb2 strain and Hyb1 males showed an intact long-term recognition memory indicates that neither the genetical background of the parental strains nor the mixture of their genes is responsible for the recognition deficit of the Hyb1 females. Upon first view, the use of juveniles of the C57BL/6JOlaHsd strain as stimulus animals might provide a critical parameter that could have contributed to the observed effect. Indeed, it was reported that brain oxytocin signalling may underlie the intra- *versus* inter-strain social recognition in a procedure that involves the presentation of the stimulus animals in small cages [54]. However, there is no evidence for a difference in the oxytocinergic signalling between 129S1/SvImJ and C57Bl/6OlaHsd animals. Further, the fact that all other groups (including Hyb1 males and both genders of Hyb2) successfully recognized the previously presented juvenile in both experimental series renders a significant contribution of the strain origin of our stimulus animals rather unlikely. A more likely explanation for the observed differences might provide the challenge to match the volatile information from sampling with the complete olfactory signature collected during choice. Other than the protocol in which the experimental subjects get direct access to the stimulus animal's body the outcome of the VFC testing is more challenging [24] and obviously more sensitive to the differences in the genetic background and/or sex. The inaccessibility to the non-volatile components, thought to provide the motivation for extensive investigatory behaviour [41], may increase the sensitivity to interfering effects that are difficult to exclude from this test procedure (e.g. handling procedures to place the juveniles in the VFC; see [26]). Therefore, testing in the VFC requires a particularly careful selection of control animals (e.g. wildtype) to avoid false positive or false negative interpretations.

Taken together, we show here that during direct encounter between the experimental subject and the stimulus animal(s) in the social discrimination procedure, mice of the 129S1/SvImJ strain and 129S1/SvImJ×C57BL/6JOlaHsd hybrids do not differ in their social recognition performance from C57BL/6JOlaHsd mice at an exposure interval of 24 h. If only the volatile components of the 'olfactory signature' are presented during sampling, female Hyb1 fail to show an intact long-term recognition memory. Thus, a social discrimination procedure based on direct exposure of the stimulus animals during *both* sampling and choice is suitable to investigate the relevance of distinct brain areas and/or intra- and extracellular signalling systems for long-term recognition memory. This procedure can be used with transgenic mouse mutants (including with a 129S1 genetic background), that may be corrupted in other learning and (long-term) memory tasks.

Supporting Information

Figure S1 Parameters obtained in the open field of the different genotypes and sex (white bars: male; black bars: female; means+SEM). A illustrates the time spent in the inner, unprotected part of a 1 m×1 m wide open field with 50 cm high walls. **B** shows the numbers of lines crossed (of a virtual grid of 10 cm×10 cm) in the open field. Two-way ANOVA (genotype×sex) followed by Scheffé's post-hoc test. a: $p<0.01$ vs. b, b' and b", a': $p<0.05$ vs. b", Hyb1 = F1-hybrid line (♀129S1/SvImJ×♂C57BL/6JOlaHsd), Hyb2 = F1-hybrid line (♀C57BL/6JOlaHsd×♂129S1/SvImJ), C57-Ola = C57BL/6JOlaHsd, 129/S1 = 129S1/SvImJ; numbers in bars = animals/group.

Table S1 Time the experimental subjects spent grooming and rearing in the open field (means ± SEM; n = 18–20 per group).

Acknowledgments

The authors thank Rita Murau for expert technical assistance and Julia Noack and PD Dr. Carsten T. Wotjak (Munich) for critical reading of the manuscript.

Author Contributions

Conceived and designed the experiments: JH ME. Performed the experiments: JH. Analyzed the data: JH ME. Wrote the paper: JH ME.

References

1. Balogh SA, McDowell CS, Stavnezer AJ, Denenberg VH (1999) A behavioral and neuroanatomical assessment of an inbred substrain of 129 mice with behavioral comparisons to C57BL/6J mice. Brain Research 836: 38–48.
2. Contet C, Rawlins JN, Deacon RM (2001) A comparison of 129S2/SvHsd and C57BL/6JOlaHsd mice on a test battery assessing sensorimotor, affective and cognitive behaviours: implications for the study of genetically modified mice. Behavioural Brain Research 124: 33–46.
3. Homanics GE, Quinlan JJ, Firestone LL (1999) Pharmacologic and behavioral responses of inbred C57BL/6J and strain 129/SvJ mouse lines. Pharmacology, Biochemistry, and Behavior 63: 21–26.
4. Logue SF, Owen EH, Rasmussen DL, Wehner JM (1997) Assessment of locomotor activity, acoustic and tactile startle, and prepulse inhibition of startle in inbred mouse strains and F1 hybrids: implications of genetic background for single gene and quantitative trait loci analyses. Neuroscience 80: 1075–1086.
5. Owen EH, Logue SF, Rasmussen DL, Wehner JM (1997) Assessment of learning by the Morris water task and fear conditioning in inbred mouse strains and F1 hybrids: implications of genetic background for single gene mutations and quantitative trait loci analyses. Neuroscience 80: 1087–1099.
6. Hefner K, Whittle N, Juhasz J, Norcross M, Karlsson R-M, et al. (2008) Impaired fear extinction learning and cortico-amygdala circuit abnormalities in a common genetic mouse strain. The Journal of Neuroscience: The Official Journal of the Society for Neuroscience 28: 8074–8085.
7. Montkowski A, Poettig M, Mederer A, Holsboer F (1997) Behavioural performance in three substrains of mouse strain 129. Brain Research 762: 12–18.
8. Hefner K, Whittle N, Juhasz J, Norcross M, Karlsson RM, et al. (2008) Impaired fear extinction learning and cortico-amygdala circuit abnormalities in a common genetic mouse strain. Journal of Neuroscience 28: 8074–8085.
9. Whittle N, Hauschild M, Lubec G, Holmes A, Singewald N (2010) Rescue of impaired fear extinction and normalization of cortico-amygdala circuit dysfunction in a genetic mouse model by dietary zinc restriction. J Neurosci 30: 13586–13596.
10. Crawley JN (1996) Unusual behavioral phenotypes of inbred mouse strains. Trends in Neurosciences 19: 181–182; discussion 188–189–181–182; discussion 188–189.
11. Crusio WE (2004) Flanking gene and genetic background problems in genetically manipulated mice. Biological Psychiatry 56: 381–385.
12. Crusio WE (1996) Gene-targeting studies: new methods, old problems. Trends in Neurosciences 19: 186–187; discussion 188–189–186–187; discussion 188–189.
13. Gerlai R (1996) Gene-targeting studies of mammalian behavior: is it the mutation or the background genotype? Trends in Neurosciences 19: 177–181.
14. Skarnes WC, Rosen B, West AP, Koutsourakis M, Bushell W, et al. (2011) A conditional knockout resource for the genome-wide study of mouse gene function. Nature 474: 337–342.
15. Carr WJ, Yee L, Gable D, Marasco E (1976) Olfactory recognition of conspecifics by domestic Norway rats. Journal of Comparative and Physiological Psychology 90: 821–828.
16. Matochik JA (1988) Role of the main olfactory system in recognition between individual spiny mice. Physiology & Behavior 42: 217–222.
17. Sawyer TF, Hengehold AK, Perez WA (1984) Chemosensory and hormonal mediation of social memory in male rats. Behavioral Neuroscience 98: 908–913.
18. Engelmann M, Wotjak CT, Landgraf R (1995) Social discrimination procedure: an alternative method to investigate juvenile recognition abilities in rats. Physiology & Behavior 58: 315–321.
19. Sinai L, Mathew R, Roder JC (2012) Impaired social memories in 129P2 inbred mice are rescued by reduced Csk expression. Genes Brain Behav 11: 559–567.
20. Dantzer R, Bluthe RM, Koob GF, Le Moal M (1987) Modulation of social memory in male rats by neurohypophyseal peptides. Psychopharmacology (Berl) 91: 363–368.
21. Engelmann M, Wotjak CT, Landgraf R (1995) Social discrimination procedure: an alternative method to investigate juvenile recognition abilities in rats. Physiol Behav 58: 315–321.
22. Richter K, Wolf G, Engelmann M (2005) Social recognition memory requires two stages of protein synthesis in mice. Learn Mem 12: 407–413.
23. Engelmann M, Hädicke J, Noack J (2011) Testing declarative memory in laboratory rats and mice using the non-conditioned social discrimination procedure. Nature Protocols 6: 1152–1162.
24. Noack J, Richter K, Laube G, Haghgoo HA, Veh RW, et al. (2010) Different importance of the volatile and non-volatile fractions of an olfactory signature for individual social recognition in rats versus mice and short-term versus long-term memory. Neurobiol Learn Mem 94: 568–575.
25. Chesler EJ, Plitt A, Fisher D, Hurd B, Lederle L, et al. (2012) Quantitative trait loci for sensitivity to ethanol intoxication in a C57BL/6Jx129S1/SvImJ inbred mouse cross. Mamm Genome 23: 305–321.
26. Engelmann M, Hädicke J, Noack J (2011) Testing declarative memory in laboratory rats and mice using the nonconditioned social discrimination procedure. Nat Protocols 6: 1152–1162.
27. Guevara-Guzman R, Buzo E, Larrazolo A, de la Riva C, Da Costa AP, et al. (2001) Vaginocervical stimulation-induced release of classical neurotransmitters and nitric oxide in the nucleus of the solitary tract varies as a function of the oestrus cycle. Brain Research 898: 303–313.
28. Larrazolo-Lopez A, Kendrick KM, Aburto-Arciniega M, Arriaga-Avila V, Morimoto S, et al. (2008) Vaginocervical stimulation enhances social recognition memory in rats via oxytocin release in the olfactory bulb. Neuroscience 152: 585–593.
29. Otsuka T, Ishii K, Osako Y, Okutani F, Taniguchi M, et al. (2001) Modulation of dendrodendritic interactions and mitral cell excitability in the mouse accessory olfactory bulb by vaginocervical stimulation. The European Journal of Neuroscience 13: 1833–1838.
30. Sanchez-Andrade G, Kendrick KM (2009) The main olfactory system and social learning in mammals. Behavioural Brain Research 200: 323–335.
31. Silberstein GB, Van Horn K, Hrabeta-Robinson E, Compton J (2006) Estrogen-triggered delays in mammary gland gene expression during the estrous cycle: evidence for a novel timing system. The Journal of Endocrinology 190: 225–239.
32. Tobin VA, Hashimoto H, Wacker DW, Takayanagi Y, Langnaese K, et al. (2010) An intrinsic vasopressin system in the olfactory bulb is involved in social recognition. Nature 464: 413–417.
33. Warren JM (1988) Age, incentives, and maze learning by hybrid mice. Behav Genet 18: 167–173.
34. Davies W, Isles A, Smith R, Karunadasa D, Burrmann D, et al. (2005) Xlr3b is a new imprinted candidate for X-linked parent-of-origin effects on cognitive function in mice. Nat Genet 37: 625–629.
35. Gregg C, Zhang J, Butler JE, Haig D, Dulac C (2010) Sex-specific parent-of-origin allelic expression in the mouse brain. Science 329: 682–685.
36. Wilkins JF (2005) Genomic imprinting and methylation: epigenetic canalization and conflict. Trends Genet 21: 356–365.
37. Bredy TW, Humpartzoomian RA, Cain DP, Meaney MJ (2003) Partial reversal of the effect of maternal care on cognitive function through environmental enrichment. Neuroscience 118: 571–576.
38. Francis DD, Meaney MJ (1999) Maternal care and the development of stress responses. Current Opinion in Neurobiology 9: 128–134.
39. Liu D, Diorio J, Day JC, Francis DD, Meaney MJ (2000) Maternal care, hippocampal synaptogenesis and cognitive development in rats. Nature Neuroscience 3: 799–806.
40. Bluthe RM, Dantzer R (1990) Social recognition does not involve vasopressinergic neurotransmission in female rats. Brain Res 535: 301–304.
41. Baum MJ (2009) Sexual differentiation of pheromone processing: links to male-typical mating behavior and partner preference. Hormones and Behavior 55: 579–588.
42. Keller M, Pierman S, Douhard Q, Baum MJ, Bakker J (2006) The vomeronasal organ is required for the expression of lordosis behaviour, but not sex discrimination in female mice. The European Journal of Neuroscience 23: 521–530.
43. Engelmann M, Ebner K, Wotjak CT, Landgraf R (1998) Endogenous oxytocin is involved in short-term olfactory memory in female rats. Behavioural Brain Research 90: 89–94.
44. Bothe GWM, Bolivar VJ, Vedder MJ, Geistfeld JG (2005) Behavioral differences among fourteen inbred mouse strains commonly used as disease models. Comparative Medicine 55: 326–334.
45. Camp M, Norcross M, Whittle N, Feyder M, D'Hanis W, et al. (2009) Impaired Pavlovian fear extinction is a common phenotype across genetic lineages of the 129 inbred mouse strain. Genes, Brain, and Behavior 8: 744–752.
46. Cook MN, Bolivar VJ, McFadyen MP, Flaherty L (2002) Behavioral differences among 129 substrains: implications for knockout and transgenic mice. Behavioral Neuroscience 116: 600–611.
47. Paulus MP, Dulawa SC, Ralph RJ, Mark AG (1999) Behavioral organization is independent of locomotor activity in 129 and C57 mouse strains. Brain Research 835: 27–36.
48. Frye CA, Duffy CK, Walf AA (2007) Estrogens and progestins enhance spatial learning of intact and ovariectomized rats in the object placement task. Neurobiology of Learning & Memory 88: 208–216.
49. Walf AA, Koonce C, Manley K, Frye CA (2009) Proestrous compared to diestrous wildtype, but not estrogen receptor beta knockout, mice have better performance in the spontaneous alternation and object recognition tasks and reduced anxiety-like behavior in the elevated plus and mirror maze. Behavioural Brain Research 196: 254–260.
50. Korol DL, Kolo LL (2002) Estrogen-induced changes in place and response learning in young adult female rats. Behavioral Neuroscience 116: 411–420.

51. Markham JA, Juraska JM (2007) Social recognition memory: influence of age, sex, and ovarian hormonal status. Physiology & Behavior 92: 881–888.
52. Maggi A, Ciana P, Belcredito S, Vegeto E (2004) Estrogens in the nervous system: mechanisms and nonreproductive functions. Annual Review of Physiology 66: 291–313.
53. McEwen BS (2001) Invited review: Estrogens effects on the brain: multiple sites and molecular mechanisms. Journal of Applied Physiology (Bethesda, Md: 1985) 91: 2785–2801.
54. Macbeth AH, Lee HJ, Edds J, Young WS (2009) Oxytocin and the oxytocin receptor underlie intrastrain, but not interstrain, social recognition. Genes Brain and Behavior 8: 558–567.

20

A Biphasic and Brain-Region Selective Down-Regulation of Cyclic Adenosine Monophosphate Concentrations Supports Object Recognition in the Rat

Maïte Hotte[1,2], François Dauphin[1], Thomas Freret[1], Michel Boulouard[1], Guenaëlle Levallet[1,3]*

1 Université de Caen Basse-Normandie, Groupe Mémoire et Plasticité comportementale (GMPc), EA4259, IFR 146, Caen, France, **2** Université de Rouen, NeoVasc, EA 4309, IFRMP23, IHURBM, Rouen, France, **3** CHU de Caen, Service d'Anatomie Pathologie, Caen, France

Abstract

Background: We aimed to further understand the relationship between cAMP concentration and mnesic performance.

Methods and Findings: Rats were injected with milrinone (PDE3 inhibitor, 0.3 mg/kg, i.p.), rolipram (PDE4 inhibitor, 0.3 mg/kg, i.p.) and/or the selective 5-HT4R agonist RS 67333 (1 mg/kg, i.p.) before testing in the object recognition paradigm. Cyclic AMP concentrations were measured in brain structures linked to episodic-like memory (i.e. hippocampus, prefrontal and perirhinal cortices) before or after either the sample or the testing phase. Except in the hippocampus of rolipram treated-rats, all treatment increased cAMP levels in each brain sub-region studied before the sample phase. After the sample phase, cAMP levels were significantly increased in hippocampus (1.8 fold), prefrontal (1.3 fold) and perirhinal (1.3 fold) cortices from controls rat while decreased in prefrontal cortex (~0.83 to 0.62 fold) from drug-treated rats (except for milrinone+RS 67333 treatment). After the testing phase, cAMP concentrations were still increased in both the hippocampus (2.76 fold) and the perirhinal cortex (2.1 fold) from controls animals. Minor increase were reported in hippocampus and perirhinal cortex from both rolipram (respectively, 1.44 fold and 1.70 fold) and milrinone (respectively 1.46 fold and 1.56 fold)-treated rat. Following the paradigm, cAMP levels were significantly lower in the hippocampus, prefrontal and perirhinal cortices from drug-treated rat when compared to controls animals, however, only drug-treated rats spent longer time exploring the novel object during the testing phase (inter-phase interval of 4 h).

Conclusions: Our results strongly suggest that a "pre-sample" early increase in cAMP levels followed by a specific lowering of cAMP concentrations in each brain sub-region linked to the object recognition paradigm support learning efficacy after a middle-term delay.

Editor: Nicoletta Landsberger, University of Insubria, Italy

Funding: The authors have no support or funding to report.

Competing Interests: The authors have declared that no competing interests exist.

* E-mail: levallet-g@chu-caen.fr

Introduction

Most modern theories of learning and memory postulate that memory processes require cyclic adenosine monophosphate (cAMP) synthesis [1]; however, there is little evidence concerning the mechanisms by which memory affects adenylyl cyclase activity (cAMP synthesis) and/or phosphodiesterase (PDE) activity (cAMP degradation). Literature reports indicate that activation of the cAMP-PKA pathway cascade by memory processes triggers activation of transcription factors such as CREB [2], leading to neural processes that underlie learning and memory [1,3–5]. Consequently, several studies argue that artificial cAMP-PKA cascade activation through intra-hippocampal infusion of 8Br-cAMP, adenylyl cyclase or PKA activation improves memory performance [4,6–9] whereas pharmacological inhibition of PKA disrupts hippocampal long term potentiation and hippocampus-based long-term memory [6,10,11]. Memory efficiency seems, however, to require a restricted or selective cAMP production; high cAMP levels do not necessarily improve memory. Indeed, studies on flies and mice show that increases in adenylyl cyclase activity can result in memory deficits [12,13]. Similarly, increasing PKA activity impairs prefrontal cortex-dependent memory in mice and expression of a constitutively active isoform of the G-protein subunit Gαs impairs mice behavioural performance in a fear-conditioning task [14]. These works clearly demonstrate the complexity of cAMP-dependent responses.

Mnesic mechanisms may be investigated through the use of an object recognition memory task, a one-phase task based on spontaneous activity and the natural preference that rodents display to explore a novel object rather than a familiar one [15]. With this paradigm, memory performances were demonstrated to be enhanced by the activation of serotonin 5-HT4 receptors (5-HT4R) [16–19], receptors that have been also demonstrated to be implicated in short- and long-term memory processes in laboratory animals [20–30] (for review see [31]). Activation of 5-HT4R, positively coupled to adenylyl cyclase, induce increases in cAMP concentrations that can be regulated by activation of cAMP phosphodiesterases (PDE) isoforms from families 1, 2, 3, and 4

[18]. Involvement of PDE4 inhibitors in working and reference memory [5,32–34] has already been well investigated through the use of rolipram, a selective PDE4 inhibitor. In fact, several studies have already reported a positive effect of PDE4 inhibition on spatial memory [3,5,33–37], inhibitory avoidance learning [5,33,38], contextual fear conditioning [4,39], and object recognition [40–42]. The PDE4 isoenzymes are encoded by four independent genes (Pde4a to Pde4d), which generate more than 25 splice variants [43,44]. Each splice variant exhibits unique properties leading to specific control of cAMP levels [45,46]. Few studies deal with the involvement of each PDE4 isoforms in memory performance as few studies have also investigated the effect of others cAMP-PDE families on memory performance, especially in the object recognition task [42,47]. Thus, from the literature, little is known about the behavioural consequence and especially memory performance following PDE4 [18], PDE3 [48] or PDE2 [47] inhibition although these PDE families accounts for a major part of the total cAMP-PDE-hydrolysing activities in the hippocampus, the prefrontal and perirhinal cortices, brain structures involved in recognition memory [18].

A way to enhance cAMP signaling and consequently influence the pathways involved in object recognition (episodic-like) memory, is to stimulate 5-HT4Rs and/or inhibit PDE enzymes, especially PDE3 and 4 [18]. Here, we further characterize the respective role of PDE3 and PDE4 in the processes of recognition memory and assayed the relationship between cAMP concentrations and mnesic performance. With this aim, we injected rats before the acquisition phase, with milrinone (a selective PDE3 inhibitor currently used in heart failure studies [49], with a half-life of 1.5–2.3 h [50,51], 0.3 mg/kg), or rolipram (a selective PDE4 inhibitor with good brain penetration and a relatively short half-life of 1–3 h [52], 0.3 mg/kg) both alone and in combination or not with the selective 5-HT4R agonist RS 67333 (1 mg/kg) [17,18]. The half-life of RS 67333 in the rat is not reported in the literature, but the data from behavioral studies [17,18] suggest that this is comparable (1–3 hours) to those of the two PDE inhibitors.

Before or after either the sample or the testing phase, cAMP concentrations were measured in the hippocampus, prefrontal and perirhinal cortices. We show that a "pre-sample" early increase in cAMP levels followed by a specific lowering of cAMP concentrations in each brain sub-region linked to the object recognition paradigm support learning efficacy after a middle-term delay. Following the different treatments and at the end of the testing trial, we also examined in these brain sub-regions i) the PDE activities to validate efficiency of PDE3 or PDE4 inhibition by their respective inhibitor and ii) the phosphoprotein phosphatase (PP) type 2 activities since cAMP concentrations have been shown to be transiently up- or down-regulated by PP2A activation in various cell types [53–55]. In fact, as cAMP-PDE limit excessive cAMP production by catalysing its hydrolysis; PP1 and PP2 (accounting for more than 90% of total phosphatase activity in brain [56]) limit PDE-induced excessive catabolism of cAMP by reversing PP2A phosphorylation of phosphorylated PPE3B [57] or particulate PDE4 activities [58]. We also demonstrate here, that milrinone alters type PP2 activities in anatomical structures linked to object recognition memory in rat.

Materials and Methods

1. Subjects

A total of 172 adult male Sprague-Dawley rats (300–350 g, René Janvier, France) were used in these experiments. Rats were housed in groups of three in a temperature controlled room under a 12L:12D cycle (lights on at 8:00 pm), with food and water provided ad libitum. All procedures were performed in conformity with National (JO 887–848) and European (86/609/EEC) legislations on animal experimentation. Behavioural procedures received approval from the Ethics Committee for Animal Experimentation of Normandy (Approval number 1009-01).

2. Behavioural experiments

2.1 Apparatus. The apparatus consisted of an open-box (100×100×60 cm) made of wood with the inside painted in black. The objects to be discriminated were made of plastic, or glass (all 5 cm height) and were available in four copies. The objects were fixed (Patafix) on the floor in the box, to ensure that they could not be displaced by the rats.

2.2 Handling and habituation. Rats were handled daily for one week prior to the study and then habituated to the apparatus and the test room. The first two days, rats were put together as a group of 3 to explore the empty arena for 10 min. On the third day, rats were put individually in the empty box for 3 min and the next two days, in the presence of an object that will not be used for the experimental task. Testing began on day 6.

2.3 Object recognition task. Animals were tested in the object recognition task as described previously [15,18]. The test session consisted of two phases with a duration of 3 min each on day 6. During the sample phase, each rat was placed in the box with two identical objects (placed close to the corners). After a delay of 4 h, during which the animal returned to its cage and both objects were replaced (one by its identical copy, the other by a new object in the same locations), the rat was returned to the box (testing phase). From rat to rat, the role (familiar or new object) as well as the relative position of the two objects were counterbalanced and randomly permuted.

The number of animals in each group was: saline-treated (n = 32), RS 67333 1 mg/kg (n = 32), milrinone 0.3 mg/kg (n = 27), milrinone+RS 67333 (n = 27), rolipram 0.3 mg/kg (n = 27), rolipram+RS 67333 (n = 27).

3. Drugs and drug administration

In all experiments, each rat was given an i.p injection of either saline (NaCl 0.9%) or RS 67333 (1 mg/kg) 30 minutes prior to the sample phase as previously described [18]. We have not tested other steps of memorization or lower doses of RS 67333 because i) RS6733-induced enhancement was reported only for the acquisition phase of information processing and ii) doses of 0.001 or 0.01 mg/kg were ineffective to enhance recognition memory [16,17]. Milrinone (PDE3 inhibitor) or rolipram (PDE4 inhibitor) was injected each at the dose of 0.3 mg/kg i.p. 45 min prior to the sample phase. Higher doses of PDE inhibitors were not tested since at high dosage, milrinone can have vasodilatory and arrhythmogenic effects [59] and rolipram can have sedative side-effects [60,61]. Efficiency of PDE3 or PDE4 inhibition was confirmed by specific PDE3 or 4 activity measures in hippocampus, prefrontal and perirhinal cortices at the end of the behavioural task. Moreover, since object recognition performance can only be determined if the animals show sufficient exploration [62], we concomitantly evaluated the exploration levels of the animals. In our experiment, a dose of 0.3 mg/kg rolipram or 0.3 mg/kg milrinone, given 45 min before the sample phase, resulted in a substantial decrease in locomotor activity but not in exploratory behaviour. A total of 148 rats were used to perform the object recognition task (tested animals), and 24 rats received the different injections, without being subjected to the behavioural task (untested animals).

4. Biochemical analysis

4.1 Cyclic AMP measurement. Cyclic AMP extraction was performed according to a procedure adapted from Rodriguez [63]. Immediately after the testing phase or at the corresponding delay following the different injections (for untested animals), rats were subjected to euthanasia by decapitation without prior anesthesia. Intact brains were dissected on ice into prefrontal cortex, perirhinal cortex and hippocampus, taken systematically in this order and in less than 3 min following euthanasia. Brain sub-regions were rinsed with cold saline and dried. Each sub-region was homogenized in cold 100% ethanol in an ice bath and the homogenate centrifuged at 20,000 g for 15 min. The supernatant was recovered and, the pellet resuspended in 1 ml of 2:1 ethanol:water solution and centrifuged as before. The combined supernatants were evaporated to dryness in a 60°C bath under a stream of nitrogen gas. The final residue was dissolved in 0.5 ml of assay buffer (0.05 M sodium acetate, pH 5.8, containing sodium azide). Cyclic AMP levels were determined with a radioimmunoassay (Amersham). This assay measures the competitive binding of ^{3}H-labeled cAMP to a cAMP-specific antibody.

4.2 Preparation of rat brain membranes and soluble fractions. Sub-cellular fractionation of the brain regions was also performed immediately after euthanasia as detailed previously [18]. Briefly, each cerebral tissue was placed, immediately upon isolation, into ice-cold homogenization buffer (20 mMTris-HCl pH 7.2, 1 mM EDTA, 250 mM sucrose, supplemented with 0.1 mMphenylmethanesulfonyl fluoride, 2 mMbenzamidin, and a mixture of antiproteases (antipain, aprotinin, leupeptin, pepstatin A) at a final concentration of 1 microg/ml), homogenized by several passages through 25-G needle. Homogenates were centrifuged at 1,000 g, 4°C for 5 min and the supernatants decanted and centrifuged at 100,000 g, 4°C for 1 h. Each supernatant (soluble fraction) and the respective pellet (particulate fraction, re-suspended in ice-cold complete homogenization buffer) were then stored at −20°C. Protein content of each fraction was determined by the method of Bradford with BSA as a standard [64]. Purity of each subcellular fraction was assayed by both lactate deshydrogenase (soluble activity) and alkaline phosphatase (membrane-associated activities) as already reported [18].

4.3 PDE Assay. Phosphodiesterase activities were assayed according to the two-step modified procedure of Thompson and Applemann [65] as already described [18]. To discriminate PDE2, PDE3 or PDE4 activities from other PDE activities, protein from each sample were incubated either in the absence (total PDE activities) or in the presence of specific inhibitors of each family: 20 µM erythro-9-(2-hydroxy-3-nonyl)-adenine (EHNA), 20 µM milrinone or 10 µM rolipram for PDE2, PDE3 and PDE4, respectively, according to their respective IC50 described elsewhere [18]. Differences between total and selective inhibitor-insensitive PDE activities were considered as corresponding PDE activities.

4.4 Phosphatase Assay. Total PP2 activities in subcellular fractions of the different rat brain structures were determined by Serine/Threonine Phosphatase Assay (Promega, Charbonière-les-Bains, France) which used a specific substrate for PP2. Assays were conducted according to the manufacturer's procedure. Free phosphate was then quantified by a colorimetric method.

4.5 SDS-PAGE Western Blot Analysis. Subcellular fraction protein from hippocampus, prefrontal and perirhinal cortices were boiled for 5 min and separated by 8% SDS-PAGE. The proteins were transferred onto a nitrocellulose membrane (1 h at 100 V and 4°C). Western blotting was then performed using an affinity-purified goat polyclonal antibody raised against a peptide that maps near the C-terminus of the human PDE4D (Santa Cruz Biotechnology). Immunoblotting with antibody that was pre-incubated with an excess of the peptide used for immunization (Santa Cruz Biotechnology) was performed as a negative control, following the instructions of the supplier. Immunoreactive bands were detected using a donkey anti-goat IgG-horseradish peroxidase (HRP) complex and an enhanced chemiluminescence (ECL) Advance Western Blotting Detection Kit (Amersham Biosciences). For β-actin detection, the blots were stripped in a stripping buffer that contained 62.5 mM Tris-HCl (pH 6.7) 2% SDS, and 100 mM β-mercaptoethanol at 58°C for 30 min, and reprobed for actin with monoclonal mouse anti-actin antibody and goat anti-mouse IgG-HRP (Calbiochem). The immunoblots were scanned on the ProXPRESS Proteomic Imaging System (Perkin Elmer Life Science, Boston, MA) and analyzed with the TotalLab Image Analysis software (Nonlinear Dynamics Ltd., Newcastle, UK).

5. Data scoring and analysis

5.1 Behavioural analysis. The experimenter sat in front of the box. Total time spent exploring each object in both the sample and the testing phases were recorded. Exploration of an object was defined as follows: directing the nose to the object at a distance <2 cm. Overall exploration times across phases were analyzed by a two-way ANOVA (phase and treatment as factors) with repeated measures. For testing phase data, exploration of each object was analyzed using a two-way repeated-measurements ANOVA with object and treatment as factors. When appropriates, post-hoc testing was performed using Fisher's least significant difference (LSD) test. We calculated discrimination indexes as D1, which is the difference in time spent exploring the two objects in testing trial (i.e. time with novel object minus time with familiar object); and D2, the discrimination ratio, which is the difference in exploration time (D1) expressed as a ratio of the total time spent exploring the two objects in the testing trail (e.g. novel-familiar/novel+familiar). This ratio makes it possible to adjust for individual or group differences in the total amount of exploration time. Comparisons were made using one-way ANOVA with treatment as factor and post-hoc testing was performed using Fisher's least significant difference (LSD) test.

Locomotor activity was measured during the test session through videotaping. The arena was divided into 9 squares (32×32 cm). During each phase, the number of entries in each square was measured. Analysis was performed using two-way repeated-measurements ANOVA with entry and treatment as factors, followed by Fisher's LSD test when necessary.

5.2 Biochemical analysis. After construction of a standard curve, cAMP levels were determined directly from the counts (in duplicate for each brain region of each animal) in nanomoles per milligram of tissue wet weight. PDE activities (in triplicates) were expressed in pmol of cAMP hydrolyzed per min and mg of protein. PP2 activities were expressed as nmol of phosphate released per min.

Statistical differences were determined through non-parametric tests adapted to small size data (Friedman and Kruskal-Wallis, followed by a post-hoc Mann-Whitney U-test; Sigma Stat software SPSS Inc, Chicago, IL).

Results

Hippocampus, prefrontal and perirhinal cortices exhibit different patterns of particulate PDE4D isoforms

Cyclic AMP-PDE was assayed in subcellular fraction from hippocampus, prefrontal and perirhinal cortices. Here, we confirmed our previous work [18] reporting that the total cAMP-PDE-hydrolysing activities of the particulate fraction from hippocampus, prefrontal and perirhinal cortices are mainly composed by PDE3 (38,2%, 34,8% and 43,4% respectively in hippocampus, prefrontal and perirhinal cortices) and PDE4 (26,7%, 43,4% and 21,4% respectively in hippocampus, prefrontal and perirhinal cortices) (data not shown). Since each PDE4D isoform plays specific roles on the cAMP concentration feedback [45,46], we furthermore characterized by western blotting the pattern of PDE4D isoforms present in these brain sub-region (**Fig. 1**). We demonstrated that PDE4D protein expression differed according to the subcellular fraction and the brain sub-region. In the particulate fraction from prefrontal cortex, the presence of nine immunoreactive proteins suggests that all nine PDE4D isoforms (i.e. PDE4D1 to PDE4D9) are expressed whereas particulate fraction of hippocampus did not exhibit PDE4D6 and of perirhinal cortex neither particulate PDE4D8/9 nor particulate PDE4D3. Nevertheless, in the particulate fraction from hippocampus, prefrontal or perirhinal cortices, both the short PDE4D1 and the long PDE4D4 isoforms are the mainly PDE4D isoforms expressed. Finally, PDE4D1, PDE4D2, PDE4D4 and PDE4D6 were the isoforms revealed in the soluble fraction from prefrontal or perirhinal cortices. A similar panel of PDE4D isoform was revealed in the soluble fraction from the hippocampus, except that no immunoreactive band matches with PDE4D2. As reflected by the densitometric analyses (right panel of **Fig. 1**), the patterns of the putative PDE4D isoforms did not display any significant structure-related differences.

RS 67333 enhances particulate PDE3 activity from the perirhinal cortex in rat

To identify which cAMP-PDE family support the rolipram-insensitive PDE activities increased in the perirhinal cortex following the selective activation of 5-HT4R (RS 67333; [18]), rats were injected with a saline solution or RS 67333 (1 mg/kg, i.p.) before the object recognition paradigm (inter-phase interval of 4 h). Immediately after the testing phase, rats were subjected to euthanasia and PDE activities were assayed. As shown in **Fig. 2**, RS 67333 elevated PDE3 activities by 71% ($P<0.01$) in the perirhinal cortex lightening this family as a key regulator of cAMP concentration in this structure linked to object recognition. No variation was measured in the supernatant fraction (data not shown).

5-HT4 receptor stimulation, PDE3- or PDE4-inhibition improve familiar object recognition after a 4-h delay in rat

Rats were then injected with milrinone (PDE3 inhibitor, 0.3 mg/kg, i.p.), rolipram (PDE4 inhibitor, 0.3 mg/kg, i.p.) and/or the selective 5-HT4R agonist RS 67333 (1 mg/kg, i.p.) before the object recognition paradigm sample phase.

We first validate the efficiency of the treatments with PDE inhibitors; rats were immediately subjected to euthanasia after the testing phase, and PDE activities from the hippocampus, prefrontal and perirhinal cortices, were assessed. We especially measured PDE3 activity for milrinone-treated animals (**Fig. 3**) and PDE4 activity for rolipram-treated animals (**Fig. 4**). Concerning the measurements of particulate PDE3 activities in milrinone-treated rats (**Fig. 3**), we showed that PDE3 activity was inhibited in the hippocampus (−30%, $P<0.01$) and prefrontal cortex (−63%, $P<0.001$), but not in the perirhinal cortex, when compared to saline-treated rats. However, pre-treatment of rats by milrinone before RS 67333 prevented the RS 67333-induced increase in particulate PDE3 in the perirhinal cortex. Finally, milrinone did not affect significantly cAMP-PDE activities supported by other families than PDE3. As illustrated in **Fig. 4**, in rolipram-treated rats, particulate PDE4 activity was lower than in saline-treated rats, in the hippocampus (−60%, $P<0.05$) and the prefrontal (−42%, $P<0.05$) cortex, while tend to be lower in the perirhinal cortex (−26%). Similar decrease when compared to RS 67333-treated group was also observed for rolipram+RS 67333-treated animals in the prefrontal cortex (−42%, $P<0.05$) and, despite no significant, in the perirhinal cortex (−47%).

As shown in **Table 1**, all rats spent a similar total time exploring both objects during either the sample or the testing phase after a 4-h delay ($P>0.05$). Comparison of locomotor activities revealed an overall significant effect of treatment during the sample phase ($F(5, 112)=9.25$, $P<0.001$). Post hoc analyses showed that rolipram- or rolipram+RS 67333-treated rats had a smaller number of entries compared to saline-treated (respectively $P<0.001$, $P<0.01$), RS 67333-treated ($P<0.001$), milrinone-treated (respectively $P<0.001$, $P<0.05$), milrinone+RS 67333- treated animals (respectively $P<0.001$, $P<0.01$) (**Table 2**). Finally, we found that milrinone-treated animals explore less than RS 67333-treated ones ($P<0.01$). Analysis of the testing phase revealed no significant treatment effect on the exploratory behaviour.

The repeated-measures ANOVA revealed i) for saline-treated rats, no significant difference of novel object exploration time (**Fig. 5**); ii) for drug-treated rats, both an overall significant effect of time spent exploring each object (F(1, 112 = 109.3, $P<0.001$) and an interaction between time exploring each object and treatment (F 5, 122) = 5.4, $P<0.001$). Post hoc analyses showed that all drug-treated rats significantly spent more time exploring the novel object, when compared to saline-treated rats (RS 67333-treated, $P<0.001$, rolipram-treated, $P<0.001$; rolipram+RS 67333-treated, $P<0.01$, $P<0.001$; milrinone-treated, $P<0.001$; milrinone+RS 67333-treated, $P<0.05$) (**Fig. 5**). This result is also confirmed by analysis of discrimination indexes (**Table 3**). ANOVA performed on D1 and D2 showed a significant treatment effect [for both D1 and D2: ($F(5, 112)=4.5$, $P<0.001$)]. Post-hoc analysis revealed that all treated animals had a greater discrimination index (D1) compared to saline-treated animals (RS 67333-treated ($P<0.001$), rolipram-treated ($P<0.001$), rolipram+RS 67333 treated ($p<0.01$), milrinone-treated ($P<0.01$) milrinone+RS 67333- treated animals ($P<0.05$). These results are also confirmed by post-hoc analysis of the discrimination ratio (D2), compared to saline-treated rats (RS 67333-treated ($P<0.001$), rolipram-treated ($P<0.001$), rolipram+RS 67333 treated ($p<0.05$) milrinone-treated ($P<0.01$) milrinone+RS 67333- treated animals ($P<0.05$).

Familiar object recognition is associated with a "pre-sample" early increase in cAMP levels in hippocampus, prefrontal and perirhinal cortices

To further characterize the cellular mechanisms involved after a 4 h-delay, rats were subjected to euthanasia before or after the sample or the testing phase of the paradigm, and cAMP was measured in the anatomical structures linked to the object recognition task (i.e. hippocampus, prefrontal and perirhinal cortices) (**Fig. 6**).

As illustrated in **Fig. 6**, cAMP concentrations measured before the sample phase (white bars), demonstrate, except in the hippocampus of rolipram treated-rats, the efficiency of RS 67333 (1 mg/kg, i.p.), rolipram (0.3 mg/kg, i.p.) and milrinone (0.3 mg/kg, i.p., data not illustrated for more readability) to increase cAMP levels in the three brain structures studied, i.e. hippocampus (~1.3 fold), prefrontal (~1.5 fold) and perirhinal cortices (~1.6 fold) when compared to saline-treated rats (Mann-Whitney test *versus* saline group, $P<0.05$).

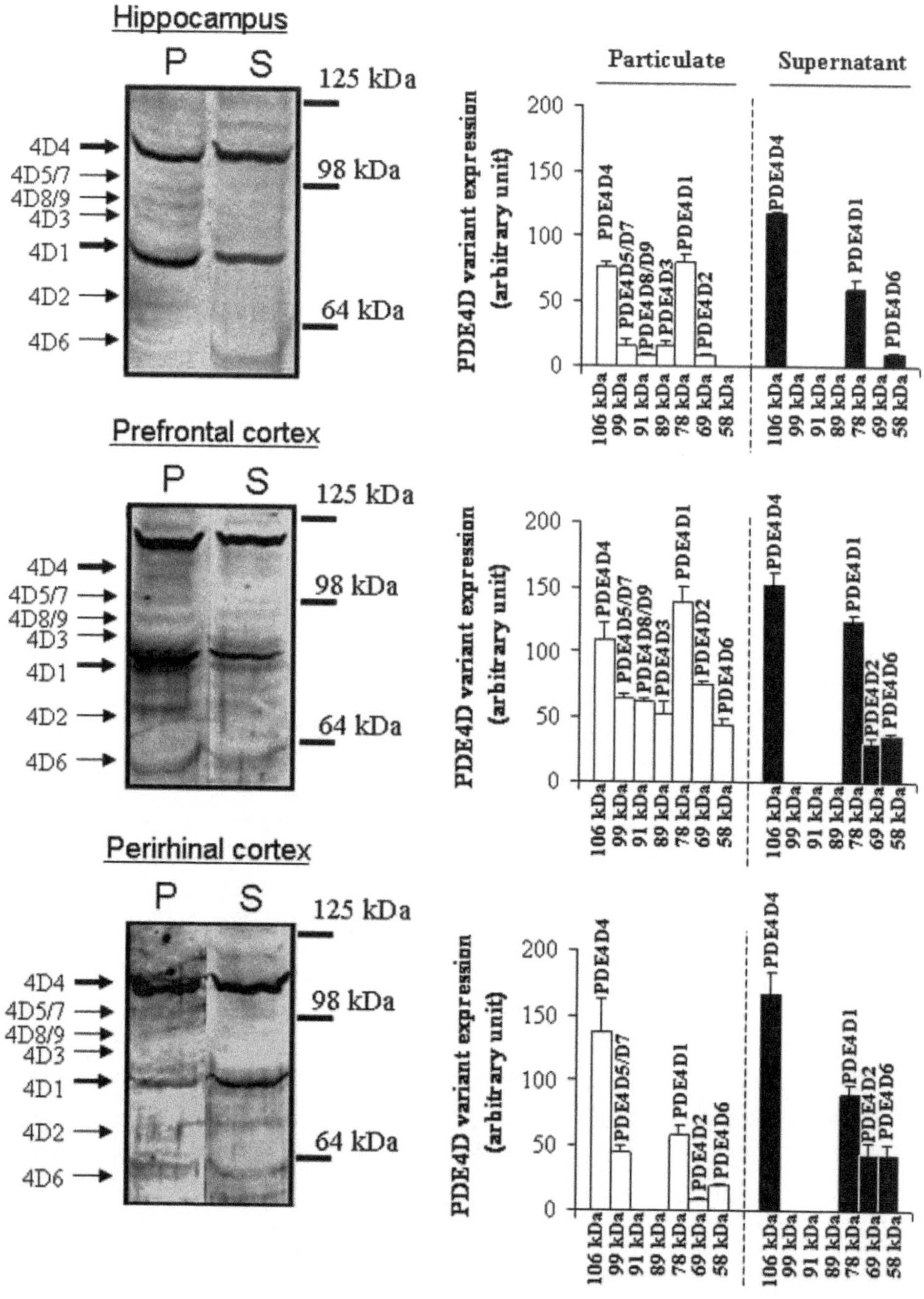

Figure 1. Expression of PDE4D proteins in the particulate and soluble fractions of rat hippocampus, prefrontal and perirhinal cortices. Particulate and soluble fractions from the rat hippocampus, the prefrontal cortex and perirhinal cortex were isolated and proteins extracted as described in Materials and Methods section. The left panel shows representative immunoblots of particulate (25 µg) and soluble (25 µg) protein fractions probed with goat polyclonal human anti-PDE4D antibody in the hippocampus, prefrontal and perirhinal cortices. Arrowheads indicate the molecular weights of the immunoreactive proteins. The right panel shows quantification; the intensities of the immunoreactive bands in the particulate and soluble fractions from hippocampus, prefrontal and perirhinal cortices were determined and normalized to those of actin. The densitometry values are the mean ± SEM (n = 3).

RS 67333, milrinone and rolipram treatments prevent the sample phase-induced increase in cAMP levels in the rat central nervous system

Analysis of the cAMP concentrations reveal that, in saline-treated rats, the sample phase induces an increase in cAMP levels in all brain regions studied (Fig. 6, grey bars versus white bars; ~1.8 fold, ~1.3 fold and ~1.3 fold for the hippocampus, prefrontal and perirhinal cortices, respectively; Mann-Whitney test, $P<0.05$). Such a sample phase-induced increase could not be observed for the drug-treated animals; a tendency to a decreased level of cAMP was even noticed in the prefrontal cortex of these animals (~0.86 to 0.62 fold according to drug treatment, Fig. 6) except those treated with milrinone+RS67333.

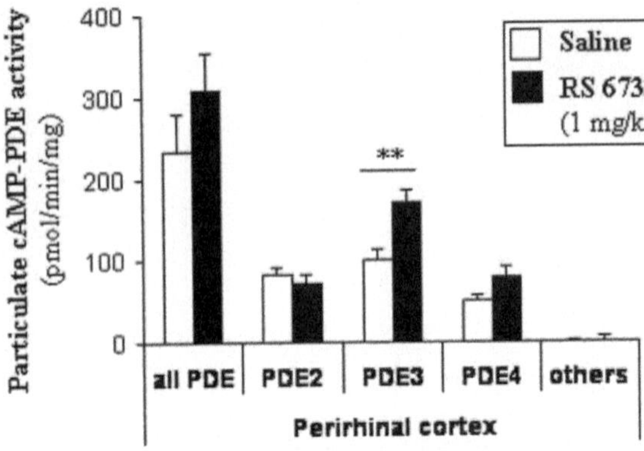

Figure 2. Effect of RS 67333 (1 mg/kg, i.p.) on PDE activities in the perirhinal cortex from rats performing the object recognition task with a 4-h delay. Rats were injected with saline or the 5-HT4 receptor agonist RS 67333 (1 mg/kg, i.p.), 30 minutes before exposure to the sample trial of he object recognition task. Immediately after the testing trial, particulate fractions from the hippocampus, prefrontal and perirhinal cortices were isolated and assayed for EHNA- (20 μM), milrinone- (20 μM) and rolipram- (10 μM) sensitive PDE activities, respective inhibitors of PDE2, PDE3 and PDE4 families. PDE inhibitor-sensitive and –insensitive PDE activities were each expressed as pmolcAMPhydrolysed/min/mg protein. Results are means ± SEM of four independent subcellular fractionations performed in triplicate. Within each subcellular compartment, * indicates a significant difference of PDE activity as compared with saline treatment within a family of PDE activity (PDE2, 3, 4 or other PDE) (**, *P*<0.01, ANOVA followed by Fisher's LSD test).

Drug treatments lower cAMP levels in hippocampus, prefrontal cortex and perirhinal cortex after the object recognition test

After the testing phase (black bars, **Fig. 6**) when compared with the situation before the testing phase (hatched bar), cAMP concentrations in saline –treated rats were once more increased in both the hippocampus (2.76 fold, P<0.05) and the perirhinal cortex (2.10 fold, P<0.05) but not in the prefrontal cortex. Nevertheless cAMP concentrations in the prefrontal cortex from saline-treated animals still tend to be higher than before the sample phase (P>0.05). Thus, between the beginning and the end of the paradigm, we reported a 4.1, 1.3 and 3.2 fold increase in cAMP concentrations, respectively in the hippocampus, prefrontal cortex and perirhinal cortex from the saline group (Mann-Whitney test, respectively P<0.05; P>0.05 and P<0.05).

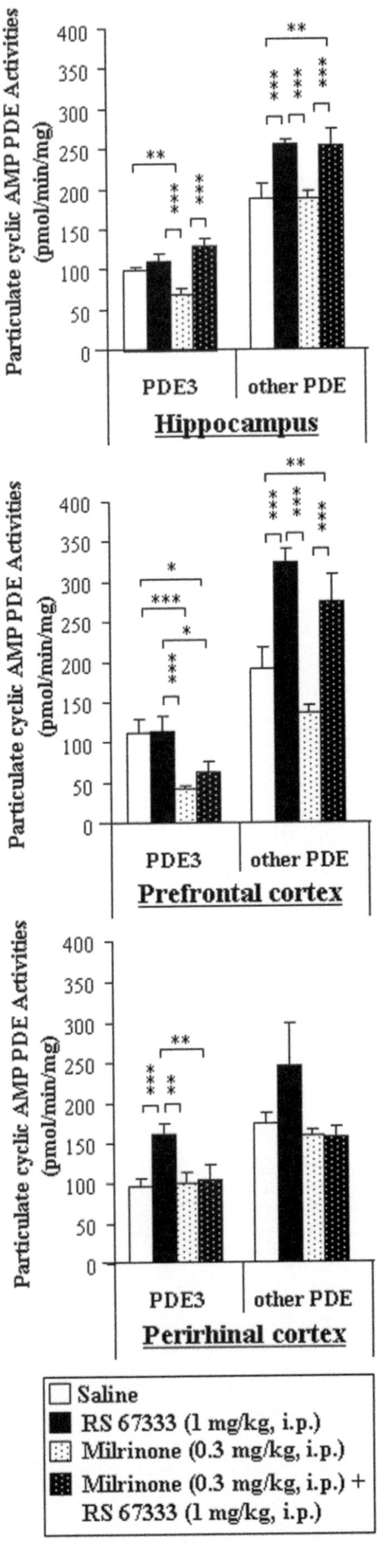

Figure 3. Milrinone (0.3 mg/kg, i.p.) specifically inhibits PDE3 activities in hippocampus, prefrontal cortex and perirhinal cortex from rats. Rats were injected with the PDE3 inhibitor (milrinone, 0.3 mg/kg, i.p.) and then with saline or the 5-HT4 receptor agonist (RS 67333, 1 mg/kg, i.p.), respectively 45 minutes and 30 minutes before the sample phase of the object recognition task. Immediately after the testing phase, both particulate and soluble fractions from the hippocampus, the prefrontal cortex and perirhinal cortex were isolated and particulate fraction was assayed for milrinone (20 μM)-sensitive PDE activities. Milrinone-sensitive and –insensitive PDE activities were expressed as pmolcAMPhydrolysed/min/mg protein. Results are means ± SEM of four independent subcellular fractionations performed in triplicate. Within each subcellular compartment, * indicated significant differences of PDE activity as compared with other treatment within a type of PDE activity (PDE3 or other PDE) (*, P<0.05, **, P<0.01, ***, P<0.01, ANOVA followed by Fisher's LSD test).

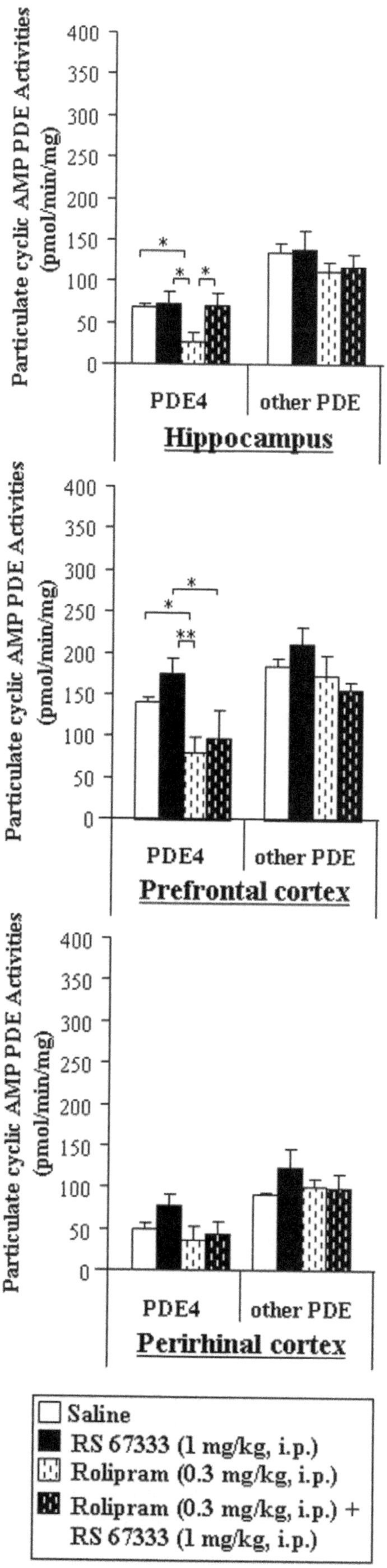

Figure 4. Rolipram (0.3 mg/kg, i.p.) specifically inhibits PDE4 activities in hippocampus, prefrontal cortex and perirhinal cortex from rats. Rats were injected with the PDE4 inhibitor (rolipram, 0.3 mg/kg), and then with saline or the 5-HT4 receptor agonist (RS 67333, 1 mg/kg), respectively 45 minutes and 30 minutes before the sample phase of the object recognition task. Immediately after the testing phase, both particulate and soluble fractions from the hippocampus, the prefrontal cortex and perirhinal cortex were isolated and the particulate fraction was assayed for rolipram (10 μM)-sensitive PDE activities. Rolipram-sensitive and –insensitive PDE activities were expressed as pmolcAMPhydrolysed/min/mg protein. Results are means ± SEM of four independent subcellular fractionations performed in triplicate. Within each subcellular compartment, * indicated significant differences of PDE activity as compared with other treatment within a type of PDE activity (PDE4 or other PDE) (*, $P<0.05$, **, $P<0.01$, ANOVA followed by Fisher's LSD test).

In the drug-treated animals, minor increases in cAMP levels (lower than in the saline-treated group) were also reported after the testing phase. Both rolipram and milrinone alone or administered before RS 67333 induced an almost 1.4 fold increase in cAMP concentrations in the hippocampus between the beginning and the end of the paradigm ($P<0.05$). Similarly, in the perirhinal cortex, a 1.6 fold increase of cAMP levels in the rolipram group (Fig. 6, $P<0.05$) and a 1.5 fold increase in the milrinone group ($P<0.05$) were reported after the testing phase. Finally, in the prefrontal cortex from drug-treated rats, cAMP levels still tend to decrease (~−8 to −22% according to drug treatment).

At the end of the paradigm, cAMP concentrations from rats injected with a PDE inhibitor alone or in combination with RS 67333 were markedly lower when compared to saline-treated rats in the hippocampus (−54 to −61% according to the treatment, $P<0.05$), the prefrontal (−13 to −31% according to the treatment) or perirhinal cortices (−13 to −35% according to the treatment) (Mann-Whitney test *versus* saline group, $P<0.05$).

Milrinone altered PP2 activities in anatomical structures linked to object recognition memory in rat

We assessed PP2 activities in both subcellular compartments (soluble and particulate) of the hippocampus, prefrontal and perirhinal cortices since excessive cAMP catabolism is limited by reversing phosphorylation of particulate PDE4 activities [53–55]. PP2 activity was measured in both the soluble and particulate fraction (**Table 4**); PP2 activity was however mainly present in the soluble fraction (68.1±2.1% to 75.3±1.7% of total PP2 activity

Table 1. Time of exploration of objects measured during the sample and the testing trails in the object recognition task.

	Time exploring objects (in s, mean ±SEM)	
	Sample	**Testing**
Saline (n = 32)	31.7±2.9	29.4±2.4
RS 67333 (1 mg/kg, i.p.) (n = 32)	33.1±3.1	34.6±2.5
Rolipram (0.3 mg/kg, i.p.) (n = 27)	29.8±3.5	32.6±2.4
Rolipram (0.3 mg/kg, i.p.)+RS 67333 (1 mg/kg, i.p.) (n = 27)	32.7±3.9	36.9±3.4
Milrinone (0.3 mg/kg, i.p.) (n = 27)	29.8±3.2	31.7±3.3
Milrinone (0.3 mg/kg, i.p.)+RS 67333 (1 mg/kg, i.p.) (n = 27)	27.3±2.7	26.7±2.4

Table 2. Locomotor activity measured during the sample and the testing trails in the object recognition task.

	Total number of entries (mean ±SEM)	
	Sample	Testing
Saline (n = 32)	50.8±2.1	48.9±3.0
RS 67333 (1 mg/kg, i.p.) (n = 32)	54.9±1.4	51.8±2.6
Rolipram (0.3 mg/kg, i.p.) (n = 27)	36.2±2.6	49.9±3.3
Rolipram (0.3 mg/kg, i.p.)+RS 67333 (1 mg/kg, i.p.) (n = 27)	37.6±3.4	53.3±3.3
Milrinone (0.3 mg/kg, i.p.) (n = 27)	46.2±2.6	46.7±3.2
Milrinone (0.3 mg/kg, i.p.)+RS 67333 (1 mg/kg, i.p.) (n = 27)	51.8±2.9	49.8±2.5

according to the structure considered). Milrinone significantly altered PP2 activities in the soluble fraction of the hippocampus (**Fig. 7a**) and perirhinal cortex (**Fig. 7c**) as well as the soluble and particulate fractions of the prefrontal cortex (**Fig. 7b**). In fact, a slight decrease was evidenced in the soluble fraction of the hippocampus from RS 67333-treated rats (−20%, P<0.01, **Fig. 7a**). Thus, in milrinone-treated rats, soluble PP2 activities

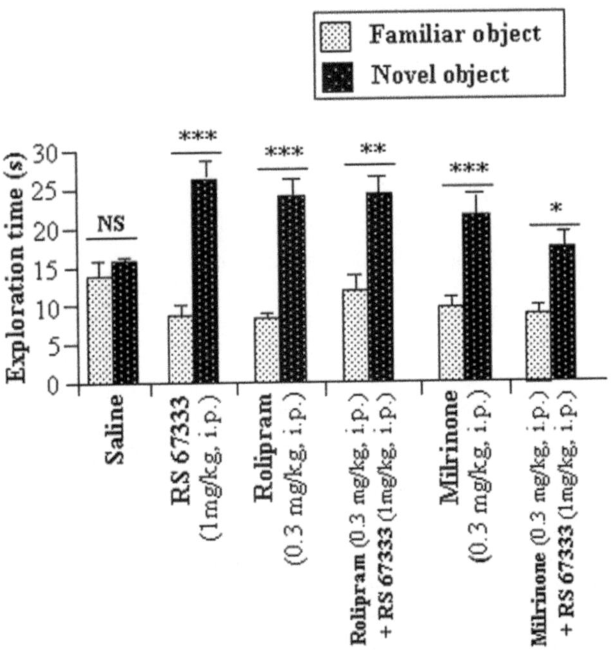

Figure 5. Object recognition task after a 4 h-delay in saline, rolipram- or milrinone- treated rats with or without RS 67333 co-treatment. Time of exploration of the familiar and novel objects during the testing phase of the object recognition memory task of saline (n = 32), RS 67333 (n = 32), rolipram (n = 27), rolipram+RS 67333 (n = 27), milrinone (n = 27), or milrinone+RS 67333 (n = 27)-treated rats. Rats were injected with the PDE inhibitor (0.3 mg/kg, i.p.) solution 45 minutes before the sample phase and with saline or RS 67333 (1 mg/kg, i.p.) solution 30 minutes before the sample phase. Object recognition was assayed after a 4 h-delay. Values are means in s ± SEM. NS: non significant, (*) indicates a significant difference in comparison with saline treatment. (*, P<0.05, **, P<0.01, ***, P<0.001, ANOVA followed by Fisher's PLSD test).

Table 3. Value of the index measures of discrimination in the object recognition task.

	D1 (mean±SEM)	D2 (mean±SEM)
Saline (n = 32)	1.4±2.3	0.1±0.08
RS 67333 (1 mg/kg, i.p.) (n = 32)	17.2±2.7***	0.5±0.06***
Rolipram (0.3 mg/kg, i.p.) (n = 27)	16.4±2.2***	0.5±0.06***
Rolipram (0.3 mg/kg, i.p.)+RS 67333 (1 mg/kg, i.p.) (n = 27)	12.9±3.6**	0.3±0.08*
Milrinone (0.3 mg/kg, i.p.) (n = 27)	12.2±2.6**	0.4±0.07**
Milrinone (0.3 mg/kg, i.p.)+RS 67333 (1 mg/kg, i.p.) (n = 27)	8.5±2.3*	0.3±0.07*

*P<0.05,
**P<0.01,
***P<0.001, one-way ANOVA followed by Fisher LSD test, comparison with saline-treated group.

decrease (−44% when compared to saline rats, P<0.001) was significantly strengthened (P<0.01) when rats were also injected with RS 67333 (PP2 activities was diminished by −70% when compared to saline rats, P<0.001). Similar observations were done in the prefrontal cortex of milrinone-treated rats (**Fig. 7b**), in which PP2 activities decrease in the soluble fraction (−53% when compared to saline rats, P<0.001) was significantly strengthened (P<0.01) when rats were also injected with RS 67333 (PP2 activities was diminished by −79% when compared to saline rats, P<0.001). PP2 activities of the particulate compartment were altered in milrinone-treated rat (−79%, P<0.001) but this decrease was not strengthened in milrinone+RS 67333 treated animals. Finally, milrinone affected soluble PP2 activities in the perirhinal cortex, when administrated alone or in association with RS 67333 (−75% when compared to saline-treated rats, P<0.001; **Fig. 7c**).

Such modifications of PP2 activities could not be evidenced in rolipram-treated animals (data not shown).

Discussion

Object recognition memory in rodents and primates is thought to be mediated, at least in part, by interactions between the perirhinal cortex, prefrontal cortex and hippocampus [66–70]. Interestingly, 5-HT4R, known to be involved in learning and memory [31], including acquisition of information [18] are widely distributed among these brain structures [71–73]. Since 5-HT4R stimulation induces an increase in cAMP that has been associated to memory processes [6,8,9,11], we first hypothesized that the drug-induced increase in cAMP might support the improvement of object recognition memory performance. Data presented here strongly suggest that the "pre-sample" early increase in cAMP levels followed by a specific lowering of cAMP concentrations in each brain sub-region involved to the object recognition paradigm improve learning efficacy after a middle-term delay.

We first confirmed the major role of PDE3 and PDE4 in the control of cAMP levels in the anatomical structures linked to the object recognition task [18]. Indeed, we reported that both the stimulation of cAMP production (RS 67333) and the inhibition of its hydrolysis (milrinone, PDE3 inhibitor or rolipram, PDE4 inhibitor) in the rat, improve familiar object recognition after a 4-h delay. Besides, we observed similar effects of PDE3 and PDE4 inhibitors. In line with previous studies, we reported a higher

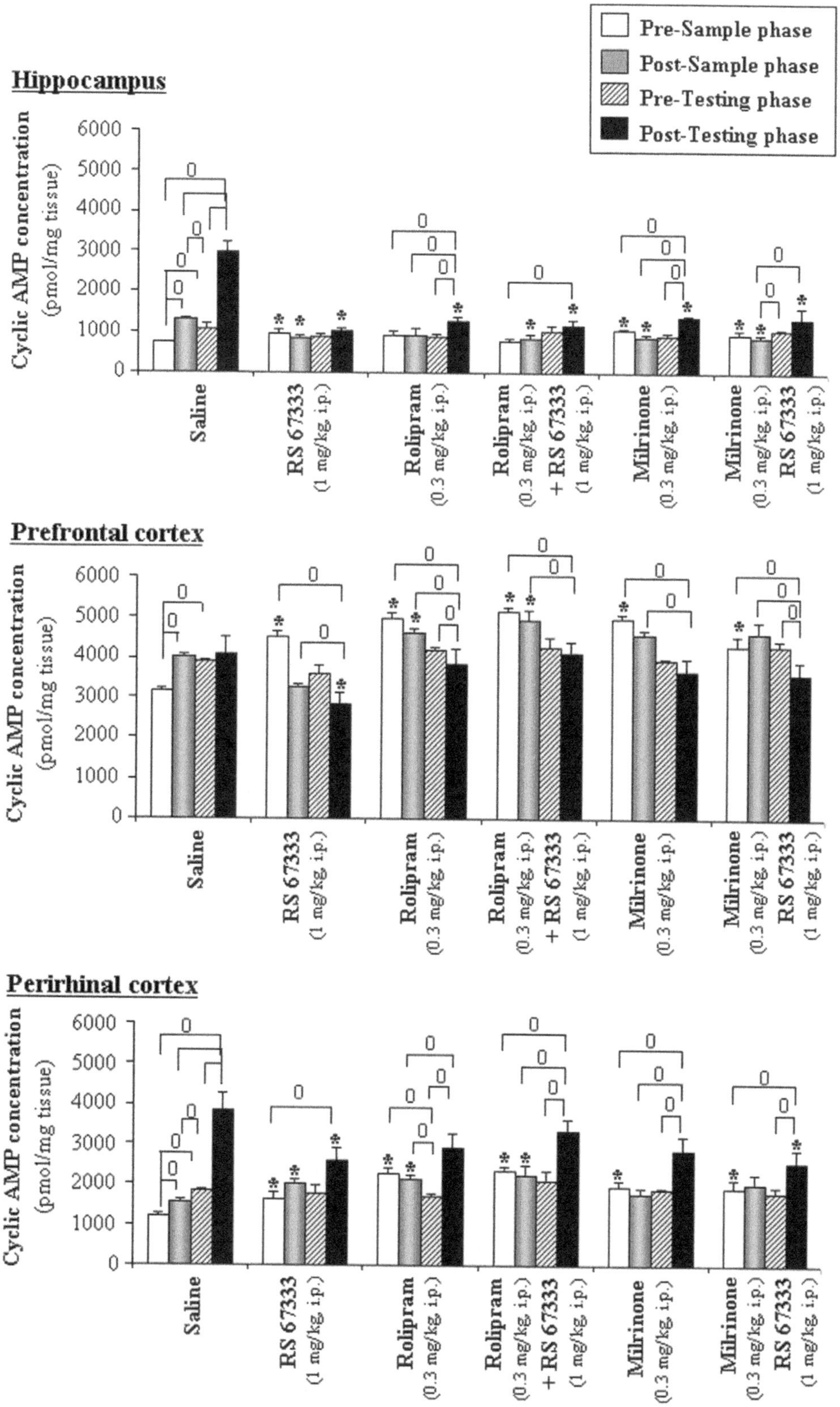

Figure 6. Rolipram and/or RS 67333 induce a biphasic modulation of cAMP concentrations in the hippocampus, prefrontal and perirhinal cortices of rats performing an object recognition task with a 4-h delay. Rats were injected with the inhibitor of PDE4 (rolipram, 0.3 mg/kg, i.p.) and then with saline or the 5-HT4 receptor agonist (RS 67333, 1 mg/kg, i.p.), respectively 45 minutes and 30 minutes before to the sample phase of the object recognition task. Rats were euthanized before or after the sample phase, or before or after the testing phase. Cyclic AMP was extracted from the hippocampus, prefrontal and perirhinal cortices and then assayed. Cyclic AMP was expressed as pmolcAMP/mg of weight tissue. Results were means ± SEM of three independent extractions performed in duplicate. (0) indicated significant differences in comparison with other steps of the paradigm in each brain sub-region, Mann-Whitney test, $P<0.05$. (*) indicated a significant difference in comparison with saline treatment in each brain sub-region. (0,*, $P<0.05$).

Table 4. Distribution of the total PP2 activities between particulate and soluble fractions from rat hippocampus, prefrontal cortex and perirhinal cortex.

	Total PP2 activities (nmol of phosphate released/min) (% of the total PP2 activities)		
	Hippocampus (n = 4)	**Prefrontal cortex (n = 4)**	**Perirhinal cortex (n = 4)**
Particulate fraction	15.83±1.18 = 24.7±1.7%	2.58±0.24 = 32.0±2.1%	3.38±0.39 = 27.8±1.7%
Soluble fraction	48.07±0.9§§§ = 75.3±1.7%	5.47±0.27§§§ = 68.1±2.1%	8.73±0.20§§§ = 72.2±1.7%

Values are means ± SEM.
§§§$p<0.001$ (ANOVA followed by Fisher's LSD test): different from the corresponding particulate fraction.

sensitivity of PDE3- than PDE4-regulated cAMP pools to PKA activation, associated to a lower efficacy of PDE3 to hydrolyze cAMP [74–77]. We hypothesized this familiar object recognition improvement to be linked to the early cAMP levels increase measured before the sample phase in the hippocampus, prefrontal and perirhinal cortices from these animals. Early cAMP activations in the prefrontal cortex have already been described to be beneficial for working memory under conditions that require hippocampal-prefrontal cortex interactions [14,78]. Moreover, the observation that the drug-induced early increase in cAMP in the brain sub-regions improves the mnesic trace is consistent with data reporting that activation of the cAMP-PKA pathway cascade improves memory processes [2,4,6,8,9] unlike inhibition of PKA [6,10,11,79]. Indeed, according to the model of Frankland and Bontempi, experience is both initially encoded in hippocampal and cortical networks. Subsequent reactivation of the hippocampal network reinstates activity in different cortical networks. This coordinated replay across hippocampal–cortical networks leads to gradual strengthening of cortico-cortical connections, progressively disengaging the memory trace from the hippocampus [80]. Thus, the higher cAMP levels induced by drug treatments could support a better acquisition of the mnesic trace that will, in turn, benefit to the animal during the testing phase.

We also observed that drug treatments induce a lowering, and/or a reduction of the awaited increases, in cAMP levels in all brains regions studied after the testing phase, when compared to saline injected animals. Hence, cAMP concentrations were systematically lower in both the hippocampus and perirhinal cortex from animals that have increased behavioural performances than in the saline-treated rats that exhibit poor object recognition performances. Appearing at a first glance as a discrepancy from data from the literature, this situation can easily be reconciled by recent studies suggesting that memory requires a restricted, or selective cAMP production rather than a large and widespread increase in cAMP levels [28,81]. Indeed, Kelly and co-workers (2008) observed an impairment of memory consolidation and/or retrieval in a fear-conditioning task in mice that express a constitutively active isoform of the G-protein subunit Gαs in the forebrain [81]. Perez-Garcia and Meneses, (2008) also demonstrate that the hippocampal production of cAMP was higher in untrained rats than in rats subjected to a behavioral task [28]. In this respect, low cAMP levels might be optimal to convert temporary memory during acquisition to long-term memory (4 h-delay) while high cAMP levels might disturb such a conversion of short-term memory to long-term memory, resulting in low performances in controls animals.

These few elements point out the complexity of cAMP-dependent responses and the putative interactions between behaviour- and drug-induced effects on cellular signaling. By rapidly degrading cAMP from selected compartments, PDEs can fix the boundaries for cAMP diffusion, shape the intracellular gradients of the second messenger and thereby modulate defined sets of PKA-mediated intracellular events. Hence, PDE alteration may affect cAMP compartmentalization, leading to untargeted cAMP signals, aberrant phosphorylation of target proteins and thus contribute to dysfunction. We report here that hippocampus, prefrontal and perirhinal cortices exhibit different patterns of particulate PDE4D isoforms. One can therefore hypothesize differential implications of PDE isoforms, keys mediators of memory and learning processes in the limitation of cAMP increase [36,37,82,83]. For example, PDE4D8 found only in the particulate fraction from the hippocampus or the prefrontal cortex, has been shown to be responsible for controlling local cAMP concentrations and PKA activity in the vicinity of β1 adrenergic receptors [46]. PDE4D3 that we only described in the particulate fraction from the hippocampus or the prefrontal cortex, was reported to bind to muscle- specific A-kinase anchoring protein (mAKAP), which in turns controls perinuclear AMP levels and recruits the MAP kinases MEK5 and ERK5 [84]. Hence, each PDE isoform plays a critical role in the specificity of cAMP-signaling, effectively creating cyclic nucleotide microdomains and/or cAMP gradients that can be sensed by the cell [46,85–88]. Accordingly, some PDE4D isoforms precisely regulate the coupling between GPCR and the Gs protein. A desensitization of 5-HT4R, through an over-stimulation by RS67333 and/or an alteration of PDE, could thus lead to a lowering of cAMP concentrations in brain regions expressing 5-HT4R. Indeed, PKA activated after a 5-HT4R stimulation, will phosphorylate 5-HT4R leading to cell membrane recruitment of GRK2, which in turn will phosphorylate i) the associated GPCR [89] inhibiting its coupling with Gs, and ii) PDE4, which locally attenuates the PKA activity by lowering local cAMP levels [90–92].

Effects induced by 5-HT4R stimulation are time-limited since PKA phosphorylation of 5-HT4R, GRK2 or PDE can be reversed by phosphatase (PP) activity [58]. PP1 and PP2, compartmentalized inside the mammalian cell [93,94], account for the major phosphatase activities [56]; PP2 is however more particularly investigated because of its ability to dephosphorylate many signaling proteins [93]. Since PP2A is activated by cAMP level increases [53–55], we suggested that phosphatase activity may have been raised in the brain sub-region structures from drug-treated rats, but it is not what we found here. While inhibition of PDE4 activity failed to alter PP2 activity, milrinone administration induced an alteration of PP2 activity, especially in the supernatant fractions from the brain regions investigated; such an alteration could in turn alter dephosphorylation of the 5-HT4R, prevent the efficiency of coupling between this receptor and the Gs protein, and thus lead to a lowered cAMP production. Such a functional contrast following selective inhibition of PDE3 and PDE4 has been already observed in many cell types [95–97]. Interestingly, the cell

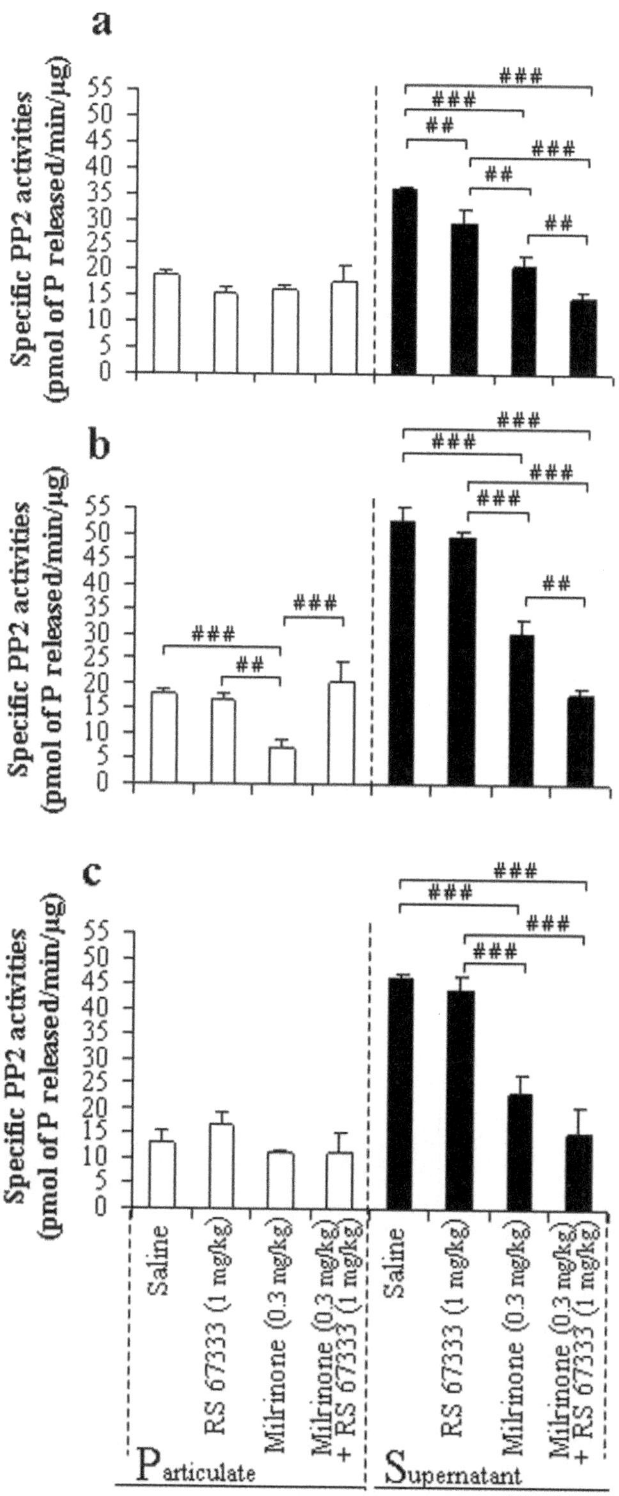

Figure 7. Effect of milrinone (0.3 mg/kg) on PP2 activities in hippocampus, prefrontal and perirhinal cortices from rats performing the object recognition task with a 4-h delay. Rats were injected with the inhibitor of PDE3 (milrinone, 0.3 mg/kg) 45 minutes before exposure then with saline or the 5-HT4 receptor agonist (RS 67333, 1 mg/kg), 30 minutes before exposure to the sample trial of the object recognition task. Immediately after the testing trial, both particulate (white bar) and soluble (black bar) fractions from the hippocampus (**a**), the prefrontal cortex (**b**) and perirhinal cortex (**c**) were isolated and were assayed for PP2 activity. PP2 activities were pmol of phosphate released by min and mg protein. Results are means ± SEM of four independent subcellular fractionations performed in triplicate. Within each subcellular compartment, # indicated significant differences of PP2 activity as compared with other treatment (#, $P<0.05$, ##, $P<0.01$, ###, $P<0.001$, ANOVA followed by Fisher's LSD test).

membrane recruited PDE4 also desensitizes the switched coupling of the β2AR to activation of Gi induced by the PKA-mediated phosphorylation [98], defining thus an appropriate coupling of the GPCR [99]. Hence, in our opinion, drugs injected before the sample phase rapidly increase cAMP levels leading to the uncoupling of 5-HT4R. However, during the test phase, further 5-HT4R stimulation does not raise cAMP level, probably because of 5-HT4R uncoupling. 5-HT4R uncoupling could thus be an adaptative mechanism to reduce cAMP levels in the presence of an excessive stimulation of 5-HT4R or an absence of PDE3 or 4 activities thus avoiding an excessive accumulation of cAMP. If such threshold of cAMP level exists and is reached by either the stimulation of 5-HT4R or inhibition of PDE3 or PDE4 alone, thus no further improvement of memory performance could be induced by the pharmacological treatments by the combination of RS 67 333 and PDE inhibitor.

Finally, early increases in cAMP levels followed by an immediate drop in cAMP concentrations have already been well described in cell differentiation, particularly in Sertoli cells [58]. Indeed, before the cAMP increase, stimulation of Sertoli cells by gonadotropin leads to an activation of the ERK pathway, while following the peak of cAMP, gonadotropin activates the PKA pathway. Interestingly, ERK pathway could prolong activation of the cAMP signaling system in cells by having both short and long term effects on PDE4D activity by respectively inactivating long PDE4D isoform (the ones to exhibits a site that allows phosphorylation by ERK) and altering PDE4D mRNA stability (for review [99]). Hence, by analogy to differentiation mechanisms, another hypothesis is that "cellular learning" may result from the crossing of a milestone, resulting in the subsequent activation of alternative intracellular signalling pathways. Increases in cAMP levels but also their subsequent declines account in mnesic performance improvement. The part of ERK pathway in these processes should be addressed in furthers works.

Our results show that a "pre-sample" early increase in cAMP levels followed by both a "post-sample" lowering of cAMP concentrations in the prefrontal cortex and a "post-test" lowering of cAMP concentrations in the hippocampus and perirhinal cortex support improved learning efficacy after a middle-term delay. If cAMP triggers a temporally defined cellular response, a major question that should be addressed in future works is to clarify how such a functionally ubiquitous signaling pathway may be involved in memory formation.

Acknowledgments

The authors thank Dr. S. Carreau (Laboratoire Œstrogènes et Reproduction, EA 2608, INRA USC 2006, University of Caen) and Dr. P. Barbey (LAMARE) for giving us access to the ultracentrifuge and the radioactivity laboratory, respectively, as well as Pr. B. Cox and Dr M. Dacher (Uniformed Service University, Bethesda, MD, USA) for english editing of the manuscript.

Author Contributions

Conceived and designed the experiments: MH GL FD MB. Performed the experiments: MH GL. Analyzed the data: MH TF FD MB GL. Contributed reagents/materials/analysis tools: MH TF GL. Wrote the paper: TF FD MB GL MH. Statistical analysis: TF.

References

1. Monti B, Berteotti C, Contestabile A (2006) Subchronic rolipram delivery activates hippocampal CREB and arc, enhances retention and slows down extinction of conditioned fear. Neuropsychopharmacology 31: 278–286.
2. Hoyer D, Pompe B, Friedrich H, Zwiener U, Baranowski R, et al. (2004) Autonomic Information Flow during awakeness, sleep, and multiple organ dysfunction syndrome assessed by mutual information function of heart rate fluctuations. Conf Proc IEEE Eng Med Biol Soc 1: 628–630.
3. Bach ME, Barad M, Son H, Zhuo M, Lu YF, et al. (1999) Age-related defects in spatial memory are correlated with defects in the late phase of hippocampal long-term potentiation in vitro and are attenuated by drugs that enhance the cAMP signaling pathway. Proc Natl Acad Sci U S A 96: 5280–5285.
4. Barad M, Bourtchouladze R, Winder DG, Golan H, Kandel E (1998) Rolipram, a type IV-specific phosphodiesterase inhibitor, facilitates the establishment of long-lasting long-term potentiation and improves memory. Proc Natl Acad Sci U S A 95: 15020–15025.
5. Zhang HT, O'Donnell JM (2000) Effects of rolipram on scopolamine-induced impairment of working and reference memory in the radial-arm maze tests in rats. Psychopharmacology (Berl) 150: 311–316.
6. Bernabeu R, Cammarota M, Izquierdo I, Medina JH (1997) Involvement of hippocampal AMPA glutamate receptor changes and the cAMP/protein kinase A/CREB-P signalling pathway in memory consolidation of an avoidance task in rats. Braz J Med Biol Res 30: 961–965.
7. Bernabeu R, Schmitz P, Faillace MP, Izquierdo I, Medina JH (1996) Hippocampal cGMP and cAMP are differentially involved in memory processing of inhibitory avoidance learning. Neuroreport 7: 585–588.
8. Prickaerts J, de Vente J, Honig W, Steinbusch HW, Blokland A (2002) cGMP, but not cAMP, in rat hippocampus is involved in early stages of object memory consolidation. Eur J Pharmacol 436: 83–87.
9. Rutten K, Prickaerts J, Hendrix M, van der Staay FJ, Sik A, et al. (2007) Time-dependent involvement of cAMP and cGMP in consolidation of object memory: Studies using selective phosphodiesterase type 2, 4 and 5 inhibitors. Eur J Pharmacol 558: 107–112.
10. Bourtchouladze R, Abel T, Berman N, Gordon R, Lapidus K, et al. (1998) Different training procedures recruit either one or two critical periods for contextual memory consolidation, each of which requires protein synthesis and PKA. Learn Mem 5: 365–374.
11. Taylor JR, Birnbaum S, Ubriani R, Arnsten AF (1999) Activation of cAMP-dependent protein kinase A in prefrontal cortex impairs working memory performance. J Neurosci 19: RC23.
12. Connolly DA, Hockley WE, Pratt MW (1996) A developmental evaluation of frequency memory for actions presented in lists, scripts, and stories. Memory 4: 243–263.
13. Pineda VV, Athos JI, Wang H, Celver J, Ippolito D, et al. (2004) Removal of G(ialpha1) constraints on adenylyl cyclase in the hippocampus enhances LTP and impairs memory formation. Neuron 41: 153–163.
14. Ramos BP, Birnbaum SG, Lindenmayer I, Newton SS, Duman RS, et al. (2003) Dysregulation of protein kinase a signaling in the aged prefrontal cortex: new strategy for treating age-related cognitive decline. Neuron 40: 835–845.
15. Ennaceur A, Delacour J (1988) A new one-trial test for neurobiological studies of memory in rats. 1: Behavioral data. Behav Brain Res 31: 47–59.
16. Lamirault L, Guillou C, Thal C, Simon H (2003) Combined treatment with galanthaminium bromide, a new cholinesterase inhibitor, and RS 67333, a partial agonist of 5-HT4 receptors, enhances place and object recognition in young adult and old rats. Prog Neuropsychopharmacol Biol Psychiatry 27: 185–195.
17. Lamirault L, Simon H (2001) Enhancement of place and object recognition memory in young adult and old rats by RS 67333, a partial agonist of 5-HT4 receptors. Neuropharmacology 41: 844–853.
18. Levallet G, Hotte M, Boulouard M, Dauphin F (2009) Increased particulate phosphodiesterase 4 in the prefrontal cortex supports 5-HT4 receptor-induced improvement of object recognition memory in the rat. Psychopharmacology (Berl) 202: 125–139.
19. Moser PC, Bergis OE, Jegham S, Lochead A, Duconseille E, et al. (2002) SL65.0155, a novel 5-hydroxytryptamine(4) receptor partial agonist with potent cognition-enhancing properties. J Pharmacol Exp Ther 302: 731–741.
20. Fontana DJ, Daniels SE, Wong EH, Clark RD, Eglen RM (1997) The effects of novel, selective 5-hydroxytryptamine (5-HT)4 receptor ligands in rat spatial navigation. Neuropharmacology 36: 689–696.
21. Galeotti N, Ghelardini C, Bartolini A (1998) Role of 5-HT4 receptors in the mouse passive avoidance test. J Pharmacol Exp Ther 286: 1115–1121.
22. Lelong V, Dauphin F, Boulouard M (2001) RS 67333 and D-cycloserine accelerate learning acquisition in the rat. Neuropharmacology 41: 517–522.
23. Letty S, Child R, Dumuis A, Pantaloni A, Bockaert J, et al. (1997) 5-HT4 receptors improve social olfactory memory in the rat. Neuropharmacology 36: 681–687.
24. Marchetti E, Dumuis A, Bockaert J, Soumireu-Mourat B, Roman FS (2000) Differential modulation of the 5-HT(4) receptor agonists and antagonist on rat learning and memory. Neuropharmacology 39: 2017–2027.
25. Marchetti E, Jacquet M, Jeltsch H, Migliorati M, Nivet E, et al. (2008) Complete recovery of olfactory associative learning by activation of 5-HT4 receptors after dentate granule cell damage in rats. Neurobiol Learn Mem 90: 185–191.
26. Meneses A (2007) Stimulation of 5-HT1A, 5-HT1B, 5-HT2A/2C, 5-HT3 and 5-HT4 receptors or 5-HT uptake inhibition: short- and long-term memory. Behav Brain Res 184: 81–90.
27. Orsetti M, Dellarole A, Ferri S, Ghi P (2003) Acquisition, retention, and recall of memory after injection of RS67333, a 5-HT(4) receptor agonist, into the nucleus basalis magnocellularis of the rat. Learn Mem 10: 420–426.
28. Perez-Garcia G, Meneses A (2008) Ex vivo study of 5-HT(1A) and 5-HT(7) receptor agonists and antagonists on cAMP accumulation during memory formation and amnesia. Behav Brain Res 195: 139–146.
29. Restivo L, Roman F, Dumuis A, Bockaert J, Marchetti E, et al. (2008) The promnesic effect of G-protein-coupled 5-HT4 receptors activation is mediated by a potentiation of learning-induced spine growth in the mouse hippocampus. Neuropsychopharmacology 33: 2427–2434.
30. Terry AV, Jr., Buccafusco JJ, Jackson WJ, Prendergast MA, Fontana DJ, et al. (1998) Enhanced delayed matching performance in younger and older macaques administered the 5-HT4 receptor agonist, RS 17017. Psychopharmacology (Berl) 135: 407–415.
31. King MV, Marsden CA, Fone KC (2008) A role for the 5-HT(1A), 5-HT4 and 5-HT6 receptors in learning and memory. Trends Pharmacol Sci 29: 482–492.
32. Egawa T, Mishima K, Matsumoto Y, Iwasaki K, Fujiwara M (1997) Rolipram and its optical isomers, phosphodiesterase 4 inhibitors, attenuated the scopolamine-induced impairments of learning and memory in rats. Jpn J Pharmacol 75: 275–281.
33. Zhang HT, Crissman AM, Dorairaj NR, Chandler LJ, O'Donnell JM (2000) Inhibition of cyclic AMP phosphodiesterase (PDE4) reverses memory deficits associated with NMDA receptor antagonism. Neuropsychopharmacology 23: 198–204.
34. Zhang HT, Zhao Y, Huang Y, Dorairaj NR, Chandler LJ, et al. (2004) Inhibition of the phosphodiesterase 4 (PDE4) enzyme reverses memory deficits produced by infusion of the MEK inhibitor U0126 into the CA1 subregion of the rat hippocampus. Neuropsychopharmacology 29: 1432–1439.
35. Rutten K, Wallace TL, Works M, Prickaerts J, Blokland A, et al. (2011) Enhanced long-term depression and impaired reversal learning in phosphodiesterase 4B-knockout (PDE4B(−/−)) mice. Neuropharmacology 61: 138–147.
36. Burgin AB, Magnusson OT, Singh J, Witte P, Staker BL, et al. (2010) Design of phosphodiesterase 4D (PDE4D) allosteric modulators for enhancing cognition with improved safety. Nat Biotechnol 28: 63–70.
37. Li YF, Cheng YF, Huang Y, Conti M, Wilson SP, et al. (2011) Phosphodiesterase-4D knock-out and RNA interference-mediated knock-down enhance memory and increase hippocampal neurogenesis via increased cAMP signaling. J Neurosci 31: 172–183.
38. Randt CT, Judge ME, Bonnet KA, Quartermain D (1982) Brain cyclic AMP and memory in mice. Pharmacol Biochem Behav 17: 677–680.
39. Gong B, Vitolo OV, Trinchese F, Liu S, Shelanski M, et al. (2004) Persistent improvement in synaptic and cognitive functions in an Alzheimer mouse model after rolipram treatment. J Clin Invest 114: 1624–1634.
40. Bourtchouladze R, Lidge R, Catapano R, Stanley J, Gossweiler S, et al. (2003) A mouse model of Rubinstein-Taybi syndrome: defective long-term memory is ameliorated by inhibitors of phosphodiesterase 4. Proc Natl Acad Sci U S A 100: 10518–10522.
41. Rutten K, Prickaerts J, Blokland A (2006) Rolipram reverses scopolamine-induced and time-dependent memory deficits in object recognition by different mechanisms of action. Neurobiol Learn Mem 85: 132–138.
42. Rutten K, Van Donkelaar EL, Ferrington L, Blokland A, Bollen E, et al. (2009) Phosphodiesterase inhibitors enhance object memory independent of cerebral blood flow and glucose utilization in rats. Neuropsychopharmacology 34: 1914–1925.
43. Houslay MD (2001) PDE4 cAMP-specific phosphodiesterases. Prog Nucleic Acid Res Mol Biol 69: 249–315.
44. Zhang HT (2009) Cyclic AMP-specific phosphodiesterase-4 as a target for the development of antidepressant drugs. Curr Pharm Des 15: 1688–1698.
45. Hoffmann R, Baillie GS, MacKenzie SJ, Yarwood SJ, Houslay MD (1999) The MAP kinase ERK2 inhibits the cyclic AMP-specific phosphodiesterase HSPDE4D3 by phosphorylating it at Ser579. EMBO J 18: 893–903.
46. Richter W, Day P, Agrawal R, Bruss MD, Granier S, et al. (2008) Signaling from beta1- and beta2-adrenergic receptors is defined by differential interactions with PDE4. EMBO J 27: 384–393.
47. van Donkelaar EL, Rutten K, Blokland A, Akkerman S, Steinbusch HW, et al. (2008) Phosphodiesterase 2 and 5 inhibition attenuates the object memory deficit induced by acute tryptophan depletion. Eur J Pharmacol 600: 98–104.
48. Zhao J, Harada N, Kurihara H, Nakagata N, Okajima K (2010) Cilostazol improves cognitive function in mice by increasing the production of insulin-like growth factor-I in the hippocampus. Neuropharmacology 58: 774–783.
49. O'Donnell JM, Frith S (1999) Behavioral effects of family-selective inhibitors of cyclic nucleotide phosphodiesterases. Pharmacol Biochem Behav 63: 185–192.
50. Edelson J, Stroshane R, Benziger DP, Cody R, Benotti J, et al. (1986) Pharmacokinetics of the bipyridines amrinone and milrinone. Circulation 73: III145–152.
51. Lindsay CA, Barton P, Lawless S, Kitchen L, Zorka A, et al. (1998) Pharmacokinetics and pharmacodynamics of milrinone lactate in pediatric patients with septic shock. J Pediatr 132: 329–334.

52. Krause W, Kuhne G (1988) Pharmacokinetics of rolipram in the rhesus and cynomolgus monkeys, the rat and the rabbit. Studies on species differences. Xenobiotica 18: 561–571.
53. Feschenko MS, Stevenson E, Nairn AC, Sweadner KJ (2002) A novel cAMP-stimulated pathway in protein phosphatase 2A activation. J Pharmacol Exp Ther 302: 111–118.
54. Moon EY, Lerner A (2003) PDE4 inhibitors activate a mitochondrial apoptotic pathway in chronic lymphocytic leukemia cells that is regulated by protein phosphatase 2A. Blood 101: 4122–4130.
55. Pullar CE, Grahn JC, Liu W, Isseroff RR (2006) Beta2-adrenergic receptor activation delays wound healing. FASEB J 20: 76–86.
56. Oliver CJ, Shenolikar S (1998) Physiologic importance of protein phosphatase inhibitors. Front Biosci 3: D961–972.
57. Resjo S, Oknianska A, Zolnierowicz S, Manganiello V, Degerman E (1999) Phosphorylation and activation of phosphodiesterase type 3B (PDE3B) in adipocytes in response to serine/threonine phosphatase inhibitors: deactivation of PDE3B in vitro by protein phosphatase type 2A. Biochem J 341(Pt 3): 839–845.
58. Levallet G, Levallet J, Bonnamy PJ (2008) FSH-induced phosphoprotein phosphatase 2A-mediated deactivation of particulate phosphodiesterase-4 activities is abolished after alteration in proteoglycan synthesis in immature rat Sertoli cells. J Endocrinol 197: 45–54.
59. Desjardins S, Cauchy MJ (1995) Comparative cardiac effects of milrinone and sodium nitroprusside in conscious rats. Drug Chem Toxicol 18: 43–59.
60. Griebel G, Misslin R, Vogel E, Bourguignon JJ (1991) Behavioral effects of rolipram and structurally related compounds in mice: behavioral sedation of cAMP phosphodiesterase inhibitors. Pharmacol Biochem Behav 39: 321–323.
61. Silvestre JS, Fernandez AG, Palacios JM (1999) Effects of rolipram on the elevated plus-maze test in rats: a preliminary study. J Psychopharmacol 13: 274–277.
62. Sik A, van Nieuwehuyzen P, Prickaerts J, Blokland A (2003) Performance of different mouse strains in an object recognition task. Behav Brain Res 147: 49–54.
63. Rodriguez R, Molino B, Weiss HR, Scholz PM (2004) Negative metabolic and coronary flow effects of decreases in cAMP and increases in cGMP in control and renal hypertensive rabbit hearts. J Appl Physiol 97: 439–445.
64. Bradford MM (1976) A rapid and sensitive method for the quantitation of microgram quantities of protein utilizing the principle of protein-dye binding. Anal Biochem 72: 248–254.
65. Thompson WJ, Appleman MM (1971) Multiple cyclic nucleotide phosphodiesterase activities from rat brain. Biochemistry 10: 311–316.
66. Bussey TJ, Duck J, Muir JL, Aggleton JP (2000) Distinct patterns of behavioural impairments resulting from fornix transection or neurotoxic lesions of the perirhinal and postrhinal cortices in the rat. Behav Brain Res 111: 187–202.
67. Ennaceur A, Neave N, Aggleton JP (1997) Spontaneous object recognition and object location memory in rats: the effects of lesions in the cingulate cortices, the medial prefrontal cortex, the cingulum bundle and the fornix. Exp Brain Res 113: 509–519.
68. Hotte M, Thuault S, Lachaise F, Dineley KT, Hemmings HC, et al. (2006) D1 receptor modulation of memory retrieval performance is associated with changes in pCREB and pDARPP-32 in rat prefrontal cortex. Behav Brain Res 171: 127–133.
69. Parker A, Gaffan D (1998) Interaction of frontal and perirhinal cortices in visual object recognition memory in monkeys. Eur J Neurosci 10: 3044–3057.
70. Warburton EC, Brown MW (2010) Findings from animals concerning when interactions between perirhinal cortex, hippocampus and medial prefrontal cortex are necessary for recognition memory. Neuropsychologia 48: 2262–2272.
71. Bonaventure P, Hall H, Gommeren W, Cras P, Langlois X, et al. (2000) Mapping of serotonin 5-HT(4) receptor mRNA and ligand binding sites in the post-mortem human brain. Synapse 36: 35–46.
72. Eglen RM, Bonhaus DW, Johnson LG, Leung E, Clark RD (1995) Pharmacological characterization of two novel and potent 5-HT4 receptor agonists, RS 67333 and RS 67506, in vitro and in vivo. Br J Pharmacol 115: 1387–1392.
73. Vilaro MT, Cortes R, Gerald C, Branchek TA, Palacios JM, et al. (1996) Localization of 5-HT4 receptor mRNA in rat brain by in situ hybridization histochemistry. Brain Res Mol Brain Res 43: 356–360.
74. Chini CC, Grande JP, Chini EN, Dousa TP (1997) Compartmentalization of cAMP signaling in mesangial cells by phosphodiesterase isozymes PDE3 and PDE4. Regulation of superoxidation and mitogenesis. J Biol Chem 272: 9854–9859.
75. Dousa TP (1998) Signaling role of PDE isozymes in pathobiology of glomerular mesangial cells. Studies in vitro and in vivo. Cell Biochem Biophys 29: 19–34.
76. Dousa TP (1999) Cyclic-3′,5′-nucleotide phosphodiesterase isozymes in cell biology and pathophysiology of the kidney. Kidney Int 55: 29–62.
77. Matousovic K, Grande JP, Chini CC, Chini EN, Dousa TP (1995) Inhibitors of cyclic nucleotide phosphodiesterase isozymes type-III and type-IV suppress mitogenesis of rat mesangial cells. J Clin Invest 96: 401–410.
78. Aujla H, Beninger RJ (2001) Hippocampal-prefrontocortical circuits: PKA inhibition in the prefrontal cortex impairs delayed nonmatching in the radial maze in rats. Behav Neurosci 115: 1204–1211.
79. Abel T, Nguyen PV, Barad M, Deuel TA, Kandel ER, et al. (1997) Genetic demonstration of a role for PKA in the late phase of LTP and in hippocampus-based long-term memory. Cell 88: 615–626.
80. Frankland PW, Bontempi B (2005) The organization of recent and remote memories. Nat Rev Neurosci 6: 119–130.
81. Kelly MP, Cheung YF, Favilla C, Siegel SJ, Kanes SJ, et al. (2008) Constitutive activation of the G-protein subunit Galphas within forebrain neurons causes PKA-dependent alterations in fear conditioning and cortical Arc mRNA expression. Learn Mem 15: 75–83.
82. Giorgi M, Modica A, Pompili A, Pacitti C, Gasbarri A (2004) The induction of cyclic nucleotide phosphodiesterase 4 gene (PDE4D) impairs memory in a water maze task. Behav Brain Res 154: 99–106.
83. Rutten K, Basile JL, Prickaerts J, Blokland A, Vivian JA (2008) Selective PDE inhibitors rolipram and sildenafil improve object retrieval performance in adult cynomolgus macaques. Psychopharmacology (Berl) 196: 643–648.
84. Dodge-Kafka KL, Bauman A, Mayer N, Henson E, Heredia L, et al. (2010) cAMP-stimulated protein phosphatase 2A activity associated with muscle A kinase-anchoring protein (mAKAP) signaling complexes inhibits the phosphorylation and activity of the cAMP-specific phosphodiesterase PDE4D3. J Biol Chem 285: 11078–11086.
85. Besheer J, Jensen HC, Bevins RA (1999) Dopamine antagonism in a novel-object recognition and a novel-object place conditioning preparation with rats. Behav Brain Res 103: 35–44.
86. Fischmeister R, Castro LR, Abi-Gerges A, Rochais F, Jurevicius J, et al. (2006) Compartmentation of cyclic nucleotide signaling in the heart: the role of cyclic nucleotide phosphodiesterases. Circ Res 99: 816–828.
87. Xiang Y, Naro F, Zoudilova M, Jin SL, Conti M, et al. (2005) Phosphodiesterase 4D is required for beta2 adrenoceptor subtype-specific signaling in cardiac myocytes. Proc Natl Acad Sci U S A 102: 909–914.
88. Zaccolo M, Pozzan T (2002) Discrete microdomains with high concentration of cAMP in stimulated rat neonatal cardiac myocytes. Science 295: 1711–1715.
89. Cong M, Perry SJ, Lin FT, Fraser ID, Hu LA, et al. (2001) Regulation of membrane targeting of the G protein-coupled receptor kinase 2 by protein kinase A and its anchoring protein AKAP79. J Biol Chem 276: 15192–15199.
90. Baillie GS, Sood A, McPhee I, Gall I, Perry SJ, et al. (2003) beta-Arrestin-mediated PDE4 cAMP phosphodiesterase recruitment regulates beta-adrenoceptor switching from Gs to Gi. Proc Natl Acad Sci U S A 100: 940–945.
91. Li X, Huston E, Lynch MJ, Houslay MD, Baillie GS (2006) Phosphodiesterase-4 influences the PKA phosphorylation status and membrane translocation of G-protein receptor kinase 2 (GRK2) in HEK-293beta2 cells and cardiac myocytes. Biochem J 394: 427–435.
92. Xin W, Tran TM, Richter W, Clark RB, Rich TC (2008) Roles of GRK and PDE4 activities in the regulation of beta2 adrenergic signaling. J Gen Physiol 131: 349–364.
93. Janssens V, Goris J (2001) Protein phosphatase 2A: a highly regulated family of serine/threonine phosphatases implicated in cell growth and signalling. Biochem J 353: 417–439.
94. Sim AT, Ludowyke RI, Verrills NM (2006) Mast cell function: regulation of degranulation by serine/threonine phosphatases. Pharmacol Ther 112: 425–439.
95. Huang Z, Ducharme Y, Macdonald D, Robichaud A (2001) The next generation of PDE4 inhibitors. Curr Opin Chem Biol 5: 432–438.
96. Lugnier C (2006) Cyclic nucleotide phosphodiesterase (PDE) superfamily: a new target for the development of specific therapeutic agents. Pharmacol Ther 109: 366–398.
97. Maurice DH, Palmer D, Tilley DG, Dunkerley HA, Netherton SJ, et al. (2003) Cyclic nucleotide phosphodiesterase activity, expression, and targeting in cells of the cardiovascular system. Mol Pharmacol 64: 533–546.
98. Zamah AM, Delahunty M, Luttrell LM, Lefkowitz RJ (2002) Protein kinase A-mediated phosphorylation of the beta 2-adrenergic receptor regulates its coupling to Gs and Gi. Demonstration in a reconstituted system. J Biol Chem 277: 31249–31256.
99. Houslay MD, Baillie GS (2003) The role of ERK2 docking and phosphorylation of PDE4 cAMP phosphodiesterase isoforms in mediating cross-talk between the cAMP and ERK signalling pathways. Biochem Soc Trans 31: 1186–1190.

21

Discovery of Bile Salt Hydrolase Inhibitors Using an Efficient High-Throughput Screening System

Katie Smith, Ximin Zeng, Jun Lin*

Department of Animal Science, University of Tennessee, Knoxville, Tennessee, United States of America

Abstract

The global trend of restricting the use of antibiotic growth promoters (**AGP**) in animal production necessitates the need to develop valid alternatives to maintain productivity and sustainability of food animals. Previous studies suggest inhibition of bile salt hydrolase (**BSH**), an intestinal bacteria-produced enzyme that exerts negative impact on host fat digestion and utilization, is a promising approach to promote animal growth performance. To achieve the long term goal of developing novel alternatives to AGPs, in this study, a rapid and convenient high-throughput screening (**HTS**) system was developed and successfully used for identification of BSH inhibitors. With the aid of a high-purity BSH from a chicken *Lactobacillus salivarius* strain, we optimized various screening conditions (e.g. BSH concentration, reaction buffer pH, incubation temperature and length, substrate type and concentration) and establish a precipitation-based screening approach to identify BSH inhibitors using 96-well or 384-well microplates. A pilot HTS was performed using a small compound library comprised of 2,240 biologically active and structurally diverse compounds. Among the 107 hits, several promising and potent BSH inhibitors (e.g. riboflavin and phenethyl caffeate) were selected and validated by standard BSH activity assay. Interestingly, the HTS also identified a panel of antibiotics as BSH inhibitor; in particular, various tetracycline antibiotics and roxarsone, the widely used AGP, have been demonstrated to display potent inhibitory effect on BSH. Together, this study developed an efficient HTS system and identified several BSH inhibitors with potential as alternatives to AGP. In addition, the findings from this study also suggest a new mode of action of AGP for promoting animal growth.

Editor: Huaijun Zhou, University of California, Davis, United States of America

Funding: No current external funding sources for this study. The work was supported by internal grant award (University of Tennessee AgResearch Innovation Grant). The funders had no role in study design, data collection and analysis, decision to publish, or preparation of the manuscript.

Competing Interests: The authors have declared that no competing interests exist.

* E-mail: jlin6@utk.edu

Introduction

One of the primary means that food animal producers seek to enhance growth performance is through the use of antibiotic growth promoters (**AGP**). Typically, AGP are defined as subtherapeutic quantities of antibiotics that enhance weight gain and feed conversion ratio [1,2]. Although this is a long-established technique with benefits to production that are still evident, concern has increased over the last several decades because AGP exert selection pressures for the emergence and persistence of drug-resistant bacteria that threaten food safety and public health [1,3]. Consequently, groups such as the World Health Organization have strongly urged proactive limitation on AGP use whereas others have banned them outright, as the European Union did in 2006 [1]. Recent suggestions by the Food and Drug Administration also support phasing out antimicrobials used for growth promotion in food animals [4]. Clearly there is an impetus to discontinue AGP use as an agricultural practice, but concerns regarding animal welfare and economic feasibility remain a concern. For this reason, AGP alternatives which could offset such negative impacts must be investigated.

Targeting the mechanism of how AGP exert their growth promoting effects is a central focus when considering what alternative strategy may be an adequate substitute. Although there is no one all-encompassing means by which AGP improve animal performance, the general scientific consensus is that AGP mediate enhanced growth performance by altering intestinal microbiota. Recent studies using poultry and swine have helped us to understand the relationships between AGP supplementation and gastrointestinal bacterial composition [5–13]. The results of such studies prove that AGP create bacterial shifts and alter the microbial diversity of the intestine, suggesting that certain populations may be more related to animal growth than others.

Although the definitive gut microbial community required for AGP-mediated optimal growth promotion is still largely unknown, previous studies have shown that the ability of AGP to promote growth is highly correlated with a decrease in activity of bile salt hydrolase (**BSH**) [14–16]. BSH is an enzyme produced by commensal bacteria in the intestine whose main function is to convert conjugated bile salts into unconjugated bile salts [17]. Unconjugated bile acids are amphipathic and able to solubilize lipids for micelle formation; however, when the amide bond is hydrolyzed by BSH, the resulting unconjugated form is much less efficient at doing so. Consistent with this finding, independent chicken studies have demonstrated that AGP usage significantly reduced population of *Lactobacillus* species, the major BSH-producers in the chicken intestine; in particular, *L. salivarius*, the dominant lactic acid bacterium present in the chicken intestine, was reduced in response to AGP treatment [5,7,16,17,18,19]. Thus, AGP may promote animal growth by reducing the activity of BSH, an enzyme exerting negative impact on host fat digestion and utilization.

Based on these findings, inhibition of BSH activity using specific inhibitors is likely a promising approach to improve growth performance of food animals. This hypothesis is supported by our recent study [20] in which a BSH enzyme with broad substrate specificity from a chicken *L. salivarius* strain [21] was identified and used for evaluating a panel of dietary compounds. In this study [20], discovery of copper and zinc compounds as potent BSH inhibitors offered a potential explanation as to why adding high concentrations of dietary copper and zinc can improve growth performance and feed efficiency of poultry [22–25] and swine [26–29]. To further test our hypothesis and develop alternatives to AGP, a significant technical hurdle is to identify potent, safe, and cost-effective BSH inhibitors. Modern computational approaches, such as homology modeling and molecular docking, would be helpful for this purpose. However, success of such structure-based computations in the discovery of BSH inhibitor relies on the availability of the defined structures of major BSH enzymes, which is still lacking at present. Since hydrolysis of soluble unconjugated bile salts by BSH generates insoluble unconjugated bile salts that could form significant precipitations [17], we took advantage of this unique hydrolysis feature and developed a high-throughput screening (**HTS**) method to rapidly and efficiently identify BSH inhibitors in this study. Subsequently, a pilot HTS using a diverse compound library identified several promising BSH inhibitors with potential as alternatives to AGP. Notably, the findings from our HTS and quantitative BSH assay also suggest a new mode of action of AGP; specifically, some AGP (e.g. tetracyclines and roxarsone) can directly inhibit intestinal BSH, likely leading to enhanced lipid metabolism and growth performance.

Materials and Methods

Purification of Recombinant BSH (rBSH)

The rBSH used in this study was purified from an *E. coli* construct that produces His-tagged BSH with broad substrate specificity from a chicken *L. salivarius* [20]. Expression of the His-tagged rBSH and purification by nickel-nitrilotriacetic acid (Ni^{2+} -NTA) affinity chromatography was carried out as described in our previous publication [20,30,31].

Standard BSH Activity Assay

Determination of BSH activity was performed in microcentrifuge tube by following a two-step procedure to ascertain the amount of liberated amino acids from sodium glycocholate as detailed in our recent publication [20]. All experiments were performed in triplicate. The BSH activity was expressed as 1 μmol of amino acids liberated from the substrate per minute per mg of BSH.

To determine inhibitory effect of a specific compound on the activity of BSH, the rBSH was incubated with or without specific compound for 30 min at 37°C prior to the addition of substrate sodium glycocholate. The standard BSH activity assay was performed to determine the residual activity. A final concentration of specific compound in a BSH activity assay was routinely 5 mM; if needed, specific compound was also serially diluted for dosing experiment. A control without compound added was set up in each independent experiment. All assays were performed in triplicate. Percentage inhibition was calculated by dividing the inhibited activity (mean activity of control – mean residual activity of presence of a compound) relative to the mean activity of control and then multiplied by 100.

Development of Microplate Precipitation-based BSH Activity Assay

In order to move towards creating a rapid method of screening for BSH inhibitors, a microtiter plate-based BSH activity assay was developed based on the publication by Tanaka *et al.* [32] with modifications. This is a precipitation-based assay that is based on the fact that hydrolysis of conjugated bile acid substrates will produce deconjugated bile acids which are insoluble at the reaction pH and can easily be visualized as a white precipitate (Fig. 1A).

For initial assays, 10 μl of rBSH (1 μg/μl) was added to the bottom of a clear 96-well microtiter plate with a round bottom. To this, 190 μl of reaction mix (178 μl of reaction buffer [0.1 M sodium-phosphate, pH 6.0], 10 μl of glycocholic acid [100 mM], and 2 μl of 1 M DTT) was added for a total reaction volume of 200 μl. Plates were incubated at 37°C for up to 6 hours. Precipitation of insoluble unconjugated bile salts was monitored every 30 minutes by visual observation concomitant to absorbance measurement at 600 nm (A_{600}) using a microplate reader (model Multiskan EX; Thermo Fisher Scientific, Vantaa, Finland). With this general assay procedure, different parameters were altered to determine the optimal assay conditions, which include the stock concentration of the rBSH (from 8 μg/μl to 0.027 μg/μl), pH of reaction buffer (6.0 or 6.5), incubation temperature (20°C, 37°C, or 40°C), conjugated bile salt substrate (glycocholic acid, glycodeoxycholic acid, taucholic acid, and taurodeoxycholic acid), and the final concentration of conjugated bile salt (ranging from 0 mM to 50 mM).

After the assay conditions were optimized for the 200 μl reaction in a 96-well plate, a proof of concept assay was performed to demonstrate the feasibility of this procedure to screen compounds for BSH inhibition by using known BSH inhibitors that were previously determined [20]. The tested inhibitors include KIO_3, $NaIO_4$, $NaIO_3$, $CuCl_2$, and Na_2SeO_3. Before adding the reaction mix, 10 μl of inhibitor was added to the 10 μl of rBSH in the bottom of each well and mixed gently by pipetting. Inhibitors were added at a final concentration of 0.5 mM in column 1 with twofold serial dilutions through column 10 of the plate. As a control, two columns on each plate served as controls: a positive control with BSH and reaction mix only and a negative control with no BSH and inhibitor added. Assays were performed in duplicate.

After all conditions were optimized, including use of inhibitors, the effectiveness and reproducibility of the microplate precipitation-based assay were further evaluated with the entire reaction scaled down to a 50 μl total volume with modification of the volume of specific component to meet the needs of HTS using 384-well microplate. For this purpose, an additional assay was performed using dimethyl sulfoxide (DMSO), the library compound solvent, as a control to rule out possible background noise caused by DMSO.

High-throughput Screening of BSH Inhibitors

High-throughput screening (**HTS**) of BSH inhibitors was performed in the HTS facility at the Vanderbilt Institute of Chemical Biology (Nashville, TN). The library (2,240 compounds in 7 source plates from Spectrum) includes biologically active and structurally diverse compounds of known drugs, experimental bioactives, and pure natural products. All compounds tested have been dissolved in DMSO at a concentration of 10 mM. Briefly, 0.25 μl of a specific compound from the Spectrum Collection library was transferred into the bottom of a clear 384-well microplate (Greiner cat # 781182) with flat bottom by using the Echo 550/555 Liquid Handlers (Labcyte). Next, 12.5 μl of BSH

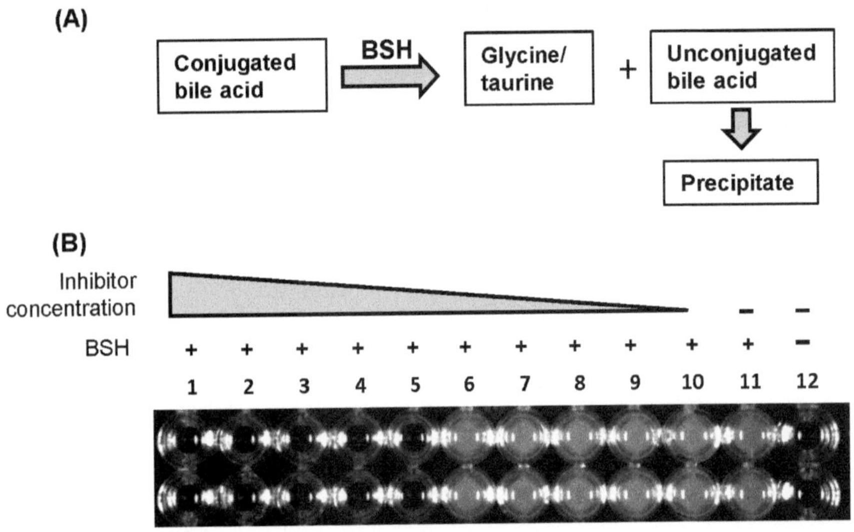

Figure 1. Optimization of conditions for high-throughput screening of BSH inhibitors. (A) Principal of the strategy for HTS screening of BSH inhibitors. (B) Proof of concept experiment to demonstrate the feasibility of the HTS methodology in 96-well plate. Using 200 µl of total reaction volume, BSH inhibitor was added at a final concentration of 0.5 mM in column 1 with 2-fold serial dilution through column 10. KIO_3 is used as an example in duplicate rows. The purified BSH was added at final concentration of 8 µg/ml in column wells 1–11. Column 12 served as a negative control with no enzyme and BSH inhibitor added.

(0.16 µg/µl prepared with reaction buffer) was added using the Multidrop Combi reagent dispenser (Thermo Scientific) and shaken for 5 minutes. Finally, 37.5 µl of reaction mix (32 µl of reaction buffer [0.1 M sodium-phosphate, pH 6.0], 2.5 µl of taurodeoxycholic acid [200 mM], and 0.5 µl of 1 M DTT) was added using the Multidrop Combi for a total reaction volume of 50 µl. The plates were subsequently shaken for 5 minutes to insure thorough mixing and spun to pull any reaction mixture back into the bottom of the well before incubation. Plates were incubated at 37°C with humidity and 5% CO_2 and absorbance measured every hour at 600 nm (A_{600}) for 4 hours using a SpectraMax M5 (Molecular Devices) with temperature controlled at 37°C; visual observations were concomitantly documented for precipitation. The reader on the system is connected via intranet to the facility's network, enabling automated data acquisition, analysis, visualization, and archival using a powerful management and chemiinformatics tool, Accelrys Pipeline Pilot; all data are stored in an Oracle database. Each plate contained three controls manually added with a multichannel pipette in a predetermined pattern to the side wells. Control 1 consisted of BSH, DMSO solvent, and reaction mix; control 2 contained a known inhibitor, BSH, and reaction mix; control 3 contained DMSO and reaction mix only.

Prior to screening the compound library, the above HTS protocol was validated twice in a full 384-well plate with checkerboard design containing activity controls (BSH, reaction mix containing substrate, and solvent DMSO) and inhibition controls (BSH, reaction mix containing substrate, and $NaIO_3$).

Selection and Validation of Identified BSH Inhibitors

A preliminary list of BSH inhibitor hits was generated after the above HTS; the wells that were completely clear after 4 hours of incubation are deemed as hits. Extensive review of relevant material safety data sheet and literature were performed for the hits with emphasis on availability, stability, toxicity, cost, and environmental impact.

The selected BSH inhibitor candidates were subjected to further *in vitro* validation using the 96-well microplate assay as well as the standard 2-step BSH assay as described above. The standard BSH assay is essential to confirm if the identified compounds are indeed real BSH inhibitors because we cannot completely rule out a possibility that certain compounds may cause less or no precipitation in reaction mix in a BSH activity-independent manner. The stock solutions of selected BSH inhibitors were purchased directly from the HTS facility at the Vanderbilt Institute of Chemical Biology, which had been solubilized in DMSO; compounds were stored at −20°C after arrival. If needed, dosing experiment was further performed to evaluate the potency of specific inhibitor.

Although the hit compounds belonging to an antibiotic family were not considered as potential alternatives to AGP, the following antibiotics were also chosen from the hit list and subjected to standard BSH assay to confirm if they had a direct inhibitory effect on BSH activity: cloxacillin sodium, cephradine, oxytetracycline, demeclocycline hydrochloride, and methacycline hydrochloride. In addition, following antibiotics that have been used as AGP were selected for evaluating their inhibitory effect on BSH activity using the standard 2-step BSH assay: doxycycline hydrochloride, roxarsone, tylosin, bacitracin, erythromycin, lincomycin, streptomycin, and sulfamethazine.

Results

Optimal Conditions for Microplate Precipitation-based BSH Activity Assay

For initial assay in 96-well plate with final volume of 200 µl, rBSH concentration, reaction buffer pH, incubation temperature and length, substrate type and concentration were altered in separate assays for assessment. As expected, a pH of 6.0 and an incubation temperature of 37°C, both of which are used in the standard 2-step assay, were conducive to efficient BSH hydrolysis. An incubation length of 2 to 4 hours was targeted in order to get several progressive measurements during HTS to monitor the dynamics of precipitation development. Given that using large quantities of rBSH could cause the reaction to proceed too quickly, rBSH was serially diluted for the plate assay; it was determined that using 10 μl of rBSH at the concentration approximately 0.16 μg/μl elicited adequate hydrolysis of substrate with maximal precipitation occurring after 2 hours of incubation. Regarding the choice of substrate, taurodeoxycholic acid was better than other tested conjugated bile salts because it elicited most observable BSH activity response (generation of white precipitation), which was also able to be corroborated by measurements using a microplate reader (data not shown).

With the above parameters optimized, a proof-of-concept experiment using known BSH inhibitors was performed. As shown in Fig. 1B, in the presence of high levels of a specific BSH inhibitor (KIO_3), the activity of BSH was inhibited, which was reflected by displaying clear, transparent wells (columns 1–5). In the presence of low concentrations of the BSH inhibitor (columns 6–10) or in the absence of the BSH inhibitor (columns 11, Fig. 1B), the reaction mix became turbid due to the precipitation of unconjugated bile salts resulting from active BSH enzyme. Other selected BSH inhibitors also displayed similar pattern as shown in Fig. 1B (data not shown), demonstrating the feasibility of the optimized microplate precipitation-based methodology for discovery of BSH inhibitors using HTS.

When the assay was further scaled down to a 50 μl total reaction volume, a prolonged incubation length (3–4 hours) was needed for precipitation to fully develop. DMSO, the solvent used for HTS library compounds, had little effect on precipitation formation with final concentration as high as 50 mM (data not shown).

HTS Discovery of BSH Inhibitors

At the HTS facility at the Vanderbilt Institute of Chemical Biology, all assays were performed in 384-well plate with 50 μl total reaction volume. Prior to screening of the compound libraries, validation assays containing different controls were carried out to ensure the system was optimal for actual inhibitor screening and that the optimized conditions translated well to this HTS format. Notably, to ensure consistent and accurate delivery of rBSH into 384-well plate using dispenser, the optimized 50-ul reaction mix setting was further modified by using larger volume of diluted rBSH (12.5 μl rather than 2.5 μl) despite final concentration of rBSH in reaction mix unchanged. With this minor on-site modification, the results of validation assays indicated that the assay could clearly distinguish between wells with or without precipitation as a result of BSH hydrolysis. Both negative controls (no BSH) and those using BSH inhibitor ($NaIO_4$ or KIO_3) remained clear in stark contrast to positive controls (no inhibitor) that displayed obvious precipitation; the visual results were also consistent with spectrometer reading (data not shown).

With the successful validation assays using HTS system, a pilot HTS using a 2,240-compound library was performed; plate setting was shown in Fig. 2A. A total of 7 plates were used for this screening, which led to 107 hits that were considered potent inhibitors. Hits were selected based on visual observation (the clear wells from lane 3 to lane 22 in a representative plate as shown in Fig. 2B) and corroborated via spectrophotometric measurement.

Validation of Selected HTS Hits with Potential as Feed Additives

Preliminary review of biochemical information of the corresponding hits eliminated most of compounds and identified 10 compounds that potentially could be used as feed additives for animal industry (Table 1). These 10 compounds were subjected to further validation using the plate assay as well as the 2-step activity assay to ascertain the quantitative inhibitory effect. The standard

Table 1. Effect of selected HTS hits on BSH activity[a].

Compound	Formula (MW)	% Inhibition
Caffeic Acid Phenethyl Ester (CAPE)	$C_{17}H_{16}O_4$ (284.31)	99.7
Riboflavin[b]	$C_{17}H_{20}N_4O_6$ (376.37)	99.3
Epicatechin monogallate	$C_{20}H_{28}O_4$ (332.44)	98.6
Gossypetin	$C_{15}H_{10}O_8$ (318.24)	97.3
Carnosic Acid	$C_{20}H_{28}O_4$ (332.44)	96.8
Menadione	$C_{11}H_8O_2$ (172.18)	87.2
Purpurogallin	$C_{11}H_8O_5$ (220.18)	83.0
Theaflavanin	$C_{20}H_{16}O_8$ (384.34)	18.1
Chrysophanol[c]	$C_{15}H_{10}O_4$ (254.24)	3.7
Folic Acid[b]	$C_{19}H_{19}N_7O_6$ (441.40)	−10.8

[a]Unless specified, the final concentration of compound in the reaction mix was 5 mM to achieve optimal resolution with the quantitative BSH activity assay.
[b]The final concentration of riboflavin in reaction mix was 1 mM.
[c]The final concentration of chrysophanol in reaction mix was 1.25 mM.
[d]The final concentration of folic acid in reaction mix was 1.5 mM.

2-step BSH activity assays identified two compounds as false positive (chrysophanol and folic acid); these two compounds also failed to inhibit BSH using 96-well plate assay.

Majority of the selected hits displayed potent inhibitory effect on BSH (Table 1). Notably, caffeic acid phenethyl ester (CAPE), riboflavin, epicatechin monogallate, gossypetin, and carnosic acid consistently inhibited activity by over 95% in independent experiments (data not shown). In particular, despite its relative lower final concentration (1 mM) in reaction mix due to solubility issue, riboflavin still exerted strong inhibition on BSH activity (>99% of inhibition) (Table 1). Because we are particularly interested in CAPE and riboflavin as AGP alternatives, dosing experiments were conducted to examine if they could inhibit BSH activity at lower concentrations. CAPE still inhibited rBSH activity by more than 50% at a final concentration of 0.25 mM (Fig. 3A) and riboflavin by more than 50% at a final concentration as low as 0.00625 mM (Fig. 3B).

Direct Inhibitory Effect of Antibiotics on BSH

Interestingly, a group of antibiotics were also HTS hits, including tetracycline and β-lactam antibiotics that have been widely used as AGP. This finding prompted us to speculate that direct inhibition of BSH activity by some antibiotics may be a new mode of action of AGP. To partly test this, we further examined a panel of antibiotics that have been used as AGP in food animals and determined if different classes of antibiotic make a difference in inhibition potential. As shown in Table 2, tetracycline antibiotics were consistently potent inhibitors. Additionally, roxarsone, an arsenical drug that has been extensively used in poultry production, also displayed high inhibitory effect on BSH activity. Both roxarsone and oxytetracycline inhibited BSH activity by over 95% when tested at a final concentration of 5 mM (Table 2). In an additional dosing experiment, oxytetracycline and roxarsone still inhibited enzyme activity by approximately 86% and 52%, respectively, at a final concentration of 0.625 mM (Fig. 4). Other tested antibiotics, such as β-lactam and lincosamide, displayed strong but relatively lower inhibitory effect on BSH activity (Table 2). However, the macrolide and bacitracin antibiotics had weak or no inhibitory effect on BSH.

(A)

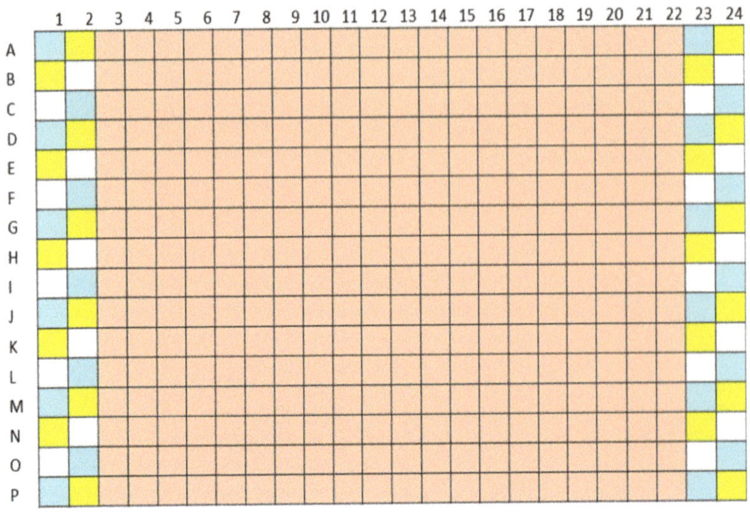

(B)

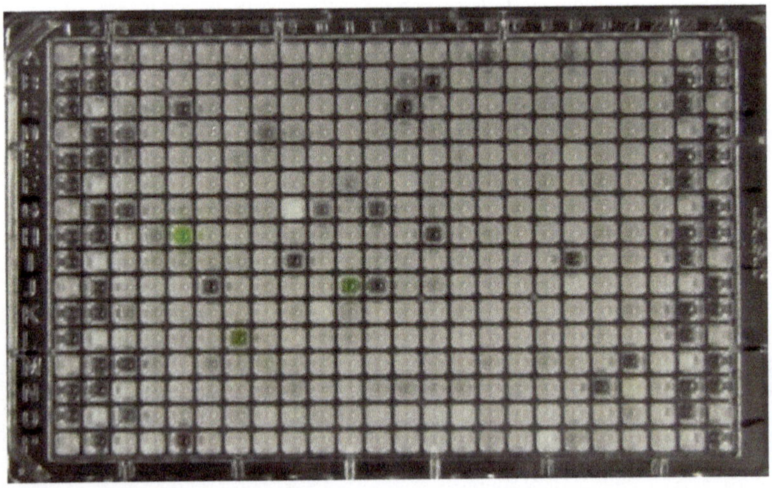

Figure 2. Representative result of high-throughput screening of BSH inhibitors. (A) Plate layout for screening BSH inhibitors. Pink boxes (columns 3–22) indicate test wells that contain library compounds of interest, BSH, and reaction mix containing substrate. Library compounds were shot into the well bottom using Echo 550/555 and enzyme and reaction mix were added using Multidrop Combi. Controls were added manually to the side wells (columns 1–2 and 23–24): blue boxes indicate activity controls (BSH, reaction mix containing substrate, and solvent DMSO), yellow boxes correspond to inhibition controls (BSH, reaction mix containing substrate, and $NaIO_3$), and white boxes are negative controls with no BSH added but include reaction mix as well as substrate. (B) The HTS results represented by one 384-well plate. Control wells indicate the assay proceeded normally. The wells in columns 3–22 that appeared clear, regardless of alternative color due to compound, and had low absorbance readings were considered hits (putative BSH inhibitors).

Discussion

Antibiotic growth promoters (**AGP**) undoubtedly have a positive influence on animal production in terms of the improvements they confer on feed utilization efficiency, animal welfare as well as the resulting financial gains. However, antibiotic resistance of zoonotic origin threatens the supply of food animals and public safety and, thus, is a compelling reason to discontinue their inclusion as feed additives. Likewise, elimination of dietary AGP will necessitate alternative strategies to compensate for setbacks to production, as have occurred in some areas where bans have been instituted [33]. Consequently, different products have been proposed as alternatives for AGP, including essential oils, prebiotics, probiotics, exogenous enzymes, and organic acids, among others, to change gut microbiota for enhanced animal health and growth performance [34]. However, very limited data is available to justify the choice of specific bacterial species or products for such microbiota manipulation despite the link between AGP induced growth promotion and altered gastrointestinal microflora [1,2].

A promising strategy for developing AGP alternative is targeting general metabolic pathway that is influenced by AGP usage. Specifically, the activity of microbial bile salt hydrolase (**BSH**) in the intestine is a valid object for manipulation since studies repeatedly have observed that AGP use is associated with reduced BSH activity and improved animal growth performance [14–16]. Conjugated bile acids play a critical role in lipid solubilization and micelle formation and are the target substrates of BSH [2,17]. However, deconjugation of bile acids by BSH compromises lipid metabolism and results in attenuated energy harvest. Therefore, we have proposed to develop BSH inhibitor-based alternatives to

Table 2. Effect of different classes of antibiotics on BSH activity.

Antibiotic class	Antibiotic name	% Inhibition
Tetracycline	Oxytetracycline*	97.4
	Demeclocycline Hydrochloride*	92.0
	Methacycline Hydrochloride*	90.2
	Doxycycline Hydrochloride	92.9
β-lactam	Cloxacillin*	61.2
	Cephradine*	61.8
Arsenicals	Roxarsone	98.4
Lincosamide	Lincomycin	67.3
Sulfonamide	Sulfamethazine	48.7
Macrolide	Tylosin	23.8
	Erythromycin	36.5
Peptides	Bacitracin	−17.8

*The antibiotic hits by HTS.

AGP for enhanced animal production and sustainability [20]; this recent study established a solid platform for us to discover novel BSH inhibitors.

In this study, we improved the capacity to screen large quantities of compounds in a time efficient manner by developing and validating a HTS assay to rapidly identify BSH inhibitors. Overall, the ability of the HTS system to identify BSH inhibitors proved to be a reliable means of inhibitor detection. Validation work was able to confirm that many of our chosen HTS hit compounds displayed high levels of BSH inhibition as expected. However, false positives were also present; chrysophanol and folic acid, both recognized as inhibitors in the library screen, exhibited minimal enzyme inhibition or even slight enhancement (Table 1), which is likely due to solution stability or chemical properties [35]. For this pilot screening, we only had the time and resources to do one all-encompassing screen. A second re-screen, including only a subset of compounds of interest, would have been beneficial in confirming our suspected compounds as legitimate inhibitors. Thus, in future HTS using a much larger compound library, it is essential to select all positive hits individually from corresponding source plates – a so-called 'cherry picking' procedure – for confirmation using the same protocol.

The Spectrum Collection library that was screened contained a wide range of structurally diverse compounds, including drug components, natural products, and other bioactive constituents. Several potent inhibitors confirmed by validation assay possess interesting beneficial qualities. Carnosic acid and epicatechin monogallate are both nutraceuticals with noted antioxidant and neuroprotective effects [36–40]. Additionally, carnosic acid is an anti-inflammatory [41] and epicatechin monogallate displays some chemopreventive [42] and antimicrobial properties [43]. Similarly, gossypetin was among the powerful inhibitors and also is noted to be both anti-inflammatory [44] and an antioxidant [45]. Each of these compounds warrants further animal research, and should availability permit, growth performance studies could reveal more as to their potential use. However, two particular compounds stood out as potent inhibitors with definite high potential as novel alternative to AGP: riboflavin and caffeic acid phenethyl ester (CAPE, also called 'phenethyl caffeate'). Riboflavin, or vitamin B2, plays a key role in energy metabolism and is a conenzyme in numerous redox reactions [46]. Aside from contributions to animal physiology, it is water-soluble and causes no known toxicities from supplementation at upper limits [47], although even at the lowest concentrations tested, riboflavin was still an extremely potent BSH inhibitor (Fig. 3B). Because it is an FDA-approved feed additive with well-established metabolic function and would be readily available to incorporate into feed, riboflavin likely may be an acceptable candidate to improve growth performance for food animals. CAPE, on the other hand, is an emerging bioactive studied because it has antioxidant [48–50], anticarcinogenic [51,52], anti-inflammatory [50,53,54], immunomodulatory [52], and antimicrobial [54] effects. It is a phenolic component of propolis and can be directly extracted or artificially synthesized by several methods [55]. Although much research has been done to characterize the aforementioned effects of CAPE, no published data exist concerning its effects on growth performance of food animals. Because CAPE can inhibit BSH at low concentrations, a trial of this natural product as a feed supplement is also highly warranted.

For initial evaluation of HTS hits, we eliminated the hits that fall in antimicrobial category when selecting the BSH inhibitors with potential as alternative to AGP. However, the finding that some antibiotic hits, such as tetracyclines that have been widely used as AGP, called our attention to a possible new mechanism of AGP: AGP may have a direct effect on the function of gut BSH enzymes [17,56] aside from modulating gut microbial ecology for enhanced growth performance. Using standard BSH assay, in this study, we have confirmed that the tested tetracycline and β-lactam antibiotics could directly inhibit BSH activity. Notably, roxarsone, an arsenical drug that has been used in the poultry industry to improve weight gain, feed conversion, and pigmentation as well as to control coccidiosis [57,58], also displayed potent inhibitory effect on BSH. Currently little is known about how roxarsone is metabolized by chickens to mediate growth promotion [58] although it is suspected it may be due to its angiogenic potential [59]. Our results suggest that roxarsone as well as some other AGP (e.g. tetracycline) may also exert their growth-promoting effect on food animals by direct inhibition of intestinal BSH enzymes for enhanced lipid metabolism and energy harvest. Thus, this unexpected finding suggests a new mode of action of AGP for promoting animal growth.

The pilot HTS in this study confirmed that our HTS protocol is a valid BSH inhibitor detection method. Future screens using larger compound libraries may reveal more novel BSH inhibitors. Having a greater number of applicable compounds will allow us to choose those with the most redeemable qualities that would include animal and environmental safety, cost, and availability. In addition, large animal trials are needed to determine the effect of 'champion' BSH inhibitors on growth performance of food animals. While *in vitro* studies are absolutely essential to justify potential AGP alternatives, intestinal *in vivo* conditions are certainly more complex. The *in vivo* stability and bioavailability must be addressed in order to establish the validity and practicality of utilizing such inhibitors as a legitimate alternative to AGP. Moreover, growth performance parameters like body weight gain, feed intake, and feed conversion ratio will need to be measured as well as morphological characteristics of the intestinal tract and meat and carcass quality. This will help to rule out or identify any negative physiological consequences associated with prolonged use of a particular BSH inhibitor. For example, because the BSH inhibitors may improve lipid metabolism, it is important to determine that energy harvest and weight gain is partitioned adequately and not skewed toward excess fat deposition, which would be undesirable for industry and consumers. Thus, comprehensive animal trials are needed to address the above

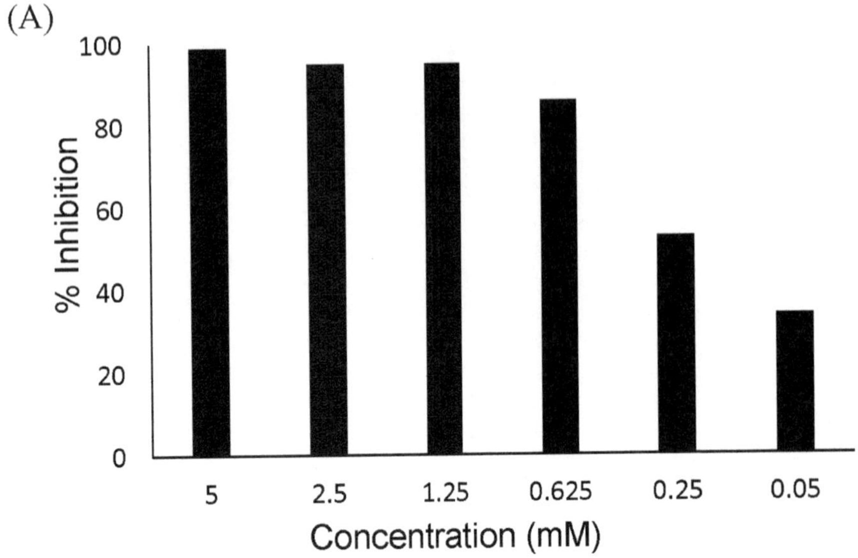

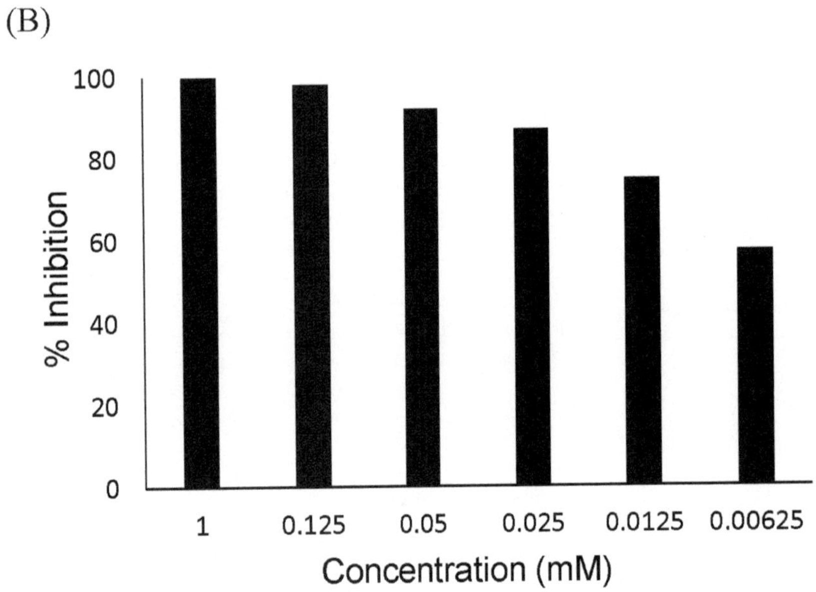

Figure 3. Dosing effects of selected BSH inhibitors on BSH activity. (A) Inhibition of BSH activity by caffeic acid phenethyl ester (CAPE). (B) Inhibition of BSH activity by riboflavin. All assays were performed in triplicate. The procedure is detailed in Materials and Methods.

issues and corroborate the performance-boosting benefits of novel BSH inhibitors.

Discovery of BSH inhibitors may help us develop 'negative-traits-mitigation' strategy to optimize probiotic products for enhanced growth performance of food animals and profitability of feed additive industry. It is no doubt that dietary probiotics, the normal commensal microorganisms, could exert various beneficial effects on food animals. However, probiotics do impose a variety of potential costs (or detrimental effects) to the animal host as well, which include the production of toxic metabolites, decreased fat digestability due to production of BSH, and the increase of mucus secretion and gut epithelial cell turn-over [2]. Since lactobacilli are dominant BSH-producers in the small intestine, dietary lactobacilli treatment may negatively affect lipid metabolism and energy harvest, consequently imposing negative impact on body weight gain. Recently, two different groups have reported that probiotic supplementation to diets significantly reduced body weight gain, fat digestibility, and feed conversion in broilers [60,61]. Based on these findings, the investigators have proposed that the detrimental effects of the probiotics on chicken growth are likely attributed to the production of intestinal BSH by lactobacilli probiotics. These recent studies provided physiological evidence for relationship between lipid metabolism and the BSH activity/BSH-producers in the chicken intestine. Therefore, the overall beneficial effects associated with specific probiotics should be carefully evaluated. The BSH inhibitors may be used together with certain probiotics to maximize the beneficial effect of the probiotics by mitigating potential negative impact of probiotics on host fat digestion due to the production of BSH enzyme.

The efficient HTS technology developed in this study should have the broad appeal to the community of researchers who study BSH and develop BSH-based practical applications. It has been

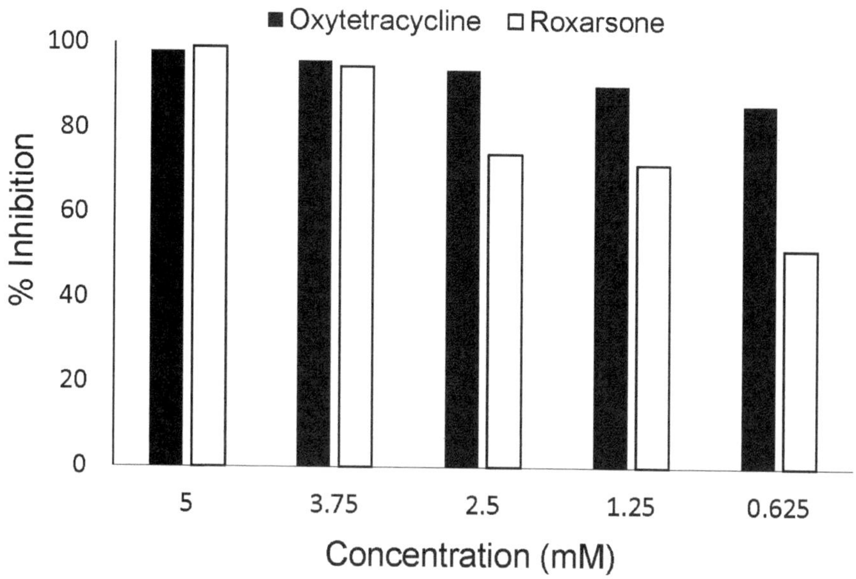

Figure 4. Dosing effects of oxytetracycline and roxarsone on BSH activity. Black bars represent Inhibition of BSH activity by oxytetracyline and white bars by roxarsone. All assays were performed in triplicate. The procedure is detailed in Materials and Methods.

increasingly recognized that intestinal BSH plays an important role in host metabolism and energy harvest [17,56]. The BSH activity has significant impact on host physiology by disturbing conjugated bile salts-mediated fat metabolism and endocrine functions, consequently affecting body weight gain and even cholesterol level in humans [17,56]. Despite the significance of gut BSH, research on BSH is still in its infancy [62]. Not surprisingly, little information exists concerning the compounds that inhibit or enhance the activity of BSH. Notably, the HTS method described in this manuscript could be easily modified to make it suitable for the discovery of BSH enhancers. Specifically, during our optimization experiments, we have observed that changing some parameters, such as using less rBSH, would prolong incubation time for full precipitation development. With the control of desired reaction time (e.g. 8 hr), the dynamics and magnitude of precipitation formation in each well (reflected by A_{600} value) could be monitored and used for screening BSH enhancers. Since enhanced BSH activity in the intestine could lower cholesterol level in humans [17,56], we speculate that BSH enhancers have a potential application in human health. Together, this study successfully developed a timely HTS technology and established a solid platform for future large-scale discovery of BSH inhibitors as well as BSH enhancers.

Acknowledgments

We are grateful to Dr. Debbie Mi (Vanderbilt HTS facility) for providing training and technical assistance to perform high-throughput screening.

Author Contributions

Conceived and designed the experiments: KS XZ JL. Performed the experiments: KS XZ JL. Analyzed the data: KS XZ JL. Contributed reagents/materials/analysis tools: JL. Wrote the paper: KS JL.

References

1. Marshall BM, Levy SB (2011) Food animals and antimicrobials: impacts on human health. Clin Microbiol Rev 24: 718–733.
2. Dibner JJ, Richards JD (2005) Antibiotic growth promoters in agriculture: history and mode of action. Poult Sci 84: 634–643.
3. van den Bogaard AE, Stobberingh EE (1999) Antibiotic usage in animals: impact on bacterial resistance and public health. Drugs 58: 589–607.
4. US FDA takes steps to reduce use of antibiotic growth promoters (2012) Vet Rec 170: 404.
5. Knarreborg A, Simon MA, Engberg RM, Jensen BB, Tannock GW (2002) Effects of dietary fat source and subtherapeutic levels of antibiotic on the bacterial community in the ileum of broiler chickens at various ages. Appl Environ Microbiol 68: 5918–5924.
6. Collier CT, Smiricky-Tjardes MR, Albin DM, Wubben JE, Gabert VM, et al. (2003) Molecular ecological analysis of porcine ileal microbiota responses to antimicrobial growth promoters. J Anim Sci 81: 3035–3045.
7. Dumonceaux TJ, Hill JE, Hemmingsen SM, Van Kessel AG (2006) Characterization of intestinal microbiota and response to dietary virginiamycin supplementation in the broiler chicken. Appl Environ Microbiol 72: 2815–2823.
8. Wise MG, Siragusa GR (2007) Quantitative analysis of the intestinal bacterial community in one- to three-week-old commercially reared broiler chickens fed conventional or antibiotic-free vegetable-based diets. J Appl Microbiol 102: 1138–1149.
9. Rettedal E, Vilain S, Lindblom S, Lehnert K, Scofield C, et al. (2009) Alteration of the ileal microbiota of weanling piglets by the growth-promoting antibiotic chlortetracycline. Appl Environ Microbiol 75: 5489–5495.
10. Danzeisen JL, Kim HB, Isaacson RE, Tu ZJ, Johnson TJ (2011) Modulations of the chicken cecal microbiome and metagenome in response to anticoccidial and growth promoter treatment. PLoS One 6: e27949.
11. La-ongkhum O, Pungsungvorn N, Amornthewaphat N, Nitisinprasert S (2011) Effect of the antibiotic avilamycin on the structure of the microbial community in the jejunal intestinal tract of broiler chickens. Poult Sci 90: 1532–1538.
12. Kim HB, Borewicz K, White BA, Singer RS, Sreevatsan S, et al. (2012) Microbial shifts in the swine distal gut in response to the treatment with antimicrobial growth promoter, tylosin. Proc Natl Acad Sci U S A 109: 15485–15490.
13. Lin J, Hunkapiller AA, Layton AC, Chang YJ, Robbins KR (2013) Response of intestinal microbiota to antibiotic growth promoters in chickens. Foodborne Pathog Dis 10: 331–337.
14. Feighner SD, Dashkevicz MP (1987) Subtherapeutic levels of antibiotics in poultry feeds and their effects on weight gain, feed efficiency, and bacterial cholyltaurine hydrolase activity. Appl Environ Microbiol 53: 331–336.
15. Knarreborg A, Lauridsen C, Engberg RM, Jensen SK (2004) Dietary antibiotic growth promoters enhance the bioavailability of alpha-tocopheryl acetate in broilers by altering lipid absorption. J Nutr 134: 1487–1492.
16. Guban J, Korver DR, Allison GE, Tannock GW (2006) Relationship of dietary antimicrobial drug administration with broiler performance, decreased population levels of Lactobacillus salivarius, and reduced bile salt deconjugation in the ileum of broiler chickens. Poult Sci 85: 2186–2194.
17. Begley M, Hill C, Gahan CG (2006) Bile salt hydrolase activity in probiotics. Appl Environ Microbiol 72: 1729–1738.

18. Engberg RM, Hedemann MS, Leser TD, Jensen BB (2000) Effect of zinc bacitracin and salinomycin on intestinal microflora and performance of broilers. Poult Sci 79: 1311–1319.
19. Zhou H, Gong J, Brisbin JT, Yu H, Sanei B, et al. (2007) Appropriate chicken sample size for identifying the composition of broiler intestinal microbiota affected by dietary antibiotics, using the polymerase chain reaction-denaturing gradient gel electrophoresis technique. Poult Sci 86: 2541–2549.
20. Wang Z, Zeng X, Mo Y, Smith K, Guo Y, et al. (2012) Identification and characterization of a bile salt hydrolase from Lactobacillus salivarius for development of novel alternatives to antibiotic growth promoters. Appl Environ Microbiol 78: 8795–8802.
21. Stern NJ, Svetoch EA, Eruslanov BV, Perelygin VV, Mitsevich EV, et al. (2006) Isolation of a Lactobacillus salivarius strain and purification of its bacteriocin, which is inhibitory to Campylobacter jejuni in the chicken gastrointestinal system. Antimicrob Agents Chemother 50: 3111–3116.
22. Ewing HP, Pesti GM, Bakalli RI, Menten JF (1998) Studies on the feeding of cupric sulfate pentahydrate, cupric citrate, and copper oxychloride to broiler chickens. Poult Sci 77: 445–448.
23. Miles RD, O'Keefe SF, Henry PR, Ammerman CB, Luo XG (1998) The effect of dietary supplementation with copper sulfate or tribasic copper chloride on broiler performance, relative copper bioavailability, and dietary prooxidant activity. Poult Sci 77: 416–425.
24. Arias VJ, Koutsos EA (2006) Effects of copper source and level on intestinal physiology and growth of broiler chickens. Poult Sci 85: 999–1007.
25. Liu ZH, Lu L, Li SF, Zhang LY, Xi L, et al. (2011) Effects of supplemental zinc source and level on growth performance, carcass traits, and meat quality of broilers. Poult Sci 90: 1782–1790.
26. Smith JW, 2nd, Tokach MD, Goodband RD, Nelssen JL, Richert BT (1997) Effects of the interrelationship between zinc oxide and copper sulfate on growth performance of early-weaned pigs. J Anim Sci 75: 1861–1866.
27. Armstrong TA, Cook DR, Ward MM, Williams CM, Spears JW (2004) Effect of dietary copper source (cupric citrate and cupric sulfate) and concentration on growth performance and fecal copper excretion in weanling pigs. J Anim Sci 82: 1234–1240.
28. Jacela JY, DeRouchey JM, Tokach MD, Goodband RD, Nelssen JL, et al. (2010) Feed additives for swine: Fact sheets - high dietary levels of copper and zinc for young pigs, and phytase. Journal of Swine Health and Production 18: 87–89.
29. Shelton NW, Tokach MD, Nelssen JL, Goodband RD, Dritz SS, et al. (2011) Effects of copper sulfate, tri-basic copper chloride, and zinc oxide on weanling pig performance. J Anim Sci 89: 2440–2451.
30. Lin J, Akiba M, Sahin O, Zhang Q (2005) CmeR functions as a transcriptional repressor for the multidrug efflux pump CmeABC in Campylobacter jejuni. Antimicrob Agents Chemother 49: 1067–1075.
31. Lin J, Michel LO, Zhang Q (2002) CmeABC functions as a multidrug efflux system in Campylobacter jejuni. Antimicrob Agents Chemother 46: 2124–2131.
32. Tanaka H, Hashiba H, Kok J, Mierau I (2000) Bile salt hydrolase of Bifidobacterium longum-biochemical and genetic characterization. Appl Environ Microbiol 66: 2502–2512.
33. Casewell M, Friis C, Marco E, McMullin P, Phillips I (2003) The European ban on growth-promoting antibiotics and emerging consequences for human and animal health. J Antimicrob Chemother 52: 159–161.
34. Huyghebaert G, Ducatelle R, Van Immerseel F (2011) An update on alternatives to antimicrobial growth promoters for broilers. Vet J 187: 182–188.
35. Oldenburg K, Pooler D, Scudder K, Lipinski C, Kelly M (2005) High throughput sonication: evaluation for compound solubilization. Comb Chem High Throughput Screen 8: 499–512.
36. Terao J, Piskula M, Yao Q (1994) Protective effect of epicatechin, epicatechin gallate, and quercetin on lipid peroxidation in phospholipid bilayers. Arch Biochem Biophys 308: 278–284.
37. Azad N, Rasoolijazi H, Joghataie MT, Soleimani S (2011) Neuroprotective effects of carnosic Acid in an experimental model of Alzheimer's disease in rats. Cell J 13: 39–44.
38. Kelsey NA, Wilkins HM, Linseman DA (2010) Nutraceutical antioxidants as novel neuroprotective agents. Molecules 15: 7792–7814.
39. Moran L, Andres S, Bodas R, Prieto N, Giraldez FJ (2012) Meat texture and antioxidant status are improved when carnosic acid is included in the diet of fattening lambs. Meat Sci 91: 430–434.
40. Xiang Q, Liu Z, Wang Y, Xiao H, Wu W, et al. (2013) Carnosic acid attenuates lipopolysaccharide-induced liver injury in rats via fortifying cellular antioxidant defense system. Food Chem Toxicol 53: 1–9.
41. Kuo CF, Su JD, Chiu CH, Peng CC, Chang CH, et al. (2011) Anti-inflammatory effects of supercritical carbon dioxide extract and its isolated carnosic acid from Rosmarinus officinalis leaves. J Agric Food Chem 59: 3674–3685.
42. Du GJ, Zhang Z, Wen XD, Yu C, Calway T, et al. (2012) Epigallocatechin Gallate (EGCG) is the most effective cancer chemopreventive polyphenol in green tea. Nutrients 4: 1679–1691.
43. Hamilton-Miller JM, Shah S (2000) Activity of the tea component epicatechin gallate and analogues against methicillin-resistant Staphylococcus aureus. J Antimicrob Chemother 46: 852–853.
44. Mounnissamy VM, Gopal V, Gunasegaran R, Saraswathy A (2002) Antiinflammatory activity of gossypetin isolated from Hibiscus sabdariffa. Indian Journal of Heterocyclic Chemistry 12: 85–86.
45. Trendafilova A, Todorova M, Nikolova M, Gavrilova A, Vitkova A (2011) Flavonoid constituents and free radical scavenging activity of Alchemilla mollis. Nat Prod Commun 6: 1851–1854.
46. Combs Jr GF (2012) Chapter 11 - Riboflavin. The Vitamins (Fourth Edition). San Diego: Academic Press. 277–289.
47. (1987) Vitamin Tolerance of Animals: The National Academies Press.
48. Altug ME, Serarslan Y, Bal R, Kontas T, Ekici F, et al. (2008) Caffeic acid phenethyl ester protects rabbit brains against permanent focal ischemia by antioxidant action: a biochemical and planimetric study. Brain Res 1201: 135–142.
49. Lee KJ, Choi JH, Hwang YP, Chung YC, Jeong HG (2008) Protective effect of caffeic acid phenethyl ester on tert-butyl hydroperoxide-induced oxidative hepatotoxicity and DNA damage. Food Chem Toxicol 46: 2445–2450.
50. Dos Santos JS, Monte-Alto-Costa A (2013) Caffeic Acid Phenethyl Ester Improves Burn Healing in Rats Through Anti-Inflammatory and Antioxidant Effects. J Burn Care Res.
51. Chung TW, Moon SK, Chang YC, Ko JH, Lee YC, et al. (2004) Novel and therapeutic effect of caffeic acid and caffeic acid phenyl ester on hepatocarcinoma cells: complete regression of hepatoma growth and metastasis by dual mechanism. FASEB J 18: 1670–1681.
52. Chan GC, Cheung KW, Sze DM (2013) The immunomodulatory and anticancer properties of propolis. Clin Rev Allergy Immunol 44: 262–273.
53. da Cunha FM, Duma D, Assreuy J, Buzzi FC, Niero R, et al. (2004) Caffeic acid derivatives: in vitro and in vivo anti-inflammatory properties. Free Radic Res 38: 1241–1253.
54. Yilmaz A, Yildirim O, Tamer L, Oz O, Cinel L, et al. (2005) Effects of caffeic acid phenethyl ester on endotoxin-induced uveitis in rats. Curr Eye Res 30: 755–762.
55. Akyol S, Ginis Z, Armutcu F, Ozturk G, Yigitoglu MR, et al. (2012) The potential usage of caffeic acid phenethyl ester (CAPE) against chemotherapy-induced and radiotherapy-induced toxicity. Cell Biochem Funct 30: 438–443.
56. Jones BV, Begley M, Hill C, Gahan CG, Marchesi JR (2008) Functional and comparative metagenomic analysis of bile salt hydrolase activity in the human gut microbiome. Proc Natl Acad Sci U S A 105: 13580–13585.
57. Chapman HD, Johnson ZB (2002) Use of antibiotics and roxarsone in broiler chickens in the USA: analysis for the years 1995 to 2000. Poult Sci 81: 356–364.
58. Nachman KE, Baron PA, Raber G, Francesconi KA, Navas-Acien A, et al. (2013) Roxarsone, inorganic arsenic, and other arsenic species in chicken: a u.s.-Based market basket sample. Environ Health Perspect 121: 818–824.
59. Basu P, Ghosh RN, Grove LE, Klei L, Barchowsky A (2008) Angiogenic potential of 3-nitro-4-hydroxy benzene arsonic acid (roxarsone). Environ Health Perspect 116: 520–523.
60. Mountzouris KC, Tsitrsikos P, Palamidi I, Arvaniti A, Mohnl M, et al. (2010) Effects of probiotic inclusion levels in broiler nutrition on growth performance, nutrient digestibility, plasma immunoglobulins, and cecal microflora composition. Poult Sci 89: 58–67.
61. Sharifi SD, Dibamehr A, Lotfollahian H, Baurhoo B (2012) Effects of flavomycin and probiotic supplementation to diets containing different sources of fat on growth performance, intestinal morphology, apparent metabolizable energy, and fat digestibility in broiler chickens. Poult Sci 91: 918–927.
62. Patel AK, Singhania RR, Pandey A, Chincholkar SB (2010) Probiotic bile salt hydrolase: current developments and perspectives. Appl Biochem Biotechnol 162: 166–180.

22

Energetic Extremes in Aquatic Locomotion by Coral Reef Fishes

Christopher J. Fulton[1]*, Jacob L. Johansen[2], John F. Steffensen[3]

1 ARC Centre of Excellence for Coral Reef Studies, Research School of Biology, The Australian National University, Canberra, Australian Capital Territory, Australia, **2** ARC Centre of Excellence for Coral Reef Studies, School of Marine and Tropical Biology, James Cook University, Townsville, Queensland, Australia, **3** Marine Biological Section, Department of Biology, University of Copenhagen, Helsingør, Denmark

Abstract

Underwater locomotion is challenging due to the high friction and resistance imposed on a body moving through water and energy lost in the wake during undulatory propulsion. While aquatic organisms have evolved streamlined shapes to overcome such resistance, underwater locomotion has long been considered a costly exercise. Recent evidence for a range of swimming vertebrates, however, has suggested that flapping paired appendages around a rigid body may be an extremely efficient means of aquatic locomotion. Using intermittent flow-through respirometry, we found exceptional energetic performance in the Bluelined wrasse *Stethojulis bandanensis*, which maintains tuna-like optimum cruising speeds (up to 1 metre s^{-1}) while using 40% less energy than expected for their body size. Displaying an exceptional aerobic scope (22-fold above resting), streamlined rigid-body posture, and wing-like fins that generate lift-based thrust, *S. bandanensis* literally flies underwater to efficiently maintain high optimum swimming speeds. Extreme energetic performance may be key to the colonization of highly variable environments, such as the wave-swept habitats where *S. bandanensis* and other wing-finned species tend to occur. Challenging preconceived notions of how best to power aquatic locomotion, biomimicry of such lift-based fin movements could yield dramatic reductions in the power needed to propel underwater vehicles at high speed.

Editor: Athanassios C. Tsikliras, University of Thessaly, Greece

Funding: Funding was provided by the Australian Research Council (to CJF) and the Danish Agency for Science, Technology and Innovation (to JFS). The funders had no role in study design, data collection and analysis, decision to publish, or preparation of the manuscript.

* E-mail: christopher.fulton@anu.edu.au

Introduction

Underwater locomotion is challenging due to the high friction and resistance imposed on a body moving through water and energy lost in the wake during undulatory propulsion [1–7]. While aquatic organisms have evolved streamlined shapes to overcome such resistance, aquatic locomotion is still considered a costly exercise [1–7]. Indeed, in tunas and other pelagic fishes that cruise the open ocean using tail-powered swimming, we see some of the fastest measured speeds for underwater swimming, but at the cost of some of the highest known rates of aerobic energy consumption during locomotion [5,6]. However, biomechanical explorations on swimming fish, birds, mammals and reptiles have suggested that flapping paired appendages (fins or flippers) around a rigid-body could be an extremely efficient form of aquatic locomotion at high sustained speeds [7–11]. Indeed, recent research on coral reef fishes swimming in this way have shown they can maintain high swimming speeds during everyday activities [8,12,13]. Such evidence rivals the paradigm that tail-powered swimming by tuna and other pelagic fishes is the pinnacle of underwater cruising locomotion [5,6,13,14].

Reef fish predominantly swim via the labriform gait, which involves moving solely the pectoral fins for propulsion [12,13,14,15]. Interestingly, these reef fishes display a wide range of pectoral fin shapes, which has been linked to differences in how they move their fins to produce thrust. Coral reef fish with wing-like fins (high aspect-ratio, AR) produce lift-based thrust via figure-eight flapping fin movements, while sister taxa with paddle-shaped fins (low AR) tend to produce resistance-based thrust via a rowing action [7]. Performance studies have indicated such modes translate to different swimming speeds according to the mechanical efficiency and energetic cost of using each form of labriform propulsion [7,12,15,16,17], with high aspect-ratio species adopting lift-based thrust being the faster swimmers. However, there is currently no empirical comparison of the energetic performance of fishes using these two styles of fin propulsion (hereafter referred to as resistance-based and lift-based labriform swimming). Given that labriform swimming is adopted by a diversity of vertebrates spanning fish to turtles and penguins [9,10,15], the underlying efficiency of this mode could help explain the widespread adoption of this form of underwater locomotion.

Integrating new data into a meta-analysis of comparative swimming energetics in bony fishes, we aimed to test the energetic efficiency of resistance- and lift-based labriform swimming for high-speed underwater locomotion. Firstly, we used intermittent flow-through respirometry to measure aerobic energy use during prolonged swimming activity in two species of coral reef fishes from the Great Barrier Reef, Australia: *Stethojulis bandanensis* and *Cheilinus fasciatus*. Chosen for their pectoral fin morphologies (indicated by the pectoral fin AR, which is measured as length of

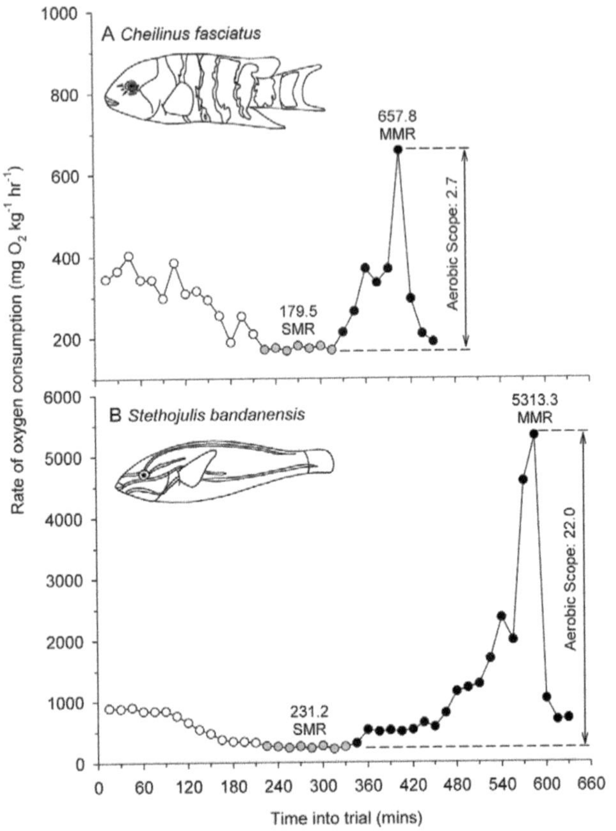

Figure 1. Rate of oxygen consumption against time in a swimming speed trial for (A) Redbreasted wrasse *Cheilinus fasciatus* (11.5 cm total body length) and (B) Bluelined wrasse *Stethojulis bandanensis* (10.5 cm). Factorial aerobic scopes are provided, which is the difference between maximum metabolic rate (MMR) during the trial (highest black dot) and the standard metabolic rate (SMR, grey dots), divided by SMR. Note the different y-axis scales, and rapid MO_2 drop on trial completion (last three dots) indicating minimal excessive post-exercise oxygen consumption (EPOC).

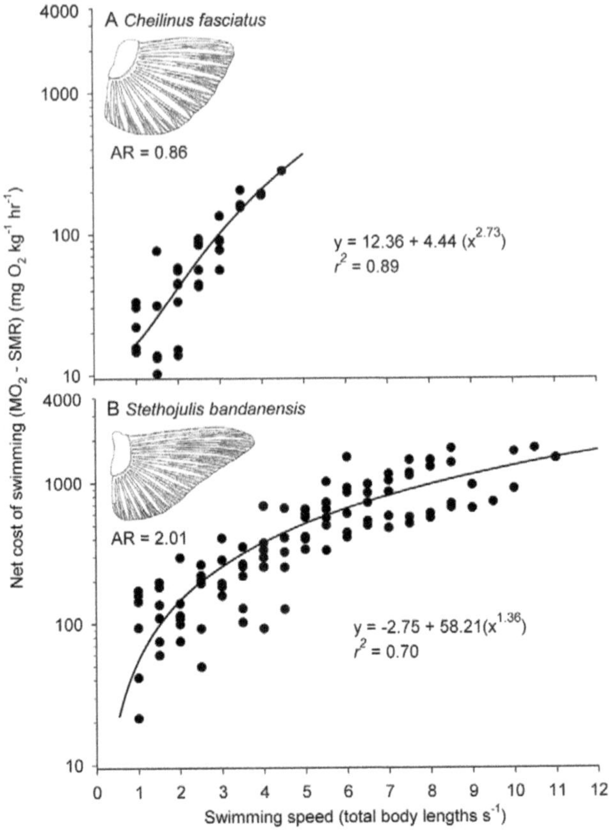

Figure 2. Net cost of swimming for two species of reef fish with alternate pectoral fin shapes, as indicated by the aspect ratio (AR), displayed different increases in oxygen consumption with speed as indicated by the semi-log plots for (A) *Cheilinus fasciatus* and (B) *Stethojulis bandanensis* fitted with the hydrodynamics-based power equation (MO_2 – SMR = a + b U ^{c}) [3]. A low exponent value (c) for *S. bandanensis* indicates a very high efficiency of locomotion with increasing speed compared to other bony fishes (Table 1).

the leading fin edge squared, divided by the fin area taken from digitized fin images following [7,15,16]) and similar demersal (reef-associated) microcarnivore lifestyle, these two species span the paddle-shaped (aspect-ratio of 0.86 in *C. fasciatus*) to wing-shaped (2.01 in *S. bandanensis*) fin shape extremes that have previously been linked to the distinct forms of resistance-based and lift-based labriform swimming (e.g. as seen in *Pseudocheilinus octotaenia* with AR of 0.76, and *Gomphosus varius* with AR of 1.75), respectively [7,12,15,16,17]. Metrics of speed and energetic performance for these two reef fish species were then compared with published values for other non-scombrid and scombrid (tuna-like) fishes to contextualize labriform swimming within the high speed performance of pelagic fishes such as tunas. Performance metrics for this comparison included the rate of increased energy consumption with speed, the gross cost of transporting their body weight one metre per second (GCOT in Joules of energy, [3,5,6]), and the optimal swimming speed (U_{opt}) at which GCOT is minimized for each species [3,5,6].

Materials and Methods

This study was carried out in strict accordance with the protocols approved by The Australian National University Animal Experimentation Ethics Committee (F.BTZ.03.06) for this specific project. All efforts were made to minimize animal suffering through careful collection, handling, and swimming trials based upon the natural rheotaxic behaviour and self-motivation of individuals. Bluelined wrasse *Stethojulis bandanensis* (n = 7, mean ± SE mass = 15.8±1.1 g, total body length 10.1±0.2 cm, body depth 2.6±0.01 cm, body width 1.4±0.01 cm) and Redbreasted wrasse *Cheilinus fasciatus* (n = 7, 34.2±7.7 g, 12±0.9 cm, 3.8±0.03 cm, 1.6±0.01 cm) were collected from the wild and tested for their swimming performance at the Lizard Island Research Station. Individuals were hand-collected during September 2008 from reefs around Lizard Island, Great Barrier Reef by divers on SCUBA (Self-Contained Underwater Breathing Apparatus) using ultra-fine barrier nets. Transported to the Lizard Island Research Station within 2 hours of capture, fish were maintained in flow-through seawater aquaria at ambient temperatures (27–28°C) and fasted for 36 hours prior to their swimming trial to maximize energy available for swimming and minimize potential respirometer fouling. All individuals were swum within 3 days of capture.

Rate of oxygen consumption (MO_2) by each fish was measured with a computerized, intermittent-flow respirometry system [18,19]. The clear Perspex respirometer consisted of an 8.3-Litre recirculating flow tank entirely submerged within a 31-Litre

Table 1. Comparative energetic performance of bony fish swimming via pectoral (families Labridae & Scaridae) and caudal (Scombridae, Carangidae, Salmonidae) propulsion, ranked by relative efficiency.

Species	Common name	Family	Factorial Aerobic Scope	Relative Efficiency
Stethojulis bandanensis	Bluelined wrasse	Labridae	22.0	1.00
Thunnus albacares [6,22]	Yellowfin tuna	Scombridae	9.0	0.83
Scarus schlegeli [8]	Yellowband parrotfish	Scaridae	4.3	0.82
Trachurus trachurus [23]	Horse mackerel	Carangidae	9.7	0.53
Onchorynchus nerka [6,24]	Sockeye salmon	Salmonidae	8.2	0.53
Cheilinus fasciatus	Redbreasted wrasse	Labridae	2.7	0.50
Dicentrarchus labrax [25]	European seabass	Moronidae	2.8	0.50

Temperatures reflected wild conditions and ranged over 20–28°C for these measurements. Published data sources are indicated.

aerated seawater bath maintained at ambient temperature (27–28°C). The flow tank could be alternately sealed or flushed with water from the bath via a computer-actuated pump, which allowed high oxygen levels (>80% saturation) to be maintained in the flow tank throughout each swimming trial. The body dimensions and fin spans of all fish (mean ± SE fin span of 4.7±0.03 cm for *S. bandanensis* and 4.6±0.01 cm for *C. fasciatus*) fitted well within the dimensions of the flow tank working section (width 9 cm, depth 11 cm, length 26 cm), which was calibrated by vane-wheel flow probe (Höntzsch GmbH, Waiblingen, Germany) against the voltage output of the propeller motor controller. Honeycomb collimators and curved baffles were used to produce a smooth laminar flow in the working section of the flow tank. Fish were monitored continuously throughout each trial to ensure they swam in this calibrated flow without any use of the walls or corners. Temperature inside the flow tank was maintained at 27.7±0.1°C via a computer-actuated cooling coil housed downstream of the working section. Oxygen partial pressure (Po_2) in the flow tank was measured with an optic-fibre sensor and oxygen meter (Fibox 3, Presens GmbH, Regensburg, Germany) linked to a computer to continuously record Po_2 measurements in real time. The respirometer was set to periodically flush the flow tank with aerated water from the bath for 6 minutes, followed by a 2-minute closed mixing period then 7 minutes of closed respirometry when oxygen measurements were taken at a rate of 1 s^{-1} to yield an MO_2 at each set flow speed.

Each fish was measured for body depth, width, mass and total length before placement in the respirometer flow tank to calculate solid blocking effects, where they were allowed to acclimate for a minimum of 5 hours at a flow speed of 0.5 total body lengths s^{-1} (4.7–7.9 cm s^{-1}). This initial speed provided adequate water mixing whilst allowing the fish to rest on the flow tank bottom without swimming. Oxygen measurements began immediately at the set flushing cycle described above, with these MO_2 measurements at 0.5 total body lengths s^{-1} used to calculate standard metabolic rate (SMR, MO_2 at zero swim speed). Flow speed was then incrementally increased by 0.5 total body lengths s^{-1} every 15 minutes and MO_2 recorded for each set speed. Incremental speed increases were continued until the fish could no longer hold position and became impinged on the downstream grid. To ensure trials measured aerobic locomotion, at the conclusion of each trial, flow speed was reduced to 1 total body length s^{-1} and MO_2 was recorded through an additional two flushing cycles. Any post-trial repayments of excessive post-exercise oxygen consumption (EPOC, required to repay any oxygen debt incurred from anaerobic activity in a trial) were measured by comparison of post-trial MO_2 against MO_2 at 1 total body length s^{-1} at the beginning of the trial. After the fish was removed, the flow tank was resealed and oxygen consumption in the empty respirometer measured to determine background levels, which were subtracted from the MO_2 values for the fish swum previously (average ± s.d. backgrounds were 116.7±80.7 mg O_2 h^{-1}, n = 14).

Oxygen consumption rate (MO_2) was determined from the slope of the linear regression of the Po_2 decline over time for each measurement cycle, using the formula [8]: $MO_2 = sV_{resp}\alpha$ where s is the slope, V_{resp} the volume of the respirometer minus the volume of the fish (calculated from body mass), and α is the solubility of oxygen in seawater. Only measurements with a regression coefficient of determination (r^2) greater than 0.94 were used in the analyses. SMR was calculated from a frequency histogram of the raw MO_2 data within the acclimation period (excluding the first four values immediately after the fish was introduced), by fitting two normal curves to separate the SMR peak, when the fish was at rest, from higher MO_2 observations that corresponded with spontaneous activity [20]. Swimming speeds were corrected for solid-blocking effects (minimal for both species, with all fish occupying less than 7% of the flow tank cross-sectional area) following [21] and calculated on-line in real time so that corrected swim speeds were used in the swimming trial settings. The relationship between swimming speed and net cost of swimming was then described using the hydrodynamics-based power equation [3]: Net cost of swimming (MO_2 − SMR) = a + b U^c, where a, b and c are constants and U is corrected swimming speed in total body lengths per second. This power equation was then transformed following [3] to estimate the optimal swimming speed (U_{opt}): $U_{opt} = [(a+SMR)/(c-1)b]^{1/c}$, where U_{opt} is the speed that minimizes energy expenditure per unit of travel distance. Finally, the gross cost of transport (GCOT, total joules of energy used to move one Newton of body weight one metre) was calculated for each individual by first multiplying total MO_2 consumption at U_{opt} by the conversion factor 14.1 J mg^{-1} O_2 and then using the equation [3,5,6]: GCOT = $MO_{2(opt)}\,(g\,U_{opt})^{-1}$, where $MO_{2(opt)}$ is the gross rate of oxygen consumption at U_{opt} (converted to units of J kg^{-1} hr^{-1}), g the acceleration due to gravity (9.8 m s^{-2}), and U_{opt} as calculated above (but converted to units of m hr^{-1}).

Comparative performance of our study species was examined in the context of a range of bony fishes using published energetic estimates, either taken directly from reported figures in each paper (which often required conversion to the same units stated above) or calculated from fitted equations. Firstly, the exponent in the hydrodynamics-based power equation was used to provide a size-independent measure locomotion efficiency (rate of increase in

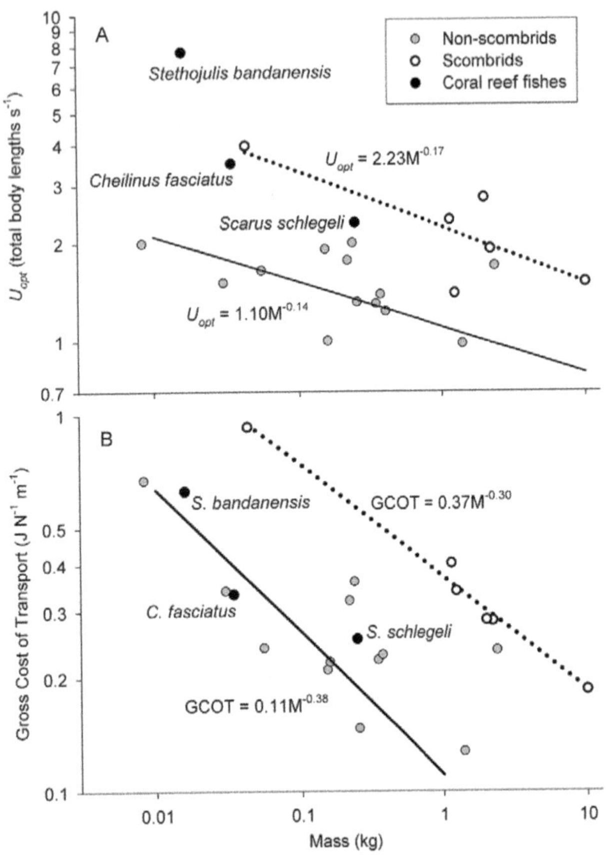

Figure 3. Comparative energetic swimming performance of bony fishes. Three coral reef and sixteen other non-scombrid fishes are presented alongside six scombrid (tuna-like) fishes of varying body mass on log-log plots of (A) Optimum swimming speed (U_{opt}) and (B) Gross cost of transport (GCOT) incurred by each species at their optimum swimming speed (U_{opt}). Dotted and solid lines denote mass-U_{opt} and mass-GCOT power functions [after 3, 6] for scombrid and non-scombrid fishes, respectively (note the range of temperatures encompassed within the underlying data in Table S1 [5,8,22–24,26–32], which reflect the conditions experienced by each species performing in the wild). Note *S. bandanensis* is well above the scombrid U_{opt} trend (A, dotted line), but has the same (much lower) GCOT as similar-sized non-scombrid fishes swimming four times slower (B, solid line).

energy consumption with increasing swimming speed) among a range of fishes. Relative efficiency was calculated by taking the lowest exponent value for a species (1.36 for *S. bandanensis*) and denoting this as 1.00, then diving this by the exponent value for each other species. Secondly, U_{opt} for our study species was placed within a log-log plot of U_{opt} versus body mass for a range of six scombrid (tuna-like) and seventeen other non-scombrid fishes (including one coral reef fish) following [3] and [6] (Table S1). Finally, the relationship between gross cost of transport (GCOT, J N^{-1} m^{-1}) and body mass was used to examine the overall energetic cost incurred by each species swimming at U_{opt} following [3].

Results

Our study species displayed swimming speeds and energetic profiles at two opposite extremes of the possible performance spectrum for labriform locomotion. At the high extreme, *S. bandanensis* displayed a rate of oxygen consumption ranging from 231 mg O_2 kg^{-1} hr^{-1} at rest, to over 5,300 mg O_2 kg^{-1} hr^{-1} at their maximum swimming speed (over 1 metre s^{-1}, Fig. 1), which is a factorial aerobic scope 22-fold above resting. By comparison, *C. fasciatus* displayed a factorial aerobic scope of just 2.7-fold (Fig. 1). Moreover, *S. bandanensis* exhibited a very low rate of increase in aerobic metabolism with increasing swimming speed when compared to the related *C. fasciatus* (Fig. 2). As such, *S. bandanensis* displayed the most efficient swimming speed performance profile known for any bony fish swimming via either a labriform or body-caudal gait (represented by the exponent c in Table 1). Indeed, when embedded within the energetic performance of other bony fishes examined to date (23 species from 11 families), *S. bandanensis* is capable of maintaining fast cruising speeds without the high energetic cost incurred by high-speed tunas and other scombrid fishes of similar body size (Table 1 and Fig. 3). When swimming at a speed that minimizes cost of transport, *S. bandanensis* can maintain 7.7 total body lengths s^{-1}, which is more than 70% faster than the optimum speeds displayed by other fishes of similar body size (2.0 to 4.5 body lengths s^{-1}, Table S1, Fig. 3A). Moreover, *S. bandanensis* can maintain such high optimum swimming speeds without an increase in the gross cost of transport beyond what is seen in a bony fish of similar body mass; tuna-like fishes swimming at similarly high cruising speeds incur up to 40% higher costs of transport than *S. bandanensis* (Fig. 3B). More generally, we see the three labriform-swimming coral reef species examined to date all display faster than expected optimum swimming speeds for their size and cost of transport (Fig. 3).

Discussion

Challenging long-held notions about the costs of underwater locomotion [1–4], we find coral reef fishes using labriform locomotion can maintain fast swimming speeds without the elevated cost of transport that has been seen in tunas and other scombrid fishes swimming at such high cruising speeds [5,6]. Notably, this performance is reflective of their daily activities in the wild, where *S. bandanensis* and other wing-finned species swim at similar or faster speeds (up to 1 metre s^{-1}) while they forage across the reef [12,15,16]. Understanding the traits that drive such energetic performance in these coral reef fishes can yield important insights for their response to changing environmental conditions, and the technological benefits that could be gained from biomimicry of flapping fin propulsion.

Extreme performance in *S. bandanensis* seems to arise from mechanical efficiency; the comparative performance of *S. bandanensis* and *C. fasciatus* matched predictions based on the relative thrust-producing efficiency of their different pectoral fin shapes and movements [7,9–11,13,15]. Such biomechanical and kinematic evidence indicates that *S. bandanensis* achieves high-speed efficiency by sweeping their wing-like (high AR) fins in a figure-eight pattern to produce lift-based thrust on all fin strokes. Rowing paddle-shaped pectoral fins (low AR) in the power-recovery stroke cycle means that *C. fasciatus* produces thrust only half the time [7,9].

Energetic performance by *S. bandanensis* is brought into extreme relief by comparison with a range of vertebrates spanning multiple families and forms of locomotion. For instance, *S. bandanensis* displayed the widest factorial scope of active metabolism ever seen in a bony fish (up to 22 times above resting, cf. 9 times in Yellowfin tuna [6]) [3,6,12,13,33], and is comparable to the metabolic scope of active endothermic birds and mammals (typically ranging from 8 to 36 times above resting, [34]). Akin to the high-performing hummingbirds [34], the enormous metabolic scope of this coral reef fish allows maintenance of a wide range of swimming speeds for prolonged time periods.

Furthermore, *S. bandanensis* increased their swimming speed with the lowest rate of increased energy consumption seen in a swimming fish. Ultimately, it means this species can optimally travel (i.e. incur a minimum cost of transport) at a speed four times faster than non-scombrid fishes of similar size, without incurring the increased energy use typically seen in tuna-like fishes swimming at this speed. It is interesting to note that other coral reef fishes using labriform swimming also display faster than expected optimal swimming speeds for their size, with similarly low costs of transport, suggesting this may be a generalized feature of pectoral-based propulsion. Collectively, these coral reef fishes adopt oscillatory fin strokes around a rigid-body posture, which is thought to bring hydrodynamic drag down to minimum [1,8,14]. When combined with the mechanical efficiency of using a flapping fin stroke to produce lift-based propulsion [1,7,9–11,13,15], the result is very fast sustained speeds in species with wing-like, high aspect-ratio fins. Matching flexibility with efficiency, *S. bandanensis* currently stands out as the highest performing swimmers for their size with respect to measures of scope, optimum swimming speed and energy consumption.

Extremes in energetic performance appear to be linked with ecological extremes in these coral reef fishes. For species with low speed efficiency, such as *C. fasciatus*, we tend to find these fish occupying calm water habitats sheltered from incident wave energy and storms. In contrast, *S. bandanensis* and other wing-finned fish species tend to be in great abundance in habitats subject to highly variable and extreme water flows generated by direct exposure to wave energy [13,15,16]. Under such hydrodynamic conditions, high efficiency across a wide range of speeds may be a physiological imperative for species to occupy these challenging, but food-rich habitats [13,15,16]. Such extremes may also place these species in good stead for future changes in the marine environment arising from climate change. Both thermal and hydrodynamic conditions appear to be increasingly intense and variable in marine habitats around the globe [35,36]. With a flexible and wide scope of aerobic metabolism, *S. bandanensis* appears to be equipped with the physiological traits needed to maintain their metabolic delivery of energy across a wide range of activities and hydrodynamic conditions. While *S. bandanensis* may be able to ride through such challenges, species such as *C. fasciatus* and other coral reef fishes of low aerobic scope [8,33,37] may require rapid adaptation in order to survive any increases in the intensity and variability of environmental conditions [33–37].

Such functional innovation in swimming prompts a rethink of the possibilities and limits to the cost of high speed performance in underwater locomotion. In fishes, we see that relative swimming performance is not necessarily related to their perceived lifestyle, with reef fishes capable of exceptional swimming performance despite their supposed sedentary, reef-associated existence [12–16]. Achieved through the use of drag-minimizing rigid-body posture, and exploiting lift-based forces from oscillating fins, these species provide some key lessons for the field of biomimetics. We are just starting to see man-made submersibles with propulsive fins reminiscent of the pectoral fins in coral reef fishes, turtles, marine birds and mammals (e.g. *Madeleine* robot turtle [9–11,38]). By incorporating a lift-based oscillatory fin movement into such technology, dramatic reductions could be achieved in the power needed to propel autonomous underwater vehicles of similar size to the fish and other aquatic animals that use this mechanism [9,11–13,38].

Supporting Information

Table S1 Meta-data for comparative analysis of the energetic swimming performance of reef fishes, scombrid, and non-scombrid fishes.

Acknowledgments

This study was conducted at *Jiigurru*, traditional sea country of the Dingaal people. We thank the Lizard Island Research Station staff for field and laboratory assistance, M. Westneat and an anonymous reviewer for helpful comments, and M. Noble for illustrations.

Author Contributions

Manuscript editing: CJF JLJ JFS. Conceived and designed the experiments: CJF JLJ JFS. Performed the experiments: CJF JLJ JFS. Analyzed the data: CJF JFS. Contributed reagents/materials/analysis tools: CJF JFS. Wrote the paper: CJF.

References

1. Schmidt-Nielson K (1972) Locomotion: energy cost of swimming, flying, and running. Science 177: 222–228.
2. Vogel S (1994) Life in moving fluids: the physical biology of flow. Princeton: Princeton University Press.
3. Videler JJ (1993) Fish swimming. New York: Chapman and Hall.
4. Alexander RM (2005) Models and the scaling of energy costs for locomotion. J Exp Biol 208: 1645–1652.
5. Videler JJ, Nolet BA (1990) Costs of swimming measured at optimum speed: scale effects, differences between swimming styles, taxonomic groups and submerged and surface swimming. Comp Biochem Physiol 97A: 91–99.
6. Korsmeyer KE, Dewar H (2001) Tuna metabolism and energetics. In: Block B, Stevens ED, editors. Tuna: physiology, ecology and evolution. London: Academic Press. 35–78.
7. Walker JA, Westneat MW (2002) Kinematics, dynamics, and energetics of rowing and flapping propulsion in fishes. Integr Comp Biol 42: 1032–1043.
8. Korsmeyer KE, Steffensen JF, Herskin J (2002) Energetics of median and paired fin swimming, body and caudal fin swimming, and gait transition in parrotfish (*Scarus schlegeli*) and triggerfish (*Rhinecanthus aculeatus*). J Exp Biol 205: 1253–1263.
9. Fish FE (1996) Transitions from drag-based to lift-based propulsion in mammalian swimming. Amer Zool 36: 628–641.
10. Baudinette RV, Gill P (1985) The energetics of 'flying' and 'paddling' in water: locomotion in penguins and ducks. J Comp Physiol B 155: 373–380.
11. Wyneken J (1997) Sea turtle locomotion: mechanics, behaviour, and energetics. In: Lutz PL, Musick JA, editors. The biology of sea turtles. Boca Raton: CRC Press. 165–198.
12. Fulton CJ (2007) Swimming speed performance in coral reef fishes: field validations reveal distinct functional groups. Coral Reefs 26: 217–228.
13. Fulton CJ (2010) The role of swimming in reef fish ecology. In: Domenici P, Kapoor BG, editors. Fish locomotion: an eco-ethological perspective. Enfield: Science Publishers. 333–373.
14. Webb PW (1994) The biology of fish swimming. In: Maddock L, Bone Q, Rayner JM, editors. Mechanics and physiology of animal swimming. Cambridge: Cambridge University Press. 45–62.
15. Wainwright PC, Bellwood DR, Westneat MW (2002) Ecomorphology of locomotion in labrid fishes. Environ Biol Fish 65: 47–62.
16. Fulton CJ, Bellwood DR, Wainwright PC (2005) Wave energy and swimming performance shape coral reef fish assemblages. Proc Roy Soc Lond B 272: 827–832.
17. Walker JA, Westneat MW (2002) Performance limits of labriform propulsion and correlates with fin shape and motion. J Exp Biol 205:177–187.
18. Steffensen JF, Johansen K, Bushnell PG (1984) An automated swimming respirometer. Comp Biochem Physiol 79A: 437–440.
19. Steffensen JF (1989) Some errors in respirometry of aquatic breathers: how to avoid and correct for them. Fish Physiol Biochem 6: 49–59.
20. Steffensen JF, Bushnell PG, Schurmann H (1994) Oxygen consumption in four species of teleosts from Greenland: no evidence of metabolic cold adaptation. Polar Biol 14: 49–54.
21. Bell WH, Terhune LDB (1970) Water tunnel design for fisheries research. J Fish Res Bd Can Tech Rep No. 195.
22. Dewar H, Graham JB (1994) Studies of tropical tuna swimming performance in a large water tunnel. I. Energetics. J Exp Biol 192: 13–31.
23. Wardle CS, Soofiani NM, O'Neill FG, Glass CW, Johnstone ADF (1996) Measurements of aerobic metabolism of a school of horse mackerel at different swimming speeds. J Fish Biol 49: 854–862.

24. Brett JR (1965) The relation of size to rate of oxygen consumption and sustained swimming speed of sockeye salmon (*Oncorhynchus nerka*). J Fish Res Bd Can 22: 1491–1501.
25. Claireaux G, Couturier C, Groison AL (2006) Effect of temperature on maximum swimming speed and cost of transport in juvenile European sea bass (*Dicentrarchus labrax*). J Exp Biol 209: 3420–3428.
26. Clark TD, Seymour RS (2006) Cardiorespiratory physiology and swimming energetics of a high-energy-demand teleost, the yellowtail kingfish (*Seriola lalandi*). J Exp Biol 209: 3940–3951.
27. Fitzgibbon QP, Strawbridge A, Seymour RS (2007) Metabolic scope, swimming performance and the effects of hypoxia in the mulloway, *Argyrosomus japonicus* (Pisces: Sciaenidae). Aquacult 270: 358–368.
28. Herskin J, Steffensen JF (1998) Energy savings in sea bass swimming in a school: measurements of tail beat frequency and oxygen consumption at different swimming speeds. J Fish Biol 53: 366–376.
29. Sepulveda CA, Dickson KA, Graham JB (2003) Swimming performance studies on the eastern Pacific bonito *Sarda chiliensis*, a close relative of the tunas (family Scombridae). I. Energetics. J Exp Biol 206: 2739–2748.
30. Gooding RM, Neill WH, Dizon AE (1981) Respiration rates and low-oxygen tolerance limits in skipjack tuna, *Katsuwonus pelamis*. Fish Bull 79: 31–48.
31. Graham JB, Lowell WR, Lai NC, Laurs RM (1989) O_2 tension, swimming velocity, and thermal effects on the metabolic rate of the Pacific albacore *Thunnus alalunga*. Exp Biol 48: 89–94.
32. Sepulveda CA, Dickson KD (2000) Maximum sustainable speeds and cost of swimming in juvenile Kawakawa tuna (*Euthynnus affinis*) and chub mackerel (*Scomber japonicus*). J Exp Biol 203: 3089–3101.
33. Johansen JL, Jones GP (2011) Increasing ocean temperature reduces the metabolic performance and swimming ability of coral reef damselfishes. Glob Change Biol 17: 2971–2979.
34. Bishop CM (1999) The maximum oxygen consumption and aerobic scope of birds and mammals: getting to the heart of the matter. Proc Roy Soc B 266: 2275–2281.
35. Hein AM, Keirsted KJ (2012) The rising cost of warming waters: effects of temperature on the cost of swimming in fishes. Biol Lett 8: 266–269.
36. Young IR, Zieger S, Babanin AV (2011) Global trends in wind speed and wave height. Science 332: 451–455.
37. Donelson JM, Munday PL, McCormick MI, Nilsson GE (2011) Acclimation to predicted ocean warming through developmental plasticity. Glob Change Biol 17: 1712–1719.
38. Long JHJr, Schumacher J, Livingston N, Kemp M (2006) Four flippers or two? Tetrapodal swimming with an aquatic robot. Bioinsp Biomim 1: 20–29.

Effect of Ensiled Mulberry Leaves and Sun-Dried Mulberry Fruit Pomace on Finishing Steer Growth Performance, Blood Biochemical Parameters, and Carcass Characteristics

Zhenming Zhou[1,2], Bo Zhou[1,2], Liping Ren[1,2], Qingxiang Meng[1,2]*

1 State Key Laboratory of Animal Nutrition, China Agricultural University, Beijing, P. R. China, 2 College of Animal Science and Technology, China Agricultural University, Beijing, P. R. China

Abstract

Fifty-one Simmental crossbred steers (357.0±16.5 kg) were used to compare a standard total mix ration (TMR) with variants on animal performance, ruminal fermentation, blood biochemical parameters, and carcass characteristics. Corn grain and cotton seed meal were partially replaced by ensiled mulberry leaves (EML) or sun-dried mulberry fruit pomace (SMFP). Experimental diets had similar amounts of crude protein (CP), acid detergent fiber (ADF), and metabolizable energy (ME). Animals were divided into three groups: control group (CONT), 8% EML group, and 6.3% SMFP group. Performance, including average daily weight gain (ADG), and dry matter intake (DMI), was measured. Blood and rumen samples were collected at the end of the experiment (16 weeks). There were no differences in final body weight ($P=0.743$), ADG ($P=0.425$), DMI ($P=0.642$), or ADG/DMI ($P=0.236$) between the groups. There were no differences ($P=0.2024$) in rumen pH values; ammonia N was lower ($P=0.0076$) in SMFP than in the EML and CONT groups. There were differences in the concentrations of total and individual volatile fatty acids, while no differences were determined in blood biochemical parameters (i.e., plasma glucose, urea concentrations, triglycerides, total protein, insulin, IgG, alanine transaminase, and aspartate aminotransferase, $P \geq 0.098$). No differences were observed in carcass characteristics ($P \geq 0.513$), tenderness ($P=0.844$), adipose and lean color values ($P \geq 0.149$), and chemical composition ($P \geq 0.400$); however, intramuscular fat was lower in the EML and SMFP groups compared to the CONT animals ($P=0.034$). In conclusion, diets supplemented with these two mulberry products in an isocaloric and isonitrogenous manner have similar effects to corn grain and cotton seed meals on steer performance, blood biochemical parameters and carcass characteristics, with the exception of ruminal VFA concentrations and lower intramuscular fat content.

Editor: Gunnar Loh, German Institute of Human Nutrition Potsdam-Rehbrücke, Germany

Funding: This study was funded by the National Nature Science Foundation of China (31101738), China Agricultural Research System (CARS-38), Chinese Universities Scientific Fund (2013QJ071, 2012 QJ108) and National Scientific Programme (2011CB302705). The funders had no role in study design, data collection and analysis, decision to publish, or preparation of the manuscript.

Competing Interests: For this study Jinwei Xumu Ltd. fed the animals and grew the mulberry leaves for the conduct of these experiments. There are no patents, products in development or marketed products to declare.

* E-mail: qxmeng@cau.edu.cn

Introduction

There is an increased demand for animal products in China and other developing countries, thereby increasing the demand for adequate and inexpensive animal feed. Scientists have developed feeding strategies that guarantee the sustainability of livestock production systems based on cost-effective alternatives and local feed sources. Several reports have shown that food and agricultural byproducts can be used as animal feed to replace cereal-based concentrates without negatively affecting animal production performance, such as ensiled and dried apple pomace which is used for finishing lambs [1], wet tomato pomace in lactating dairy cows [2], and tomato and cucumber waste fruits in dairy goats [3]. Mulberry trees are fast-growing deciduous plants, which are cultivated in many regions of the world. In China, mulberry cultivation areas are estimated to cover more than 10^6 ha [4], and the biomass yield of fresh mulberry leaves is approximately 25~30 tonnes/ha/year. Mulberry has been used as feed for silk worms for hundreds of years. Furthermore, it is used in Chinese herbal medicine for hyperlipemia and diabetes treatments due to its antioxidant, antimicrobial, and anti-hyperlipidemic properties [5–8]. Studies have also shown that mulberry trees are potential protein sources for herbivores [9,10]. The addition of mulberry leaves to lamb feed reduces the need for expensive protein supplements [11]. One previous study on the rumen and gastrointestinal tract digestibility of mulberry leaves showed that it had similar digestible energy and crude protein values as alfalfa hay [12]. Compared with dry processing, mulberry leaf silage has several advantages for animal feeding. For example, the weather during harvesting need not be as fair and dry as when dried mulberry leaves are harvested. During fermentation, the silage bacteria acts on carbohydrates in the leaves to produce lactic, acetic, propionic, and butyric acid, which act as natural preservatives for mulberry leaves. This preservative method is particularly important during winters, when green forage is unavailable.

Mulberry pomace, a byproduct of the production of mulberry juice, which consists mainly of peels and stems, accounts for approximately 8% of the fresh weight of the mulberry. Typically, mulberry pomace is used as a soil enhancer or is otherwise dumped in landfills, which contributes to environmental problems. The results of our previous study revealed that mulberry fruit pomace is an excellent potential feed source for ruminal animals [13]. Even though several studies have focused on the nutritive and antioxidant value of mulberry leaves and mulberry fruit pomace, little is known about the effects of mulberry leaves and mulberry fruits on the blood biochemical parameters and carcass characteristics of finishing steers.

Our hypothesis is that EML and SMFP can be used in the diets for finishing steers without negatively affecting their productive performance. To validate this assumption, we evaluated the effects of feed supplementation with these two mulberry products in an otherwise isocaloric and isonitrogenous diet. Corn grain and cotton seed meal were partially replaced by EML or SMFP, and the performance, rumen fermentation, blood biochemical parameters, and carcass characteristics of finishing steers was assessed.

Materials and Methods

Animals and Diets

The animals used in this study were handled in strict accordance with the Regulations for Laboratory Animals of Beijing. The protocol was approved by the Animal Welfare Committee of the China Agricultural University (Permit Number: DK1008). In this study, 51 medium-frame crossbred Simmental steers (357.0±16.5 kg; 15 months old) were used and housed in a tie-stall facility. An adjustment period of 2 weeks allowed the steers to become acclimated to routine feeding and to allot time for proper diet adjustment before the experiment. During the adjustment period, experimental diets were gradually fed to the animals. At the end of the 2-week period, the animals were fed the experimental diets. The basal diet met the NRC (2000) requirements for 350-kg beef cattle with weight gain of 1.2 kg/d.

Mulberry leaves were harvested from a farm (Beijing, China) during July, 2010. The harvested mulberry material was ensiled without additives after chopping. The mulberry silage was then used for the feeding experiment after being stored for 50 days; mulberry fruit pomace was purchased from a local company (Guosen Co., Beijing, China). The chemical composition and silage fermentation characteristics of EML or SMFP are shown in Table S1. The animals were divided into three treatment groups: the CONT group (n = 17), which received standard TMR; the EML group (n = 17), which received a dietary ingredient of ensiled mulberry leaves; or the SMFP group (n = 17), which received a dietary ingredient of sun-dried mulberry fruit pomace (Table 1). Animals were fed twice daily at approximately 0800 and 1700 and had ad libitum access to water and a trace mineral salt block. Diet refusals were recorded every day; the amounts of feed were adjusted every 3 to 4 d to maintain the preferred daily diet refusal rate of 5~10%. Steer body weight was monitored to assess growth performance during the 16-week trial period.

Table 1. Ingredients and nutrient composition of the experimental diets.

Item	Experimental diet[1]		
	CONT	EML	SMFP
Ingredient composition, %(DM basis)			
Corn silage	40.0	40.0	40.0
Brewer's grain	20.0	20.0	20.0
Corn	30.2	27.0	24.5
Cotton seed meal	7.8	3.0	7.2
Ensiled mulberry leaves	0.0	8.0	0.0
Sun-dried mulberry fruit pomace	0.0	0.0	6.3
Limestone powder	0.7	0.7	0.7
Calcium hydrogen phosphate	0.1	0.1	0.1
Sodium bicarbonate	0.5	0.5	0.5
Salt	0.5	0.5	0.5
Premix[2]	0.2	0.2	0.2
Nutrient composition, DM basis			
ME (Mcal/kg)	2.69	2.69	2.70
CP (%)	14.01	14.06	13.98
Ca (%)	0.56	0.57	0.55
P (%)	0.33	0.33	0.33
ADF (%)	21.42	22.56	21.51

[1]CONT = control; EML = ensiled mulberry leaves; SMFP = sun-dried mulberry fruit pomace.
[2]Supplied per kilogram of dietary DM: 15 mg of Cu, 65 mg of Zn, 28 mg of Mn, 0.7 mg of I, 0.2 mg of Co, 0.3 mg of Se, 6,000 IU of vitamin A, 600 IU of vitamin D, and 47 IU of vitamin E.
[3]ME = metabolizable energy; CP = crude protein; ADF = acid detergent fiber.

Sample Collection and Laboratory Analysis

Samples of dietary ingredients (EML and SMFP), three experimental diets, and diet refusals were collected every 2 weeks, dried at 72°C in an air oven, and ground in a Wiley Mill (A. H. Thomas Co., Philadelphia, PA) through a 1-mm screen. The dried samples were stored for chemical and nutrient content analysis. Briefly, samples were analyzed for amylase-treated ADF using the Ankom200 Fiber Analyzer (Ankom Technology Corp., Macedon, NY) according to previous study [14]; heat stable amylase (Ankom Technology Corp.) and sodium sulfite (Fisher Scientific, Waltham, MA) were used. Nitrogen content of three treatments and diet refusals were analyzed with a Nitrogen Analyzer (Model Rapid N III, Elementar, Analysis Systems GmbH, Germany); 6.25 was used as the conversion factor to obtain crude protein (CP) values. When there were differences in the nutritional values of the diet between the measured and initial values, the dietary formula was adjusted according to the measured nutrient values.

Blood samples (10 mL) were collected 2 hours after morning feeding at 10:00 on the final feeding week by jugular venipuncture into a single 10-mL vacuum tube containing Na-heparin. Samples were placed on ice for transport to the laboratory and centrifuged at 5,000 × g for 20 min at 4°C, and collected plasma were frozen at –20°C until analysis. Plasma concentrations of glucose, urea, triglycerides, total protein, alanine transaminase, and aspartate aminotransferase were determined using an automated enzymatic colorimetric principle on a Cobas Integra 400 Instrument (Roche Diagnostics) as described in our previous study [15].

After the steers were slaughtered, samples (approximately 500 mL, mix of liquid and solid) from the dorsal, central, and ventral regions of the rumen were collected to form one composite sample, and then strained the sample through four layers of cheesecloth for rumen fluid collection. The pH of the rumen fluid was measured immediately using a Model PHS-3C pH meter (Shanghai Leici Scientific Instrument Co., Ltd., China). To the filtrates (10 mL), 1 mL of 50% (v/v) HCl was added for ammonia N analysis, and 1 mL of a metaphosphoric acid (187.5 g/L) and formic acid (250 mL) solution was added for volatile fatty acids or VFA determination [16]. Ammonia N concentration in the filtrates was determined with a UV-VIS 8500 spectrophotometer (Shanghai Tianmei Scientific Instrument Co., Ltd., China). VFAs were quantified using a gas liquid chromatograph (GLC, 6890 N model, Agilent Technologies) with a capillary column (30 m×0.32 mm *i.d.*; 0.50 μm phase thickness) [17].

Steer Slaughter and Carcass Measurements

At the end of feeding trail, steers were starved for 24 h before they were loaded (at 0600 h) and transported 5 km to a commercial slaughterhouse. Efforts were made to minimize the suffering of steers. Briefly, wide ramps with slopes (11°) were used for the loading and unloading steers, and transport vehicles were equipped with non-slip flooring. Once a vehicle arrived at slaughterhouse, efficient scheduling procedures were implemented to ensure that steers were unloaded quickly. Pre-slaughter handling systems were used to encourage the smooth movement of stock. When steers were slaughtered, they were restrained in an upright position with their head held fast and the neck exposed in a suitable position for the incision of the throat. The knife used for cattle has a long, extremely sharp, and undamaged blade. The intention was to produce the immediate outpouring of blood by severing both jugular veins and both carotid arteries. Steer final BW was recorded prior to slaughter to determine dressing percentage. Hot carcass weight was obtained on the day of slaughter; carcasses were subsequently stored at 4°C for 48 h. Cold carcass weights and carcass shrink weights were measured. Carcasses were cut between the 12th and 13th ribs for assessment of carcass quality attributes and yield grade parameters. Carcass quality attributes included 12th-rib fat thickness, longissimus muscle (LM) area, intramuscular fat, and ultimate pH values. Carcass color was assessed by measuring L*, a*, and b* color values using a portable Minolta chromameter (Minolta Chroma Meter CR-400 colorimeter, Minolta Corp., Osaka, Japan) on the cut lean surface and carcass external fat along the lateral side of the carcass. Color readings were recorded in the L* (0 = black, 100 = white), a* (negative values = green, positive values = red), and b* (negative values = blue, positive values = yellow) color space (CIELAB); large pieces of connective tissue and intramuscular fat were avoided. Color saturation was calculated as described by the operational manual [18].

Meat Measurements

Warner-Bratzler shear force (WBSF) values were determined according to the AMSA (1995) guidelines. Following carcass data collection, strip loins were excised from the left side of each carcass, vacuum-packaged, and aged for 14 d at 4°C. After the aging period, 2.54-cm-thick steaks were made, vacuum packaged and frozen (−20°C) for later analysis. Steaks for WBSF were thawed for 24 h at 4°C and cooked to an internal temperature of 71°C in an open-hearth electric broiler (George Foreman GGR50, Salton Inc., Lake Forest, IL) with thermocouples inserted to approximately the geometric center of the steak. Cooking loss, expressed as a percentage of weight loss, was calculated from the ratio between the initial weight (before cooking) and the final weight (after cooking). At least six 1.27-cm-diameter pieces were removed from each steak parallel to the muscle fiber orientation [19]. The pieces were sheared perpendicular to the muscle fiber orientation using a shear device (Warner-Bratzler Meat Shear 2356X). Peak shear force values were expressed in kg; six pieces were analyzed.

Meat samples were trimmed of external fat and connective tissue for determination of ash, protein, moisture, and ether-extractable lipid. Briefly, crude protein content was calculated from the nitrogen content (%N × 6.25) analyzed by a Nitrogen Analyzer (Model Rapid N III, Elementar, Analysis Systems GmbH, Germany); ash content was determined by heating the steak sample at 550°C for 15 h; moisture content was determined by weight loss after freeze-drying at −55°C for 5 d; and lipid content was determined by the Ankom procedure (Am 5-04, AOCS, 2004). Analyses were performed in duplicate and mean values were used for statistical analysis.

Statistical Analyses

Data for animal growth performance, rumen fermentation, blood biochemical parameters, and carcass characteristics were analyzed using the GLM procedure (SAS Inst. Inc., Cary, NC). Diet was considered as a fixed effect in this model, and the animal was considered as random effect. Statistical differences between the mean values were tested using Duncan's multiple range test. Significant effects of the treatment were declared at $P<0.05$.

Results and Discussion

Feed Intake and Growth Performance

The chemical composition of the experimental diets is shown in Table 1. Crude protein, energy, and ADF contents were similar in all experimental diets. The low NDF and high crude protein contents in mulberry leaves make this forage a palatable feed for ruminants [20]. Therefore, silage mulberry leaves could improve livestock productivity by improving digestibility. Previous studies have shown that mulberry leaves benefit animal health and growth performance. Several bioactive compounds in mulberry leaves, including 1-deoxynojirimycin, γ-aminobutyric acid, and phenolic compounds, have strong antioxidant properties [21]. Additionally, mulberry fruits have antioxidative and anti-inflammatory properties [22], which suggests that mulberry silage and sun-dried mulberry fruits may be useful as a functional feed for improving the growth performance and health status of beef cattle.

The DMI, ADG, and FCR of the Simmental crossbred steers are summarized in Table 2. Ensilaged mulberry leaves had the same effect as the corn grain and cotton seed meal they replaced on animal growth performance ($P > 0.1$) after 16 weeks, which indicated that EML and SMFP could be used in finishing steers without negative effects on animal growth performance. Our results are similar to previous study, which reported that the partial replacement of cereals with mulberry leaves ($P<5\%$) had no effects on the DMI of ruminant animals [23]. Furthermore, there were no negative effects on ADG in growing cattle when cotton seed was partially replaced with fresh mulberry leaves [9]. Taking into account the feed cereal amount, average feed intakes, and current costs, a reduction of 5.65% and 5.94% in feeding costs may be achieved by using diets that consist of EML or SMFP, respectively (unpublished data).

Rumen Fermentation

The variables of rumen fermentation are summarized in Table 3. There were no differences ($P = 0.2024$) in the pH values of

Table 2. Growth performance of finishing steers fed a total mixed ration supplemented with ensiled mulberry leaves (EML) or sun-dried mulberry fruit pomace (SMFP).

Item	Experimental diet[1]			SEM	P-value
	CONT	EML	SMFP		
Animal number	17	17	17		
Initial body weight (kg)	353.6	360.5	356.75	5.60	0.812
Final body weight (kg)	500.2	495	491	11.68	0.743
ADG (kg/d)	1.22	1.19	1.20	0.14	0.425
DMI (kg/d)	7.96	8.15	8.29	0.211	0.642
ADG/DMI	0.153	0.141	0.145	0.083	0.236

[1]CONT = control; EML = ensiled mulberry leaves; SMFP = sun-dried mulberry fruit pomace.

Table 3. Rumen fermentation of finishing steers fed a total mixed ration supplemented with ensiled mulberry leaves (EML) or sun-dried mulberry fruit pomace (SMFP).

Item	Experimental diet[1]			SEM	P-value
	CONT	EML	SMFP		
pH value	5.51	5.65	5.71	0.08	0.2024
Ammonia N (mg/dL)	17.82^{a}	17.93^{a}	11.95^{b}	1.44	0.0076
Total VFA (mM)	83.16^{a}	83.87^{a}	70.65^{b}	3.88	0.0427
Individual VFA (mM/100 mM))					
Acetate	68.26^{b}	69.13^{b}	70.78^{a}	0.43	0.0015
Propionate	20.07^{a}	19.40^{a,b}	18.57^{b}	0.30	0.0085
Isobutyrate	0.77^{a,b}	0.80^{a}	0.63^{b}	0.05	0.0466
Butyrate	8.14	7.99	7.86	0.25	0.7361
Isovalerate	1.38^{a}	1.43^{a}	1.12^{b}	0.06	0.0053
Valerate	1.38^{a}	1.25^{a}	1.04^{b}	0.05	0.0008
Acetate/Propionate	3.41^{b}	3.58^{b}	3.82^{a}	0.07	0.0022

[1]CONT = control; EML = ensiled mulberry leaves; SMFP = sun-dried mulberry fruit pomace.
[a,b]Values in the same row with different superscripts are significantly different ($P<0.05$).

rumen fluid between the treatment groups. The rumen ammonia N concentration was lower ($P = 0.0076$) in the SMFP group than in the EML and CON groups. The VFA concentration in the rumen fluid was higher ($P = 0.0427$) in the EML and CON groups than in the SMFP group. There were differences in the concentrations of individual fatty acids between the treatment groups. The concentration of acetate was higher ($P = 0.0015$) in the SMFP group than in the CONT or EML groups. The concentration of propionate was lower ($P = 0.0085$) in SMFP than in CONT. The ratio of acetate to propionate was higher ($P = 0.0022$) in SMFP than in CONT and EML; however, the difference between CONT and EML was not statistically significant. There were no significant differences between the treatment groups in the concentrations of butyrate ($P = 0.7361$). The concentrations of isobutyrate, valerate, and isovalerate were lower ($P \leq 0.0466$) in SMFP than in CONT and EML; however, the difference between CONT and EML was not statistically significant. When animals are fed high concentrate diets, the amount of propionate produced increases relative to the acetate concentration, which results in increased glucose production and more available net energy for the animals. Increased propionate production also increases insulin secretion, which increases fat and protein synthesis and decreases fat and protein degradation [24].

SMFP and EML provided carbohydrates for ruminal bacteria, which utilized more ammonia N for microbial growth, and consequently reduced the amount of free ammonia N in the rumen (Table 3). No information is available on the effect of EML or SMFP on rumen fermentation. Our results revealed that SMFP decreased ruminal ammonia N concentrations and increased acetate/propionate ratios, indicating that SMFP has the ability to modify rumen fermentation and bacterial composition, and improve the efficiency of N metabolism in the rumen. Studies have reported that mulberry fruits are rich in polyphenols [25]. The presence of condensed tannins in the diet reduce soluble protein in the rumen and rumen ammonia N concentration, but increase non-ammonia N outflow from sheep rumen [26,27]. We proposed that polyphenols in SMFP had similar effects on protein degradation and non-ammonia N outflow.

Blood Biochemical Parameters

The effects of EML and SMFP on the blood biochemical parameters are shown in Table 4. There were no differences in plasma urea nitrogen between EML, SMFP, and CONT ($P > 0.05$). Plasma urea nitrogen is the end product of proteolysis and amino acid metabolism; its concentration is dependent on the crude protein level. Additionally, plasma urea nitrogen is negatively correlated with body nitrogen deposition and the utilization rate of proteins. High levels of casein in the diet increase plasma urea-N concentrations, whereas supplemental energy or glucose decrease plasma urea-N concentrations [28-30]. The lack of differences in plasma urea-N concentrations between the treatment groups indicates that there was probably no amino acid deficiency or imbalance in the EML and SMFP groups. Insulin is an anabolic hormone with key roles in glucose [31,32], lipid [33], and protein metabolism [34]. The stabilization of plasma insulin reflects a balance in protein and amino acid metabolism. There were no significant differences in plasma insulin concentrations between the three groups ($P = 0.491$; Table 4), indicating that EML and SMFP had the similar effect as corn grain and cotton seed meal on amino acid metabolism in steers. A previous study reported that insulin regulates feed intake and nutrient partitioning in ruminants [35]. Our result of the same level of insulin concentration is consistent, with no differences found in the DMI between the three treatments. Plasma glucose, albumin, cholesterol, triglycerides, HDL, LDL, and VLDL were not affected by the supplemented feeds ($P > 0.05$). Additionally, alanine transaminase and aspartate aminotransferase concentrations were not affected by treatments ($P > 0.05$), which indicated the absence of hepatic or cardiac dysfunction in the animals.

The addition of agricultural byproducts, including bioactive compounds, to animal feed affects the metabolic activity of finishing steers. For example, grape pomace modulate metabolic syndrome [36], wine byproducts lower plasma triacylglycerol and phospholipid concentrations [37]. The chronic intake of bioactive compounds can affect animal performance. Silage mulberry leaves and fruits contain several bioactive compounds, such as phenolic and flavonoid compounds [22,25], which increase humoral immunity, cell mediated immunity, and activate macrophage activity [38]. However, the results of our study showed no

Table 4. Blood biochemical parameters of finishing steers fed a total mixed ration supplemented with ensiled mulberry leaves (EML) or sun-dried mulberry fruit pomace (SMFP).

Item	Experimental diet[1]			SEM	P-value
	CONT	EML	SMFP		
Blood urea N (mmol/L)	3.85	3.41	3.67	0.055	0.469
Insulin (μU/mL)	1.35	1.46	1.45	0.021	0.491
Total protein (g/L)	78.34	79.57	74.21	0.044	0.856
Albumin (g/L)	32.87	34.89	32.42	0.190	0.696
Glucose (g/L)	5.21	5.35	5.11	0.456	0.143
Cholesterol (mmol/L)	4.21	4.02	4.34	0.267	0.456
Triglyceride (mmol/L)	0.33	0.31	0.34	0.493	0.098
HDL (mmol/L) [2]	2.64	2.41	2.35	0.562	0.251
LDL (mmol/L) [2]	0.95	1.13	0.98	0.345	0.667
VLDL (mmol/L) [2]	1.45	1.62	1.51	0.876	0.811
Alanine transaminase (U/L)	30.64	29.96	30.11	0.454	0.349
Aspartate aminotransferase(U/L)	35.11	36.23	35.64	0.659	0.289
IgG (g/L)	11.50	12.34	12.62	0.516	0.538

[1]CONT = control; EML = ensiled mulberry leaves; SMFP = sun-dried mulberry fruit pomace.
[2]HDL: High density lipoprotein; LDL: Low density lipoprotein; VLDL: Very low density lipoprotein.

differences in blood metabolism characteristics between the treatment groups could be attributed to two possible reasons: i) processing conditions of fermentation and sun-drying could reduce or degrade the bioactive compounds; and ii) the supplementation of <10% of the feed could not initiate obvious downstream effects in animal physiological systems.

Carcass Characteristics

The carcass characteristics of the three experimental groups are shown in Table 5. There were no differences between the EML, SMFP, and CONT groups ($P \geq 0.513$ to $P \geq 0.964$). Mean LM area ranged from 76.48 cm^2 (CONT) to 60 cm^2 (SMF); mean fat thickness over the 12th rib ranged from 0.29 cm (SMF) to 0.31 cm (CONT). Similar carcass characteristics revealed that there were no adverse effects by EML or SMFP. There were no differences in dressing percentage between EML, SMFP, and CONT (P = 0.973). Dressing percentage was 54.14% in EML and 54.89% in SMFP. Hot carcass weight and cold carcass weight were not affected by the treatments. Switching from a conventional diet to a partially substituted diet can change the amount of fat deposited in finishing steers, thereby affecting profitability and consumer acceptance. Energy density and feeding intensity are the main factors that affect carcass and intramuscular fatty acid composition [39]. Modifying diet energy density and increasing the supply of unsaturated fatty acids affect muscle fatty acid composition [40]. Although the three treatment groups had similar metabolizable energy (ME) in their diets, intramuscular fat content was lower in the EML and SMFP groups than in the CONT group (P = 0.034). The lower propionate concentrations in the rumen of the SMFP group probably contributed to the lower intramuscular fat content. Different fat contents between the treatment groups could be attributed to differences in feed chemical compositions. Therefore, the EML and SMFP groups may have received plant secondary compounds with anti-lipogenesis properties.

Table 5. Carcass characteristics of finishing steers fed a total mixed ration supplemented with ensiled mulberry leaves (EML) or sun-dried mulberry fruit pomace (SMFP).

Item	Experimental diet[1]			SEM	P-value
	CONT	EML	SMFP		
BW (kg)	500.2	495	491	11.68	0.743
HCW (kg)	272.0	269.5	271.3	10.56	0.591
Cold carcass weight (kg)	271.1	268.0	270.2	9.69	0.513
Dressing percentage (%)	54.19	54.14	54.89	0.58	0.973
12th-rib fat thickness (cm)	0.31	0.28	0.29	1.28	0.713
LMA[2] (cm^2)	76.48	78.64	77.45	3.69	0.532
pH value	6.15	6.04	6.11	0.31	0.26
External fat color[3,4]					
L*	34.11	40.23	38.32	2.56	0.754
a*	15.14	18.05	27.68	1.19	0.895
b*	8.73	7.45	7.29	1.05	0.149
c*	19.52	20.07	20.10	0.93	0.256
12th-rib lean color[4]					
L*	81.98	80.14	80.88	2.265	0.783
a*	3.69	3.56	3.84	0.862	0.861
b*	7.73	7.98	8,07	1.614	0.715
c*	8.34	8.70	8.42	0.123	0.532

[1]CONT = control; EML = ensiled mulberry leaves; SMFP = sun-dried mulberry fruit pomace.
[2]LMA = LM area measured at 12th rib.
[3]Fat color measurements obtained approximately 20 cm ventrally to the lateral process of the split carcass adjacent the 13th rib.
[4]CIE color measurements: L* = lightness, black (0) to white (100); positive a* = red; negative a* = green; positive b* = yellow; negative b* = blue; c* = color saturation = $[(a^*)^2 + (b^*)^2]^{1/2}$ whereby a large number is considered to be more vivid.

To our knowledge, this study is the first to evaluate the carcass characteristics of steers. Future studies should assess the effect of different levels of EML and SMFP on animal growth performance. The supplementation of finishing steer diets with either EML or SMFP did not affect WBSF values (P = 0.844). Carcass color and other meat quality characteristics are shown in Table 5. There were no differences in lean L* values (P = 0.783), a* values (P = 0.861), b* values (P = 0.715), or c* values (P=0.532) between treatment groups. Similarly, there were no differences in adipose L* values (P=0.754), a* values (P=0.895), b* values (P=0.149), or c* values (P=0.256) between treatment groups. There were no differences in pH values of the 48-h LM between the treatment groups; pH values were within the normal pH range (Table 5). There were no significant differences in the color values of adipose and lean tissues; therefore, EML and SMFP did not affect adipose and lean tissue color in steers.

Table 6 shows the effect of the different treatment groups on moisture, protein, intramuscular fat, and ash percentages of steaks. Moisture, protein, and ash percentages were similar between the three treatment groups; however, intramuscular fat content was significantly higher in the CONT group than in the EML and SMFP groups. A previous study showed that WBSF values are negatively correlated with marbling scores and intramuscular fat

Table 6. WBSF and chemical composition of steak samples of finishing steer fed a total mixed ration supplemented with ensiled mulberry leaves (EML) or sun-dried mulberry fruit pomace (SMFP).

Item	Experimental diet[1]			SEM	*P*-value
	CONT	EML	SMFP		
WBSF[2] (kg)	4.78	4.75	4.34	0.61	0.844
Cooked rate (%)	65.72	66.89	68.13	2.53	0.961
Drip loss (%)	6.89	7.12	6.93	1.24	0.922
Moisture (%)	75.49	75.82	75.65	0.37	0.146
Protein (%, DM basis)	87.24	87.62	86.82	0.42	0.276
Intramuscular fat (%, DM basis)	5.62[a]	4.53[b]	4.62[b]	0.34	0.034
Ash (%, DM basis)	5.36	5.74	5.78	0.21	0.467

[1]CONT = control; EML = ensiled mulberry leaves; SMFP = sun-dried mulberry fruit pomace.
[2]WBSF = Warner-Bratzler shear force values of tenderness.
[a,b]Values in the same row with different superscripts are significantly different ($P<0.05$).

content, indicating that as intramuscular fat decreases, WBSF values increase and steaks become tougher [19]. However, our results revealed that there were no differences in WBSF values, even at the lowest intramuscular fat content, in the EML and SMFP groups. Intramuscular fat plays an integral role in steak tenderness; high intramuscular fat contents are associated with increased beef tenderness. Even though there were no differences in tenderness or flavor, low intramuscular fat contents affect juiciness, flavor, and sensory attributes.

Conclusion

This study compared steer performance variables between those fed with a standard TMR and those fed with diets in which corn grain and cotton seed meal were partially replaced by EML or SMFP. Diets that included these two mulberry products had similar effects as the corn grain and cotton seed meal on general performance, blood biochemical parameters, and carcass characteristics. The supplemented feeds did produce variable levels of ruminal VFAs and lower intramuscular fat content.

Supporting Information

Table S1 Chemical composition and silage fermentation characteristics of ensiled mulberry leaves (EML) or sun-dried mulberry fruit pomace (SMFP).

Acknowledgments

We thank the staff of the Jinwei Xumu Ltd for their assistance in preparing the manuscript.

Author Contributions

Conceived and designed the experiments: QM ZZ. Performed the experiments: ZZ BZ. Analyzed the data: BZ LR. Contributed reagents/materials/analysis tools: BZ LR. Wrote the paper: ZZ QM.

References

1. Taasoli G, Kafilzadeh F (2008) Effects of dried and ensiled apple pomace from puree making on performance of finishing lambs. Pak J Biol Sci 11: 294–297.
2. Weiss WP, Frobose DL, Koch ME (1997) Wet tomato pomace ensiled with corn plants for dairy cows. J Dairy Sci 80: 2896–2900.
3. Romero-Huelva M, Ramos-Morales E, Molina-Alcaide E (2012) Nutrient utilization, ruminal fermentation, microbial abundances, and milk yield and composition in dairy goats fed diets including tomato and cucumber waste fruits. J Dairy Sci 95: 6015–6026.
4. Liu JX, Yao J, Yan B, Yu JQ, Shi ZQ (2001) Effects of mulberry leaves to replace rapeseed meal on performance of sheep feeding on ammoniated rice straw diet. Small Rumin Res 39: 131–136.
5. Zou Y, Liao S, Shen W, Liu F, Tang C, et al. (2012) Phenolics and antioxidant activity of mulberry leaves depend on cultivar and harvest month in southern china. Int J Mol Sci 13: 16544–16553.
6. Choi J, Kang HJ, Kim SZ, Kwon TO, Jeong SI, et al. (2013) Antioxidant effect of astragalin isolated from the leaves of Morus alba L. against free radical-induced oxidative hemolysis of human red blood cells. Arch Pharm Res.
7. Wang W, Zu Y, Fu Y, Efferth T (2012) In vitro antioxidant and antimicrobial activity of extracts from Morus alba L. leaves, stems and fruits. Am J Chin Med 40: 349–356.
8. Chen J, Li X (2007) Hypolipidemic effect of flavonoids from mulberry leaves in triton WR-1339 induced hyperlipidemic mice. Asia Pac J Clin Nutr 16 Suppl 1: 290–294.
9. Vu CC, Verstegen MWA, Hendriks WH, Pham KC (2011) The Nutritive Value of Mulberry Leaves (Morus alba) and Partial Replacement of Cotton Seed in Rations on the Performance of Growing Vietnamese Cattle. Asian-Australasian Journal of Animal Sciences 24: 1233–1242.
10. Deshmukh SV, Pathak NN, Takalikar DA, Digraskar SU (1993) Nutritional effect of mulberry (Morus alba) leaves as sole ration of adult rabbits. World Rabbit Science 1: 67–69.
11. Salinas-Chavira J, Castillo-Martinez O, Ramirez-Bribiesca JE, Mellado M (2011) Effect of increasing levels of white mulberry leaves (Morus alba) on ruminal dry matter degradability in lambs. Tropical Animal Health and Production 43: 995–999.
12. Doran MP, Laca EA, Sainz RD (2007) Total tract and rumen digestibility of mulberry foliage (Morus alba), alfalfa hay and oat hay in sheep. Animal Feed Science and Technology 138: 239–253.
13. Zhou B, Meng QX, Ren LP, Shi FH, Wei Z, et al. (2012) Evaluation of chemical composition, in situ degradability and in vitro gas production of ensiled and sun-dried mulberry pomace. Journal of Animal and Feed Sciences 21: 188–197.
14. Van Soest PJ, Robertson JB, Lewis BA (1991) Methods for dietary fiber, neutral detergent fiber, and nonstarch polysaccharides in relation to animal nutrition. J Dairy Sci 74: 3583–3597.
15. Xue F, Zhou Z, Ren L, Meng Q (2011) Influence of rumen-protected lysine supplementation on growth performance and plasma amino acid concentrations in growing cattle offered the maize stalk silage/maize grain-based diet. Animal Feed Science and Technology 169: 61–67.
16. Min BR, Solaiman S, Gurung N, Behrends J, Eun JS, et al. (2012) Effects of pine bark supplementation on performance, rumen fermentation, and carcass characteristics of Kiko crossbred male goats. J Anim Sci 90: 3556–3567.
17. Li YL, Meng QX (2006) Effect of different types of fibre supplemented with sunflower oil on ruminal fermentation and production of conjugated linoleic acids in vitro. Archives of Animal Nutrition 60: 402–411.
18. Minolta (1994) Precise color communications: color control from feeling to instrumentation. Minolta Co Ltd, Ramsey, NJ.
19. Arnett EJ, Fluharty FL, Loerch SC, Zerby HN, Zinn RA, et al. (2012) Effects of forage level in feedlot finishing diets on carcass characteristics and palatability of Jersey beef. J Anim Sci 90: 960–972.
20. Todaro M, Sinacori A, Marinaro G, Alicata ML, Giaccone P (2007) Palatability and in vivo digestibility of mulberry leaves (Morus latifolia CV. Kokusou 21) in sheep feeding. Journal of Animal and Veterinary Advances 6: 509–512.
21. Cheong SH, Kim KH, Jeon BT, Park PJ, Hwang IH, et al. (2012) Effect of mulberry silage supplementation during late fattening stage of Hanwoo (Bos taurus coreanae) steer on antioxidative enzyme activity within the longissimus muscle. Animal Production Science 52: 240–247.
22. Isabelle M, Lee BL, Ong CN, Liu X, Huang D (2008) Peroxyl radical scavenging capacity, polyphenolics, and lipophilic antioxidant profiles of mulberry fruits cultivated in southern China. J Agric Food Chem 56: 9410–9416.
23. Salinas-Chavira J, Castillo-Martinez O, Ramirez-Bribiesca JE, Mellado M (2011) Effect of increasing levels of white mulberry leaves (Morus alba) on ruminal dry matter degradability in lambs. Trop Anim Health Prod 43: 995–999.
24. Johnson DD, Mitchell GE, Jr., Tucker RE, Hemken RW (1982) Plasma glucose and insulin responses to propionate in preruminating calves. J Anim Sci 55: 1224–1230.
25. Butkhup L, Samappito W, Samappito S (2013) Phenolic composition and antioxidant activity of white mulberry (Morus alba L.) fruits. International Journal of Food Science and Technology 48: 934–940.

26. Min BR, Attwood GT, Reilly K, Sun W, Peters JS, et al. (2002) Lotus corniculatus condensed tannins decrease in vivo populations of proteolytic bacteria and affect nitrogen metabolism in the rumen of sheep. Canadian Journal of Microbiology 48: 911–921.
27. Min BR, McNabb WC, Barry TN, Peters JS (2000) Solubilization and degradation of ribulose-1,5-bisphosphate carboxylase/oxygenase (EC 4.1.1.39; Rubisco) protein from white clover (Trifolium repens) and Lotus corniculatus by rumen microorganisms and the effect of condensed tannins on these processes. Journal of Agricultural Science 134: 305–317.
28. Bailey EA, Titgemeyer EC, Olson KC, Brake DW, Jones ML, et al. (2012) Effects of ruminal casein and glucose on forage digestion and urea kinetics in beef cattle. Journal of Animal Science 90: 3505–3514.
29. Lopez-Campos O, Bodas R, Prieto N, Frutos P, Andres S, et al. (2011) Vinasse added to the concentrate for fattening lambs: Intake, animal performance, and carcass and meat characteristics. Journal of Animal Science 89: 1153–1162.
30. Bailey EA, Titgemeyer EC, Olson KC, Brake DW, Jones ML, et al. (2012) Effects of supplemental energy and protein on forage digestion and urea kinetics in growing beef cattle. Journal of Animal Science 90: 3492–3504.
31. Aschenbach JR, Kristensen NB, Donkin SS, Hammon HM, Penner GB (2010) Gluconeogenesis in Dairy Cows: The Secret of Making Sweet Milk from Sour Dough. Iubmb Life 62: 869–877.
32. Winkelman LA, Overton TR (2012) The effects of increasing doses of 2 preparations of long-acting insulin on short-term plasma profiles of glucose and insulin in lactating dairy cows. Journal of Dairy Science 95: 6974–6982.
33. Mao S, Zhang R, Wang D, Zhu W (2012) The diversity of the fecal bacterial community and its relationship with the concentration of volatile fatty acids in the feces during subacute rumen acidosis in dairy cows. BMC Veterinary Research 8: (6 December 2012)-(2016 December 2012).
34. Tesseraud S, Metayer S, Duchene S, Bigot K, Grizard J, et al. (2007) Regulation of protein metabolism by insulin: Value of different approaches and animal models. Domestic Animal Endocrinology 33: 123–142.
35. Laarveld B, Christensen DA, Brockman RP (1981) The effect of insulin on net metabolism of glucose and amino acids by the bovine mammary gland. Endocrinology 108: 2217–2221.
36. Sagdic O, Ozturk I, Yilmaz MT, Yetim H (2011) Effect of Grape Pomace Extracts Obtained from Different Grape Varieties on Microbial Quality of Beef Patty. Journal of Food Science 76: M515–M521.
37. Hogan S, Zhang L, Li J, Sun S, Canning C, et al. (2010) Antioxidant rich grape pomace extract suppresses postprandial hyperglycemia in diabetic mice by specifically inhibiting alpha-glucosidase. Nutr Metab (Lond) 7: 71.
38. Yang WZ, Li YL, McAllister TA, McKinnon JJ, Beauchemin KA (2012) Wheat distillers grains in feedlot cattle diets: feeding behavior, growth performance, carcass characteristics, and blood metabolites. J Anim Sci 90: 1301–1310.
39. Sami AS, Augustini C, Schwarz FJ (2004) Effect of feeding intensity and time on feed on intramuscular fatty acid composition of Simmental bulls. Journal of Animal Physiology and Animal Nutrition 88: 179–187.
40. Price BD, Garmyn AJ, Derington HM, Galyean ML, Jackson SP, et al. (2011) Effects of high-oil corn on feedlot performance, carcass characteristics, fatty acid profiles, beef palatability, and retail case life traits of beef top loin steaks. Journal of Animal Science 89: 809–816.

24

Successful Cognitive Aging in Rats: A Role for mGluR5 Glutamate Receptors, Homer 1 Proteins and Downstream Signaling Pathways

Caroline Ménard*, **Rémi Quirion**

Douglas Mental Health University Institute, Department of Psychiatry, McGill University, Montréal, Canada

Abstract

Normal aging is associated with impairments in cognition, especially learning and memory. However, major individual differences are known to exist. Using the classical Morris Water Maze (MWM) task, we discriminated a population of 24-months old Long Evans aged rats in two groups - memory-impaired (AI) and memory-unimpaired (AU) in comparison with 6-months old adult animals. AI rats presented deficits in learning, reverse memory and retention. At the molecular level, an increase in metabotropic glutamate receptors 5 (mGluR5) was observed in post-synaptic densities (PSD) in the hippocampus of AU rats after training. Scaffolding Homer 1b/c proteins binding to group 1 mGluR facilitate coupling with its signaling effectors while Homer 1a reduces it. Both Homer 1a and 1b/c levels were up-regulated in the hippocampus PSD of AU animals following MWM task. Using immunohistochemistry we further demonstrated that mGluR5 as well as Homer 1b/c stainings were enhanced in the CA1 hippocampus sub-field of AU animals. In fact mGluR5 and Homer 1 isoforms were more abundant and co-localized in the hippocampal dendrites in AU rats. However, the ratio of Homer 1a/Homer 1b/c bound to mGluR5 in the PSD was four times lower for AU animals compared to AI rats. Consequently, AU animals presented higher PKCγ, ERK, p70S6K, mTOR and CREB activation. Finally the expression of immediate early gene Arc/Arg3.1 was shown to be higher in AU rats in accordance with its role in spatial memory consolidation. On the basis of these results, a model of successful cognitive aging with a critical role for mGluR5, Homer 1 proteins and downstream signalling pathways is proposed here.

Editor: Mark P. Mattson, National Institute on Aging Intramural Research Program, United States of America

Funding: This research was supported by a grant from the Canadian Institutes of Health Research (CIHR) to Rémi Quirion (Grant #MOP-8580). Caroline Menard is the recipient of a fellowship from CIHR (http://www.cihr-irsc.gc.ca/e/193.html). The funders had no role in study design, data collection and analysis, decision to publish, or preparation of the manuscript.

Competing Interests: The authors have declared that no competing interests exist.

* E-mail: caroline.menard@douglas.mcgill.ca

Introduction

Aging is a natural process characterized by various physical alterations: slower reaction time, cognitive alterations and modification of the muscles density [1]. However, major individual differences are observed. The concept of successful aging consists of three components namely, low probability of disease and disability, high cognitive and physical capacity and active engagement in general [1]. In the present study, we explored variability in cognitive abilities associated with normal aging using Long Evans rats as model. Similar to humans, a certain proportion of rats becomes memory-impaired with aging (AI) while others maintain spatial memory abilities comparable to young animals, and are classified as aged memory-unimpaired (AU) [2,3,4,5]. Alterations in gene expression have been proposed to at least partly explain this phenomenon [6,7]. Another hypothesis relates to the incapacity of AI subjects to adapt and properly encode new information [8]. Hippocampus-dependent spatial memory is not the only behavioural profile associated to Long Evans AI and AU sub-groups. A reduced reactivity to novelty in exploratory paradigms, gustatory/olfactory stimulus and an increased reaction to pain has also been reported in AI animals whereas in AU animals all these behaviours were similar to young rats [3].

Glutamate and its receptors are closely involved in mechanisms underlying spatial learning and hippocampus-dependent memory processes [9,10]. The stimulation of ionotropic α-amino-3-hydroxy-5-methyl-4-isoxazole-proprionic acid (AMPA) and N-methyl-D-aspartate (NMDA) receptors induces synaptic plasticity leading to long-term potentiation (LTP) and long-term depression (LTD) [9,11]. Alterations in synaptic plasticity, including LTP, have been previously associated with age-related memory impairments [12]. Interestingly, Lee et al. [13] had correlated NMDA receptor-independent LTD and successful cognitive aging in Long Evans rats. This form of LTD is mediated by the post-synaptic group 1 metabotropic glutamate receptors (mGluR), mGluR1α and mGluR5 [14]. mGluR1α receptors have been associated with the post-synaptic specialization of excitatory synapses and are concentrated in perisynaptic and extrasynaptic areas [15] while mGluR5 functionally interact with NMDA receptors in the postsynaptic densities (PSD) [16]. mGluR5 receptors are abundant in the adult hippocampus and cerebral cortex [17] and are involved in the maintenance of synaptic plasticity [14,18,19,20,21]. To further investigate the cellular mechanisms involved in successful cognitive aging, we studied mGluR levels and related signaling pathways in AI versus AU animals focusing on the hippocampal formation as a key structure involved in spatial learning [22].

Higher mGluR5 receptors levels were observed in AU hippocampus PSD after training compared to both AI and 6-months old animals. Metabotropic receptors closely interact with scaffolding proteins such as the family of Homers [23,24,25,26]. The expression of all Homer 1 isoforms was only increased in the PSD of AU rats. Homer 1b/c are constitutively expressed and coupled with mGluR5 and its signaling effectors while Homer 1a is produced following synaptic activity and reduces the coupling of the receptor [26,27,28,29]. A lower Homer 1a/Homer 1b/c ratio bound to mGluR5 receptors was observed in AU animals, leading to stronger activation of protein kinases and immediate early gene expression, especially in the CA1 hippocampal sub-field. These changes were correlated with the performance of aged rats in the MWM, which led us to propose a working model of successful cognitive aging.

Materials and Methods

Animals

Male Long–Evans rats were purchased from Charles River Laboratories (St Constant, QC, Canada) at 12 months of age and were housed in our animal facility until 24-months old. Another group was purchased at 3 months of age and housed in our animal facility until they were 6 months old. Animals were housed two per cage and maintained on a 12 h light/dark cycle with *ad libitum* access to food (Purina Lab Chow; Mondou, Montreal, QC, Canada) and water. Animal care, surgery and handling procedures were approved by the McGill University Animal Care Committee (protocol number 3589) and the Canadian Council for Animal Care.

Morris Water Maze (MWM)

The long-term reference memory version of the MWM, a hippocampus-dependant behavioral task [30] was used to investigate spatial memory in Long–Evans rats. Rats were required to find, in a 1.4 m diameter pool, a submerged platform (14 cm in diameter) located 2 cm below the surface of water (24°C), rendered opaque by the addition of white paint. Animals were pseudo-randomly started from a different position on each trial and used distal visuo-spatial cues to find the hidden escape platform that remained in the centre of the same quadrant throughout training [30]. Rats were given three trials of 90 s per day for five consecutive days. Animals were guided to the platform if it was not located within 90 s. All the rats remained on the platform for 15 s before removal. After the acquisition phase on day 5, rats were given one probe trial of 60 s for which the platform was removed from the pool. The number of times the animal swim over the platform location was evaluated.

Motor/Visual Controls

Swimming speed (see results section) and distance swam (data not shown) were used as controls of motor function, a parameter potentially altered with age. The probe test was followed by a cued trial of 60 s in which the platform was visible to assess visual deficits and motivation to escape from water.

Classification Criteria

The learning curves and probe tests of aged rats were compared to those of 6-months old adult rats to classify them as AI or AU [3]. An animal was considered unimpaired when the learning curve and learning probe test results were less than two statistical variations to the average of adult rats. Aged rats were classified as AI when animals swam for over 25 seconds to find the platform on both days 4 and 5 (average of three trials per day). Memory impairment was confirmed with platform crossings and time spent in the target and opposite quadrants during the learning probe test. Few animals showing inconsistent results, for example latencies on days 4 and 5 similar to AI rats but the probe tests results comparable to the AU group were removed from the study and used to optimize experimental conditions.

Following the first week of training, all animals were submitted to a reverse memory paradigm to assess inhibitory learning [31]. Briefly, the platform was moved to the opposite quadrant while the position of visual cues stayed the same. Rats were then trained to find the new platform location for four consecutive days (three trials per day). A second probe test was conducted at the end of the last day (day 11). Finally, a third probe test was done seven days later to evaluate memory consolidation and retrieval. Animals were sacrificed 2–3 h after the last MWM probe test to allow immediate early gene (IEG) expression (day 18).

After each trial, rats were immediately placed under a heat lamp to dry and prevent hypothermia. To control for possible effects due to circadian cycles, all trials were performed at approximately the same time of the day between 10–15 h. Data derived from the MWM task were recorded on computer using a video tracking system (HVS, Buckingham, UK).

Antibodies

mGluR5 antibody was procured from Millipore (Billerica, MA, USA) and used at a 1/1000 dilution for Western Blot. Homer 1 (1/200) and actin (1/2000) antibodies were purchased from Santa Cruz Biotechnology (Santa Cruz, CA, USA). The PSD95, p42/44 (Thr202/Tyr204), p42/p44, p70S6K (Thr389), p70S6K (Thr412/Ser424), p70S6K, mTOR (Ser2448), mTOR, CREB (Ser133) and CREB antibodies were all used at a 1/1000 dilution for Western Blot and purchased from Cell Signaling (Danvers, MA, USA). All other antibodies were procured from Abcam (Cambridge, MA, USA) and also used at a 1/1000 dilution. For immunohistochemistry experiments, mGluR5 and Homer 1b/c antibodies were incubated at 1/1000 and 1/200 dilution, respectively. For immunofluorescence co-localization, mGluR5, Homer 1a, Homer 1b/c, PKCγ (Thr674) and mTOR (Ser2448) were incubated at 1/100 dilution while Arc and MAP2 were used at 1/250 dilution.

Specificity of mGluR5 and Homer 1 antibodies has been established using Western Blot and immunohistochemistry experiments. As shown in the supplemental figure S1A, a band around the expected molecular weight (132 kDa) is strongly reactive to mGluR5 antibody. Incubation of this antibody with the mGluR5 peptide (provided by the company) completely blocked the interaction in both Western Blot and immunohistochemistry analyses. Specificity of the anti-rabbit secondary antibody was verified as well with the omission of the primary antibody. Incubation of the tissue preparation with Homer 1a primary antibody produced non-specific bands (Fig. S1B). Preadsorption with Homer 1a peptide blocked the interaction of only one band (mol. weight between 30 and 36 kDa), which was close to the expected molecular weight (30 kDa). Again, the specificity of the anti-goat secondary antibody was verified with omission of the primary antibody. Homer 1b/c primary antibody was highly specific with only one reactive band around 45 kDa (Fig. S1C). Incubation with the peptide completely blocked the interaction in both Western Blot and immunohistochemistry.

Subcellular fractionation and PSD isolation

Animals were sacrificed by decapitation and brains were dissected to collect the hippocampal formation and adjacent cortex and the tissue was quickly frozen in isopentane at −80°C.

The preparation included the hippocampus, entorhinal, perirhinal and portions of adjacent neocortices [32]. To compare protein levels, tissues were collected in 7 ml of Tris-acetate buffer (50 mM, pH 7.4) containing 100 μM EGTA, 0.32 M sucrose and several protease inhibitors (leupeptin 5 μM, phenylmethylsulfonyl fluoride 200 μM, N-tosyl-L-phenylalanine chloromethyl ketone 1 μg/ml, Sigma-Aldrich Canada, Oakville, Ont.). Subsequently, samples were homogenized with a polytron and total protein concentrations were determined using bicinchoninic acid protein assay kit (Pierce, Rockford, IL, USA). Two ml of each homogenate sample was frozen for Western Blot and immunoprecipitation experiments.

To obtain membrane fractions, 5 ml of sample homogenates were purified by differential centrifugation as reported previously [33]. Briefly, homogenates were centrifuged at 1,000×g for 10 min. Supernatants were then centrifuged at 14,000×g for 20 min and the resulting pellet was defined as the crude synaptosomal fraction. Synaptosomes were resuspended in Tris-acetate buffer (50 mM, pH 7.4, 100 μM EGTA, 0.32 M sucrose and proteases inhibitors).

To purify the PSD, the synaptosomal fraction was diluted with 20 mM Tris-HCl, pH 6.0, 0.1 mM $CaCl_2$ containing 1% Triton X-100, mixed for 20 min at 4°C with agitation and then centrifuged at 40,000×g for 20 min at 4°C [34]. The pellet containing the isolated synaptic junctions was collected. To separate presynaptic components from the PSD, the pellet was resuspended in 20 mM Tris-HCl, pH 8.0, 0.1 mM $CaCl_2$ containing 1% Triton X-100. Samples were mixed again for 20 min at 4°C with agitation and then centrifuged at 40,000×g for 20 min at 4°C. The insoluble pellet containing the PSD fraction was suspended in Tris-acetate buffer (50 mM, pH 7.4, 100 μM EGTA, 0.32 M sucrose and proteases inhibitors) and stored at −80°C until use. Purification of the PSD was controlled by Western Blot using PSD95 to assess the protein enrichment (Fig. S1D).

Electrophoresis and immunoblotting

Western blot analysis was carried out on aliquots of homogenates and PSD fractions obtained from AI, AU and 6-months old adult rat brains. Aliquots were subjected to sodium dodecyl sulfate polyacrylamide gel electrophoresis (SDS-PAGE) on denaturing NuPAGE® Novex 4–20% Bis–Tris gel (Invitrogen, Carlsbad, CA, USA). Proteins were transferred onto Hybond-C nitrocellulose membranes (Amersham Biosciences, Little Chalfont, UK). To block nonspecific sites, membranes were first incubated for 1 h at room temperature in phosphate-buffered saline (PBS) containing 2% bovine serum albumin (BSA). Membranes were next incubated with primary antibodies directed against group 1 mGluR (mGluR1α and mGluR5), Homer proteins (Homer 1a, Homer 1b/c, Homer 2, Homer 3) and signaling pathways enzymes - PKCγ (Thr674), PKCγ, p42/44 MAPK (Thr202/Tyr204), p42/44 MAPK, p70S6K (Thr389), p70S6K (Thr421/Ser424), p70S6K, mTOR (Ser2448), mTOR, CREB (Ser133), CREB - in PBS containing 2% BSA. Bands corresponding to proteins were detected with peroxidase-conjugated secondary antibody (Santa Cruz Biotechnology, CA, USA) and Western Lightning Chemiluminescence Reagent Plus (Perkin Elmer, Boston, MA, USA) on Kodak BioMax MS film (Amersham Biosciences). The same labelling procedure was used for negative controls but primary antibody was omitted. Actin level was used as a loading control. Immunoblots were placed on a Northern light illuminator and computer-generated images were analyzed semiquantitatively by densitometry with a microcomputer imaging device (Imaging Research, MCID, St. Catharines, ON, Canada).

Immunohistochemistry

Coronal sections (20-μm) of perfused brains (fixed with 4% paraformaldehyde and cryoprotected in 30% sucrose solution) at the level of dorsal hippocampus were washed in PBS (pH 7.4) for 5 min then immersed in methanol containing 0.3% H_2O_2 for 30 min at room temperature to quench endogenous peroxidases. After three washes of 5 min each in PBS, sections were preincubated with 10% normal goat serum (NGS) at room temperature for 1 h, followed by incubation with polyclonal mGluR5 or Homer 1b/c primary antibodies (1/1000 or 1/200 in PBS with 1% NGS) overnight at 4°C. Subsequently, sections were washed 3 times in PBS and incubated in biotinylated anti-rabbit IgG (1:200, Vector Laboratories, Burlingame, CA, USA) in 1% NGS for 30 min. After washing in PBS, sections were finally incubated with avidin biotinylated enzyme complex (ABC reagent, Vector Laboratories, Burlington, ON, Canada) diluted in PBS for 30 min. The peroxidase reaction was carried out with 0.02% hydrogen peroxide and 3,3′-diaminobenzidine tetrahydrochloride (0.1% in 100 mM Tris-HCl buffer, pH 7.4). Sections were then washed in tap water, cleaned and mounted on Superfrost Plus slides (Thermo Scientific, Portsmouth, NH, USA). Immunohistochemical staining was visualized using bright-field microscopy at 2× and 10× magnification. Controls were prepared using the same labelling procedure, but primary antibody was omitted (Fig. S1).

Immunofluorescence

Coronal 20-μm sections of perfused brains at the level of the dorsal hippocampus were washed in PBS (pH 7.4) for 5 min. Sections were permeabilized with 0.2% Tween 20 in PBS for 10 min at room temperature then processed for immunofluorescence labelling. In brief, sections were first incubated in 10% NGS (or normal horse serum for Homer 1a) diluted in 0.1 M PBS with 0.05% Tween 20 (PBST) and 1% BSA for 60 min at room temperature, followed by overnight incubation with primary antibodies at 4°C in a solution of 1% serum and 1% BSA in 0.1 M PBST. After 3 washes in PBS, sections were incubated with corresponding secondary antibodies (1:500, Invitrogen, Carlsbad, CA, USA) conjugated with Alexa Fluor 488 or Alexa Fluor 568 in 1% BSA in PBST for 2 h at room temperature in dark. Sections were washed three times with PBS for 5 min each in dark. Nuclei were stained with Hoechst solution (2 μg/ml, Invitrogen, Carlsbad, CA, USA) for 5 min and subsequently the sections were washed and coverslipped with Fluoromount-G (Southern Biotech, Birmingham, AL, USA). Pictures were taken at 40× magnification with an Axio Observer microscope with Apotome (Carl Zeiss MicroImaging GmbH, Germany).

Immunoprecipitation

20 μg of PSD and synaptic membrane proteins were immunoprecipitated to investigate mGluR1α/5-Homer protein interactions. Samples were suspended in 150 μl of lysis buffer (100 mM Tris/HCl, 100 μM EGTA, pH 7.4) containing 1% SDS and heated at 100°C for 5 min, followed by dilution in 600 μl of cold lysis buffer containing 2% Triton X-100. Insoluble materials were removed by centrifugation at 15,000×g for 30 min. Protein A-agarose beads (Millipore, Billerica, MA, USA) and antibody (3 μl) mixtures were added to the supernatant and incubated overnight with agitation at 4°C. Immunoreactive complexes were recovered by brief centrifugation and pellets were washed with lysis buffer-2% Triton X-100. Proteins were finally eluted in 2× Laemmli sample buffer containing 4% SDS, 20% glycerol, 10% 2-mercaptoethanol, 0.004% bromophenol blue and 0.125 M Tris HCl, pH 6.8. (Sigma-Aldrich Canada, Oakville, Ontario) and

heated at 100°C for 10 min. Immunoprecipitated proteins were subjected to SDS-PAGE and immunoblotting (see preceding section).

Statistical analysis

In all cases, data are presented as means ± SEM. MWM latencies and platform crossings were analyzed by two-way ANOVA followed by Bonferroni *post hoc* analysis with GraphPad Prism software. Swim speed, time spent in target and opposite quadrants and protein levels were analyzed by one-way ANOVA followed by Bonferroni *post hoc* analysis. Behavioural data correlations were evaluated with a two-tailed Pearson's test. Significance level was set at $p<0.05$.

Results

Alterations in cognitive abilities with normal aging

The effect of normal aging on spatial memory was evaluated on the basis of MWM performance in 24-months old rats. We compared learning curves of aged rats with those of young animals (6-months old). The behavioural training protocol is presented in Figure 1A. As shown in Fig. 1B, the average learning curve of the AI (N = 31) animals was significantly different from the AU (N = 32) and adult rats (N = 26) for the last 4 days (**$p<0.01$, ***$p<0.001$). A considerable difference was also observed between the performance of adult and AU rat on day 2 (***$p<0.001$). The average performance of the AI animals was much lower compared to the AU and adult rats for the first probe test (*** $p<0.001$). The adult animals also crossed the platform more often than the AU rats (*$p<0.05$), which is probably due to a slightly faster swimming speed (*$p<0.05$) for these animals (Fig. 1C). Motor activity as shown by the swim speed was comparable for both groups of aged rats (Fig. 1C, 0.167±0.009 m/s for AI vs 0.169±0.006 m/s for AU). Time spent in the target and opposite quadrants was evaluated to confirm memory status. As depicted in Fig. 1D, adult and AU rats spent significantly more time (38.06±1.83%, ***$p<0.001$ and 33.76±2.34%, *$p<0.05$ respectively) in the platform quadrant (target) while AI animals swam randomly during the learning probe test; 24.90±2.23% of time in the target quadrant versus 26.12±1.84% in the opposite quadrant.

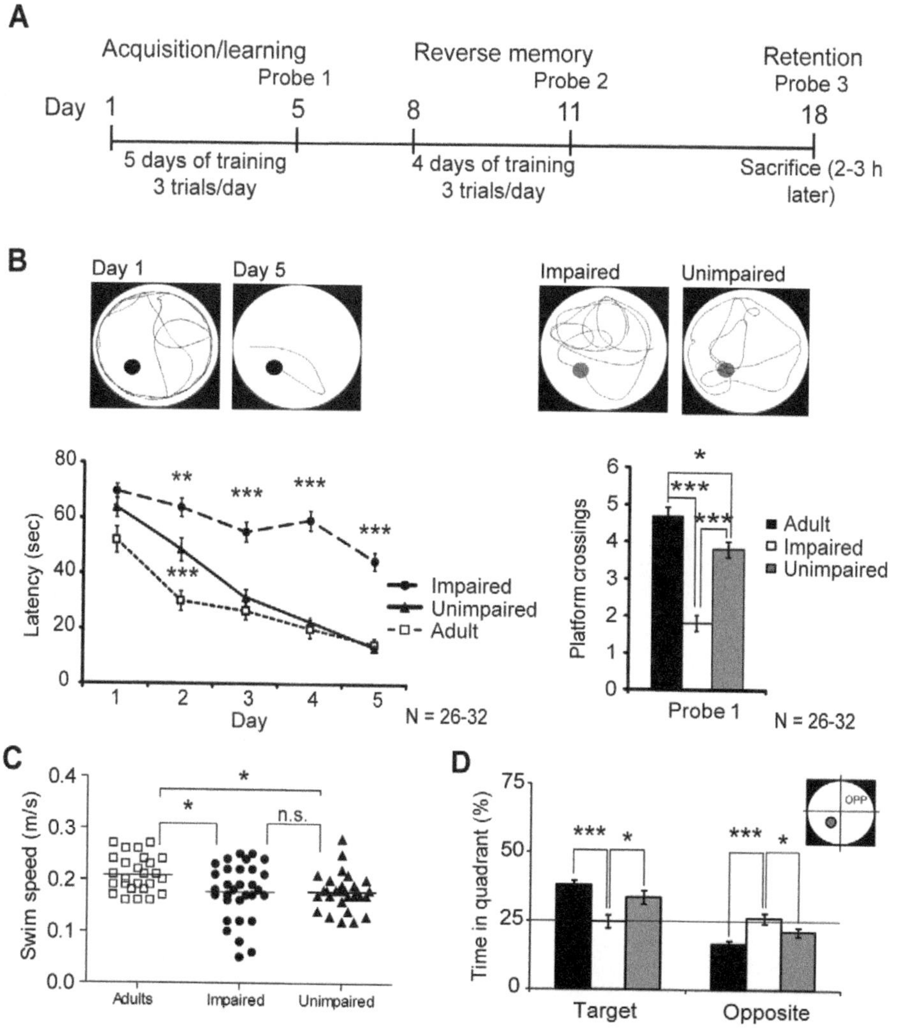

Figure 1. Effect of aging on the MWM acquisition task. A) Behavioral protocol schedule. B) Learning curves of the AI (Impaired), AU (Unimpaired) and 6 months old (Adult) rats for week 1 as well as average probe 1 test results (platform crossings) for all groups. Representative tracking pictures for day 1, day 5 and AI vs AU probe tests are presented. C) Swim speed distribution in all groups during probe 1 test. D) Time spent in target and opposite quadrants during learning of probe test. Data represent mean ± SEM for all animals of each group (N = 26–32), *$p<0.05$, **$p<0.01$, ***$p<0.001$.

Aging has been shown to alter synaptic plasticity associated with new memories [8]. This hypothesis was evaluated using a reverse memory task. The platform position was modified and rats were trained again, 3 times a day, for 4 consecutive days. As shown in Figure 2A, the average learning curve of AU rats overlapped with that of adult animals while the performance of AI animals was getting worse (***p<0.001). A second probe test was conducted on day 11 and similar to the first test, AI animals crossed the platform fewer times (Fig. 2A ***p<0.001). No significant change was observed for young adult versus AU animals. Motor activity was evaluated by swim speed and results were identical to the first probe test with adult rat group swimming slightly faster (Fig. 2B, *p<0.05). Time spent in the new target and opposite quadrants was assessed to evaluate if aged animals were going back to the initial platform position (i.e. learning), suggesting an impairment of new memories formation. Interestingly, adult and AU animals spent similar percentage of time in the new target quadrant (Fig. 2C, 39.36±1.75 vs 35.42±2.32% respectively, *p<0.05). Similar to the first learning probe test, AI rats spent 25% of the time swimming in both target and opposite quadrants. Finally, an evaluation of memory consolidation was done with a third probe test one week later (probe 3). The AU animals again crossed the previous platform position (reverse) as often as the adult rats while the AI rats struggled to remember it (Fig. 2D, **p<0.01, *p<0.05). None of the groups went back to the learning platform quadrant significantly (Fig. 2E); adult and AU rats were focused on the reverse platform quadrant (*p<0.05) while AI animals still swam randomly (25.57±1.80%).

As mentioned in the Materials and Methods section, the main criteria to classify the aged rats in either impaired or unimpaired sub-groups was day 5 latency. Individual performance of aged rats was compared to the mean value of adult animals and as shown in Fig. 3A, the criterion was set at 25 seconds (***p<0.001). Learning and reverse probe test results were also taken into account for classification. The distribution cleavage is less evident but still significantly different among the groups (Fig. 3B, ***p<0.001). Correlation between day 5 latency and average probe test performance for each individual was determined to confirm the criterion validity. As presented in Fig. 3C, day 5 latencies were correlated with the mean of learning and reverse probe tests for adult animals (**p<0.01), with correlation being stronger for aged rats (Fig. 3C, ***p<0.001). Finally, correlation between day 5 latencies and time spent in overall target quadrant was evaluated to confirm memory adaptative abilities. While adult animal performances show a tendency towards adaptative abilities, correlation for aged rats was also strongly significant (Fig. 3D, ***p<0.001), confirming that the 24-months old animals spending more time swimming in the target quadrant during learning and reverse probe tests found the platform faster on day 5 as well.

Decrease in metabotropic glutamate receptor levels in aged memory-impaired animals

Considering the importance of metabotropic glutamate receptors LTD for successful aging and cognition [13], we investigated group 1 mGluR receptor expression in the adult, AI and AU hippocampal formation following MWM training to explain the performance discrepancies. The mGluR1α protein level was unaltered in the homogenates of all groups (Fig. 4A). However,

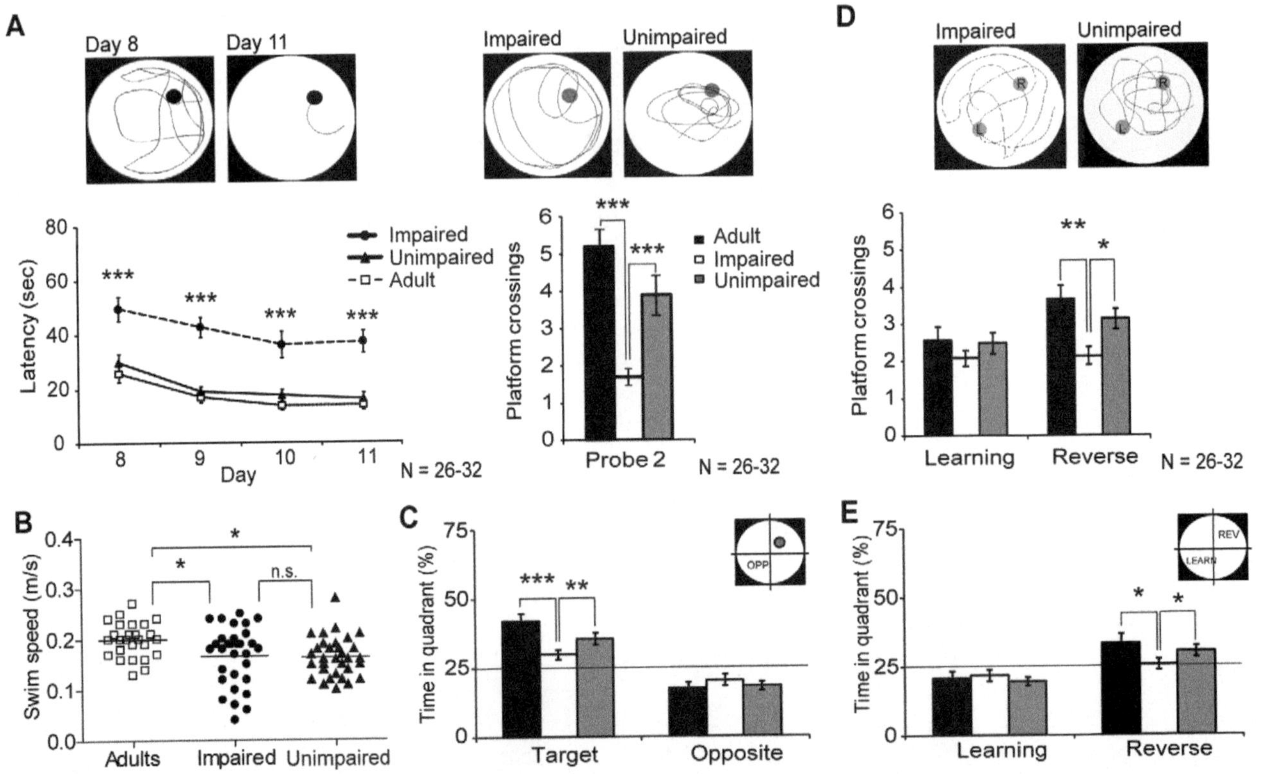

Figure 2. Effect of aging on reverse memory and retention. A) Learning curves and representative tracking pictures for week 2 as well as average results for the reverse probe test on day 11 (probe 2). B) Swim speed distribution in all groups during probe 2 test. C) Time spent in target and opposite quadrants during the reverse probe test. D) Results of the retention probe test conducted one week after the reverse probe test. Representative tracking pictures for AI and AU groups are presented. E) Time spent in learning and reverse platform quadrants during the retention probe test. Data represent mean ± SEM for all animals of each group (N = 26–32), *p<0.05, **p<0.01, ***p<0.001.

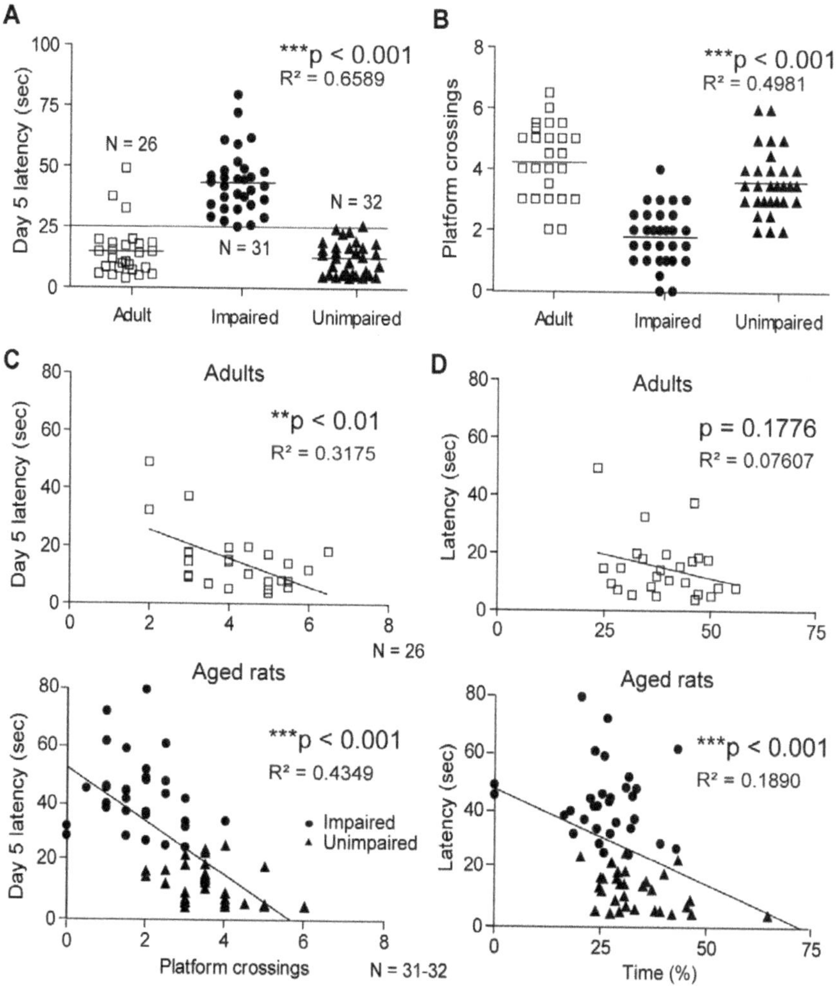

Figure 3. Correlation between day 5 latency and cognitive abilities. A) Distribution of day 5 latency for all 3 groups; this parameter is used to correlate protein levels with MWM performance. B) Number of platform crossings distribution during learning and reverse probe tests (mean of probes 1 and 2) for all groups. C) Day 5 latency correlation with the mean of learning and reverse probe test results for both adult and aged rats. D) Correlation of day 5 latency with the mean of time spent in target quadrant during learning and reverse probe tests for both adult and aged rats. Data represent mean ± SEM for all animals of each group (N=26–32), $^{**}p<0.01$, $^{***}p<0.001$.

mGluR5 was decreased in the AI homogenate samples (Fig. 4A, 56%±4% vs 97%±9% AI vs AU, $^{**}p<0.01$ and 100%±17% AI vs adult rats, $^{*}p<0.05$). As presented in Fig. 3, day 5 latency is a good indicator of cognitive faculties. Thus, the correlation between this parameter and molecular changes was evaluated for each individual animal. Protein level of mGluR5 correlates with aged rats MWM performance on day five (Fig. 4B, $^{*}p<0.05$). Since synaptic transmission is concentrated at synapses [9], PSD fraction was purified and mGluRs expression level was evaluated. Receptor level of mGluR1α was reduced in the PSD of AI animals (61%±9%, $^{***}p<0.001$, Fig. 4C), while the level of mGluR5 receptor was significantly increased in the PSD of the AU group (Fig. 4C, 158%±12%), not only compared to AI animals (96%±7%, $^{**}p<0.01$) but also with young adults (100%±21%, $^{*}p<0.05$). The correlation between the behavioural data and mGluR1α in aged rats was significant (Fig. 4D, $^{*}p<0.05$) but was stronger for mGluR5 receptor ($^{***}p<0.001$). Changes in mGluR5 receptor levels were restricted to a specific area of the hippocampal formation. Indeed, mGluR5 staining was predominant in the dendrites of the CA1 sub-field and reduced in the AI group (Fig. 4E).

mGluRs and Homer proteins interactions are altered in the hippocampal formation of aged memory-impaired animal

Group 1 metabotropic glutamate receptors are associated with Homer scaffolding proteins [23,24,25,26]. The Homer 1b/c, Homer 2 and Homer 3 long isoforms are constitutively expressed and mediate mGluR signaling. An increase in synaptic activity induces the transcription of the dominant negative Homer 1a short isoform [26,27,28,29]. A significant up-regulation in the level of Homer 3 protein was observed in hippocampal homogenates of the AU compared to AI animals (Fig. 5A, 145%±13% vs 94%±8%, $^{***}p<0.001$) while AI animals demonstrated an increase in Homer 2 (134%±5%, $^{*}p<0.05$, $^{**}p<0.01$). The AU animals demonstrated strong increase in the levels of Homer 1a (Fig. 5B, 168%±10%, $^{**}p<0.01$, $^{***}p<0.001$) and Homer 1b/c (171%±12%, $^{**}p<0.01$, $^{***}p<0.001$) protein in the PSD compared to both AI and young adult rats, while Homer 3 level was decreased in the AI group (54%±6%, $^{**}p<0.01$, $^{***}p<0.001$). MWM performance were significantly correlated with protein levels of Homer 1a (Fig. 5C, $^{***}p<0.001$), Homer 1 b/c ($^{***}p<0.001$) and Homer 3 ($^{**}p<0.01$) in the PSD. Similar to

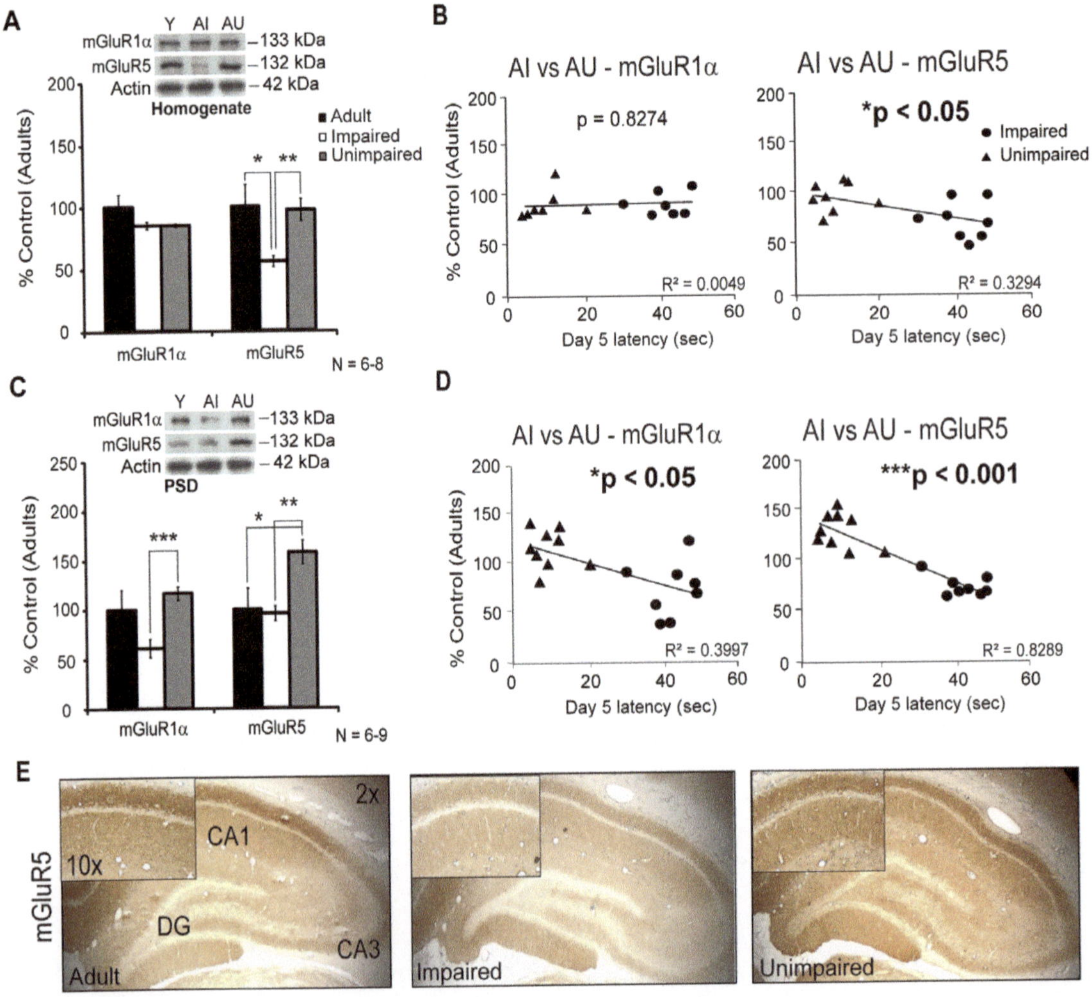

Figure 4. Comparison of group 1 mGluR levels in adult (Y), AI and AU hippocampal formations. A) mGluR5 protein level was decreased in the hippocampal homogenates of AI animals and B) this modification was correlated to the MWM performance for aged rats. C) mGluR1α receptor level in the PSD was decreased for AI rats while AU animals showed an increase in mGluR5 level. D) Both alterations were correlated to the MWM day 5 latencies for aged rats. E) mGluR5 immunostaining was strongly reduced in the AI hippocampus particularly in the dendrites of the CA1 region. AU displayed small increase in mGluR5 level in CA1 and DG. Western Blot values represent mean ± SEM of three separate experiments (N = 6–9) and the data are expressed as percentage of control (6-months adult rats), *$p<0.05$, **$p<0.01$, ***$p<0.001$.

mGluR5, Homer 1 b/c immunostaining was more intense in dendrites of the CA1 hippocampal sub-field particularly in AU animals (Fig. 5D).

In addition, we investigated the regional specificity of the expression of mGluR5 and Homer 1 isoforms using double immunofluorescence. As Homer 1a is an IEG translated following synaptic activity [35], it is mainly co-localized with mGluR5 in the nuclear area of the cells (Fig. 6A). IEG expression was expected since the animals were sacrificed 2–3 hours after the last probe test. Homer 1a expression was more abundant in CA1 neurons of AU animals and occasionally co-localized with mGluR5 in dendrites (Fig. 6A). Homer 1 b/c expression was strongly co-localized with mGluR5 in CA1 neurons and especially in dendrites, particularly in AU rats (Fig. 6B).

Homer interactions have previously been shown to regulate mGluR dependent synaptic plasticity [35]. Moreover, Van Keuren-Jensen & Cline [36] demonstrated the importance of the Homer 1a/Homer 1b/c ratio for mGluR-mediated plasticity in optic tectal cells, where Homer 1a acts as a dominant negative modulator. We observed an increase in the Homer 1 ratio for mGluR1α in synaptosomes but no significant change at the PSD level in AU animals compared to AI rats (Fig. 7A, 1.38±0.38 vs 0.71±0.04, *$p<0.05$). Indeed in the PSD, both Homer 1a and Homer 1b/c protein levels bound to mGluR1α were significantly decreased with memory impairments resulting in a similar ratio (Fig. 7B, top graph, *$p<0.05$ and **$p<0.01$ respectively). The Homer 1 ratio in synaptosomes was significantly correlated with MWM performance (Fig. 7C, **$p<0.01$). Following mGluR5 immunoprecipitation in the PSD, we observed a change in the Homer 1a/Homer 1b/c ratio for AI animals (Fig. 7D, 4.21±0.48, ***$p<0.001$). Indeed, a reduced amount of Homer 1b/c proteins was bound to mGluR5 receptors for this group (Fig. 7E, top graph, 32%±24%, *$p<0.05$, **$p<0.01$), whereas the level remained the same for Homer 1a (99%±11%) compared to adult and AU rats. The ratio in PSD for the AU group (1.37±0.22) was higher but not significantly different from the adult animals (Fig. 7D, 0.89±0.35). However it was considerably decreased in synaptosomes compared to both adult and AI rats (Fig. 7D, 0.51±0.06, **$p<0.01$, ***$p<0.001$). The MWM performance correlates strongly with Homer 1a/Homer 1b/c proteins ratio bound to

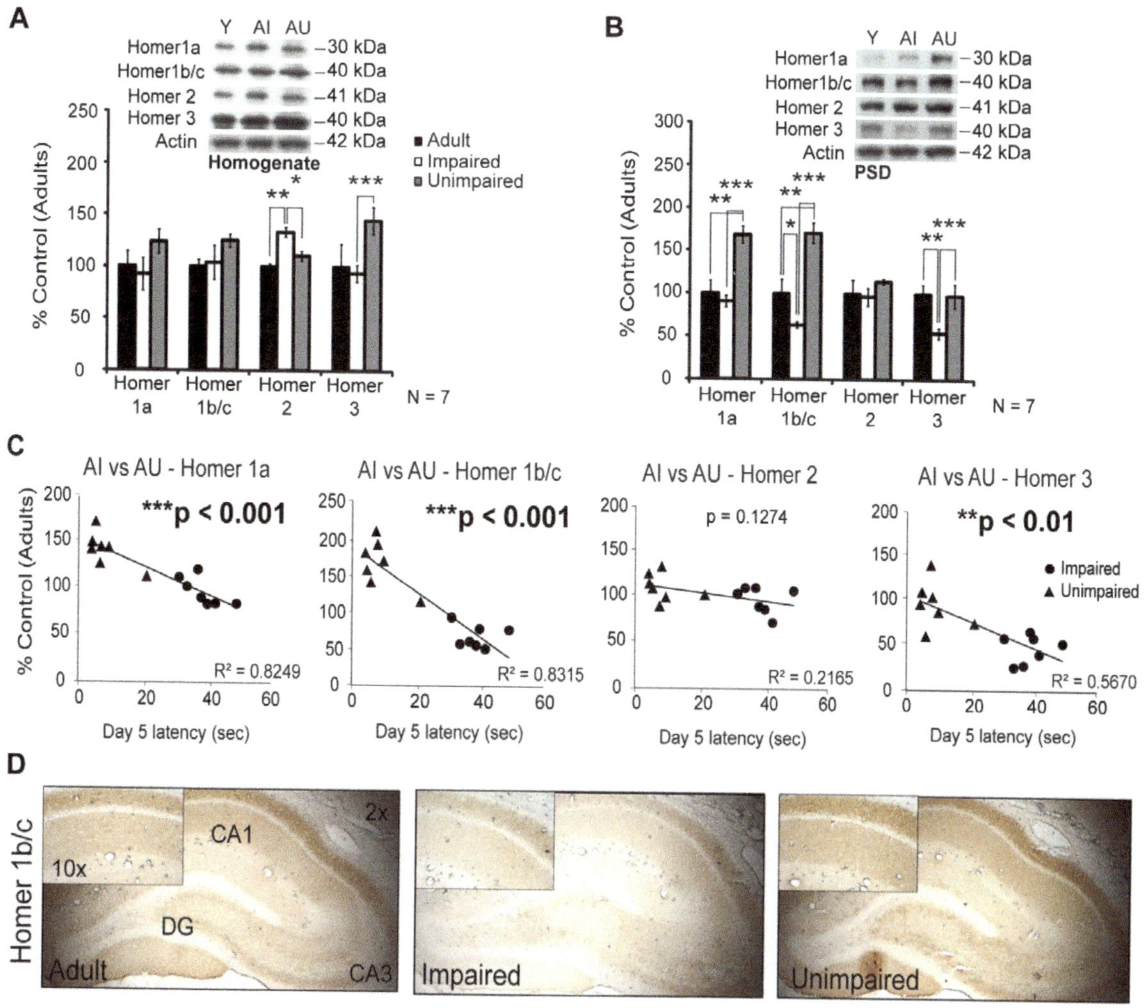

Figure 5. Comparison of Homer protein levels in adult (Y), AI and AU hippocampal formations. A) Homer 3 protein level was increased in AU homogenates, while Homer 2 level was increased in AI hippocampal formation. B) Homer 1a and 1 b/c protein levels were considerably increased in AU PSD. The AI animals displayed reduced levels of Homer 3 in the PSD. C) All these alterations in the PSD were correlated to MWM day 5 latencies. D) Similar to mGluR5, Homer 1 b/c immunostaining was reduced in the AI hippocampus, mainly in the CA1 area. Western Blot values represent mean ± SEM of three separate experiments (N = 7) and the data are expressed as percentage of control (6-months adult rats), *p<0.05, **p<0.01, ***p<0.001.

mGluR5 receptors in both PSD (Fig. 7F, ***p<0.001) and synaptosomes (***p<0.001).

Signaling machinery and immediate early gene expression is more efficient in the hippocampal formation of aged memory-unimpaired rats

Homer 1a uncouples mGluR5 from the translation machinery [35,37] and many enzymes are associated with group 1 mGluRs receptor activation [38,39,40,41]. Since alterations in signaling pathways underlying learning processes could explain memory impairments, next we evaluated the activation and protein levels of several enzymes (Fig. 8A). Protein kinase C gamma (PKCγ) is specific to neurons and colocalized with mGluR5 [42]. It has also been previously associated with learning and memory processes [43,44,45,46]. PKCγ total protein level was similar in all groups (Fig. 8B); however the phosphorylation of threonine 674 residue was significantly increased in AU animals (Fig. 8B, 135%±15% vs 78%±9% for AI, **p<0.01). Extracellular signal-regulated kinase 1/2 (ERK1/2) activation and protein level were also evaluated [38,40,41]. ERK1/2 phosphorylation (Thr202/Tyr204) was significantly lower while the protein level remained unchanged for AI rats (Fig. 8B, 56%±24%, **p<0.01, ***p<0.001). Group 1 mGluRs activation has been previously coupled to the translation regulatory factor p70S6 kinase (p70S6K) via ERK1/2 and the mammalian target of rapamycin (mTOR) [40]. The phosphorylation of threonine 389 residue was lower in AI animals compared to adult rats but not significantly different from AU rats (Fig. 8C, 65%±18% vs 100%±11%, *p<0.05 and 92%±16%). Conversely, Thr421/Ser424 phosphorylation was specifically enhanced in the AU animals (Fig. 8C, 161%±20%, *p<0.05) but total protein levels remained unchanged. A dramatic decrease in mTOR phosphorylation (Ser2448) was observed for the AI group (Fig. 8D, 31%±22%, **p<0.01, ***p<0.001) and alteration in mTOR translational control has been previously associated with learning and memory deficits [47]. Finally, the inhibition of c-AMP response element binding protein (CREB) has been shown to disrupt mice reactivated spatial memory [48]. An increase in CREB Ser133 phosphorylation with no change in CREB expression was observed in the AU group (Fig. 8D, 126%±9%, *p<0.05). The activation of signaling enzymes PKCγ and mTOR

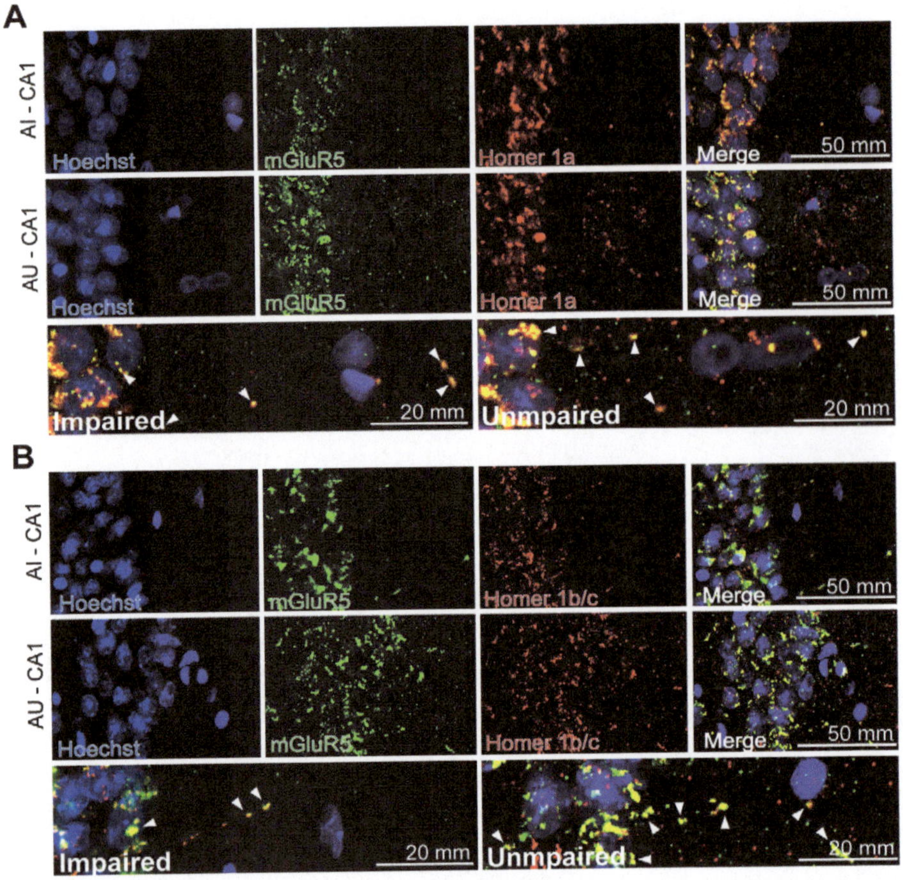

Figure 6. Co-localization of mGluR5 and Homer 1 isoforms. A) Homer 1a immunofluorescence and co-localization with mGluR5 was overall stronger in the cell bodies and dendrites of CA1 area in AU animals. At higher magnification (bottom), both mGluR5 and Homer 1a appear to be expressed more in the dendrites of AU animals but the co-localization seems to be similar for both AI and AU rats. B) mGluR5 and Homer 1 b/c co-localization was also overall higher in the cell bodies and dendrites of the CA1 region in AU animals. Both mGluR5 and Homer 1b/c are more expressed and co-localized in the AU dendrites compared to AI (bottom).

was significantly correlated with MWM day 5 latencies (Fig. 8E, *$p<0.05$, ***$p<0.001$, respectively). Using double immunofluorescence with the neuron-specific microtubule-associated protein 2 (MAP2), we confirmed the higher level of PKCγ activation in AU hippocampus. As shown in Figure 8F, PKCγ (Thr674) level is increased in the neurons of the CA1 sub-field for AU animals, particularly in the dendrites. Unsurprisingly, mTOR (Ser2448) is present close to the nuclei and again more abundant in CA1 neurons of the AU animals (Fig. 8G). Only a few proteins were observed in the dendrites of AU animals.

Transcription of the IEG Arc/Arg 3.1 modulates LTP consolidation, LTD and homeostatic plasticity in the hippocampus [49]. Considering the lower activation of mGluR5 signaling pathways in the AI rat hippocampal formation following the MWM training, Arc/Arg 3.1 expression and localization was investigated. As expected, Arc/Arg 3.1 level was decreased in AI animals (Fig. 9A, 79%±3%, *$p<0.05$) compared to AU and young adult rats. This change was correlated with the MWM performance (Fig. 9B, *$p<0.05$). Another IEG, Zif 268 is important for synaptic plasticity [50] and the expression of LTP and long-term memory [51]. No significant change was observed for this IEG (Fig. 9A and Fig. 9B). Higher Arc/Arg 3.1 level for AU compared to AI animals was confirmed by double immunofluorescence with MAP2 (Fig. 9C). Most of the proteins were detected in the nuclei of the CA1 region neuronal cells.

Discussion

In the present study, we report for the first time an involvement of mGluR5 receptors, Homer 1 proteins, and their downstream signaling pathways in hippocampus-dependent spatial memory for successful cognitive aging. Using the MWM, two distinct groups of aged animals were characterized by different learning abilities in a Long Evans rat population. Following the MWM task, we observed an up-regulation of mGluR5 protein level in homogenates and PSD of AU compared to the AI group, while mGluR1α levels were altered only in the PSD. These changes were significantly correlated with MWM performances of day 5. Interestingly, the scaffolding proteins, Homer 1a and Homer 1b/c were increased in PSD of AU animals. However, a lower ratio of Homer 1a/1b/c proteins was co-immunoprecipitated with mGluR5 receptors in the AU group, suggesting strengthened signaling pathways. This hypothesis was confirmed by Western Blot analysis of PKCγ (Thr674), ERK 1/2 (Thr202/Tyr204), p70S6K (Thr421/Ser424), mTOR (Ser2448) and CREB (Ser133). Higher PKCγ (Thr674) and mTOR (Ser2448) levels in AU hippocampus CA1 sub-field were confirmed in the dendrites and cell bodies, respectively, by double immunofluorescence with the neuronal marker MAP2. Memory consolidation is associated with the expression of IEG and a lower Arc level observed in the AI group, suggests an altered synaptic plasticity. Hence, mGluR5 and

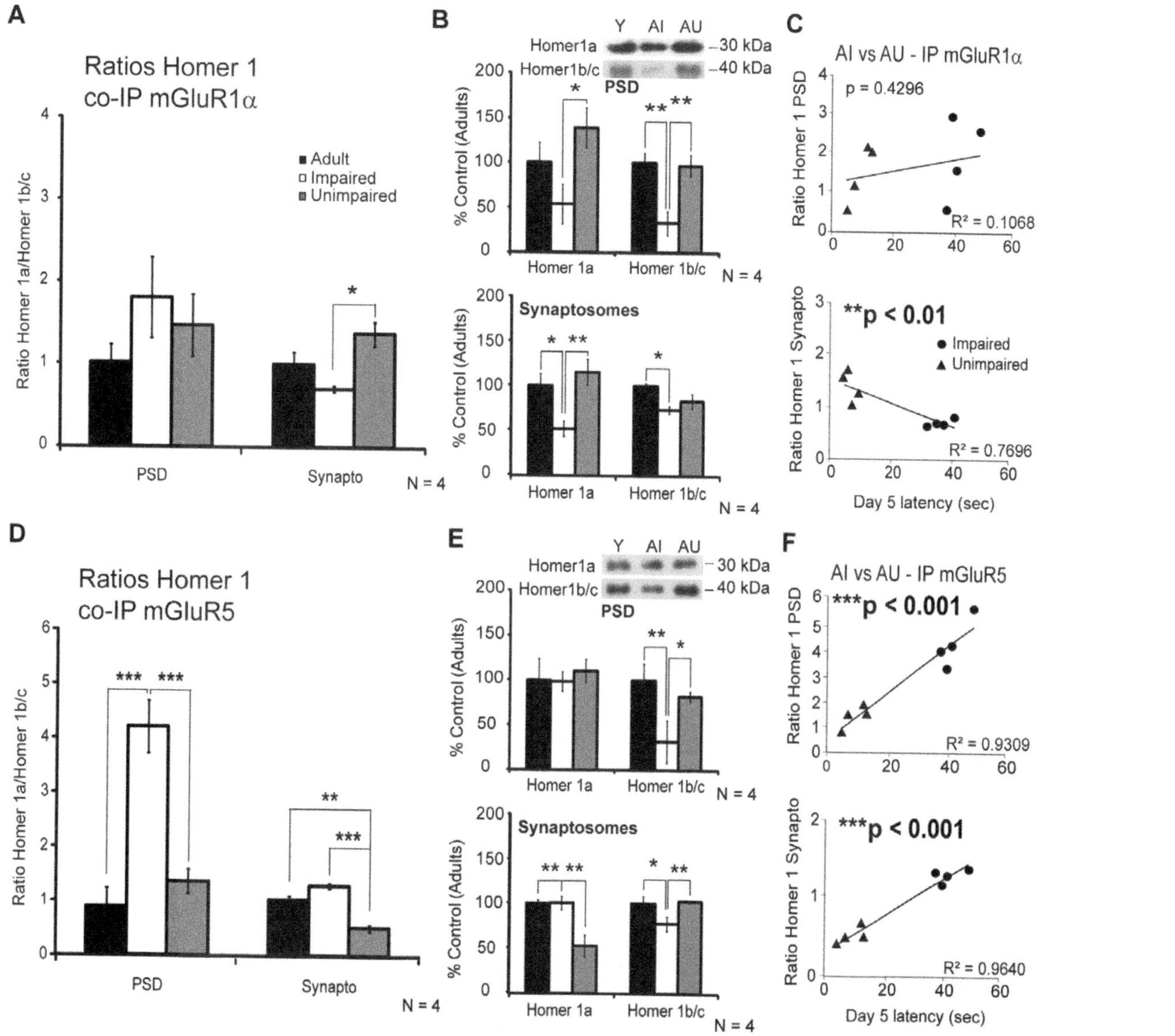

Figure 7. Comparison of Homer 1 isoforms ratio in adult (Y), AI and AU hippocampal formation. A) mGluR1α receptors were immunoprecipitated for PSD and synaptosomes preparations and Homer protein levels were determined by Western Blot. AU showed an increase in Homer 1a/Homer 1 b/c ratio but only in synaptosomes. B) Graphs for average Homer 1 protein levels bound to mGluR1α in the PSD and synaptosomes of each group. C) The aged rats MWM day 5 performances were correlated with the ratio of Homer 1 isoforms and mGluR1α in synaptosomes. D) The ratio of Homer 1 isoform protein levels co-immunoprecipitated with mGluR5 receptors was significantly greater in the PSD of AI rats. Conversely, this ratio was decreased in synaptosomes of AU animals. E) Graphs for average Homer 1 protein levels bound to mGluR5 in the PSD and synaptosomes of each group. F) The Homer 1 isoform ratio co-immunoprecipitated with mGluR5 receptors was strongly correlated with aged rats MWM performances. Data represent mean ± SEM of three separate experiments (N = 4) and are expressed as percentage of control (6-months adult rats), **p*<0.05, ***p*<0.01, ****p*<0.001.

Homer 1 proteins expressed in the PSD may play a key role in the regulation of IEG.

LTD associated with Group 1 mGluR has been previously related to memory processing in Long Evans rats. While NMDA-dependent LTD was correlated with MWM performance in adult rats, NMDA-independent LTD was linked to successful aging [13]. In the present study, we modified the protocol to assess not only acquisition of spatial memory but also inhibitory learning (reverse memory) and retention (remote memory). The average latency during the reverse memory task was minimally improved in AI animals, suggesting a severe impairment to encode new spatial information. In agreement with this result, Xu et al. [31] observed significant deficits in MWM reverse memory task in mGluR5 knockout mice and proposed a role of this receptor in inhibitory learning. Increased level of mGluR5 receptor protein in the PSD following MWM training was not observed in AI and adult rats, highlighting the critical role of this receptor in normal cognition during aging. A decrease in mGluRs-mediated phosphoinositide signal transduction has already been reported in AI animals without significant changes in mGluR5 receptor protein [52]. Injection of pharmacological agents related to mGluR5 activity could be an intriguing future set of experiments to confirm or infirm the conclusions presented here. Unfortunately the survival rate of Long Evans rat population decreases exponentially

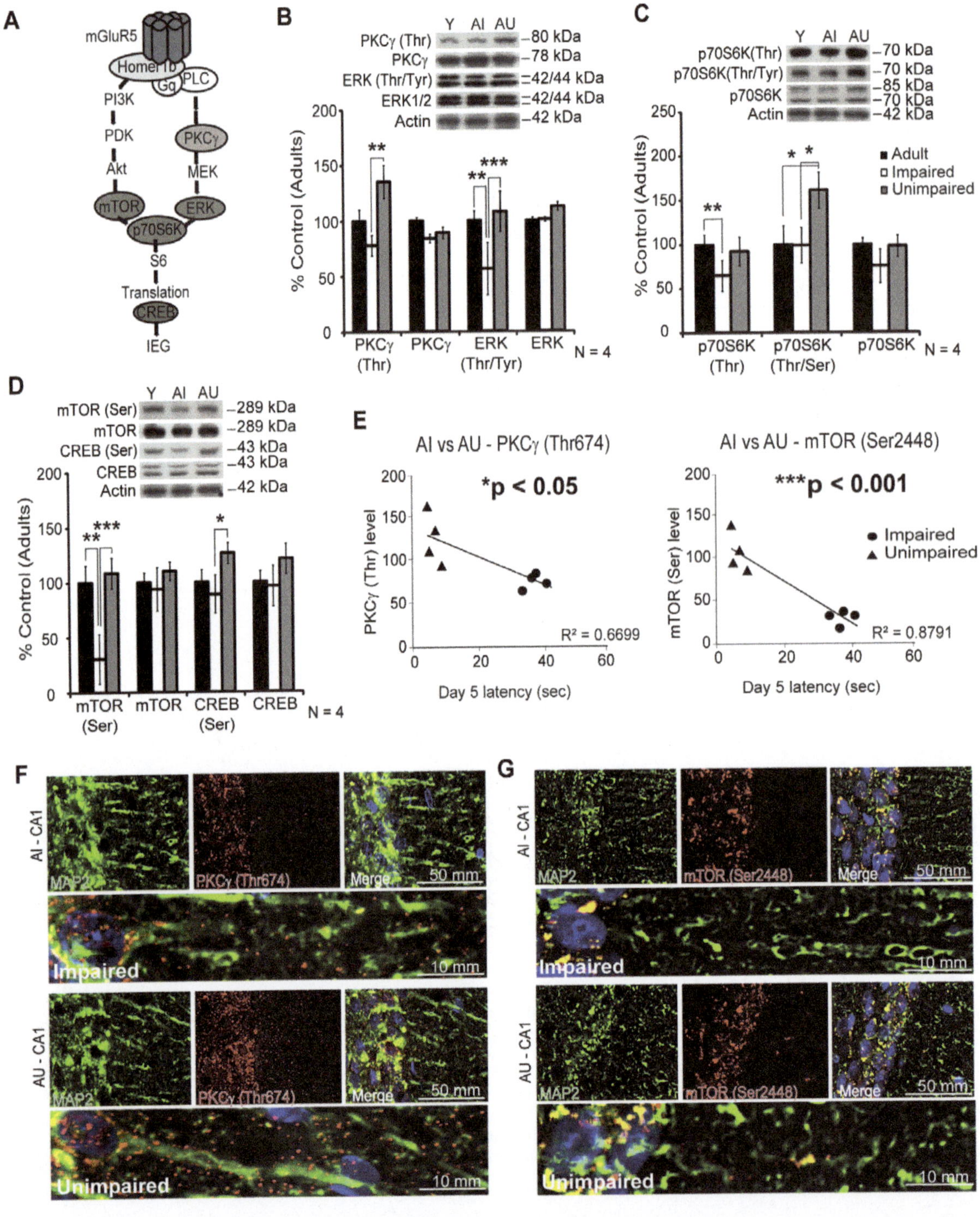

Figure 8. Activation of signaling pathways related to mGluR5 in adult (Y), AI and AU animals. A) Representative sketch of the multiple enzymes involved in mGluR5 activation. B) PKCγ Thr674 phosphorylation level is higher in AU homogenates while ERK 1/2 phosphorylation was decreased in AI animals. Total protein levels were unaffected in all groups. C) Thr421/Ser424 sites of p70S6 kinase enzyme were significantly more phosphorylated in AU homogenates. Total protein level remained constant in all groups. D) The strongest variation of phosphorylation level was observed for mTOR (Ser2448) in the AI group. A slight increase of CREB (Ser133) phosphorylation was also detected in AU homogenates. E) PKCγ (Thr674) and mTOR (Ser2448) phosphorylation levels were correlated with the MWM day 5 latencies in aged rats. As shown in Fig. 6F the expression of PKCγ (Thr674) was higher in dendrites of the CA1 sub-field in AU rats using double immunofluorescence with the neuronal marker MAP2. G) For mTOR (Ser2448), the increase was mostly localized in the CA1 cell bodies for AU rats. Western Blot results are represented as mean ± SEM of three separate experiments (N = 4) and the data are expressed as percentage of control (6-months adult rats), *$p<0.05$, **$p<0.01$, ***$p<0.001$.

with age and reaches to only 12% at 24 months (mean of 3 cohorts of 300 rats each at 12 months of age), making pharmacological studies in this animal model very difficult and expensive to pursue.

Other rat strains also present alterations in cognitive abilities with aging. While 20–40% of Long Evans rats become memory-impaired at 24–26 months of age in the MWM task, 45–100% of

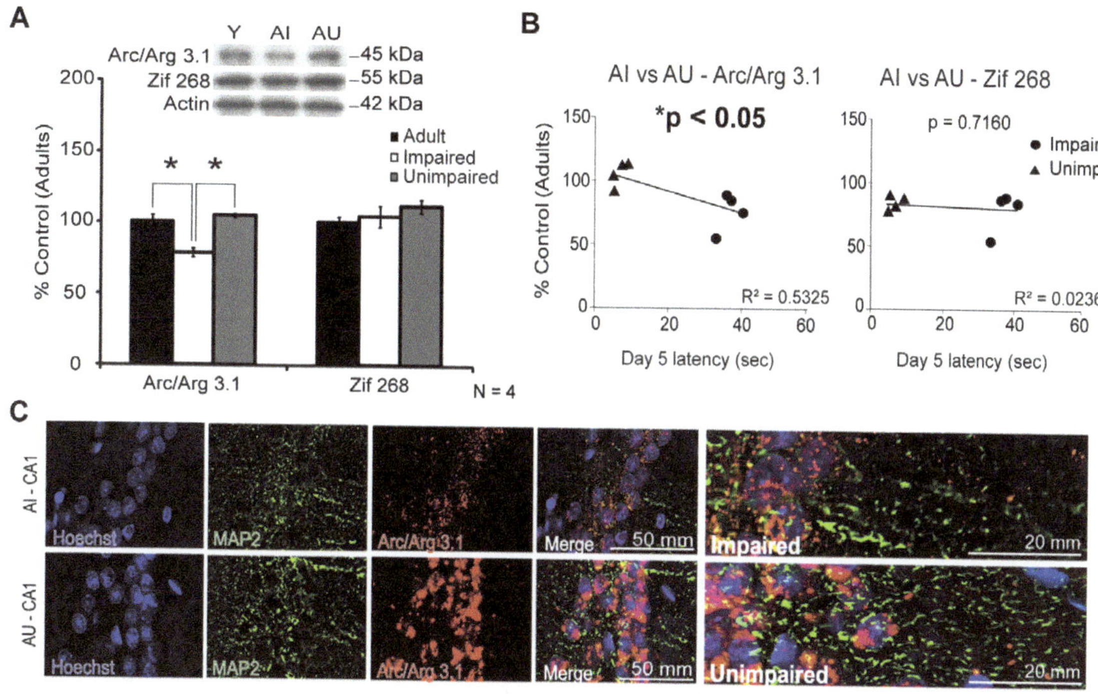

Figure 9. IEG Arc/Arg 3.1 and Zif 268 levels in adult (Y), AI and AU hippocampal homogenates. A) Arc/Arg 3.1 level was significantly decreased in AI homogenates while Zif 268 was unaltered. B) Arc/Arg 3.1 expression level was correlated to aged rats MWM performance. C) Arc staining was higher for the AU animals in the CA1 sub-field nuclei. Western Blot values represent mean ± SEM of three separate experiments (N = 4) and the data are expressed as percentage of control (6-months adult rats), *$p<0.05$.

Fisher rats population (depending on the study) are classified as memory-impaired at the same age [5]. Sprague Dawley rats also show memory impairments in short- and long-term memory recognition tasks at 24 months [53]. Age-related decrease of NMDA-mediated synaptic plasticity has been associated with such impairments [53]. mGluR5 and Homer 1 protein levels have never been investigated in these animal models after learning/training. Thus this could be of a great interest to confirm our data in the Long Evans rat model.

Homer proteins act as scaffolding support for the group 1 metabotropic glutamate receptors [24,25]. Since AU animals demonstrated higher levels of mGluR5 receptors, increase in the associated Homer proteins was expected. Both Homer 1a and 1b/c levels were increased in the PSD of those animals. However the ratio of Homer 1a/Homer 1b/c bound to mGluR5 was higher in the PSD of AI group. Based on the results obtained in this study for Long Evans AU rats, here we propose a model (Fig. 10) to summarize the mechanisms implicated in successful aging after learning/training. First, a basal high level of mGluR5 binding to the long Homer protein isoforms in the PSD is required. Following MWM training, mGluR5-related signaling pathways are activated and PKCγ is phosphorylated. The role of this enzymatic activity has been reported in learning and spatial memory [46] and the levels of PKCγ increase with age [43]. In the present study, an increase in the phosphorylation of PKCγ (Thr674) was seen only in AU animals whereas ERK1/2 (Thr202/Tyr204) and mTOR (Ser2448) phosphorylation levels were similar in both AU and young adult rats. The phosphorylation level of ERK1/2 was previously correlated to cognitive impairments related to aging [54]. Moreover, ERK plays a key role in the hippocampal CA1/CA2 regions in long-term spatial memory following multiple training trials in the MWM [55]. While AU animals optimally use mGluR5-related pathways, younger animals might favour NMDA receptors-mediated signaling. Indeed, Gong et al. [56] suggested a critical role for NMDA receptors and mTOR to induce dendritic protein synthesis in hippocampal neurons. Furthermore, spatial memory performance strongly correlates with the magnitude of NMDA-dependent LTP in young rats while NMDA-independent LTP was closely associated to cognition in aged rats [57].

Alterations in the level and phosphorylation of CREB have been reported in cognitively-impaired aged rats [58,59]. Following MWM training, AU rats displayed high levels of phosphorylated CREB, suggesting an activation of the transcription machinery, which was confirmed by a higher level of IEG Arc especially in the CA1 area of the hippocampus. Zif268 levels were not significantly different among the groups. Since, this IEG seems particularly important for plasticity underlying the first training days [50]; it may not be relevant after three weeks. Synaptic activity also induced the synthesis of Homer 1a [23] but its sustained expression in the dorsal hippocampus of transgenic mice impairs LTP and spatial working memory [60]. Interestingly, recombinant adeno-associated virus-mediated overexpression of Homer 1a impairs hippocampal-dependent memory while the overexpression of Homer 1c slightly facilitates memory performance [61]. Thus perturbations of synaptic group 1 mGluR equilibrium by Homer 1 isoforms may occur in AI animals. Indeed these receptors could more easily be uncoupled from its downstream signaling pathways, leading to long-term lower protein synthesis required for synaptic plasticity.

Metabotropic mGluR5 receptors dynamically regulate NMDA receptor function through Homer-Shank multi-protein scaffolding and this interaction can be disrupted by Homer 1a [62]. On the other hand, the stimulation of NMDA receptors induces a positive feedback regulation of mGluR1α via calpain-mediated truncation [63]. Blocking NMDA-dependent LTD with specific NR2B antagonists impairs spatial memory consolidation [64]. Therefore,

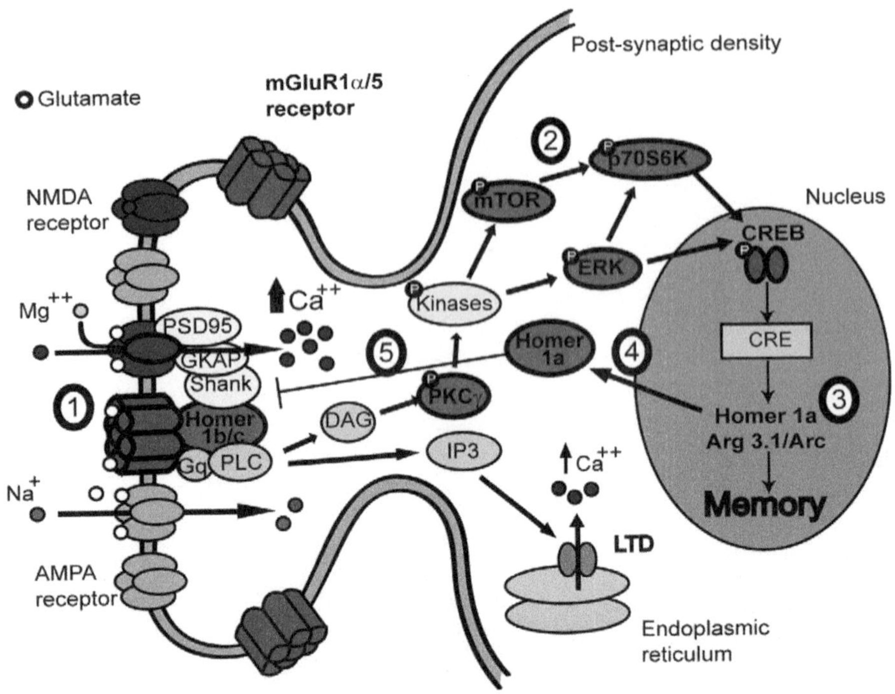

Figure 10. Possible mechanisms activated in the aged memory-unimpaired PSD following MWM training. MWM training induces release of glutamate which then activates group 1 mGluRs on the plasma membrane (1). Metabotropic mGluR5 are linked to ionotropic glutamate receptors by multiple scaffolding proteins (Homer 1 b/c, Shank, GKAP, PSD95). Following mGluR activation, protein kinases PKCγ, ERK1/2, mTOR, p70S6K and CREB become phosphorylated (2) and induce IEG expression (3). Homer 1a is released in the cytosol and translocated to plasma membrane (4) uncoupling the receptors with its signaling effectors (5). High levels of mGluR5 receptors bound to Homer long isoforms in the PSD of the AU rats maintain the activation of crucial signaling pathways and IEG expression following spatial memory training.

it would be of interest to evaluate possible interactions between metabotropic and ionotropic glutamate receptors in our model. Another intriguing hypothesis involving ionotropic glutamate receptors is the homeostatic scaling of the sensitivity of hippocampal synapses by group 1 mGluR and Homer 1a [65]. This form of synaptic plasticity adjusts the strength of excitatory synapses to stabilize firing [66]. Network activity induces transient Homer 1a expression and decreases the level of synaptic AMPA receptors [65]. Moreover, synapse-specific homeostatic plasticity is also observed in mature synapses and requires IEG Arc [67]. Aging likely affect these refined synaptic processes [68] and consequently information storage in the brain.

Most of the changes observed in the AI rat brains were in the CA1 hippocampal region in accordance with its role in the organization of recent memories [69]. Gene expression profiles have been investigated in the CA1, CA3 and DG areas in adult, AI and AU groups [5,70]. Interestingly, changes observed in the CA3 region were better correlated to memory impairments [70]. However, changes in gene expression associated to lysosome, immune function and mRNA translation, were also correlated to cognitive deficits without being specific to the CA3 region [70]. Our group previously reported a reduction in the expression level of various genes in the dorsal hippocampus following scopolamine-induced amnesia including Homer 1a, GABA B receptor and Zif268 [71]. Finally, normal aging seems to affect the pre-synaptic modulation of glutamatergic synapses. Interestingly, Zhang et al. [72] reported a decrease in extracellular glutamate levels in the hippocampus of 24-months old cognitively-impaired Fischer-344 rats. It would thus be of interest to compare the integrity of pre-synaptic components in AI and AU Long Evans rats.

Excessive protein synthesis downstream mGluR5 activation has been proposed to explain the fragile X syndrome, a genetic syndrome characterized by multiple symptoms including intellectual dysfunction, social anxiety and working memory deficits. Many symptoms of this disease can be reduced by modulating mGluR5 in animal models [73,74]. For example, the acute blockade of mGluR5 and ERK1/2 decreased protein synthesis to wild-type level in a fragile X mental retardation protein (FMRP) knockout mice model [75]. FMRP is a cytoplasmic mRNA binding protein able to regulate the translation of the amyloid precursor protein through mGluR [76]. Interestingly, the accumulation of beta-amyloid (Aβ) at synapses observed in Alzheimer's brain reduces mGluR5 lateral diffusion in the membrane leading to an aberrant accumulation of glutamate receptors, elevated intracellular Ca+ levels and the synapse dysfunction [77]. The clustering and lateral mobility of mGluRs are influenced by Homer interactions [78] and Aβ has been reported to decrease synaptic levels of Homer 1b [79]. In this perspective, it could be attractive to investigate mGluR and Homer 1 protein levels and their interactions with aging in pathological mouse models.

In this study we identified for the first time, mGluR5 glutamate receptors, Homer 1 proteins and their downstream signalling pathways as mediators of successful cognitive aging in a population of Long Evans rats. After training, more changes were observed in the hippocampal formation of aged memory-unimpaired compared to memory-impaired rats, suggesting a better adaptation of their brain plasticity to aging. Some of those variations were not observed in the 6-months adult animals, indicating a specific mechanism underlying spatial learning and memory in the aged hippocampus. Hopefully, our work will help to understand the individual differences observed with aging not only in rodents but also in humans and provide novel hypothesis to understand pathological aging.

Supporting Information

Figure S1 Control experiments for the specificity of mGluR5 and Homer 1 antibodies and PSD purification. A) mGluR5 antibody strongly reacted to one specific band (132 kDa). Immunostaining was blocked by peptide preadsorption in both Western Blot and immunohistochemistry. Specificity of the anti-rabbit secondary antibody was assessed by primary antibody omission. B) Homer 1a primary antibody shows several nonspecific bands. One band between 30 and 36 kDa was blocked by pre-incubation with peptide. Specificity of the anti-goat secondary antibody was assessed by primary antibody omission. C) Homer 1b/c antibody was highly specific with only one immunoreactive band. No staining was observed following preadsorption with the peptide or when the primary antibody was omitted. D) Purification of the PSD by subcellular fractionation and Triton X-100 treatment was assessed by Western Blot using PSD95 as a marker. Actin was used as a loading control.

Acknowledgments

The authors would like to thank Drs Jean-Guy Chabot, Salah El Mestikawy and Gregory Dal-Bo for their technical assistance and Mira Thakur for editing this manuscript.

Author Contributions

Conceived and designed the experiments: CM RQ. Performed the experiments: CM. Analyzed the data: CM. Contributed reagents/materials/analysis tools: CM RQ. Wrote the paper: CM RQ.

References

1. Rowe JW, Kahn RL (1997) Successful aging. The Gerontologist 37: 433–440.
2. Aubert I, Rowe W, Meaney MJ, Gauthier S, Quirion R (1995) Cholinergic markers in aged cognitively impaired Long-Evans rats. Neuroscience 67: 277–292.
3. Rowe WB, Spreekmeester E, Meaney MJ, Quirion R, Rochford J (1998) Reactivity to novelty in cognitively-impaired and cognitively-unimpaired aged rats and young rats. Neuroscience 83: 669–680.
4. Gallagher M, Bizon JL, Hoyt EC, Helm KA, Lund PK (2003) Effects of aging on the hippocampal formation in a naturally occurring animal model of mild cognitive impairment. Experimental gerontology 38: 71–77.
5. Benoit CE, Rowe WB, Menard C, Sarret P, Quirion R (2011) Genomic and proteomic strategies to identify novel targets potentially involved in learning and memory. Trends in pharmacological sciences 32: 43–52.
6. Brouillette J, Quirion R (2008) Transthyretin: a key gene involved in the maintenance of memory capacities during aging. Neurobiology of aging 29: 1721–1732.
7. Rowe WB, Blalock EM, Chen KC, Kadish I, Wang D, et al. (2007) Hippocampal expression analyses reveal selective association of immediate-early, neuroenergetic, and myelinogenic pathways with cognitive impairment in aged rats. The Journal of neuroscience: the official journal of the Society for Neuroscience 27: 3098–3110.
8. Wilson IA, Gallagher M, Eichenbaum H, Tanila H (2006) Neurocognitive aging: prior memories hinder new hippocampal encoding. Trends in neurosciences 29: 662–670.
9. Bredt DS, Nicoll RA (2003) AMPA receptor trafficking at excitatory synapses. Neuron 40: 361–379.
10. Lee YS, Silva AJ (2009) The molecular and cellular biology of enhanced cognition. Nature reviews Neuroscience 10: 126–140.
11. Lamprecht R, LeDoux J (2004) Structural plasticity and memory. Nature reviews Neuroscience 5: 45–54.
12. Barnes CA (2003) Long-term potentiation and the ageing brain. Philosophical transactions of the Royal Society of London Series B, Biological sciences 358: 765–772.
13. Lee HK, Min SS, Gallagher M, Kirkwood A (2005) NMDA receptor-independent long-term depression correlates with successful aging in rats. Nature neuroscience 8: 1657–1659.
14. Palmer MJ, Irving AJ, Seabrook GR, Jane DE, Collingridge GL (1997) The group I mGlu receptor agonist DHPG induces a novel form of LTD in the CA1 region of the hippocampus. Neuropharmacology 36: 1517–1532.
15. Nicoletti F, Bockaert J, Collingridge GL, Conn PJ, Ferraguti F, et al. (2011) Metabotropic glutamate receptors: from the workbench to the bedside. Neuropharmacology 60: 1017–1041.
16. Collett VJ, Collingridge GL (2004) Interactions between NMDA receptors and mGlu5 receptors expressed in HEK293 cells. British journal of pharmacology 142: 991–1001.
17. Romano C, van den Pol AN, O'Malley KL (1996) Enhanced early developmental expression of the metabotropic glutamate receptor mGluR5 in rat brain: protein, mRNA splice variants, and regional distribution. The Journal of comparative neurology 367: 403–412.
18. Bortolotto ZA, Collett VJ, Conquet F, Jia Z, van der Putten H, et al. (2005) The regulation of hippocampal LTP by the molecular switch, a form of metaplasticity, requires mGlu5 receptors. Neuropharmacology 49 Suppl 1: 13–25.
19. Manahan-Vaughan D, Braunewell KH (2005) The metabotropic glutamate receptor, mGluR5, is a key determinant of good and bad spatial learning performance and hippocampal synaptic plasticity. Cerebral cortex 15: 1703–1713.
20. Bikbaev A, Neyman S, Ngomba RT, Conn PJ, Nicoletti F, et al. (2008) MGluR5 mediates the interaction between late-LTP, network activity, and learning. PloS one 3: e2155.
21. Neyman S, Manahan-Vaughan D (2008) Metabotropic glutamate receptor 1 (mGluR1) and 5 (mGluR5) regulate late phases of LTP and LTD in the hippocampal CA1 region in vitro. The European journal of neuroscience 27: 1345–1352.
22. Bird CM, Burgess N (2008) The hippocampus and memory: insights from spatial processing. Nature reviews Neuroscience 9: 182–194.
23. Brakeman PR, Lanahan AA, O'Brien R, Roche K, Barnes CA, et al. (1997) Homer: a protein that selectively binds metabotropic glutamate receptors. Nature 386: 284–288.
24. Xiao B, Tu JC, Petralia RS, Yuan JP, Doan A, et al. (1998) Homer regulates the association of group 1 metabotropic glutamate receptors with multivalent complexes of homer-related, synaptic proteins. Neuron 21: 707–716.
25. Xiao B, Tu JC, Worley PF (2000) Homer: a link between neural activity and glutamate receptor function. Current opinion in neurobiology 10: 370–374.
26. Fagni L, Worley PF, Ango F (2002) Homer as both a scaffold and transduction molecule. Science's STKE: signal transduction knowledge environment 2002: re8.
27. Yang L, Mao L, Tang Q, Samdani S, Liu Z, et al. (2004) A novel Ca2+-independent signaling pathway to extracellular signal-regulated protein kinase by coactivation of NMDA receptors and metabotropic glutamate receptor 5 in neurons. The Journal of neuroscience: the official journal of the Society for Neuroscience 24: 10846–10857.
28. Duncan RS, Hwang SY, Koulen P (2005) Effects of Vesl/Homer proteins on intracellular signaling. Experimental biology and medicine 230: 527–535.
29. Mao L, Yang L, Tang Q, Samdani S, Zhang G, et al. (2005) The scaffold protein Homer1b/c links metabotropic glutamate receptor 5 to extracellular signal-regulated protein kinase cascades in neurons. The Journal of neuroscience: the official journal of the Society for Neuroscience 25: 2741–2752.
30. Morris R (1984) Developments of a water-maze procedure for studying spatial learning in the rat. Journal of neuroscience methods 11: 47–60.
31. Xu J, Zhu Y, Contractor A, Heinemann SF (2009) mGluR5 has a critical role in inhibitory learning. The Journal of neuroscience: the official journal of the Society for Neuroscience 29: 3676–3684.
32. Lauterborn JC, Lynch G, Vanderklish P, Arai A, Gall CM (2000) Positive modulation of AMPA receptors increases neurotrophin expression by hippocampal and cortical neurons. The Journal of neuroscience: the official journal of the Society for Neuroscience 20: 8–21.
33. Henley JM (1995) Subcellular localization and molecular pharmacology of distinct populations of [3H]-AMPA binding sites in rat hippocampus. British Journal of Pharmacology 115: 295–301.
34. Moron JA, Abul-Husn NS, Rozenfeld R, Dolios G, Wang R, et al. (2007) Morphine administration alters the profile of hippocampal postsynaptic density-associated proteins: a proteomics study focusing on endocytic proteins. Molecular & cellular proteomics: MCP 6: 29–42.
35. Kammermeier PJ, Worley PF (2007) Homer 1a uncouples metabotropic glutamate receptor 5 from postsynaptic effectors. Proceedings of the National Academy of Sciences of the United States of America 104: 6055–6060.
36. Van Keuren-Jensen K, Cline HT (2006) Visual experience regulates metabotropic glutamate receptor-mediated plasticity of AMPA receptor synaptic transmission by homer1a induction. The Journal of neuroscience: the official journal of the Society for Neuroscience 26: 7575–7580.
37. Ronesi JA, Huber KM (2008) Homer interactions are necessary for metabotropic glutamate receptor-induced long-term depression and translational activation. The Journal of neuroscience: the official journal of the Society for Neuroscience 28: 543–547.

38. Gallagher SM, Daly CA, Bear MF, Huber KM (2004) Extracellular signal-regulated protein kinase activation is required for metabotropic glutamate receptor-dependent long-term depression in hippocampal area CA1. The Journal of neuroscience: the official journal of the Society for Neuroscience 24: 4859–4864.
39. Hou L, Klann E (2004) Activation of the phosphoinositide 3-kinase-Akt-mammalian target of rapamycin signaling pathway is required for metabotropic glutamate receptor-dependent long-term depression. The Journal of neuroscience: the official journal of the Society for Neuroscience 24: 6352–6361.
40. Page G, Khidir FA, Pain S, Barrier L, Fauconneau B, et al. (2006) Group I metabotropic glutamate receptors activate the p70S6 kinase via both mammalian target of rapamycin (mTOR) and extracellular signal-regulated kinase (ERK 1/2) signaling pathways in rat striatal and hippocampal synaptoneurosomes. Neurochemistry international 49: 413–421.
41. Antion MD, Hou L, Wong H, Hoeffer CA, Klann E (2008) mGluR-dependent long-term depression is associated with increased phosphorylation of S6 and synthesis of elongation factor 1A but remains expressed in S6K-deficient mice. Molecular and cellular biology 28: 2996–3007.
42. Liu JX, Tang YC, Liu Y, Tang FR (2008) mGluR5-PLCbeta4-PKCbeta2/PKCgamma pathways in hippocampal CA1 pyramidal neurons in pilocarpine model of status epilepticus in mGluR5+/+ mice. Epilepsy research 82: 111–123.
43. Colombo PJ, Wetsel WC, Gallagher M (1997) Spatial memory is related to hippocampal subcellular concentrations of calcium-dependent protein kinase C isoforms in young and aged rats. Proceedings of the National Academy of Sciences of the United States of America 94: 14195–14199.
44. Colombo PJ, Gallagher M (2002) Individual differences in spatial memory among aged rats are related to hippocampal PKCgamma immunoreactivity. Hippocampus 12: 285–289.
45. Saito N, Shirai Y (2002) Protein kinase C gamma (PKC gamma): function of neuron specific isotype. Journal of biochemistry 132: 683–687.
46. Nithianantharajah J, Murphy M (2009) Experience on the Barnes spatial maze influences PKCgamma levels in the hippocampus. The International journal of neuroscience 119: 1014–1030.
47. Parsons RG, Gafford GM, Helmstetter FJ (2006) Translational control via the mammalian target of rapamycin pathway is critical for the formation and stability of long-term fear memory in amygdala neurons. The Journal of neuroscience: the official journal of the Society for Neuroscience 26: 12977–12983.
48. Kim R, Moki R, Kida S (2011) Molecular mechanisms for the destabilization and restabilization of reactivated spatial memory in the Morris water maze. Molecular brain 4: 9.
49. Bramham CR, Worley PF, Moore MJ, Guzowski JF (2008) The immediate early gene arc/arg3.1: regulation, mechanisms, and function. The Journal of neuroscience: the official journal of the Society for Neuroscience 28: 11760–11767.
50. Alberini CM (2009) Transcription factors in long-term memory and synaptic plasticity. Physiological reviews 89: 121–145.
51. Jones MW, Errington ML, French PJ, Fine A, Bliss TV, et al. (2001) A requirement for the immediate early gene Zif268 in the expression of late LTP and long-term memories. Nature neuroscience 4: 289–296.
52. Nicolle MM, Colombo PJ, Gallagher M, McKinney M (1999) Metabotropic glutamate receptor-mediated hippocampal phosphoinositide turnover is blunted in spatial learning-impaired aged rats. The Journal of neuroscience: the official journal of the Society for Neuroscience 19: 9604–9610.
53. Kollen M, Stephan A, Faivre-Bauman A, Loudes C, Sinet PM, et al. (2010) Preserved memory capacities in aged Lou/C/Jall rats. Neurobiology of aging 31: 129–142.
54. Williams BJ, Bimonte-Nelson HA, Granholm-Bentley AC (2006) ERK-mediated NGF signaling in the rat septo-hippocampal pathway diminishes with age. Psychopharmacology 188: 605–618.
55. Blum S, Moore AN, Adams F, Dash PK (1999) A mitogen-activated protein kinase cascade in the CA1/CA2 subfield of the dorsal hippocampus is essential for long-term spatial memory. The Journal of neuroscience: the official journal of the Society for Neuroscience 19: 3535–3544.
56. Gong R, Park CS, Abbassi NR, Tang SJ (2006) Roles of glutamate receptors and the mammalian target of rapamycin (mTOR) signaling pathway in activity-dependent dendritic protein synthesis in hippocampal neurons. The Journal of biological chemistry 281: 18802–18815.
57. Boric K, Munoz P, Gallagher M, Kirkwood A (2008) Potential adaptive function for altered long-term potentiation mechanisms in aging hippocampus. The Journal of neuroscience: the official journal of the Society for Neuroscience 28: 8034–8039.
58. Brightwell JJ, Gallagher M, Colombo PJ (2004) Hippocampal CREB1 but not CREB2 is decreased in aged rats with spatial memory impairments. Neurobiology of learning and memory 81: 19–26.
59. Monti B, Berteotti C, Contestabile A (2005) Dysregulation of memory-related proteins in the hippocampus of aged rats and their relation with cognitive impairment. Hippocampus 15: 1041–1049.
60. Celikel T, Marx V, Freudenberg F, Zivkovic A, Resnik E, et al. (2007) Select overexpression of homer1a in dorsal hippocampus impairs spatial working memory. Frontiers in neuroscience 1: 97–110.
61. Klugmann M, Symes CW, Leichtlein CB, Klaussner BK, Dunning J, et al. (2005) AAV-mediated hippocampal expression of short and long Homer 1 proteins differentially affect cognition and seizure activity in adult rats. Molecular and cellular neurosciences 28: 347–360.
62. Bertaso F, Roussignol G, Worley P, Bockaert J, Fagni L, et al. (2010) Homer1a-dependent crosstalk between NMDA and metabotropic glutamate receptors in mouse neurons. PloS one 5: e9755.
63. Xu W, Wong TP, Chery N, Gaertner T, Wang YT, et al. (2007) Calpain-mediated mGluR1alpha truncation: a key step in excitotoxicity. Neuron 53: 399–412.
64. Ge Y, Dong Z, Bagot RC, Howland JG, Phillips AG, et al. (2010) Hippocampal long-term depression is required for the consolidation of spatial memory. Proceedings of the National Academy of Sciences of the United States of America 107: 16697–16702.
65. Hu JH, Park JM, Park S, Xiao B, Dehoff MH, et al. (2010) Homeostatic scaling requires group I mGluR activation mediated by Homer1a. Neuron 68: 1128–1142.
66. Turrigiano GG (2008) The self-tuning neuron: synaptic scaling of excitatory synapses. Cell 135: 422–435.
67. Beique JC, Na Y, Kuhl D, Worley PF, Huganir RL (2011) Arc-dependent synapse-specific homeostatic plasticity. Proceedings of the National Academy of Sciences of the United States of America 108: 816–821.
68. Burke SN, Barnes CA (2010) Senescent synapses and hippocampal circuit dynamics. Trends in neurosciences 33: 153–161.
69. Frankland PW, Bontempi B (2005) The organization of recent and remote memories. Nature reviews Neuroscience 6: 119–130.
70. Haberman RP, Colantuoni C, Stocker AM, Schmidt AC, Pedersen JT, et al. (2011) Prominent hippocampal CA3 gene expression profile in neurocognitive aging. Neurobiology of aging 32: 1678–1692.
71. Brouillette J, Young D, During MJ, Quirion R (2007) Hippocampal gene expression profiling reveals the possible involvement of Homer1 and GABA(B) receptors in scopolamine-induced amnesia. Journal of neurochemistry 102: 1978–1989.
72. Zhang WQ, Mundy WR, Thai L, Hudson PM, Gallagher M, et al. (1991) Decreased glutamate release correlates with elevated dynorphin content in the hippocampus of aged rats with spatial learning deficits. Hippocampus 1: 391–397.
73. Dolen G, Bear MF (2008) Role for metabotropic glutamate receptor 5 (mGluR5) in the pathogenesis of fragile X syndrome. The Journal of physiology 586: 1503–1508.
74. Krueger DD, Bear MF (2011) Toward fulfilling the promise of molecular medicine in fragile X syndrome. Annual review of medicine 62: 411–429.
75. Osterweil EK, Krueger DD, Reinhold K, Bear MF (2010) Hypersensitivity to mGluR5 and ERK1/2 leads to excessive protein synthesis in the hippocampus of a mouse model of fragile X syndrome. The Journal of neuroscience: the official journal of the Society for Neuroscience 30: 15616–15627.
76. Westmark CJ, Malter JS (2007) FMRP mediates mGluR5-dependent translation of amyloid precursor protein. PLoS biology 5: e52.
77. Renner M, Lacor PN, Velasco PT, Xu J, Contractor A, et al. (2010) Deleterious effects of amyloid beta oligomers acting as an extracellular scaffold for mGluR5. Neuron 66: 739–754.
78. Serge A, Fourgeaud L, Hemar A, Choquet D (2002) Receptor activation and homer differentially control the lateral mobility of metabotropic glutamate receptor 5 in the neuronal membrane. The Journal of neuroscience: the official journal of the Society for Neuroscience 22: 3910–3920.
79. Roselli F, Hutzler P, Wegerich Y, Livrea P, Almeida OF (2009) Disassembly of shank and homer synaptic clusters is driven by soluble beta-amyloid(1–40) through divergent NMDAR-dependent signalling pathways. PloS one 4: e6011.

PERMISSIONS

All chapters in this book were first published in PLOS ONE, by The Public Library of Science; hereby published with permission under the Creative Commons Attribution License or equivalent. Every chapter published in this book has been scrutinized by our experts. Their significance has been extensively debated. The topics covered herein carry significant findings which will fuel the growth of the discipline. They may even be implemented as practical applications or may be referred to as a beginning point for another development.

The contributors of this book come from diverse backgrounds, making this book a truly international effort. This book will bring forth new frontiers with its revolutionizing research information and detailed analysis of the nascent developments around the world.

We would like to thank all the contributing authors for lending their expertise to make the book truly unique. They have played a crucial role in the development of this book. Without their invaluable contributions this book wouldn't have been possible. They have made vital efforts to compile up to date information on the varied aspects of this subject to make this book a valuable addition to the collection of many professionals and students.

This book was conceptualized with the vision of imparting up-to-date information and advanced data in this field. To ensure the same, a matchless editorial board was set up. Every individual on the board went through rigorous rounds of assessment to prove their worth. After which they invested a large part of their time researching and compiling the most relevant data for our readers.

The editorial board has been involved in producing this book since its inception. They have spent rigorous hours researching and exploring the diverse topics which have resulted in the successful publishing of this book. They have passed on their knowledge of decades through this book. To expedite this challenging task, the publisher supported the team at every step. A small team of assistant editors was also appointed to further simplify the editing procedure and attain best results for the readers.

Apart from the editorial board, the designing team has also invested a significant amount of their time in understanding the subject and creating the most relevant covers. They scrutinized every image to scout for the most suitable representation of the subject and create an appropriate cover for the book.

The publishing team has been an ardent support to the editorial, designing and production team. Their endless efforts to recruit the best for this project, has resulted in the accomplishment of this book. They are a veteran in the field of academics and their pool of knowledge is as vast as their experience in printing. Their expertise and guidance has proved useful at every step. Their uncompromising quality standards have made this book an exceptional effort. Their encouragement from time to time has been an inspiration for everyone.

The publisher and the editorial board hope that this book will prove to be a valuable piece of knowledge for researchers, students, practitioners and scholars across the globe.

LIST OF CONTRIBUTORS

Ghazaleh Samoudi and Hans Nissbrandt
Department of Pharmacology, Institute of Neuroscience and Physiology, The Sahlgrenska Academy, University of Gothenburg, Göteborg, Sweden

Mayank B. Dutia
Centre for Integrative Physiology, University of Edinburgh, Edinburgh, United Kingdom

Filip Bergquist
Department of Neurology, Sahlgrenska University Hospital, Göteborg, Sweden

Hui-Chun Ku and Ming-Jai Su
Institute of Pharmacology, College of Medicine, National Taiwan University, Taipei, Taiwan

Clévio Nóbrega
CNC - Center for Neurosciences & Cell Biology, University of Coimbra, Coimbra, Portugal

Isabel Nascimento-Ferreira, Isabel Onofre and Luiís Pereira de Almeida
CNC - Center for Neurosciences & Cell Biology, University of Coimbra, Coimbra, Portugal
Faculty of Pharmacy, University of Coimbra, Coimbra, Portugal

David Albuquerque
Faculty of Sciences and Technology, University of Coimbra, Coimbra, Portugal

Hirokazu Hirai
Department of Neurophysiology, Gunma University Graduate School of Medicine, Maebashi, Gunma, Japan

Nicole Déglon
Lausanne University Hospital, Department of Clinical Neurosciences, Laboratory of Cellular and Molecular Neurotherapies, Lausanne, Switzerland

Jean A. Hall and William R. Vorachek
Department of Biomedical Sciences, College of Veterinary Medicine, Oregon State University, Corvallis, Oregon, United States of America

Janice K. Huntera, Whitney C. Stewartb, Wayne D. Mosher and Gene J. Pirelli
Department of Animal and Rangeland Sciences, College of Agricultural Sciences, Oregon State University, Corvallis, Oregon, United States of America

Gerd Bobe
Department of Animal and Rangeland Sciences, College of Agricultural Sciences, Oregon State University, Corvallis, Oregon, United States of America Linus Pauling Institute, Oregon State University, Corvallis, Oregon, United States of America

Charles T. Estill
Department of Animal and Rangeland Sciences, College of Agricultural Sciences, Oregon State University, Corvallis, Oregon, United States of America
Department of Clinical Sciences, College of Veterinary Medicine, Oregon State University, Corvallis, Oregon, United States of America

Jorge A. Vanegas
Department of Clinical Sciences, College of Veterinary Medicine, Oregon State University, Corvallis, Oregon, United States of Ameri

Jean A. Hall and William R. Vorachek
Department of Biomedical Sciences, College of Veterinary Medicine, Oregon State University, Corvallis, Oregon, United States of America

Janice K. Huntera, Wayne D. Mosher, Gene J. Pirelli and Whitney C. Stewartb
Department of Animal and Rangeland Sciences, College of Agricultural Sciences, Oregon State University, Corvallis, Oregon, United States of America

Gerd Bobe
Department of Animal and Rangeland Sciences, College of Agricultural Sciences, Oregon State University, Corvallis, Oregon, United States of America
Linus Pauling Institute, Oregon State University, Corvallis, Oregon, United States of America

Charles T. Estill
Department of Animal and Rangeland Sciences, College of Agricultural Sciences, Oregon State University, Corvallis, Oregon, United States of America
Department of Clinical Sciences, College of Veterinary Medicine, Oregon State University, Corvallis, Oregon, United States of America

Jorge A. Vanegas
Department of Clinical Sciences, College of Veterinary Medicine, Oregon State University, Corvallis, Oregon, United States of America

José L. Quesado and José P. Andrade
Departamento de Anatomia, Faculdade de Medicina da Universidade do Porto, Porto, Portugal

Gisela H. Maia and Nikolai V. Lukoyanov
Departamento de Anatomia, Faculdade de Medicina da Universidade do Porto, Porto, Portugal
Neural Networks Group, Instituto de Biologia Molecular e Celular da Universidade do Porto, Porto, Portugal
Programa Doutoral em Neurociências, Universidade do Porto, Porto, Portugal

Joana I. Soares
Neural Networks Group, Instituto de Biologia Molecular e Celular da
Universidade do Porto, Porto, Portugal
Programa Doutoral em Neurociências, Universidade do Porto, Porto, Portugal

Joana M. do Carmo and Pedro A. Andrade
Programa Doutoral em Neurociências, Universidade do Porto, Porto, Portugal

Lara Moumné and Gillian P. Bates
Department of Medical and Molecular Genetics, King's College London, London, United Kingdom

Ken Campbell and Yingbin Ouyang
Taconic, Cranbury, New Jersey, United States of America

David Howland
CHDI Management/CHDI Foundation, Princeton, New Jersey, United States of America

Motohide Nishio and Masahiro Satoh
NARO Institute of Livestock and Grassland Science, Tsukuba, Japan

J. Jeffry Howbert, Vanessa Ruedebusch, Jaideep Mavoori, Kent Leyde and W. Douglas Sheffield
NeuroVista Corp., Seattle, Washington, United States of America

Edward E. Patterson
Veterinary Medical Center, University of Minnesota, St. Paul, Minnesota, United States of America

S. Matt Stead3, Ben Brinkmann, Vincent Vasoli, Daniel Crepeau and Gregory A. Worrell
Mayo Systems Electrophysiology Laboratory, Mayo Clinic, Rochester, Minnesota, United States of America

Charles H. Vite
School of Veterinary Medicine, University of Pennsylvania, Philadelphia, Pennsylvania, United States of America

Beverly Sturges
Veterinary School, University of California Davis, Davis, California, United States of America

Brian Litt
Department of Bioengineering, University of Pennsylvania, Philadelphia, Pennsylvania, United States of America

Christian Agrillo, Laura Piffer and Angelo Bisazza
Department of General Psychology, University of Padova, Padova, Italy

Brian Butterworth
Institute of Cognitive Neuroscience, University College of London, London, United Kingdom

Dhananjay R. Namjoshi, Georgina Martin, James Donkin, Anna Wilkinson, Sophie Stukas, Jianjia Fan, Michael Carr, Sepideh Tabarestani and Cheryl L. Wellington
Department of Pathology and Laboratory Medicine, The University of British Columbia, Vancouver, British Columbia, Canada

Kelli Wuerth and Robert E. W. Hancock
Department of Microbiology and Immunology, The University of British Columbia, Vancouver, British Columbia, Canada

Sandrine Thuret, Michaela Thallmair, Laura L. Horky and Fred H. Gage
Laboratory of Genetics, The Salk Institute for Biological Studies, La Jolla, California, United States of America

Maaike van Putten, Margriet Hulsker, Vishna Devi Nadarajah, Sandra H. van Heiningen, Ella van Huizen, Maarten van Iterson, Peter Admiraal, Tobias Messemaker, Johan T. den Dunnen, Peter A. C. t'Hoen and Annemieke Aartsma-Rus
Department of Human Genetics, Leiden University Medical Center, Leiden, the Netherlands

Brandon A. Baiamonte, Franklin A. Lee, Steve T. Brewer and Gerald J. LaHoste
Applied Biopsychology Program, Department of Psychology, University of New Orleans, New Orleans, Louisiana, United States of America

Daniela Spano
Dipartimento di Biochemicae Biotecnologie Mediche, Universitàdi Napoli, Naples, Italy
CEINGE Biotecnologie Avanzate, Naples, Italy

Luca Madaro, Andrea Pelle, Carmine Nicoletti, Annunziata Crupi, Valeria Marrocco and Marina Bouché
Unit of Histology, and IIM, Sapienza University, DAHFMO, Rome, Italy

Gianluca Bossi and Silvia Soddu
Department of Experimental Oncology, Regina Elena Cancer Institute, Rome, Italy

Thomas Hannagan, Johannes C. Ziegler, Stéphane Dufau, Joël Fagot and Jonathan Grainger
Laboratoire de Psychologie Cognitive, Aix-Marseille University, & CNRS (Centre National de la Recherche Scientifique), Marseille, France

Vanessa Donega, Cindy T. J. van Velthoven and Cora H. Nijboer
Laboratory of Neuroimmunology and Developmental Origins of Disease, University Medical Centre Utrecht, Utrecht, The Netherlands

Cobi J. Heijnen
Laboratory of Neuroimmunology and Developmental Origins of Disease, University Medical Centre Utrecht, Utrecht, The Netherlands
Department of Symptom Research, MD Anderson Cancer Center, Houston, Texas, United States of America

Frank van Bel
Department of Neonatology, University Medical Center Utrecht, Utrecht, The Netherlands

Martien J. H. Kas
Rudolf Magnus Institute of Neuroscience, Department of Neuroscience and Pharmacology, University Medical Center Utrecht, Utrecht, The Netherlands

Annemieke Kavelaars
Department of Symptom Research, MD Anderson Cancer Center, Houston, Texas, United States of America

Apiradee Thammanomai, Hiroshi Hamakawa, Erzsébet Bartolák-Suki and Béla Suki
Department of Biomedical Engineering, Boston University, Boston, Massachusetts, United States of America

Apiradee Thammanomai, Hiroshi Hamakawa, Erzsébet Bartolák-Suki, Béla Suki
Department of Biomedical Engineering, Boston University, Boston, Massachusetts, United States of America

Jana Hädicke
Otto-von-Guericke-Universität, Institut für Biochemie und Zellbiologie, Magdeburg, Germany

Mario Engelmann
Otto-von-Guericke-Universität, Institut für Biochemie und Zellbiologie, Magdeburg, Germany
Center for Behavioral Brain Sciences, Magdeburg, Germany

Franc¸ois Dauphin, Thomas Freret and Michel Boulouard
Universitéde Caen Basse-Normandie, Groupe Mémoire et Plasticité comportementale (GMPc), EA4259, IFR 146, Caen, France

Guenaëlle Levallet
Universitéde Caen Basse-Normandie, Groupe Mémoire et Plasticitécomportementale (GMPc), EA4259, IFR 146, Caen, France
CHU de Caen, Service d'Anatomie Pathologie, Caen, France

Maïte Hotte
Universitéde Caen Basse-Normandie, Groupe Mémoire et Plasticitécomportementale (GMPc), EA4259, IFR 146, Caen, France
Universitéde Rouen, NeoVasc, EA 4309, IFRMP23, IHURBM, Rouen, France

Katie Smith, Ximin Zeng and Jun Lin
Department of Animal Science, University of Tennessee, Knoxville, Tennessee, United States of America

Christopher J. Fulton
ARC Centre of Excellence for Coral Reef Studies, Research School of Biology, The Australian National University, Canberra, Australian Capital Territory, Australia

Jacob L. Johansen
ARC Centre of Excellence for Coral Reef Studies, School of Marine and Tropical Biology, James Cook University, Townsville, Queensland, Australia

John F. Steffensen
Marine Biological Section, Department of Biology, University of Copenhagen, Helsingør, Denmark

Zhenming Zhou, Bo Zhou, Liping Ren and Qingxiang Meng
State Key Laboratory of Animal Nutrition, China Agricultural University, Beijing, P. R. China
College of Animal Science and Technology, China Agricultural University,
Beijing, P. R. China

Caroline Ménard and Rémi Quirion
Douglas Mental Health University Institute, Department of Psychiatry, McGill University, Montréal, Canada

Index